FORMULAS

DISTANCE between (x_1, y_1) and (x_2, y_2) is $d = \sqrt{(x_2 - x_1)^2 + (y_2 - y_1)^2}$.

SLOPE of the line passing through (x_1, y_1) and (x_2, y_2) is $m = \dfrac{y_2 - y_1}{x_2 - x_1}$; if θ is the angle of elevation, then $m = \tan \theta$.

PYTHAGOREAN THEOREM
A triangle with sides a and b, and hypotenuse c (the side opposite the right angle) is a right triangle if and only if $a^2 + b^2 = c^2$.

QUADRATIC FORMULA
If $Ax^2 + Bx + C = 0$, $A \neq 0$, then $x = \dfrac{-B \pm \sqrt{B^2 - 4AC}}{2A}$

The *discriminant* is $B^2 - 4AC$; if $B^2 - 4AC < 0$, then there are *no real* solutions,

$B^2 - 4AC = 0$, then there is *one real* solution, and

$B^2 - 4AC > 0$, then there are *two real* solutions.

FACTORING RULES
$$x^2 - y^2 = (x - y)(x + y)$$
$$x^3 - y^3 = (x - y)(x^2 + xy + y^2)$$
$$x^3 + y^3 = (x + y)(x^2 - xy + y^2)$$

RELATIONSHIP BETWEEN DEGREE AND RADIAN MEASURE
$$\frac{\text{angle in degrees}}{360} = \frac{\text{angle in radians}}{2\pi}$$

ARC LENGTH FORMULA
The arc length of a circle of radius r and central angle θ (measured in radians) is $s = r\theta$.

GROWTH/DECAY FORMULA $A = A_0 e^{rt}$

FORMS OF A COMPLEX NUMBER z

Rectangular form $z = a + bi$

Polar form $z = r(\cos \theta + i \sin \theta) = r \operatorname{cis} \theta$

Conjugates $a + bi$ and $a - bi$ are conjugates. Their product is a real number $a^2 + b^2$.

Absolute value $|z| = \sqrt{a^2 + b^2}$; if $b = 0$, then $\sqrt{a^2} = |a|$; that is, $|a| = a$ if $a \geq 0$, and $|a| = -a$ if $a < 0$.

Conversion formulas   From rectangular to polar: $r = \sqrt{a^2 + b^2}$; $\theta' = \tan^{-1}\left|\dfrac{b}{a}\right|$, where θ' is the reference angle for θ.

From polar to rectangular: $a = r \cos \theta$, $b = r \sin \theta$

DeMOIVRE'S THEOREM
If n is a natural number, $(r \operatorname{cis} \theta)^n = r^n \operatorname{cis} n\theta$.

nTH ROOT THEOREM
If n is a positive integer, $(r \operatorname{cis} \theta)^{1/n} = \sqrt[n]{r} \operatorname{cis}\left[\dfrac{1}{n}(\theta + 360°k)\right]$, where $k = 0, 1, ..., n - 1$.

AREA FORMULAS
Let $K = $ area, $r = $ radius, $b = $ base, $s = $ side, $\ell = $ length, $w = $ width, and $h = $ height.

Square $K = s^2$ *Rectangle* $K = \ell w$ *Circle* $K = \pi r^2$

Trapezoid $K = h\left(\dfrac{b_1 + b_2}{2}\right)$ *Triangle* $K = \dfrac{1}{2}bh$ *Sector* $K = \dfrac{1}{2}\theta r^2$

Other triangle formulas: Two sides and an included angle: $K = \dfrac{1}{2}bc \sin \alpha$, $K = \dfrac{1}{2}ac \sin \beta$, $K = \dfrac{1}{2}ab \sin \gamma$

Two angles and an included side: $K = \dfrac{a^2 \sin \beta \sin \gamma}{2 \sin \alpha}$, $K = \dfrac{b^2 \sin \alpha \sin \gamma}{2 \sin \beta}$, $K = \dfrac{c^2 \sin \alpha \sin \beta}{2 \sin \gamma}$

Three sides where $s = \dfrac{1}{2}(a + b + c)$: $K = \sqrt{s(s - a)(s - b)(s - c)}$

VOLUME FORMULAS
For a parallelepiped (a box): $V = \ell wh$

For a cone: $V = \dfrac{1}{3}\pi r^2 h$ or $V = \dfrac{1}{3}\pi r^3 \tan \alpha$ for angle of elevation from the base to the vertex

ESSENTIALS OF TRIGONOMETRY

Third Edition

KARL J. SMITH

Santa Rosa Junior College

Brooks/Cole Publishing Company

I(T)P® An International Thomson Publishing Company

Pacific Grove • Albany • Belmont • Bonn • Boston • Cincinnati • Detroit • Johannesburg • London •
Madrid • Melbourne • Mexico City • New York • Paris • Singapore • Tokyo • Toronto • Washington

Sponsoring Editor: *Bob Pirtle/Margot Hanis*
Project Development Editor: *Elizabeth Rammel*
Marketing: *Margaret Parks/Laura Caldwell*
Editorial Assistant: *Melissa Duge/Jennifer Wilkinson*
Production Coordinator: *Tessa A. McGlasson*
Production: *Susan L. Reiland*
Interior Design: *Carolyn Deacy*

Cover Design: *Larry Molmud*
Cover Photo: *Tom Van Sant, Geosphere Project/Photo Researchers*
Illustration: *Lori Heckelman*
Typesetting: *Bi-Comp, Inc.*
Cover Printing: *Phoenix Color Corp.*
Printing and Binding: *R. R. Donnelley/Crawfordsville*

For more information, contact:

BROOKS/COLE PUBLISHING COMPANY
511 Forest Lodge Road
Pacific Grove, CA 93950
USA

International Thomson Publishing Europe
Berkshire House 168-173
High Holborn
London WC1V 7AA
England

Thomas Nelson Australia
102 Dodds Street
South Melbourne, 3205
Victoria, Australia

Nelson Canada
1120 Birchmount Road
Scarborough, Ontario
Canada M1K 5G4

International Thomson Editores
Seneca 53
Col. Polanco
11560 México, D. F., México

International Thomson Publishing GmbH
Königswinterer Strasse 418
53227 Bonn
Germany

International Thomson Publishing Asia
221 Henderson Road
#05-10 Henderson Building
Singapore 0315

International Thomson Publishing Japan
Hirakawacho Kyowa Building, 3F
2-2-1 Hirakawacho
Chiyoda-ku, Tokyo 102
Japan

You can request permission to use material from this text through the following phone and fax numbers:
Phone: 1-800-730-2214; Fax: 1-800-730-2215.

Printed in the United States of America

10   9   8   7   6   5   4   3

Library of Congress Cataloging-in-Publication Data

Smith, Karl J.
  Essentials of trigonometry / Karl J. Smith. — 3rd ed.
  Includes index.
  ISBN 0-534-34806-8
  1. Trigonometry.   I. Title.
QA531.S6527   1998
516.24'2—dc21
                        97-40526
                        CIP

PREFACE

There is perhaps nothing which so occupies, as it were, the middle position of mathematics, as trigonometry.
J. F. Herbart, 1890

What Is Trigonometry?

Trigonometry is usually considered to be a transition course between high school algebra and calculus. It changes our focus from the development of manipulative skills to an understanding of mathematical concepts. Quite simply, trigonometry is the study of six functions called the *trigonometric functions* and their relationships to one another, as well as the study of applications involving these six functions. But equally important, trigonometry occupies that middle position of mathematics that changes our mathematical framework or way of viewing the world around us.

About This Book

This text is for students preparing to take calculus or other courses that require a background in trigonometry. It is written as a two- or three-credit college course in trigonometry. The prerequisite is a second-year high school algebra course or equivalent. This text assumes that you have a calculator, but is calculator-independent in its development. Some of the assumptions regarding calculators are discussed in Appendix A. The text also assumes a familiarity with the set of real numbers including, in particular, numbers such as $\sqrt{2}$, $\sqrt{3}$, π, and e. Many of the necessary ideas from algebra, including rational expressions, radical expressions, and equation solving, are reviewed in Appendix C. Even though a course in geometry is also a desirable prerequisite, more and more students are taking trigonometry without having taken high school geometry; consequently, in addition to a review of geometry in Appendix D, the text includes all of the necessary ideas from geometry as they are needed.

New Features

This edition is much more application-oriented than the previous edition. The goal of the design is to make it functional and easy-to-use for the student. The next four pages illustrate these features.

To the Instructor

Essentials of Trigonometry, Third Edition, is a right-triangle development of trigonometry. By the end of the first chapter, the basic ideas of the cosine, sine, and tangent ratios in a right triangle, along with their respective inverses, are introduced and are used to solve both triangles and right-triangle applications.

Chapter Openers Clear, clean, and to the point. List of sections included in the chapter, a preview of the chapter, and a relevant quotation.

Road Signs have been added to help the student through the material.

Stop Signs tell the student to stop for a few moments to study the material.

Warning Signs tell the student to make a special note of some mentioned fact because it is used throughout the rest of the book.

*This is an important definition. Spend some time learning this before going on. We list the cosine first because cosine always comes first (it is the **first** component); sine is second because it is the second component (also second alphabetically). Note that, with this definition, θ is no longer restricted as it was in Section 1.3.*

This may seem like fine print to you, but spending some time on this help window now will pay dividends to you later in this book.

CHAPTER THREE

GRAPHS OF TRIGONOMETRIC FUNCTIONS

PREVIEW

3.1 Graphs of the Standard Trigonometric Functions
3.2 General Cosine, Sine, and Tangent Curves
3.3 Trigonometric Graphs
3.4 Inverse Trigonometric Functions
3.5 Chapter 3 Summary

In this chapter, we introduce the concept of graphing the trigonometric functions. You can understand these functions more easily if you form a mental image or "picture" of them. In the first section, a quick, simple method called framing is discussed and is then used in the second section to graph general cosine, sine, and tangent functions easily and efficiently in one step.

Throughout this book, we will avail ourselves of computer and calculator technology to "see" many relationships more easily, but that usage does not reduce our need to understand the nature and properties of trigonometric graphs.

SECTION 3.1

GRAPHS OF THE STANDARD TRIGONOMETRIC FUNCTIONS

x	y
0	500
1	470.7107
2	400
3	329.2893
4	300
5	329.2893
6	400
7	470.7107
8	500
9	470.7107
10	400
11	329.2893
12	300

PROBLEM OF THE DAY

In a classic study,* Lotka considered the interdependence of two species, the first of which serves as food for the other. If we consider the populations of foxes and rabbits in a certain region, we might find that their populations go up and down in cycles (but out of phase with one another). Suppose it was found that the population of the rabbits changes according to the formula

$$y = 400 + 100 \cos \frac{\pi x}{4}$$

where y is the number of rabbits after time x (measured in months). What is the graph representing the rabbit population?

How would you answer this question?

* From *Elements of Mathematical Biology* by Alfred Lotka, New York: Dover Publications, 1956.

Section 2

**UNIT CIRCLE DEFINITION O___
TRIGONOMETRIC FUNCTION___**

Let θ be an angle with verte___
θ is the positive x-axis, and ___
terminal side of θ with the *un___
are defined as follows:

Cosine function:　　cos θ ___

Sine function:　　sin θ ___

Tangent function:　　$\tan \theta = \dfrac{b}{a}, a \neq 0$　(read "tangent of θ is b/a")

Secant function:　　$\sec \theta = \dfrac{1}{a}, a \neq 0$　(read "secant of θ is $1/a$")

Cosecant function:　　$\csc \theta = \dfrac{1}{b}, b \neq 0$　(read "cosecant of θ is $1/b$")

Cotangent function:　　$\cot \theta = \dfrac{a}{b}, b \neq 0$　(read "cotangent of θ is a/b")

Help Window ? X

Trigonometry is about the study of these six functions, so it is important to remember the unit circle definition. The angle θ is an important part of this definition. The words cos, sin, tan, and so on are meaningless without θ. You can speak about the cosine function, or you can speak about cos θ, but you cannot correctly write only cos.

Also, take special note of the conditions by 0. For example, $a \neq 0$ means that sec θ and tan θ are not defined when θ is 90°, 270°, or any angle coterminal with either of these angles. If $b \neq 0$, then csc θ and cot θ are not defined when θ is 0°, 180°, or any angle coterminal with these angles. Thus, cosecant and cotangent are not defined for 0° or 180°. In this book, we assume all values that cause division by zero are excluded from the domain.

The angle θ in the unit circle definition is called the **argument** of the function. The argument, of course, does not need to be the same as the variable. For example, in $\cos(2\theta + 1)$, we say the *function* is $\cos(2\theta + 1)$, the *argument* is $2\theta + 1$, and the variable is θ.

Evaluation of Trigonometric Functions

We now can answer the Problem of the Day. In many applications, we know the argument and want to find one or more of its trigonometric functions. To

Definitions are enclosed in a box.

Problem of the Day is used to motivate each section. Note the use of real-life data.

Help Windows rephrase important ideas in everyday words, without mathematical jargon.

Table (top left)

30° (210°)			(329°) 149°		
'	Sin	Tan	Cot°	Cos	'
0	.50000	.57735	1.7321	.86603	60
1	.50025	.57774	1.7309	.86588	59
2	.50050	.57813	1.7297	.86573	58
3	.50076	.57851	1.7286	.86559	57
4	.50101	.57890	1.7274	.86544	56
5	.50126	.57929	1.7262	.86530	55
6	.50151	.57968	1.7251	.86515	54
7	.50176	.58007	1.7239	.86501	53
8	.50201	.58046	1.7228	.86486	52
9	.50227	.58085	1.7216	.86471	51
10	.50252	.58124	1.7205	.86457	50
11	.50277	.58162	1.7193	.86442	49
12	.50302	.58201	1.7182	.86427	48
13	.50327	.58240	1.7170	.86413	47
14	.50352	.58279	1.7159	.86398	46
15	.50377	.58318	1.7147	.86384	45
16	.50403	.58357	1.7136	.86369	44
17	.50428	.58396	1.7124	.86354	43
18	.50453	.58435	1.7113	.86340	42
19	.50478	.58474	1.7102	.86325	41

Historical Note

For years, the primary method for evaluating trigonometric functions used what is known as a trigonometric table. Typically, the values of the sine, cosine, and tangent in the first quadrant were given. The availability of inexpensive

Historical Notes are used to enrich the material.

TABLE 2.2 Signs of the Trigonometric Functions

	II	I
	Sine and Cosecant positive; others negative	All positive
	Tangent and Cotangent positive; others negative	Cosine and Secant positive; others negative
	III	IV

Easy-to-remember form:
A Smart **T**rig **C**lass

II	I
Smart	**A**
Trig	**C**lass
III	IV

Real-life data are used to generate a graph.

Bump Signs are used to warn of some unexpected or difficult material.

Computer Windows make references to available computer programs.

EXAMPLE 10

Graphing a Curve from Given Data Points

Let us reconsider the Problem of the Day.

The flywheel on a lawn mower has 16 evenly spaced cooling fins. If we rotate the engine through two complete revolutions, we can generate 32 data points, with the first coordinate representing time and the second representing the depth of the piston. Determine an equation of the curve generated by the data as shown in Table 3.3.

Problem of the Day questions are answered in the section after the necessary material has been presented.

Solution

As previously noted, we plot the data points, as shown in Figure 3.33, and note that the points are sinusoidal.

TABLE 3.3 Data Points for Lawn Mower Engine

x	y	x	y
0	0	16	0
1	0.2	17	0.2
2	0.8	18	0.8
3	1.7	19	1.7
4	2.5	20	2.5
5	3.4	21	3.4
6	4.0	22	4.0
7	4.4	23	4.4
8	4.5	24	4.5
9	4.4	25	4.4
10	4.0	26	4.0
11	3.5	27	3.5
12	2.2	28	2.2
13	1.7	29	1.7
14	0.9	30	0.9
15	0.2	31	0.2
		32	0

$y = 2.25$

It looks as if $x = 4$ when $y = 2.25$.

FIGURE 3.33 Data points for the Problem of the Day

The graph seems to have a period of 16 and an amplitude of 2.25. This means that for the general sine curve

$$y - k = a \sin b(x - h)$$

Thus, $a = 2.25 = \frac{9}{4}$; period $p = \frac{2\pi}{b} = 16$ so that $b = \frac{\pi}{8}$. Finally, we need to find an appropriate starting point, (h, k). There are many points we might choose, but we select a y-value halfway between the maximum and minimum, that is, $y = 2.25$. Next, we approximate the intersection of the line $y = 2.25 = \frac{9}{4}$ and the graph: it appears to be (4, 2.25).

$$y - \frac{9}{4} = \frac{9}{4} \sin \frac{\pi}{8}(x - 4)$$

---

minutes, multiply each by

on in 4 minutes (location Can you draw the location

and Sine

= 6.5, b = 8.3, and c = 12.2. a degree.

l in this text).

alues.

35°

Given triangle   To the nearest tenth of a degree, the solution is $\gamma = 110.5°$.

Computer Window [?] [X]

There are many computer programs that will help you solve triangles. One such program is *CONVERGE* from JEMware and is typical. You begin with the choice on the pull-down menu called *Alg/Trig—Solving Triangles* and then input the given values. Next, YOU must choose the correct method of solution:

1. SSS   2. SAS   3. SSA   4. AAS or ASA   5. AAA

Then, if you make the correct choice, the program forces you to select one of the following:

1. No solution   2. Infinitely many solutions   3. One solution using the law of cosines
4. One solution using the law of sines   5. Two solutions using the law of sines

Finally, after making the correct choice here (or being tutored if you do not), you can request not only the solution, but also a diagram of the given triangle.

Some problems can be worked by either the law of cosines or the law of sines. If you read the preceding Calculator Window, you might have noticed that this software uses the law of sines for the ambiguous case. Consider the following example where we use both solutions. Our goal is to show that, even though both methods may work, one method is preferable.

Calculator Window　? X

To solve a system of equations using a graphing calculator, begin by graphing the two curves:

$$y_1 = x$$
$$y_2 = \cos x$$

Since the scale is not shown on calculator screens, it would be difficult to estimate points of intersection if it were not for a built-in function called *trace*. When you press ⟦TRACE⟧, the cursor will appear on one of the graphed curves, and the coordinates of the cursor will appear at the bottom of the screen:

Y₁◼X
Y₂◼cos X
Xmin=-6　　Ymin=-2
Xmax=9　　Ymax=2
Xscl=.5　　Yscl=1

X=.70212766　Y=.70212766

On some calculator models, the numeral at the top tells you which curve the cursor will follow; the 1 here indicates the cursor will trace along the curve whose equation was entered at the y_1 prompt, namely $y = x$.

The trace values shown on your calculator may vary depending on the model and on the window set. We find the estimated solution (0.7, 0.7). Some calculators have an ISECT or INTERSECTION command. If you use such a calculator, you can find the intersection point with greater accuracy. For this example, we find a more accurate (but still not exact) intersection point, (0.7390851, 0.7390851).

Calculator Windows are included as an integral part of the text. Notice how this Calculator Window relates to the solution of Example 6, even though Example 6 does not require that the student have a graphing calculator.

Titled Examples help students follow what is happening.

Systems of Equations involving trigonometric equations lay the groundwork for solving equations in Chapter 4.

EXAMPLE 6　**Estimating Solutions to Systems of Equations**

State the number of solutions and estimate the x-value of the solutions.

a. $\begin{cases} y = \frac{1}{5}x \\ y = \cos x \end{cases}$　　b. $\begin{cases} y = \frac{1}{2} \\ y = \cos x \end{cases}$　　c. $\begin{cases} y = \frac{3}{2} \\ y = \sin x \end{cases}$

Solution　a. Graph $y_1 = \frac{1}{5}x$ and $y_2 = \cos x$ as shown in Figure 3.54. Note the use of y_1 and y_2 to indicate how you might enter these equations into a calculator.

FIGURE 3.54　Graphs of $y = \frac{1}{5}x$ and $y = \cos$

The period appears to be about 6.25, and the amplitude is about 1.4. Suppose we assume that the Problem of the Day is asking for the exact values for the period and the amplitude. Is there a formula for these properties of the graph? In this section, we find a formula that gives us exact values for the period and amplitude of $y = \sin x + \cos x$.

Product-to-Sum Identities

It is sometimes convenient, and even necessary, to write a trigonometric sum as a product or a product as a sum. These are known as the **product-to-sum identities.**

PRODUCT-TO-SUM IDENTITIES

30. $2 \cos \alpha \cos \beta = \cos(\alpha - \beta) + \cos(\alpha + \beta)$
31. $2 \sin \alpha \sin \beta = \cos(\alpha - \beta) - \cos(\alpha + \beta)$
32. $2 \sin \alpha \cos \beta = \sin(\alpha + \beta) + \sin(\alpha - \beta)$
33. $2 \cos \alpha \sin \beta = \sin(\alpha + \beta) - \sin(\alpha - \beta)$

The proofs of the product-to-sum identities involve systems of equations and the addition laws for cosine:

$$\cos(\alpha - \beta) = \cos \alpha \cos \beta + \sin \alpha \sin \beta$$
$$\cos(\alpha + \beta) = \cos \alpha \cos \beta - \sin \alpha \sin \beta$$

Add these equations:

$$\cos(\alpha - \beta) + \cos(\alpha + \beta) = 2 \cos \alpha \cos \beta \quad \text{This is identity 30.}$$

Subtract these equations:

$$\cos(\alpha - \beta) - \cos(\alpha + \beta) = 2 \sin \alpha \sin \beta \quad \text{This is identity 31.}$$

Starting over with the addition laws for sine:

$$\sin(\alpha + \beta) = \sin \alpha \cos \beta + \cos \alpha \sin \beta$$
$$\sin(\alpha - \beta) = \sin \alpha \cos \beta - \cos \alpha \sin \beta$$

Add these equations:

$$\sin(\alpha + \beta) + \sin(\alpha - \beta) = 2 \sin \alpha \cos \beta \quad \text{This is identity 32.}$$

Subtract these equations:

$$\sin(\alpha + \beta) - \sin(\alpha - \beta) = 2 \cos \alpha \sin \beta \quad \text{This is identity 33.}$$

▽ YIELD

New Terms are in **boldface** to help the student recognize them as important. These terms are included in the chapter review at the end of each chapter.

Important formulas and definitions are contained in a **box**.

Yield Signs are used to indicate that only the main idea presented is necessary (not the derivation).

CUMULATIVE REVIEW CHAPTERS 1–3

Essential Concepts

1. IN YOUR OWN WORDS Distinguish among equal angles, coterminal angle, and reference angles.

2. IN YOUR OWN WORDS Discuss the concept of arc length. What are the arc length formulas?

3. From memory, give the radian measure for each of the angles whose degree measure is stated. Also sketch the angle.

a. 30°   b. 45°   c. 60°   d. 90°
e. 180°   f. 270°   g. 360°   h. 120°
i. 135°   j. 150°   k. −45°   ℓ. −300°
m. −240°   n. −225°

4. From memory, give the degree measure for each of the angles w~~~~
sketch the a~~~~

a. $\frac{\pi}{3}$

e. π

i. $\frac{11\pi}{6}$

m. $\frac{-11\pi}{6}$

5. IN YOUR O~~~
definition of~~~
memory.

6. IN YOUR O~~~
definition of~~~
memory.

7. IN YOUR O~~~
of the trigon~~~

8. IN YOUR O~~~
principle an~~~
examples.

9. Name the re~~~
function.
a. cosine
d. secant

10. State the two ratio identities.

11. State the three Pythagorean identities.

12. Write out the table of exact values.

Problems 13–21: Find the exact value for the given expressions.

13. a. $\tan \frac{\pi}{4}$ **b.** $\cos 0$ **c.** $\cos 30°$
d. $\sin \frac{\pi}{2}$ **e.** $\tan 0$ **f.** $\cos \frac{\pi}{4}$

14. a. $\tan 180°$ **b.** $\sin 45°$ **c.** $\sin \pi$
d. $\sin 60°$ **e.** $\cos 270°$ **f.** $\tan \frac{\pi}{6}$

15. a. $\cos 60°$ **b.** $\sin \frac{\pi}{6}$ **c.** $\sin 0°$
d. $\csc 90°$ **e.** $\sec \frac{3\pi}{2}$ **f.** $\sec 90°$

Cumulative Reviews are provided to give practice for midterm examinations.

IN YOUR OWN WORDS are writing problems that ask the student to summarize important ideas. **Discussion Problems,** which call for essay-type answers, are also included.

---

14. What is the smallest force necessary to keep a 3-ton boxcar from sliding down a hill that makes an angle of 4° with the horizontal? Assume that friction is ignored.

15. A mine shaft is dug into the side of a sloping hill. The shaft is dug horizontally for 485 ft. Next, a turn is made so that the angle of elevation of the second shaft is 58.0°, thus forming a 58° angle between the shafts. The shaft is then continued for 382 ft before exiting, as shown in Figure 5.69. How far is it along a straight line from the entrance to the exit, assuming that all tunnels are in a single plane?

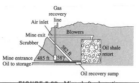

FIGURE 5.69   Mine shaft schematic

Discussion Problems

1. Describe what is meant by "the ambiguous case~~~

2. Outline your "method of attack" for solving tri~~~

3. Describe a process for finding the area of a tria~~~

4. What is meant by a "trigonometric substitution"~~~

Miscellaneous Problems—Chapters 1–~~~

Problems 1–4: Draw a unit circle and label the poi~~~ shown in the figure. Find:

a. The coordinates of A and P
b. $|OA|$ **c.** $|PA|$
d. $|OB|$ **e.** Area of $\triangle AOB$

1.   **2.**

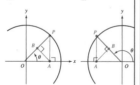

Miscellaneous Problems at the end of each chapter provide cumulative reviews for all chapters preceding and including that chapter.

Group Research Projects are included to encourage students to work together. **Individual Research Projects** are also provided.

---

63. A piston rotating at 60 rev/min is attached to a crankshaft, as shown in Figure 3.69. The connecting rod AB is 6 inches, and the radius of the crankshaft is 2 in. Draw a graph (for the first 10 seconds) indicating the height y of the point B at x seconds after it reaches the uppermost position. [See Problem 63, Section 2.4 (page 82) for an equation.]

FIGURE 3.69   Crankshaft

Group Research Projects

1. Plot the sunrise and sunset for your town's latitude. On another graph, plot the length of daylight for your own town. Can you find equations for the plotted data?

2. In calculus, it is shown that if a particle is moving in simple harmonic motion according to the equation

$$y = a \cos(\omega \tau + h)$$

then the velocity of the particle at time t is given by

$$y = a\omega \sin(\omega \tau + h)$$

Graph this velocity curve, where the frequency is 0.25 cycle/unit of time, $h = \pi/6$, and $a = 120/\pi$. Find a maximum value and a minimum value for this velocity.

3. In the 1920s, Soviet economist Nikolai Kondratieff identified what he believed to be a long-term sinusoidal cycle governing economic activity. Economists today talk about cycles of bear markets and bull markets. Do some research on economic cycles.

4. A formula for the voltage V of a charged capacitor connected to a coil is a function of time, t. The voltage (in volts) gradually diminishes to 0 over time according to the formula

$$V(t) = e^{-1.5t} \cos \pi \tau$$

Graph V for $0 \le t \le 5$, and write a paper on the relationship between voltage and trigonometry.

5. A touch-tone phone's "dial" is arranged in four rows and three columns, with each row and each column set at a different frequency, as shown in Figure 3.70.

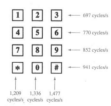

FIGURE 3.70   Touch-tone phone

When you press a particular button, the note sounded is the sum of two frequencies determined by the row and column. For example, if you press 1, then the sound emitted is

$$y = \sin 2\pi(697)x + \sin 2\pi(1,209)x$$

Graph this sound, and write a paper on touch-tone phone sounds.

One of my goals of this edition is suggested with the title; I want to present a basic trigonometric book that includes all of the essentials of trigonometry. A companion book, *Trigonometry for College Students, Seventh Edition,* presents a unit-circle approach to trigonometry and includes, in addition to the material in this edition, a discussion of coordinate systems, sets of numbers, functions, and conic sections.

Trigonometry requires a great deal of memory work for the student, so I have taken every opportunity to present the material in a manner that helps the student. For example, if you ask most professors to list the trigonometric functions, they will say, "sine, cosine, tangent, ..." However, in this book, I present them in the order cosine, sine, tangent, ... Why? My experience has shown that if the student always remembers that the cosine is first, the material will be easier to remember. Cosine comes first alphabetically, it is associated with the *x* (the *first* component), with the *first* component on a unit circle, and with the *first* two quadrants when dealing with the inverse functions. Throughout the book I have kept in mind how a student learns and have presented the processes as smoothly and naturally as possible for the student who is seeing this material for the first time.

Changes for the Third Edition

This is a significant revision from the previous edition. It has been twelve years since this book was last revised and, during the past decade, trigonometry has changed dramatically. Virtually every student today has access to a calculator to evaluate trigonometric functions, and many students have graphing calculators and software to enhance their understanding of the material. This book seeks to use those technologies to enhance the learning of the mathematics, but beyond evaluating the trigonometric functions, is not dependent on technology.

Since this is an extensive revision from the previous edition, I would like to give you a chapter-by-chapter overview.

Chapter 1 This chapter introduces angles and degree measure of angles. Arc length has been moved from the first section to the last section, which allows for a more focused presentation. The use of calculators to evaluate the trigonometric functions has caused the delay of the introduction of the secant, cosecant, and cotangent functions to Chapter 2 because knowledge of the reciprocal functions is necessary for calculator evaluation. Solving right triangles requires only the functions of cosine, sine, and tangent, and the focus of this chapter is to build a firm foundation of trigonometry in a setting that is concrete and reassuring for the students.

Chapter 2 This chapter introduces angles and radian measure of angles. Eight essential identities, which we call fundamental identities, are considered in Chapter 2. By the time the students consider solving equations and proving other identities, they will have learned these eight essential trigonometric relationships.

Chapter 3 Graphing is introduced early and given special emphasis throughout the book. You will find an innovative and unique method called *framing,* which is used to quickly and efficiently communicate the concepts of amplitude, period, and phase displacement without the usual difficulties of plotting "ugly coordinates." After teaching this method for over 25 years to both students and instructors, I am convinced that, by using this technique, your students will have a better understanding of graphing more complicated equations—with less class time.

Chapter 4 This chapter discusses trigonometric equations and identities. This edition has a much more focused presentation than the previous editions. I believe that students do much better with trigonometric identities if they are studied in the context of equation solving; that is, an *identity* is an equation that is true for all replacements of the variable. Consequently, simple equations are solved first and then identities are proved. After proving several identities, we return to solving more involved equations. Calculus students must be able to solve trigonometric equations and use trigonometric identities in a variety of contexts.

Chapter 5 This chapter presents solving oblique triangles and vectors. I have emphasized the concepts, rather than the mechanics involved in solving triangles, and have taken time in the text to help the student choose a method for solution. I think the immediate introduction to the law of cosines eliminates the confusion that sometimes arises with the law of sines and its ambiguous case. The law of cosines eliminates the need even to introduce the ambiguous case, since the quadratic formula leads directly to the proper analysis of the problem.

Chapter 6 This chapter considers complex numbers, De Moivre's theorem, and rectangular and trigonometric forms of a number. This topic leads naturally to the polar-form representation of a point and graphing in polar form.

Chapter 7 At one time, logarithms were needed in a trigonometry course to do the calculations required in solving triangles, but that is no longer the case. Over the last decade we have seen a transition in the treatment of logarithms from a computational topic to a functional topic. The treatment of logarithms in this text is not simply lifted from an algebra book, but is presented in the context of trigonometry. For example, in Section 7.1, harmonic motion of a damped vibration curve, which requires both e and a sine or cosine function, is introduced.

Appendices Topics that are needed as background for this text are included in the appendices for easy review and reference.

Supplementary Materials

Instructor's Solutions Manual This manual includes teaching suggestions, transparency masters, and answers to all problems in the text (0-534-35208-1).

ITP Tools (Macintosh and Windows) ITP Tools, a fully integrated suite of programs that includes ITP Test/Tutorial, ITP Class Manager, and ITP Test On-Line, provides you with text-specific algorithmic testing and tutorial options designed to offer you greater flexibility and your students greater continuity. The computer platforms supported include Windows 3.1, Windows95, Windows NT, Macintosh, and Power-Macintosh.

For a more complete description of each component of ITP Tools and to download a self-running demonstration, please visit Brooks/Cole's web site: http://www.brookscole.com (Mac: 0-534-35210-3; Win: 0-534-35211-1).

Westest DOS 4.01 This computerized testing system allows instructors to create, edit, store, and print exams directly to word processor format from algorithmic, text-specific test banks. Professors can create test banks by copying test items individually, as a range of test items, using numerical select, or by random selection. Includes integrated computer Gradebook (0-534-35212-X).

Printed Test Items (0-534-35213-8).

Boxer Trigonometry CD-ROM Lab Pack This multimedia Windows CD-ROM provides a step-by-step instructional approach and visual simulations that allow you to explore trigonometric concepts. Lab packs available for qualified adoptions (0-534-35685-0).

Student Solutions Manual Includes detailed step-by-step solutions for selected odd-numbered problems (0-534-35209-X).

Converge This supplement (by John Mobray) provides graphing technology for the IBM and is available from JEMware, The Kawaiahao Plaza Executive Center, 567 South King Street, Suite 178, Honolulu, HI 96813-3076 (phone 808-523-9911).

Acknowledgments

I am grateful to the following persons who reviewed the manuscript for this edition: Jay Abramson, *Amarillo College;* Mitch Anderson, *University of Hawaii at Hilo;* Kolman Brand, *Nassau Community College;* Debbie Cochener, *Austin Peay State University;* David Dubriske, *Arkansas State University;* Bonnie Hodge, *Austin Peay State University;* Jim Keefe, *Denver Institute of Technology;* Lee Minor, *Western Carolina University;* Bette L. Warren, *Eastern Michigan University;* and James F. Gordon, a friend and very meticulous reader.

I would also like to thank the reviewers of the previous editions: Martin Brown, *Jefferson Community College;* Julius Burkett, *Stephen F. Austin University;* Carroll Commons, *Tarrant County Community College;* Art Dull, *Diablo Valley College;* George Edenharder, *Milwaukee School of Engineering;* Molly Fails, *Terra Technical College;* Marjorie Freeman, *University of Houston—Downtown College;* Carolyn Funk, *Thornton Community College;* Jack Goebel, *Montana College of Mineral Science and Technology;* Mary Lou Hart, *Brevard Community College;* Bill Hinds, *Midwestern State University;* Louis Hoelzle, *Bucks Community College;* Helen Kriegsman, *Pittsburg State University;* Wendell Motter, *Florida A&M University;* Gilbert Nelson, *North Dakota State University;* Charlene Pappin, *Wright City Community College;* Vicki Schell, *Pensacola Junior College;* Dorothy Schwellenbach, *Hartnell College;* Donna Szott, *Community College of Allegheny County, South Campus;* Ann Thorne, *College of Du-Page;* Jon Weerts, *Triton College;* Jerry Wilkerson, *Missouri Western State University;* and Barry Wood, *Santa Rosa Junior College.*

My thanks also go to Michael W. Ecker and Bobby May for their careful reading of the book, for working all of the examples and problems, and for their insights and suggestions for improving the textbook. I am grateful to the people at Brooks/Cole who worked on this book, including Elizabeth Rammel, Tessa

McGlasson, and Bob Pirtle. There are many who have helped with this publication: Sue C. Howard, who researched photographs; Katherine Townes, who designed the endpapers and quick reference card; and Lori Heckelman, who rendered completely new art for this edition.

I would especially like to thank my production editor, Susan Reiland, who has the ability to find my mistakes, keep my spirits up, and produce a first-class book. Without her, this book would have been much more difficult to publish.

And last, but not least, I would like to thank my wife and partner, Linda, who has always been there for me in every possible way.

Karl J. Smith
Sebastopol, CA
e-mail: smithkjs@wco.com

TO THE STUDENT

Each new course begins with a sense of anticipation, and this course is probably no exception. You are enrolling in a course and your instructor has adopted this textbook. What can you expect? You will learn about the *trigonometric functions* and their properties. Whether or not you have previously had a right-triangle definition of the trigonometric functions, Chapter 1 will provide a solid foundation for your understanding of the material of this course. We will give (in Chapter 2) a much more general definition of a trigonometric function. It is important that you thoroughly understand this definition as it is discussed in this book.

You will find that the course will be much easier if you *read the textbook prior to discussing the material in class.* Do not assume you are finished with the day's work as soon as you have finished the day's homework assignment. Read, study, and *understand* the important ideas. There will be some terms and facts you will need to *memorize.* I have tried to make it easy for you to know what you need to do; follow the advice of the street signs. Definitions and formulas are enclosed in boxes, procedures are in screened boxes; new terms are shown in boldface and are again listed at the end of each chapter. Chapters begin with a preview and end with a list of objectives, so you should know exactly what is expected of you. If you are not sure what is important, ask your instructor, or give me a call. My home phone number is (707) 829-0606.

You will also see that answers to all the problems in the sample tests and cumulative tests are included in the answer section to help you prepare for examinations.

Use your book and make it your own. Books are expensive, and I hope you see this book as an investment that will pay off not only for this course, but also as a valuable reference book in your future mathematics courses. If you intend to continue your study in mathematics, you will want to make this book part of your professional library so that you have access to the notation and formulas developed in this course.

CONTENTS

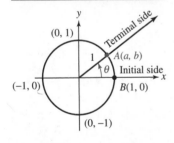
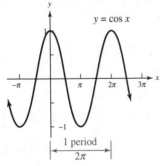

CHAPTER FOUR

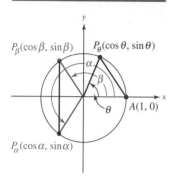

TRIGONOMETRIC EQUATIONS AND IDENTITIES 163

CHAPTER FIVE

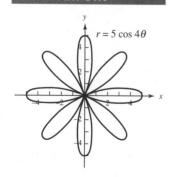

OBLIQUE TRIANGLES AND VECTORS 222

CHAPTER SIX

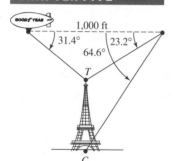

COMPLEX NUMBERS AND POLAR-FORM GRAPHING 281

CHAPTER SEVEN

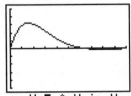

Y₁◼e^-Xsin X
Xmin=0   Ymin=-.5
Xmax=5   Ymax=.5
Xscl=.5 Yscl=.1

LOGARITHMIC FUNCTIONS 332

Y₁◼sin X/X
Xmin=-6 Ymin=-.5
Xmax=6   Ymax=1.2
Xscl=1   Yscl=.5

APPENDICES 383

INDEX 458

CHAPTER ONE

RIGHT-TRIANGLE TRIGONOMETRY

PREVIEW

The origins of trigonometry are obscure, but we know that it existed long before calculus. Its roots go back more than 2,000 years to the Mesopotamian, Babylonian, and Egyptian civilizations. The ancients used trigonometry in a very practical way to measure triangles as a means of surveying land. Today, trigonometry is used in a much more theoretical way, not only in mathematics, but also in electronics, engineering, and computer science. Trigonometry is that branch of mathematics that deals with the properties and applications of six functions: cosine, sine, tangent, secant, cosecant, and cotangent. The first three of these functions are defined in this chapter, and the last three follow in the next chapter.

We begin this chapter by investigating angle measurement. Most students are familiar with degree measure; thus, we develop this concept first. Historically, degree measure was used before the decimal numeration system we use today. For more advanced work, a more convenient system of measurement for angles, called radian measure, is used. We study radian measure in the second chapter.

In the second section, we review, from geometry, the notion of similar triangles to lay the foundation for the definition of trigonometric functions.

In the third section, the right-triangle definition of the trigonometric function ratios is given, which leads to an important application of trigonometry, called solving a right triangle. We introduce a variety of applications involving solutions of triangles or finding parts of a triangle.

The chapter concludes by finding the length of an arc. The foundation for our study of trigonometric functions in Chapter 2 is laid in the last section of this chapter by considering angles that are not angles of any triangle.

Mathematics is the instrument by which the engineer tunnels our mountains, bridges our rivers, constructs our aqueducts, erects our factories and makes them musical by the busy hum of spindles. Take away the results of the reasoning of mathematics, and there would go with it nearly all the material achievements which give convenience and glory to modern civilization.
  Edward Brooks
  Mental Science and Culture (Philadelphia: Normal Publishing, 1891, p. 255).

ANGLES AND DEGREE MEASURE

Measurement

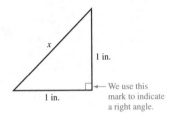

FIGURE 1.1
What is the length of side x?

PROBLEM OF THE DAY

What is the length of the segment labeled x in Figure 1.1? (Assume that the length of each side is exactly one inch.)

How would you answer this question?

***

1. Use a ruler?
If you use a ruler calibrated in inches, you might answer by saying the length of the segment is $1\frac{7}{16}$ in. If you use a ruler calibrated in centimeters, you might say the length is 3.5 cm.

2. Use a mathematical formula?
You might recognize this as a right triangle (see Appendix D) and remember the **Pythagorean theorem** from geometry. In a right triangle with legs of length a and b and hypotenuse c,

$$a^2 + b^2 = c^2$$

Thus, if legs a and b each have measure 1,

$$1^2 + 1^2 = x^2$$
$$2 = x^2$$

From this quadratic equation, we know that $x = \pm\sqrt{2}$ (square root property). However, to answer the question, recall that we are measuring the length of a segment, so the answer is

$$x = \sqrt{2}$$

An approximate value can be found with a calculator:

$$x \approx 1.414213562$$

This introductory exercise raises several issues that must be considered in this course:

1. Measurements are never exact.
2. There may be more than one possible measurement unit. For this example, we used one ruler calibrated in inches and another ruler calibrated in centimeters. We could have chosen other measuring instruments. The instrument that we use often determines the unit of measurement we use.
3. Since measurements are never exact, we need to decide how *precise* the measure should be. For example, the measurement might be to the nearest

Calculator Window **?** **X**

As noted in both the preface and Appendix A, we expect you to use a calculator with this textbook. We assume that you know how to carry out the basic operations; we will not show you the keystrokes for arithmetic operations. You can check your owner's manual, but we assume that your calculator recognizes the correct order of operations.

A NOTE TO THE STUDENT
Did you read the preface of this book? You will find some important information there. Here is a reminder of some features of this book that will ease your progress. These road signs are signals to help you understand trigonometry.

When you see the stop sign, you should stop for a few moments and study the material by the stop sign. Memorize this material.

When you see the warning sign, you should make a special note of some mentioned fact because it will be used throughout the rest of the book.

When you see the yield sign, it means that you need only remember the stated result, and that the derivation is optional.

When you see the bump sign, some unexpected or difficult material follows, and you will need to slow down to understand the discussion.

inch, foot, or even nearest mile. For this Problem of the Day, we decided to measure to the nearest sixteenth of an inch or the nearest tenth of a centimeter. The *precision* of our measurement is often determined by our purpose. For example, if we are measuring a room to lay carpet, the precision of our measurement might be different than when we are measuring an airport hangar.

4. When using a mathematical formula, we sometimes assume that the given information represents exact values (the legs are one inch long); thus, we say that the answer from the Pythagorean theorem is *exact*. That is, we would say that $\sqrt{2}$ is an exact representation of the length of the segment, whereas if we use a calculator, we would say that 1.414213562 is an *approximate* representation of the answer. Note the use of the *approximately equal to* symbol, $\approx$.

5. The *accuracy* of an answer depends on the information that we are given for a particular problem. Suppose that we measure the length of the sides of the triangle (to the nearest tenth centimeter) to be 2.5. Then we use the Pythagorean theorem to find

$$2.5^2 + 2.5^2 = x^2 \qquad \text{\textit{Write } } a = 2.5 \text{ \textit{and} } b = 2.5.$$
$$12.5 = x^2 \qquad \text{\textit{Note that } } 2.5^2 + 2.5^2 \neq (2.5 + 2.5)^2.$$
$$x = \sqrt{12.5} \qquad \text{\textit{Square root property (positive for length)}}$$

If we use a calculator to evaluate $\sqrt{12.5} \approx 3.535533906$, the answer should reflect the same degree of accuracy as the given information: $\sqrt{12.5} \approx 3.5$ (since the sides were measured to the nearest tenth of a centimeter).

We use the following rule throughout this book.

ACCURACY IN THIS BOOK

All measurements are as precise as given in the text. When we ask you to make a measurement, the precision will be specified. Carry out all calculations without rounding. When you use a calculator, use all decimal places displayed during successive computations. After you obtain a *final answer,* round this answer so that it is as accurate as the *least precise* measurement.

We will illustrate this rule in various examples throughout this book, so it is not necessary to work accuracy examples at this time. However, if you would like to practice this concept, you can refer to Appendix B.

We have mentioned the **Pythagorean theorem,** and because it is an important result in trigonometry, we repeat it here for emphasis.

> ### PYTHAGOREAN THEOREM
>
> The sum of the squares of the legs of a right triangle is equal to the square of the hypotenuse. That is,
>
> $$a^2 + b^2 = c^2$$
>
> Also, if $a^2 + b^2 = c^2$, then the triangle with sides of lengths a, b, and c is a right triangle.

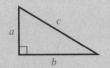

EXAMPLE 1

Finding the Lengths of Sides of a Right Triangle

If $a = 2\sqrt{3}$ and $b = 2$, find c if a, b, and c are sides of a right triangle.

Solution

$$a^2 + b^2 = c^2$$
$$(2\sqrt{3})^2 + (2)^2 = c^2$$
$$4 \cdot 3 + 4 = c^2$$
$$16 = c^2$$
$$4 = c$$

Note: The square root property tells us that $c = \pm4$, but since c is the length of a side of a triangle, the positive value is the desired length.   ∎

Angles

Essential to the definition of the trigonometric functions is the idea of an angle. In geometry, an angle is usually defined as the union of two rays with a common endpoint. In trigonometry and calculus, a more general definition is used.

> ### ANGLE
>
> An **angle** is formed by rotating a ray about its endpoint (called the **vertex**) from some initial position (called the **initial side**) to some terminal position (called the **terminal side**). The measure of an angle is the amount of rotation from the initial side to the terminal side.

If the rotation of the ray is in a counterclockwise direction, the measure of the angle is called **positive,** and if the rotation is in a clockwise direction, the measure is called **negative.** The notation $\angle ABC$ means the measure of an angle with vertex B and points A and C (different from B) on the sides; $\angle B$ denotes the measure of an angle with vertex at B, and a curved arrow is used to denote the direction and amount of rotation, as shown in Figure 1.2. If no arrow is shown, the measure of the angle is considered to be the smallest positive rotation.

TABLE 1.1 Commonly
Used Greek Letters

Symbol	Name
α	alpha
β	beta
γ	gamma
δ	delta
θ	theta
μ	mu
ϕ or φ	phi
λ	lambda
ω	omega

Note: The lowercase Greek
letter π always represents the
irrational number approximately equal to 3.141592654.

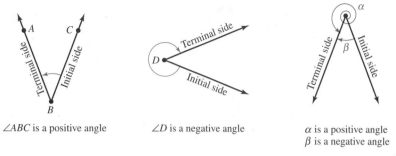

$\angle ABC$ is a positive angle $\angle D$ is a negative angle α is a positive angle
                                                                          β is a negative angle

FIGURE 1.2   Examples of angles

Lowercase Greek letters signify both the angles and the measure of angles.
For example, θ may represent the angle or the measure of the angle called
θ; you will know which is meant by the context in which it is used. Some
commonly used Greek letters are shown in Table 1.1.

Degree Measurement of Angles

Several units of measurement are used to measure angles, and we begin with
one with which you are probably familiar. Let α be an angle with a point P (not
the vertex) on the terminal side. As this side is rotated through one revolution,
the trace of the point P forms a circle. It is easy to understand how angles of
one or more revolutions are measured, but much of our work is with angles
less than one revolution, so we will define measures that are smaller than one
revolution. Historically, the most common scheme divides one revolution into
360 equal parts, with each part called a **degree.** Sometimes even finer divisions
are necessary, so a degree is divided into 60 equal parts, each called a **minute**
($1° = 60'$), which is further divided into 60 equal parts, each called a **second**
($1' = 60''$). For most applications, we write decimal degrees instead of minutes
and seconds. That is, 41.5° is preferred over 41°30′. A protractor for degree
measure is shown in Figure 1.3.

Historical Note

The degree was first
symbolized by the letter *G*
for *gradus*, which means
"degree." We see this
symbol in a calendar
compiled by Johannes
Regiomontanus.

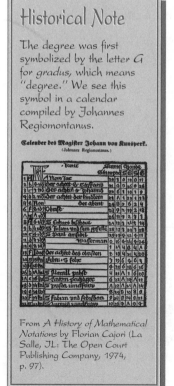

From *A History of Mathematical
Notations* by Florian Cajori (La
Salle, JL: The Open Court
Publishing Company, 1974,
p. 97).

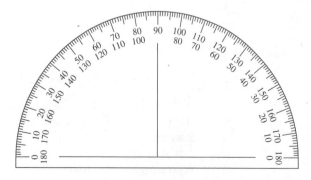

FIGURE 1.3   Protractor for degree measure

You will not need to use a protractor for the problem sets in this book, but you will need to know the approximate sizes of various angles measured in degrees, as shown in Figure 1.4.

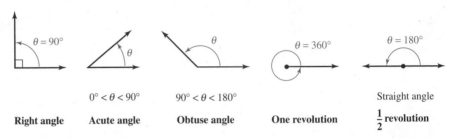

Right angle	Acute angle	Obtuse angle	One revolution	$\frac{1}{2}$ revolution

FIGURE 1.4   Types of angles

Other common angles are illustrated in Example 2. When we write $m\angle B = 45°$, we mean that the measure of angle B is 45°, and when we write $\theta = 30°$, we mean that the measure of angle θ is 30°.

EXAMPLE 2

Drawing an Angle with a Given Degree Measure

Sketch each angle from memory (without using a protractor).

a. $\alpha = 30°$ **b.** $m\angle G = 45°$ **c.** $\beta = -60°$ **d.** $\gamma = 270°$
e. $m\angle ABC = 135°$

Solution

a.   **b.**   **c.**

d. $\gamma = 270°$ **e.**

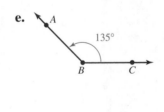

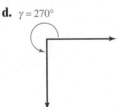

Remember the approximate sizes of the angles 0°, 30°, 45°, 60°, 90°, 180°, 270°, and 360°; these are used frequently in mathematics.

With the widespread use of calculators, it is common to represent minutes and seconds as decimal degrees. In this book, we use decimal degrees rather than minutes and seconds. However, you may still encounter an old reference that uses degrees/minutes/seconds that you may wish to convert to decimal degrees. This is illustrated in Example 3.

EXAMPLE 3   **Converting Degree Measure to Decimal Degrees**

Convert the given angles to decimal degrees.

a. 25°30′   **b.** 128°14′   **c.** 42°13′40″

Solution   **a.** Since 1° = 60′, the number of minutes should be divided by 60 to convert it to decimal degrees.

$$30' = \frac{30°}{60} = \left(\frac{1}{2}\right)° = 0.5°$$

Thus, 25°30′ = 25.5°.

b. $14' = \dfrac{14°}{60} = 0.2\overline{3}° \approx 0.23°$

If an angle is measured in minutes, then round the converted decimal to the nearest hundredth of a degree.

Thus, $128°14' = 128° + \left(\dfrac{14}{60}\right)° \approx 128.23°$.

c. Since 1° = 60′ = 3,600″, we have

$$13'40'' = \left(\frac{13}{60}\right)° + \left(\frac{40}{3,600}\right)° = \left(\frac{820}{3,600}\right)°$$

By division,

$$\frac{820}{3,600} = 0.22\overline{7}$$

If an angle is measured in seconds, then round the converted decimal to the nearest thousandth of a degree.

Thus, $42°13'40'' = 42° + \left(\dfrac{13}{60}\right)° + \left(\dfrac{40}{3,600}\right)° \approx 42.228°$. ■

Because the degree/minute/second notation is, for the most part, obsolete, it is not necessary to practice converting from decimal degrees to degrees/minutes/seconds. However, as an intellectual exercise, you might wish to describe this process, which is requested in Problem 41.

PROBLEM SET 1.1

Essential Concepts

Problems 1–2: *Name the given Greek letter.*

1. a. θ **b.** α **c.** ϕ **d.** ω **e.** μ

2. a. γ **b.** β **c.** δ **d.** λ **e.** φ

Problems 3–4: *Write the symbol for the given Greek letter.*

3. a. delta   **b.** phi   **c.** theta   **d.** omega

4. a. alpha   **b.** beta   **c.** gamma   **d.** lambda

Level 1 Drill

Problems 5–7: *Indicate the number of degrees in the angle formed by the terminal side rotating counterclockwise.*

5. a. 1 revolution   **b.** $\frac{1}{2}$ revolution

6. a. $\frac{2}{3}$ revolution   **b.** $\frac{1}{12}$ revolution

7. a. $\frac{1}{4}$ revolution   **b.** $\frac{3}{8}$ revolution

Problems 8–10: *Indicate the number of degrees in the angle formed by the terminal side rotating clockwise.*

8. a. $\frac{5}{8}$ revolution   **b.** $\frac{3}{4}$ revolution

9. a. $\frac{1}{4}$ revolution   **b.** $\frac{1}{8}$ revolution

10. a. $\frac{1}{3}$ revolution   **b.** $\frac{1}{36}$ revolution

Problems 11–14: *Without using a protractor, draw an angle that approximates each angle with the measure indicated. Classify each angle as acute, right, obtuse, straight, or none of these.*

11. a. 60°   **b.** 180°   **c.** 45°   **d.** 360°   **e.** 300°

12. a. 90°   **b.** 30°   **c.** 0°   **d.** 270°   **e.** 150°

13. a. −30°   **b.** 135°   **c.** −120°   **d.** −270°   **e.** −200°

14. a. −60°   **b.** 315°   **c.** −330°   **d.** −45°   **e.** −90°

Level 2 Drill

Problems 15–18: *Change the given measures to decimal degrees to the nearest hundredth.*

15. a. 65°40′   **b.** 146°50′   **c.** 85°20′

16. a. 127°10′   **b.** 240°30′   **c.** 315°25′

17. a. 62°55′   **b.** 315°20′   **c.** 25°25′

18. a. 242°50′   **b.** 205°18′   **c.** 325°58′

Problems 19–22: *Change the given measures to decimal degrees to the nearest thousandth.*

19. a. 128°10′40″   **b.** 13°30′50″   **c.** 48°28′10″

20. a. 281°31′36″   **b.** 210°40′15″   **c.** 94°21′31″

21. a. 16°42″   **b.** 29°17″   **c.** 143°23″

22. a. 38′42″   **b.** 12′24″   **c.** 6′7″

Problems 23–32: *Remember from geometry (see Appendix D) that the sum of the three angles of a triangle is 180°. Use Figure 1.5 to find the listed angles.*

23. β if $\theta = 60°$

24. β if $\theta = 40°$

25. γ if $\phi = 45°$

26. γ if $\phi = 30°$

27. α in terms of β and γ

28. β in terms of α and γ

29. γ in terms of α and β

30. α if $\theta = 35.5°$ and $\phi = 16.9°$

31. β if $\theta + \phi = 100°$ and $\theta = \phi$

32. γ if $\theta + \phi = 80°$ and $\phi = \frac{2}{3}\theta$

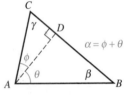

$\alpha = \phi + \theta$

FIGURE 1.5
Angles in triangles

Problems 33–40: *Assume that a and b are legs of a right triangle and c is the hypotenuse. Use the Pythagorean theorem to find the missing length(s). If all three sides are given, find x and then give the lengths of the sides.*

33. $a = 2, b = 2$ **34.** $a = 8, b = 6$

35. $a = 5, c = \sqrt{41}$ **36.** $b = 3, c = 3\sqrt{5}$

37. $a = x, b = x + 7, c = 13$

38. $a = x, b = x + 1, c = 5$

39. $a = x − 1, b = x, c = \sqrt{13}$

40. $a = x, b = x + 2, c = \sqrt{10}$

41. IN YOUR OWN WORDS Describe a process for changing from decimal degrees to degrees/minutes/seconds.

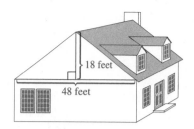

FIGURE 1.7

Level 2 Applications

42. In Figure 1.6, what is β if we want the post to be vertical and if we know that the angle the stairway makes with the horizontal is 15°?

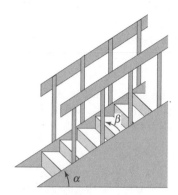

FIGURE 1.6   Stairway

43. What is the length of one side of the roof (rounded to the nearest inch) of the house shown in Figure 1.7? Assume the distance across the house (called the span) is 48 ft and the roof extends 18 ft above the horizontal.

Level 3
Individual Project

44. Historical Question The division of one revolution (a circle) into 360 equal parts (called *degrees*) is no doubt due to the sexagesimal (base 60) numeration system used by the Babylonians. Several explanations have been put forward to account for the choice of this number. (For example, see Howard Eves' *In Mathematical Circles.*) One possible explanation is based on the fact that the radius of a circle can be applied exactly six times to its circumference as a chord. Show this with a sketch. Notice that sexagesimal division (that is, division into 60 equal parts) makes this geometric construction easy. It seems natural that each chord should be further divided into 60 equal parts, resulting in the division of the circle into 360 equal parts. Write a short research paper on the history of degree measure for an angle.

SECTION 1.2

SIMILAR TRIANGLES

Measurement

Let h = height of tree

PROBLEM OF THE DAY

Suppose you want to find the height of a tall tree, and do not have the means to take the measurement directly. How can you determine the height of the tree?

How would you answer this question?

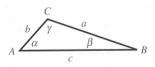

FIGURE 1.8
Correctly labeled triangle

Vertices	Sides	Angles
A	a	α
B	b	β
C	c	γ

Historical Note

The origins of trigonometry
are obscure, but we do
know that it began more
than 2,000 years ago in
Mesopotamia, Babylonia,
and Egypt. The word
trigonometry comes from
the words "triangle"
(*trigon*) and "measure-
ment" (*metry*). Some of the
early names associated
with trigonometry are
Ptolemy, known as
Claudius Ptolemy,
Hipparchus, and
Menelaus. Their goal was
to build a quantitative
astronomy that could be
used to predict the paths
of the heavenly bodies, and
to aid in telling time,
calendar reckoning,
navigating, and studying
geography.

To answer this question, we will begin with the idea of similar triangles, and then in the next section we will use this idea to motivate a definition of three trigonometric functions.

One of the most important uses of trigonometry is finding certain measurements in a triangle when other measurements in that triangle are given. Recall from geometry that every triangle has three sides and three angles, which are called the *six parts* of the triangle. A typical triangle is labeled in Figure 1.8. The vertices are labeled A, B, and C, with the sides opposite those vertices labeled a, b, and c, respectively. The angles are labeled α, β, and γ, respectively. In this section, we assume that γ denotes the right angle and c the *hypotenuse* (the longest side) of the right triangle. Also, remember that for *any* triangle, the sum of the measures of the angles is 180°.

Similar Triangles

It is possible for two figures to have the same shape, but not necessarily the same size. Such figures are called *similar figures.* We will now focus on **similar triangles.** If $\triangle ABC$ is similar to $\triangle DEF$, we write

$$\triangle ABC \sim \triangle DEF$$

Similar triangles are shown in Figure 1.9.

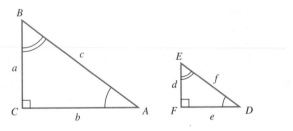

FIGURE 1.9   Similar triangles

Since similar figures have the same shape, we talk about **corresponding angles** and **corresponding sides.** The corresponding angles of similar triangles are the angles that have the same measure. It is customary to label the vertices of triangles with capital letters and the sides opposite the angles at those vertices with corresponding lowercase letters. It is easy to see that, if the triangles are similar, the corresponding sides are the sides opposite equal angles. In Figure 1.9, we see that

$\angle A$ and $\angle D$ are corresponding angles;
$\angle B$ and $\angle E$ are corresponding angles; and
$\angle C$ and $\angle F$ are corresponding angles.

Side a ($\overline{BC}$) is opposite $\angle A$, and side d ($\overline{EF}$) is opposite $\angle D$, so we say that a corresponds to d; b corresponds to e; and c corresponds to f.

Even though corresponding angles are the same size, corresponding sides do not need to have the same length. If they do have the same length, then the triangles are congruent. However, when they are not the same length, we can say they are *proportional.* As Figure 1.9 illustrates, when we say the sides are proportional, we mean

NOTE

There are six ratios that can be formed by the sides of a triangle and these ratios are the basis of the definition of the trigonometric functions, which are stated later.

$$\frac{a}{b} = \frac{d}{e} \qquad \frac{a}{c} = \frac{d}{f} \qquad \frac{b}{c} = \frac{e}{f}$$

$$\frac{b}{a} = \frac{e}{d} \qquad \frac{c}{a} = \frac{f}{d} \qquad \frac{c}{b} = \frac{f}{e}$$

SIMILAR TRIANGLES

Two triangles are **similar** if two angles of one triangle have the same measure as two angles of the other triangle. If the triangles are similar, then their corresponding sides are proportional.

EXAMPLE 1   **Similar Triangles**

Identify pairs of triangles that are similar in Figure 1.10.

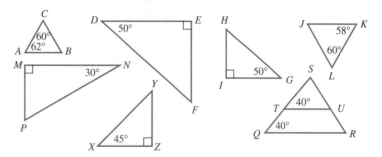

FIGURE 1.10   Identifying similar triangles

Solution $\triangle ABC \sim \triangle JKL;\ \triangle DEF \sim \triangle GIH;\ \triangle SQR \sim \triangle STU.$ ■

EXAMPLE 2   **Finding Unknown Lengths in Similar Triangles**

Given the similar triangles below, find the unknown lengths marked b' and c'.

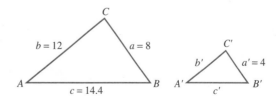

Solution Since corresponding sides are proportional (other proportions are possible), we have

$$\frac{a'}{a} = \frac{b'}{b} \qquad \frac{a}{c} = \frac{a'}{c'}$$

$$\frac{4}{8} = \frac{b'}{12} \qquad \frac{8}{14.4} = \frac{4}{c'}$$

$$b' = \frac{4(12)}{8} \qquad c' = \frac{14.4(4)}{8}$$

$$= 6 \qquad\qquad = 7.2 \qquad\qquad\blacksquare$$

EXAMPLE 3 **Finding a Perimeter Using Similar Triangles**

In equilateral $\triangle ABC$, suppose $|\overline{DE}| = 2$ and is parallel to $\overline{AB}$, as shown at the right. If $|\overline{AB}|$ is three times as long as $|\overline{DE}|$, what is the perimeter of quadrilateral $ABED$?

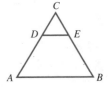

Solution $\triangle ABC \sim \triangle DEC$, so $\triangle DEC$ is equilateral. This means that $\overline{CE}$ and $\overline{DC}$ both have length 2; thus, $\overline{EB}$ and $\overline{AD}$ both have length 4. The perimeter of the quadrilateral is

$$|\overline{AB}| + |\overline{BE}| + |\overline{DE}| + |\overline{AD}| = 6 + 4 + 2 + 4 = 16 \qquad\blacksquare$$

Finding similar triangles is simplified even further if we know that the triangles are right triangles; in that case the triangles are similar if one of the acute angles has the same measure as an acute angle of the other.

EXAMPLE 4 **Problem of the Day: Using Similar Triangles to Find an Unknown Length**

Suppose that a tree and a yardstick are casting shadows as shown in Figure 1.11. If the shadow of the yardstick is 3 yards long and the shadow of the

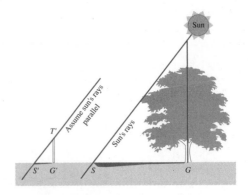

FIGURE 1.11 Height of a tree

tree is 12 yards long, use similar triangles to estimate the height of the tree if you know that angles S and S' are the same size.

Solution   Since $\angle G$ and $\angle G'$ are right angles, and since S and S' are the same size, we see that $\triangle SGT \sim \triangle S'G'T'$. Therefore, corresponding sides are proportional. Since the length of the yardstick's shadow is 3 times the length of the stick, the tree's shadow should be 3 times its height.

$$\frac{1}{3} = \frac{h}{12}$$

$$h = \frac{1(12)}{3}$$

$$= 4$$

The tree is 4 yards tall.   ∎

PROBLEM SET 1.2

Essential Concepts

1. What does it mean for a triangle to be labeled in a standard way?

2. What does it mean for triangles to be similar?

Level 1 Drill

Problems 3–6: *Name the corresponding parts of the triangles.*

3.
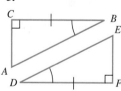

4.
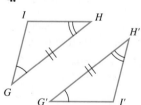

5.
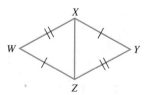

6.
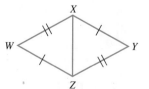

Problems 7–12: *Tell whether it is possible to conclude that the pairs of triangles are similar.*

7.
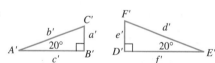

8.
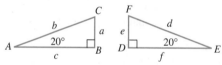

9.
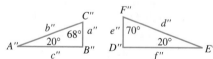

10.
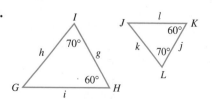

11.

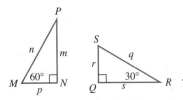

12.

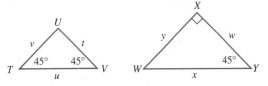

Level 2 Drill

Problems **13–18:** *Given the two similar triangles shown, find the unknown lengths.*

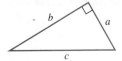

13. $a = 4$, $b = 8$, $a' = 2$; find b'.

14. $b = 5$, $c = 15$, $b' = 2$; find c'.

15. $c = 6$, $a = 4$, $c' = 8$; find a'.

16. $a' = 7$, $b' = 8$, $a = 5$; find b.

17. $b' = 8$, $c' = 12$, $c = 4$; find b.

18. $c' = 9$, $a' = 2$, $c = 5$; find a.

Level 2 Applications

19. How far from the base of a building must a 26-ft ladder be placed so that it reaches 10 ft up the wall?

20. How high up a wall does a 26-ft ladder reach if the bottom of the ladder is placed 6.0 ft from the building?

21. A carpenter wants to be sure that the corner of a building is square and measures 6.0 ft and 8.0 ft along the sides. How long should the diagonal be?

22. What is the exact length of the hypotenuse if the legs of a right triangle are 2 in. each?

23. What is the exact length of the hypotenuse if the legs of a right triangle are 3 ft each?

24. An empty lot is 40 ft by 65 ft. How many feet would you save by walking diagonally across the lot instead of walking the length and width?

25. A television antenna is to be erected and held by guy wires. If the guy wires are 15 ft from the base of the antenna and the antenna is 10 ft high, what is the exact length of each guy wire? What is the length of each guy wire rounded to the nearest foot? If three guy wires are attached, how many feet of wire should be purchased, if it cannot be bought in fractions of a foot?

26. In equilateral $\triangle ABC$, shown in Figure 1.12, D is the midpoint of segment $\overline{AB}$. What is the length of $\overline{CD}$?

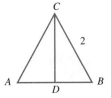

FIGURE 1.12   Height of a triangle

27. In Figure 1.13, $\overline{AB}$ and $\overline{DE}$ are parallel. What is the length of $\overline{AB}$?

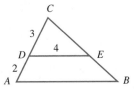

FIGURE 1.13   Length of a base

28. Ben walked diagonally across a rectangular field that measures 100 ft by 240 ft. How far did he walk?

29. Use similar triangles and a proportion to find the length of the lake shown in Figure 1.14.

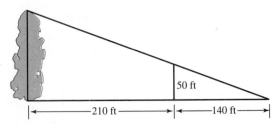

FIGURE 1.14   Finding the length of a lake

30. Use similar triangles and a proportion to find the height of the house shown in Figure 1.15.

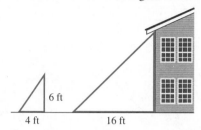

6 ft

4 ft 16 ft

FIGURE 1.15 Finding the height of a building

31. If a tree casts a shadow of 12 ft at the same time a 6-ft person casts a shadow of $2\frac{1}{2}$ ft, find the height of the tree to the nearest foot.

32. If a circular cone of height 10 cm and a radius of 4 cm contains liquid with height measuring 3.8 cm, what is the volume of the liquid?

SECTION 1.3

TRIGONOMETRIC RATIOS

PROBLEM OF THE DAY

Let h = height of tree

The Problem of the Day in Section 1.2 found the height of a tree by using similar trianglels and shadows. Now suppose it is a cloudy day and there are no shadows. What is the height of the tree?

How would you answer *this* question?

We will answer this Problem of the Day by considering a right triangle and the Pythagorean theorem, as well as a very important definition of what are called the trigonometric functions. In the previous section, you may have noticed that given any triangle, there are three ratios (fractions), along with their reciprocals, that can be formed with the three sides of a right triangle.

Right-Triangle Definition

Consider a right triangle and an acute angle θ; that is, θ is not the right angle of the right triangle. Let c be the hypotenuse, and a and b the legs. Then

$$c^2 = a^2 + b^2 \quad \textit{Pythagorean theorem}$$

Recall that the word **hypotenuse** refers to the longest side (the side opposite the right angle). The angle consists of two sides: the hypotenuse and the **adjacent side; the opposite side** is the side of the triangle that is not a side of the given angle.

This lead us to a very important definition.

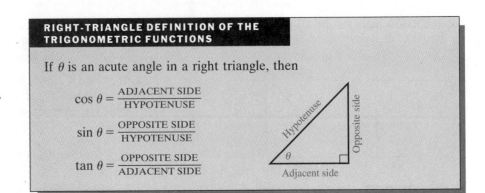

RIGHT-TRIANGLE DEFINITION OF THE TRIGONOMETRIC FUNCTIONS

If θ is an acute angle in a right triangle, then

$$\cos \theta = \frac{\text{ADJACENT SIDE}}{\text{HYPOTENUSE}}$$

$$\sin \theta = \frac{\text{OPPOSITE SIDE}}{\text{HYPOTENUSE}}$$

$$\tan \theta = \frac{\text{OPPOSITE SIDE}}{\text{ADJACENT SIDE}}$$

The symbol "cos θ" is called the **cosine function** and is read "cosine theta." Similarly, "sin θ" is called the **sine function,** and "tan θ" is called the **tangent function.** The cosine, sine, and tangent functions are the primary ratios of a triangle (later we will give names to their reciprocals), and together are known as the **trigonometric ratios** or **trigonometric functions.**

EXAMPLE 1

Finding the Trigonometric Functions for a Given Triangle

Given a triangle with sides 3, 4, and 5, as shown in Figure 1.16.

a. Find cos α, sin α, and tan α.
b. Find cos β, sin β, and tan β.

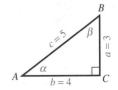

FIGURE 1.16 $\triangle ABC$

Solution

Because $3^2 + 4^2 = 5^2$, we see that $\triangle ABC$ is a right triangle.

a. $\cos \alpha = \frac{4}{5} = 0.8$; $\sin \alpha = \frac{3}{5} = 0.6$; $\tan \alpha = \frac{3}{4} = 0.75$
b. $\cos \beta = \frac{3}{5} = 0.6$; $\sin \beta = \frac{4}{5} = 0.8$; $\tan \beta = \frac{4}{3}$ ∎

Precision, Accuracy, and Significant Digits

Measurement is never exact. Therefore decisions must be made as to how **precise** the measurement should be. A measurement might be precise to the nearest inch, nearest foot, or even nearest mile. The precision of a measurement depends not only on the instrument used but also on the purpose of the measurement. For example, if you are measuring a room to lay carpet, the precision of your measurement might be different than when you are measuring an airport hangar.

Accuracy refers to your answer. Suppose an instrument is used that measures to the nearest tenth of a unit. One measurement is 4.6 and the other measurement is 2.1. If, in the process of your work, you need to multiply these numbers, the result obtained is

$$4.6 \times 2.1 = 9.66$$

This product is calculated to two decimal places. However, it does not seem quite right that an answer can be more accurate (two decimal places) than the instrument used to obtain the measurements (one decimal place). In this book, we will require that the accuracy of answers not exceed the precision of the measurement. This means that after doing your calculations, the final answer should be rounded. The principle we use was stated on page 3. This means that, to avoid round-off error, you should round only once (at the end). This is particularly important if you are using a calculator that displays 8, 10, 12, or even more decimal places. When solving triangles, we are dealing with measurements that are never exact. In "real life," measurements are made with a certain number of digits of precision, and results are not claimed to have more digits of precision than the least accurate number in the input data. This limitation is sometimes confusing when comparing linear measurements of the lengths of the sides of a triangle with angle measurements in that same triangle. In this book, we assume the following relationship between the *accuracy* of the measurement of sides and angles.

Accuracy in sides		*Equivalent accuracy in angles*
Two significant digits	⇔	Nearest degree
Three significant digits	⇔	Nearest tenth of a degree
Four significant digits	⇔	Nearest hundredth of a degree

The definition of significant digits, as well as a set of rules for working with significant digits, is given in Appendix B.

EXAMPLE 2

Significant Digits in Triangle Solutions

Round each set of triangle descriptions to the least accurate measurement. (*Note:*   We write $\alpha = 35°$ to mean the measure of angle α is 35°.)

a. $a = 46.6$, $b = 35.3$, $c = 42.7$; $\alpha = 72.7°$, $\beta = 46.3°$, $\gamma = 61°$

b. $a = 803$, $b = 455$, $c = 521.7952$;
$\alpha = 110°$, $\beta = 32.07896°$, $\gamma = 37.52103°$

c. $a = 68.123$, $b = 95.3$, $c = 128.0656$;
$\alpha = 31.3659°$, $\beta = 46.7358°$, $\gamma = 101.9025°$

d. $a = 34.6$, $b = 27.43995$, $c = 61.5$;
$\alpha = 8.5°$, $\beta = 6.731806°$, $\gamma = 164.76682°$

Solution

a. The least accurate measurement is $\gamma = 61°$, which is to the nearest degree, so the answers involving angles should be rounded to the nearest degree, and the answers involving sides should be rounded to two significant digits:

$$a = 47, \quad b = 35, \quad c = 43; \qquad \alpha = 73°, \quad \beta = 46°, \quad \gamma = 61°$$

b. The least accurate measurement is $\alpha = 110°$, which is to the nearest degree, so the answers involving angles should be rounded to the nearest degree, and the answers involving sides should be rounded to two significant digits:

$$a = 800, \quad b = 460, \quad c = 520; \qquad \alpha = 110°, \quad \beta = 32°, \quad \gamma = 38°$$

As we point out in Appendix B, the answer $a = 800$ is ambiguous regarding the number of significant digits (is it one or two?), and to be clear we would need to use scientific notation. We do not wish to require scientific notation in this book, so we allow this ambiguity in answers such as this ($a = 800$), where it is clear from the other measurements in the triangle that we are rounding to two significant digits.

c. The least accurate measurement is $b = 95.3$, which is three significant digits. This means that the angles should be rounded to the nearest tenth of a degree, and the answers involving sides should be rounded to three significant digits:

$$a = 68.1, \quad b = 95.3, \quad c = 128; \quad \alpha = 31.4°, \quad \beta = 46.7°, \quad \gamma = 101.9°$$

d. The least accurate measurements are $a = 34.6$, $c = 61.5$, and $\alpha = 8.5°$, which all indicate that we should round answers to three significant digits and to the nearest tenth of a degree:

$$a = 34.6, \quad b = 27.4, \quad c = 61.5; \quad \alpha = 8.5°, \quad \beta = 6.7°, \quad \gamma = 164.8°$$

Once again, we use this road sign to remind you of an important rounding principle discussed in Appendix B. **Do not round at the beginning or in the middle of a problem;** *round only when stating answers.* ■

Evaluating Trigonometric Functions

The process of finding the values of a trigonometric function is called **evaluation.** These values used to be found from tables, but today calculators are used.

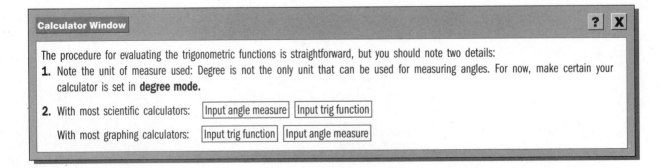

Calculator Window ? X

The procedure for evaluating the trigonometric functions is straightforward, but you should note two details:

1. Note the unit of measure used: Degree is not the only unit that can be used for measuring angles. For now, make certain your calculator is set in **degree mode.**

2. With most scientific calculators: | Input angle measure | | Input trig function |

With most graphing calculators: | Input trig function | | Input angle measure |

EXAMPLE 3   **Checking Calculator Evaluation of the Trigonometric Functions**

Evaluate the trigonometric functions using a calculator. Check your calculator answers with those shown here: **a.** cos 70° **b.** sin 70° **c.** tan 70° **d.** cos 20° **e.** sin 50° **f.** −sin 50°

Solution   **a.** $\cos 70° \approx 0.3420201433$ **b.** $\sin 70° \approx 0.9396926208$

c. $\tan 70° \approx 2.747477419$ **d.** $\cos 20° \approx 0.9396926208$

e. $\sin 50° \approx 0.7660444431$ **f.** $-\sin 50° \approx -0.7660444431$ ■

Solving Triangles

A triangle is made up of three sides and three angles. We say that **a triangle is solved** when the measurements of all six parts are known and listed. Typically, three parts will be given, or known, and it will be our task to find the other three parts to solve the triangle. Sometimes, however, we will be looking for only one part of the triangle.

EXAMPLE 4   **Solving a Right Triangle Given a Side and an Angle**

Solve the right triangle with $a = 50.0$ and $\alpha = 35°$.

Solution   Begin by drawing the triangle (not necessarily to scale), as shown in Figure 1.17.

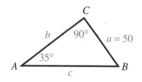

FIGURE 1.17
Given side *a* and angle α

$\alpha = 35°$ *Given*

$\beta = 55°$ *Since $\alpha + \beta = 90°$ for any right triangle with right angle C*

$\gamma = 90°$ *Given*

$a = 50$ *Given*

Note that we write the answer to the nearest unit because the given angle is measured to the nearest degree, and we use the *least accurate* of the given measurements to determine the accuracy of the answer.

b: $\tan 35° = \dfrac{50}{b}$ *Definition of tangent*

$b = \dfrac{50}{\tan 35°}$ *Solve for b.*

≈ 71.40740034

$b = 71$ *Round answer to two significant digits.*

$$c: \qquad \sin 35° = \frac{50}{c} \qquad \textit{Definition of sine}$$

$$c = \frac{50}{\sin 35°} \approx 87.17233978$$

$$c = 87 \qquad \textit{Again, round to two significant digits.}$$ ■

EXAMPLE 5

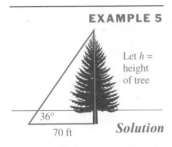

Let h = height of tree

36°

70 ft   *Solution*

Problem of the Day: Finding the Height of a Tree (Without the Sun)

Find the height of a tree without using shadows. Measure out a distance from the base of the tree, say $d = 70$ ft. Next, approximate the angle formed from the ground to the top of the tree, as shown in the figure. Suppose this angle is 36°. Now, find the height of the tree.

We use the right-triangle definition of tangent:

$$\tan 36° = \frac{h}{70}$$

$$h = 70 \tan 36°$$

$$\approx 50.85797696$$

To the correct number of significant digits (two), we say the height of the tree is 51 ft.   ■

Inverse Functions

Sometimes we know the sides of a right triangle, and we need to find the sizes of the angles. For example, suppose the legs of a right triangle are 8.0 in. and 12 in., as shown in Figure 1.18.

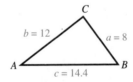

C

$b = 12$ $a = 8$

A —————— B
$c = 14.4$

FIGURE 1.18   Finding the angles of a right triangle

We know that the angles of this triangle should be labeled α, β, and γ, and we know (since it is a right triangle) that $\gamma = 90°$. We want to find α and β. We also know

$$\tan \alpha = \tfrac{8}{12} = \tfrac{2}{3}$$

We seek the angle α whose tangent is $\tfrac{2}{3}$, and we write this as

$$\alpha = \text{the angle whose tangent is } \tfrac{2}{3}$$

$$\alpha = \tan^{-1} \tfrac{2}{3}$$

In other words, we use the symbol $\tan^{-1}x$ to mean the angle whose tangent is x, and this function is called the **inverse tangent** function. We give similar definitions for $\cos^{-1}x$, called the **inverse cosine** function, and $\sin^{-1}x$, called the **inverse sine** function, as shown in the following box.

INVERSE TRIGONOMETRIC FUNCTIONS

In a right triangle,

$\cos^{-1}x$ is the angle whose cosine is x;
$\sin^{-1}x$ is the angle whose sine is x; and
$\tan^{-1}x$ is the angle whose tangent is x.

Help Window   ? X

We will generalize this definition later in the book, and at that time the values specified with this definition will be called the **principal values** for the inverse cosine, inverse sine, and inverse tangent. For now, we note that the angles we seek must be acute angles. Do not confuse $\cos^{-1}x$ and $(\cos x)^{-1}$. The symbol $\cos^{-1}x$ means the angle whose cosine is x, but the symbol $(\cos x)^{-1}$ is the reciprocal of $\cos x$, namely

$$(\cos x)^{-1} = \frac{1}{\cos x}$$

We use a calculator to find the values of $\cos^{-1}x$, $\sin^{-1}x$, and $\tan^{-1}x$. Find these keys on your calculator. We continue by finding the angles for the right triangle with sides 8.0 in. and 12 in.:

$$\tan^{-1}\tfrac{2}{3} \approx 33.69006753° \qquad \textit{Note: Be sure to enclose } \tfrac{2}{3} \textit{ in parentheses when entering this number in your calculator.}$$

To the correct number of significant digits (two for this example), we see that $\alpha = 34°$. We can find β by

NOTE

Do not forget the parentheses when entering $\frac{3}{2}$ into your calculator. $\tan^{-1}(3/2)$ means $\tan^{-1}\frac{3}{2}$ and $\tan^{-1}3/2$ means $\frac{\tan^{-1}3}{2}$.

$$\beta = \tan^{-1}\tfrac{3}{2} \qquad \textit{See Figure 1.18.}$$
$$\approx 56.30993247° \qquad \textit{You can enter } \tan^{-1}(3/2) \textit{ or } \tan^{-1}1.5.$$

To the correct number of significant digits, $\beta = 56°$. As a check, we note that the acute angles of a right triangle must be **complementary**—that is, add up to 90°:

$$\alpha + \beta \approx 33.699006753° + 56.30993247° = 90°$$

If you check by adding the rounded answers ($34° + 56°$) you will obtain an answer of approximately 90°.

EXAMPLE 6   **Solving a Right Triangle Given Two Sides**

Solve the right triangle with $a = 30.0$ and $c = 70.0$.

Solution   Begin by drawing the triangle as shown in Figure 1.19.

$$\alpha: \qquad \sin \alpha = \frac{30.0}{70.0} \qquad \textit{Definition of sine}$$

$$\alpha = \sin^{-1} \frac{3}{7} \qquad \textit{Definition of arcsine}$$

$$\approx 25.37693353°$$

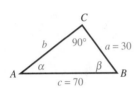

FIGURE 1.19
Given sides a and c

$\alpha = 25.4°$ *Three significant digits*

$\gamma = 90.0°$ *Given (right triangle)*

$\beta:$ $\beta = 180° - \alpha - \gamma \approx 64.62306647°$

$\beta = 64.6°$

$a = 30.0$ *Given*

$b:$ $a^2 + b^2 = c^2$ *Pythagorean theorem*

$$30.0^2 + b^2 = 70.0^2$$

$$b^2 = 4{,}000$$

$$b = 20\sqrt{10} \approx 63.2455532$$

$b = 63.2$ *Three significant digits*

$c = 70.0$ *Given*   ■

PROBLEM SET 1.3

Essential Concepts

1. State the right-triangle definition of the trigonometric functions.

2. What is the agreement in this book about the accuracy of the lengths of sides and the accuracy in the measurement of angles?

3. What is the agreement in this book about rounding your answers?

4. In a correctly labeled right triangle, $\triangle ABC$, if $|\overline{AC}| = |\overline{BC}| = 5$, evaluate:
a. $\cos \alpha$, $\sin \alpha$, and $\tan \alpha$
b. $\cos \beta$, $\sin \beta$, and $\tan \beta$

5. In a correctly labeled right triangle, $\triangle ABC$, if $|\overline{AB}| = 1$ and $|\overline{BC}| = \frac{1}{2}\sqrt{3}$, evaluate:
a. $\cos \alpha$, $\sin \alpha$, and $\tan \alpha$
b. $\cos \beta$, $\sin \beta$, and $\tan \beta$

6. IN YOUR OWN WORDS Describe each expression in words.
a. $\cos^{-1} T$
b. $\sin^{-1}(x + y)$
c. $\tan^{-1}\sqrt{1 + x^2}$

Level 1 Drill

Problems 7–18: Use a calculator to evaluate the given functions; round your answers to four decimal places.

7. a. $\cos 8°$ **b.** $\sin 24°$ **c.** $\tan 62°$

8. a. $\sin 80°$ **b.** $\tan 45°$ **c.** $\cos 60°$

9. a. $\tan 30°$ **b.** $\cos 70°$ **c.** $\sin 50°$

10. a. $\sin 40°$ **b.** $\dfrac{1}{\sin 40°}$ **c.** $\sin^2 40°$

11. a. $\cos 50°$ **b.** $\dfrac{1}{\cos 50°}$ **c.** $\cos^2 50°$

12. a. $\tan 56°$ **b.** $\dfrac{\sin 56°}{\cos 56°}$

13. a. $\cos 49°$ **b.** $\sin 41°$

14. a. $\sin 82°$ **b.** $-\sin 82°$

15. a. $\cos 25°$ **b.** $-\cos 25°$

16. a. $\tan 62°$ **b.** $-\tan 62°$

17. a. $(\sin 88°)^2 + (\cos 88°)^2$
  b. $(\sin 35°)^2 + (\cos 35°)^2$
  c. $(\sin 11°)^2 + (\cos 11°)^2$

18. a. $\dfrac{1}{(\cos 48°)^2} - (\tan 48°)^2$

  b. $\dfrac{1}{(\cos 19°)^2} - (\tan 19°)^2$

  c. $\dfrac{1}{(\cos 48°)^2} - (\tan 48°)^2$

Problems 19–26: Find the requested values correct to the nearest degree.

19. $\cos^{-1} 0.6$ **20.** $\sin^{-1} 0.6$

21. $\tan^{-1} 0.6$ **22.** $\tan^{-1} 1.9$

23. $\tan^{-1} 2.5$ **24.** $\cos^{-1} 0.125$

25. $\cos^{-1} 0.913$ **26.** $\sin^{-1} 0.0963$

Problems 27–33: Evaluate each expression and show the calculator display. Be sure to use the degree symbol if the answer is an angle.

27. a. $\cos^{-1} 0.5$ **b.** $(\cos 5°)^{-1}$

28. a. $\sin^{-1} 0.8$ **b.** $(\sin 8°)^{-1}$

29. a. $\tan^{-1} 1.2$ **b.** $(\tan 1.2°)^{-1}$

30. $\cos^{-1} 0.5 + \sin^{-1} 0.5$ **31.** $(\cos 0.5° + \sin 0.5°)^{-1}$

32. $2 \sin^{-1} 0.85$ **33.** $(2 \sin 0.85°)^{-1}$

Level 2 Drill

Problems 34–57: Solve each of the right triangles (γ is a right angle).

34. $a = 80; \beta = 60°$ **35.** $b = 37; \alpha = 69°$

36. $a = 29; \alpha = 76°$ **37.** $b = 90; \beta = 13°$

38. $a = 49; \alpha = 45°$ **39.** $b = 82; \alpha = 50°$

40. $c = 28.3; \alpha = 69.2°$ **41.** $c = 28.7; \alpha = 67.8°$

42. $c = 75.4; \alpha = 62.5°$ **43.** $c = 418.7; \beta = 61.05°$

44. $\beta = 57.4°; a = 70.0$ **45.** $\alpha = 56.00°; b = 2,350$

46. $\alpha = 42°; b = 350$ **47.** $\beta = 32.17°; c = 343.6$

48. $b = 3,200; c = 7,700$ **49.** $b = 3,100; c = 3,500$

50. $a = 145; b = 240$ **51.** $a = 85.3; b = 125.5$

52. $a = 45, b = 29$ **53.** $a = 12.5, b = 45.6$

54. $b = 86.5, c = 125.8$ **55.** $a = 135, b = 256$

56. $a = 10.06, c = 45.00$ **57.** $b = 189.4, c = 219.9$

Level 2 Applications

58. The most powerful lighthouse is on the coast of Brittany, France, and is 50 m tall. Suppose you are in a boat just off the coast, as shown in Figure 1.20. Determine your distance from the base of the lighthouse if $\theta = 12°$.

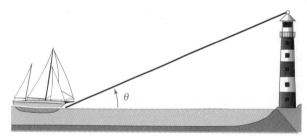

FIGURE 1.20   Distance from a boat to the *Créac'h d'Ouessant* on the coast of Brittany, France

59. Find the distance $|\overline{PA}|$ across the river shown in Figure 1.21 if you know that $\theta = 32.5°$.

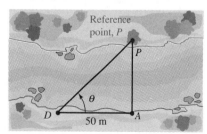

FIGURE 1.21   Distance across a river

60. The angle to the top of a flagpole from a point on the ground 40 ft from its base is 51.3°. What is the height of the flagpole?

61. A 16-ft ladder on level ground is leaning against a house. If the angle of the ladder forms with the ground is 52°, how far above the ground is the top of the ladder?

62. A 16-ft ladder on level ground is leaning against a house. If the base of the ladder is placed 5.0 ft from the house, what is the angle formed at the top of the ladder?

63. What is the size of the angle formed by two fences in a back yard if the sides measure 25 ft and 35 ft (as shown in Figure 1.22)?

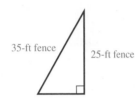

35-ft fence   25-ft fence

FIGURE 1.22   Finding an angle

64. Find the height of the Barrington Space Needle if the angle measured from the ground 1,000 ft from a point on the ground directly below the top to the top of the needle is 58.15°.

65. The world's tallest chimney is the stack at the International Nickel Company. Find its height if the

angle from a point 1,000 ft from the ground directly below the top of the stack to the top of the stack is 51.36°.

66. What is the angle of the sun above the horizon when a 6.0-ft person casts a shadow that is 8.0 ft long?

67. What is the angle of the sun above the horizon when a person 5 ft 10 in. tall casts a shadow that is 8 ft 2 in. long?

68. What is the distance from right field (point A) to left field (point C), using the dimensions of a baseball field as shown in Figure 1.23? Assume measurements are exact and give your answer to the nearest foot.

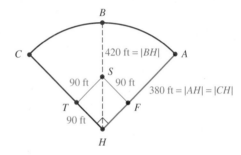

FIGURE 1.23   Baseball field

RIGHT-TRIANGLE APPLICATIONS

Sight to a landmark on the other side.

You are here

PROBLEM OF THE DAY

How can you go about measuring the distance across a canyon?

How would you answer this question?

This Problem of the Day is an application of solving triangles. From the time of the Egyptian rope stretchers in the 6th century B.C. to celestial navigation of spacecraft in the 21st century, triangles and trigonometry have been used to measure inaccessible distances. In this section, we will introduce some terminology associated with surveying, navigation, and aviation.

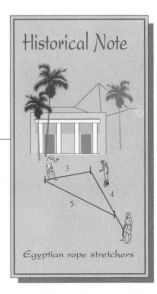

Historical Note

Egyptian rope stretchers

Angles of Depression and Elevation

The solution of right triangles is necessary in a variety of situations. The first one to be considered concerns an observer looking at an object at different elevations. The **angle of elevation** is the acute angle measured up from a horizontal line to the line of sight, whereas the **angle of depression** is the acute angle measured down from the horizontal line to the line of sight.

EXAMPLE 1

Using Angle of Elevation to Find an Inaccessible Height

We consider a problem from the previous section, this time with some new terminology. Suppose we want to find the height of a tree outside the classroom. We estimate the angle of elevation of the tree from a point on the ground 42 ft from its base to be 33°. (See Figure 1.24.) Find the height of the tree.

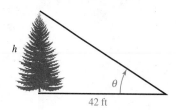

FIGURE 1.24   Height of a tree

Solution

The angle of elevation is 33°; let h = height of the tree. Then

$$\tan 33° = \frac{h}{42} \quad \textit{Definition of tangent}$$

$$h = 42 \tan 33° \approx 27.27511891$$

The tree is about 27 feet tall.   ∎

EXAMPLE 2

Finding the Distance from the Air

The distance across a canyon can be determined from an airplane. Suppose the angles of depression to the two sides of a canyon are 43° and 55°, as shown in Figure 1.25. If the altitude of the plane is 20,000 ft, how far is it across the canyon?

Solution

Label parts x, y, θ, and ϕ, as shown in Figure 1.25.

$$\theta = 90° - 55° = 35°$$
$$\phi = 90° - 43° = 47°$$

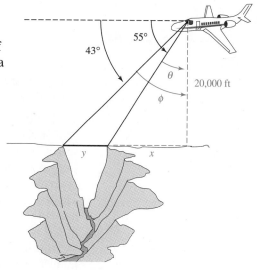

FIGURE 1.25   Distance across a canyon from the air

First find x:

$$\tan 35° = \frac{x}{20{,}000}$$

$$x = 20{,}000 \tan 35°$$

Next, find $x + y$: $\tan 47° = \dfrac{x + y}{20{,}000}$

$$x + y = 20{,}000 \tan 47°$$

Thus, $y = 20{,}000 \tan 47° - 20{,}000 \tan 35° \approx 7{,}443$. Rounding to two significant digits, we find the distance across the canyon to be 7,400 ft.   ■

Bearing

Another application of the solution of right triangles involves the **bearing** of a line, which is defined as an acute angle made with a north–south line. When giving the bearing of a line, first write N or S to determine whether to measure the angle from the north or from the south side of a point on the line. Then give the measure of the angle followed by E or W, denoting which side of the north–south line you are measuring. Some examples are shown in Figure 1.26.

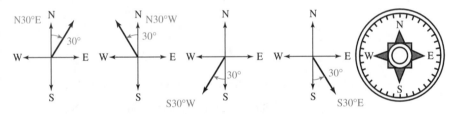

FIGURE 1.26   Measuring angles using bearing

EXAMPLE 3

Problem of the Day: Using Bearing to Find an Inaccessible Distance

To find the width $|\overline{AB}|$ of a canyon (see Figure 1.27), a surveyor measures 100 m from A in the direction of N42.6°W to locate point C. The surveyor then determines that the bearing of $\overline{CB}$ is N73.5°E. Find the width of the canyon if point B is situated so that $\angle BAC = 90.0°$.

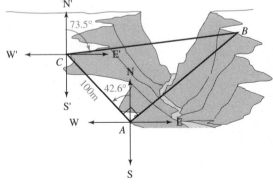

FIGURE 1.27   Surveying problem

Solution   Let $\theta = \angle BCA$ in Figure 1.27.

$$\angle BCE' = 16.5° \quad \textit{Complementary angles}$$
$$\angle ACS' = 42.6° \quad \textit{Alternate interior angles}$$
$$\angle E'CA = 47.4° \quad \textit{Complementary angles}$$
$$\theta = \angle BCA = \angle BCE' + \angle E'CA = 16.5° + 47.4° = 63.9°$$

Since $\tan \theta = \dfrac{|\overline{AB}|}{|\overline{AC}|}$, $|\overline{AB}| = |\overline{AC}| \tan \theta$

$$= 100 \tan \theta$$
$$= 100 \tan 63.9°$$
$$\approx 204.1253967$$

The canyon is 204 meters across.   ■

EXAMPLE 4   **Proving an Inverse Identity Using Right Triangles**

Prove $\sin^{-1}x + \cos^{-1}x = 90°$ for $0 < x < 1$.

Solution   Think about what this is saying: $\sin \alpha$ and $\cos \beta$ are both x, so $\sin^{-1}x$ is the angle whose sine is x; this is α.

$\sin^{-1}x$ is the angle whose sine is x; this is α.
$\cos^{-1}x$ is the angle whose cosine is x; this is β.

Since α and β are the acute angles of a right triangle, this problem is asking us to prove that the sum of the acute angles of a right triangle is 90°. This is obvious since the sum of the angles of any tringle is 180°. Here are the details that will prove this observation. Consider the right triangle shown in Figure 1.28.

Label the length of the side opposite α as x and the length of the hypotenuse as 1. Then

$$\sin \alpha = \frac{x}{1} = x;$$

therefore $\alpha = \sin^{-1}x$

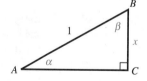

FIGURE 1.28   Right triangle

$$\cos \beta = \frac{x}{1} = x;$$

therefore $\beta = \cos^{-1}x$

Because the sum of the angles of any triangle is 180° and since C is a right angle, we see that the sum of the measures of angles α and β is 90°:

$$\alpha + \beta = 90°$$
$$\sin^{-1}x + \cos^{-1}x = 90°$$

■

PROBLEM SET 1.4

Essential Concepts

1. What is meant by angle of elevation?

2. What is meant by angle of depression?

3. What is meant by bearing? Explain the notation S38.5°W. Why do we not write W38.5°S?

4. Suppose ℓ and m are parallel lines.

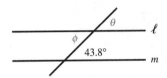

   a. What is the measure of angle θ?
   b. What is the measure of angle φ?

Level 1 Applications

5. The angle of elevation of a building from a point on the ground 150 ft from its base is 38°. Find the height of the building.

6. From a cliff 310 ft above the shoreline, the angle of depression of a ship is 31°. Find the distance of the ship from a point directly below the observer.

7. From a police helicopter flying at 1,250 ft, a stolen car is sighted at an angle of depression of 53.4°. Find the distance of the car from a point directly below the helicopter.

8. A guy wire is used to attach an antenna to the ground. If the wire is 145 ft long and is attached at a height of 120 ft, what is the angle of elevation of the antenna at the point where the wire is attached to the ground?

9. A contractor wants to put a fence around a lot in the shape of a right triangle. If the hypotenuse is 35 ft and one angle is 29.5°, how much fence is needed?

10. You need to enclose a garden in the shape of a right triangle. If one side (not the hypotenuse) is 85 ft and the angle opposite that side is 42.4°, what is the perimeter of the garden?

Level 2 Applications

11. To get the most energy from the sun, suppose you need to adjust the height, *h*, of a 12-ft solar heater panel (see Figure 1.29) so that the angle θ is 90°. What is *h* when the angle of elevation of the sun is 42°?

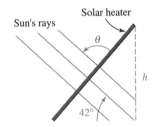

FIGURE 1.29   Solar heater panel

12. Suppose you want to plant a flowering plant that needs the direct rays of the sun, and would like to know how close to a 6-ft fence this will be possible (see Figure 1.30). After consulting an almanac, you find that the lowest angle of elevation of the sun during the plant's flowering season is 35°20′. How close to the fence (rounded to the nearest inch) should the plant be placed?

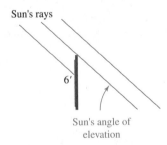

FIGURE 1.30   Tracking the sun's rays

13. Suppose you are constructing rafters with pitch angle θ = 35°. If *x* = 20 ft, what is the height *h*, as shown in Figure 1.31?

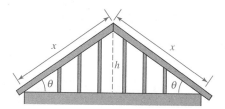

FIGURE 1.31   Rafter cross-section

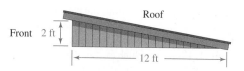

FIGURE 1.32   Shed roof angle

14. Suppose you are constructing rafters (see Figure 1.31) with a pitch angle of 40°. If $h = 4$ ft 6 in., what is the length, x, of the rafters?

15. Suppose you are constructing a shed and you need to cut a triangular piece of siding by scribing an angle marked θ. If the dimensions are shown in Figure 1.32, what is the proper angle for θ (to the nearest degree)?

16. The angle from the ground to the top of the Great Pyramid of Khufu (or Cheops) from a point on the ground 351 ft from a point directly below the top is 52.0°. Find the height of the pyramid.

Photo Researchers © Farrell Grehan

Historical Note

The angle of elevation for a pyramid is the angle between the edge of the base and the slant height, the line from the apex of the pyramid to the midpoint of any side of the base. It is the maximum possible ascent for anyone trying to climb the pyramid to the top. In an article, "Angles of Elevation to the Pyramids of Egypt," in *The Mathematics Teacher* (February 1982, pp. 124–127), author Arthur F. Smith notes that the angle of elevation of these pyramids is either about 44° or 52°. Why did the Egyptians build pyramids using these angles of elevation? (See Problem 40.)

Smith states that (according to Kurt Mendelssohn, *The Riddle of the Pyramids*, New York: Praeger Publications, 1974) Egyptians might have measured long horizontal distances by means of a circular drum with some convenient diameter such as one cubit. The circumference would then have been π cubits. To design a pyramid of convenient and attractive proportions, the Egyptians used a 4 : 1 ratio for the rise relative to revolutions of the drum. Smith then shows that the angle of elevation of the slant height is

$$\tan^{-1}\frac{4}{\pi} \approx 51.9°$$

If a small angle of elevation was desired (as in the case of the Red Pyramid), a 1 : 3 ratio might have been used. In that case, the angle of elevation of the slant height is

$$\tan^{-1}\frac{3}{\pi} \approx 43.7°$$

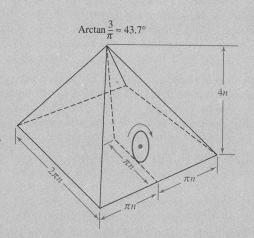

$\text{Arctan}\dfrac{3}{\pi} \approx 43.7°$

17. To find the east–west boundary of a piece of land, a surveyor must divert her path from point C on the boundary by proceeding due south for 300 ft to point A. Point B, which is due east of point C, is now found to be in the direction of N49°E from point A. What is the distance $|\overline{CB}|$?

18. To find the distance across a river that runs east–west, a surveyor locates points P and Q on a north–south line on opposites sides of the river. She then paces out 150 ft from Q due east to a point R. Next, she determines that the bearing of $\overline{RP}$ is N58°W. How far is it across the river?

19. A wheel 5.00 ft in diameter rolls up a 15.0° incline. What is the height of the bottom of the wheel above the base of the incline after the wheel has completed one revolution?

20. On top of the Empire State Building is a television tower. From a point 1,000 ft from a point on the ground directly below the top of the tower, the angle of elevation to the bottom of the tower is 51.34° and to the top of the tower is 55.81°. What is the length of the tower?

21. If the Empire State Building and the Sears Tower were situated 1,000 ft apart, the angle of depression from the top of the Sears Tower to the top of the Empire State Building would be 11.53°. The angle of depression to the foot of the Empire State Building would be 55.48°. Find the heights of the buildings.

22. To find the boundary of a piece of land, a surveyor must divert his path from a point A on the boundary for 500 ft in the direction S50°E. He then determines that the bearing of a point B located directly south of A is S40°W. Find the distance $|\overline{AB}|$.

23. To find the distance across the river, a surveyor locates points P and Q on opposite sides of the river. Next, the surveyor measures 100 m from point Q in the direction S35°E to point R. Then the surveyor determines that point P is now in the direction N25.0°E from point R and that $\angle PQR$ is a right angle. Find the distance across the river.

24. In the movie *Close Encounters of the Third Kind*, there was a scene in which the star, Richard Dreyfuss, was approaching Devil's Tower in Wyoming. He could have determined his distance from Devil's Tower by first stopping at a point P and estimating the angle P, as shown in Figure 1.33.

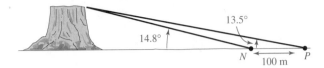

FIGURE 1.33   Devil's Tower

After moving 100 m toward Devil's Tower, he could have estimated the angle N, as shown in the figure. How far away from Devil's Tower is point N?

25. What is the height of Devil's Tower in Problem 24?

26. The distance from the earth to the sun at a particular time is 92.9 million miles and the angle formed between Venus, the earth, and the sun (as shown in Figure 1.34) is 47.0°. Find the distance from the sun to Venus.

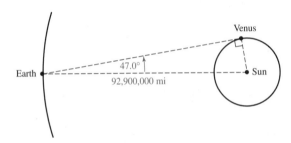

FIGURE 1.34   Distance from sun to Venus

27. Use the information in Problem 26 to find the distance from the earth to Venus.

28. To determine the height of the building shown in Figure 1.35, we select a point P and find that the angle of elevation is 59.64°. We then move out a distance of 325.4 ft (on a level plane) to point Q and find that the angle of elevation is now 41.32°. Find the height h of the building.

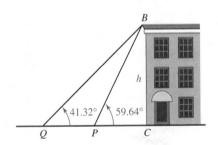

FIGURE 1.35   Height of an office building

29. Using Figure 1.35, let the angle of elevation at P be α and at Q be β, and let the distance from P to Q be d. If h is the height of the building, show that

$$h = \frac{d \tan \alpha \tan \beta}{\tan \alpha - \tan \beta}$$

Level 3 Applications

30. If the sun rises at 6:15 A.M. and is directly overhead at 12:15 P.M., what is the time when the angle of the sun above the horizon is 30.0°?

31. If the sun rises at 6:07 A.M. and is directly overhead at 12:10 P.M., what is the time when the angle of the sun above the horizon is 30.0°?

32. Rework Problem 31 for an angle of 42.0° above the horizon.

33. A 6.0-ft person is casting a shadow of 4.2 ft.
   a. What is the angle (to the nearest degree) of the sun above the horizon?
   b. What time of the morning is it if the sun rose at 6:15 A.M. and is directly overhead at 12:15 P.M.?

34. A person 5 ft 8 in. tall is casting a shadow of 3 ft 4 in.
   a. What is the angle (to the nearest degree) of the sun above the horizon?
   b. What time of the morning is it if the sun rose at 6:28 A.M. and is directly overhead at 12:14 P.M.?

35. Find the lengths (both exact and approximate, rounded to the nearest inch) of all of the pieces of the truss shown in Figure 1.36.

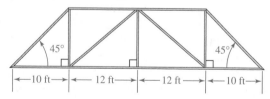

FIGURE 1.36   Building truss

36. A TI-92 or an interactive geometric program (such as *The Geometer's Sketchpad 3*) is necessary for this problem.* Consider a correctly labeled triangle such

*See "Using Interactive-Geometry Software for Right-Angle Trigonometry," by Charles Embse and Arne Engebretsen, *The Mathematics Teacher,* October 1996, pp. 602–605.

as that shown in Figure 1.8. Construct a table showing the lengths of the sides, namely $|\overline{BC}|$ and $|\overline{AB}|$ relative to $\angle BAC$. Also compare $\sin A$ and the ratio $|\overline{BC}|/|\overline{AB}|$. One possible display is shown in Figure 1.37.

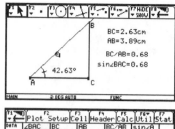

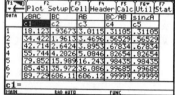

FIGURE 1.37   Sample TI-92 output

   a. What will happen if $m\angle BAC$ reaches 90°?
   b. What will happen if $m\angle BAC$ is between 90° and 180°?
   c. Reformulate the questions in parts **a** and **b** for the ratio $|\overline{AC}|/|\overline{AB}|$.
   d. Reformulate the questions in parts **a** and **b** for the ratio $|\overline{BC}|/|\overline{AC}|$.

Level 3 Theory

37. Suppose that $\triangle ABC$ is *not* a right triangle, but has an obtuse angle β. Draw $\overline{BD}$ perpendicular to $\overline{AC}$ forming right triangles $\triangle ABD$ and $\triangle BDC$ (with right angles at D). Show that

$$\frac{\sin \alpha}{a} = \frac{\sin \gamma}{c}$$

38. Suppose that $\triangle ABC$ is *not* a right triangle, but has an obtuse angle γ. Draw $\overline{CE}$ perpendicular to $\overline{AB}$ forming right triangles $\triangle ACE$ and $\triangle BEC$ (with right angles at E). Show that

$$\frac{\sin \alpha}{a} = \frac{\sin \beta}{b}$$

39. Suppose that $\triangle ABC$ is *not* a right triangle, but has an obtuse angle α. Draw $\overline{AF}$ perpendicular to $\overline{BC}$

forming right triangles $\triangle CAF$ and $\triangle ABF$ (with right angles at F). Show that

$$\frac{\sin \beta}{b} = \frac{\sin \gamma}{c}$$

**Level 3
Individual Project**

40. INDIVIDUAL RESEARCH PROJECT
   Historical Question If we define the angle of elevation of a pyramid as the angle between the edge of the base and the slant height, we find that the major Egyptian pyramids are about 44° or 52°. Write a paper explaining why the Egyptians built pyramids using these angles of elevation.
   References: Billard, Jules B., ed., *Ancient Egypt* (Washington, D.C.: National Geographic Society, 1978). Edward, I. E. S., *The Pyramids of Egypt* (New York: Penguin Books, 1961). Smith, Arthur F. "Angles of Elevation of the Pyramids of Egypt," *The Mathematics Teacher,* February 1982, pp. 124–127.

ANGLES AND ARC LENGTH

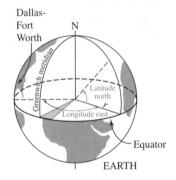

PROBLEM OF THE DAY

How far north of the equator is the Dallas–Fort Worth airport?

How would you answer this question?

You might begin by checking in an atlas or on the Internet to find that the Dallas–Fort Worth airport is located at approximately 33°N latitude and 97°W longitude. But what does that mean? In this section we will consider angles relative to a coordinate system, as well as angles that are not angles of any triangle (that is, they may be more than 180° or less than 0°). We will then define the notion of arc length and conclude by answering this Problem of the Day.

Standard-Position Angles

It is often useful to superimpose a Cartesian coordinate system on an angle so that the vertex is at the origin and the initial side is along the positive x-axis. The angle is then said to be in **standard position.** In this book, we assume that angles are in standard position unless otherwise noted. Angles in standard position that have the same terminal sides are called **coterminal angles.** Given any angle α, there is an unlimited number of angles coterminal with α (some positive and some negative; remember, angles can be more than 180°). In Figure 1.38, β is coterminal with α. Can you find other angles coterminal with α?

   If the terminal side of θ is on one of the coordinate axes, then θ is called a **quadrantal angle.** If θ is not a quadrantal angle, then it is said to be in a certain **quadrant** if its terminal side lies in that quadrant. Thus, in Figure 1.38, both α and β are called Quadrant II angles.

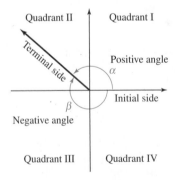

FIGURE 1.38
Standard-position angles α and β: α is a positive angle, and β is a negative angle. α and β are also coterminal angles.

EXAMPLE 1

Finding the Positive Angle Coterminal with a Given Angle

Find the positive angle less than one revolution that is coterminal with the given angle and classify the angle by quadrant.

a. −30°   **b.** 500°   **c.** −240°   **d.** −2,000°

Solution

In each case, draw the given angle in standard position and then draw the coterminal angle.

a. −30° is in Quadrant IV

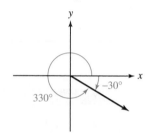

The angles −30° and 330° are coterminal.

b. 500° is in Quadrant II

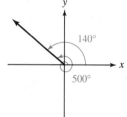

The angles 500° and 140° are coterminal.

c. −240° is in Quadrant II

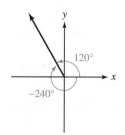

The angles −240° and 120° are coterminal.

d. −2000° is in Quadrant II

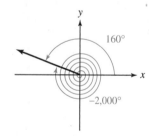

To find the number of revolutions, divide by 360°: $2,000 \div 360 = 5.\overline{5}$. The angle is between 5 and 6 revolutions. To find how much more than 5 revolutions, subtract 5 revolutions $(5 \cdot 360°)$ from 2,000°:

$$2,000° - 5(360°) = 200°$$

The angle is 5 revolutions (clockwise) plus 200° (clockwise). We see that −2,000° and 160° are coterminal.

■

Do not confuse equal angles with coterminal angles. Equal angles are angles with the same measure, so 30° and −330° are NOT equal angles, but rather coterminal angles. (In fact, how could a positive number ever be equal to a negative number?) Equal angles are always coterminal angles, but coterminal angles are not necessarily equal. For example, if we say ∠ABC is equal to ∠CBA, we mean that both name the same angle, but if we say ∠ABC is coterminal with ∠CBA, we mean that they share the same terminal side but are not necessarily equal.

Reference Angles

We now introduce still another relationship between pairs of angles. If an angle θ is not a quadrantal angle, then we refer to its *reference angle,* which is denoted by θ' throughout this book.

REFERENCE ANGLE

If an angle θ is not a quadrantal angle, then the **reference angle** θ' is defined as the acute angle the terminal side of θ makes with the x-axis.

Help Window ? X

This means that θ' is always positive (regardless of θ) and is between 0° and 90°.

For any angle θ, its reference angle θ' satisfies the inequality

$$0° < \theta' < 90°$$

The procedure for finding reference angles varies, depending on the quadrant of θ, as shown in Figure 1.39. Notice that the reference angle θ', shown in color, is always drawn to the x-axis and *never* to the y-axis. Also, remember that if θ is a quadrantal angle, there is no reference angle.

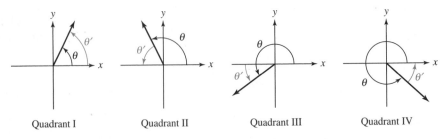

Quadrant I Quadrant II Quadrant III Quadrant IV

FIGURE 1.39 Reference angles

EXAMPLE 2 **Finding Reference Angles**

Find the reference angle for each angle, then draw both the given angle and the reference angle.

a. $-30°$ **b.** $500°$ **c.** $-135°$ **d.** $790°$

Solution **a.** Draw the angle $-30°$, as shown:

Reference angle is $30°$; remember that reference angles are always positive.

Note: $30° \neq -30°$

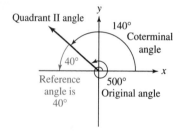

b. Draw the angle $500°$, as shown:

If the angle is more than one revolution, first find a nonnegative coterminal angle less than one revolution ($500° - 360° = 140°$). The reference angle is

$$180° - 140° = 40°$$

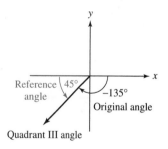

c. Draw the angle $-135°$, as shown:

The reference angle is

$$180° - 135° = 45°$$

d. Draw the angle $790°$; the coterminal angle is $70°$.

The reference angle is the same as the coterminal angle, namely $70°$.

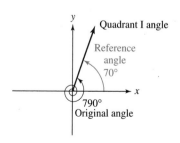

It is NOT correct to say $790° = 70°$. It IS correct to say $790°$ and $70°$ are coterminal angles.

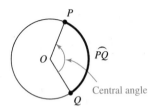

FIGURE 1.40
Central angle and an arc PQ

Arc Length

Draw a circle with any nonzero radius r and center at O. An **arc** is part of a circle, so if P and Q are any two distinct points on a circle, then the part of the circle from P to Q is an arc of the circle. We also say that $\angle POQ$ is a **central angle** that is **subtended** by arc PQ, as shown in Figure 1.40.

We can use angles to find the length of an arc of a circle, called *arc length,* and denoted by s. To do this, we note that one revolution (360°) is one **circumference** of the circle, one-half revolution (180°) is one-half circumference, and one-fourth revolution (90°) is one-fourth circumference. This leads us to the following result.

ARC LENGTH

The **arc length** s of a circle of circumference C and central angle θ is given by the proportion

$$\frac{\theta}{360°} = \frac{s}{C}$$

This is the proportion solved for s:

$$s = \frac{\theta}{360°}(C)$$

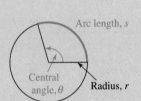

This proportion has three variables, the angle θ measured in degrees, the arc length s, and the circumference C. If we know any two of these variables, we can use this proportion to find the value of the third variable. We also know (from geometry) that $C = 2\pi r$ for a radius r, so that

$$s = \frac{\theta}{360°}(C) = \frac{\theta}{360°}(2\pi r) = \frac{\theta\pi}{180°}r$$

EXAMPLE 3

Finding the Arc Length

If a circle has a circumference of 36 in. and a central angle of 45°, what is the length of the intercepted arc?

Solution   Use the arc length formula where $C = 36$ in. and $\theta = 45°$:

$$\frac{\theta}{360°} = \frac{s}{C} \qquad \textit{Arc length formula}$$

$$\frac{45°}{360°} = \frac{s}{36 \text{ in.}} \qquad \textit{Substitute the known values.}$$

$$s = 36\left(\frac{45}{360}\right) \text{ in.} \qquad \textit{Multiply both sides by 36 in.}$$

$$= 4.5 \text{ in.} \qquad \textit{Simplify.}$$

The arc length is 4.5 inches.   ∎

We conclude this section by answering the Problem of the Day. Locations on the earth are given by *latitude* and *longitude*. **Latitude** is measured north/south of the equator by angles between 0° and 90°. **Longitude** is measured by angles between 0° and 180° either west or east of the Greenwich meridian, as shown in Figure 1.41. A **nautical mile** is defined to be s for a central angle of 1′ on the earth.

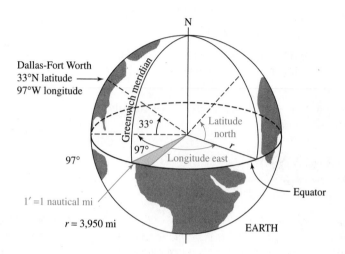

FIGURE 1.41 Latitude and longitude on earth; assume that the radius of the earth is 3,950 mi or 6,370 km (kilometer)

EXAMPLE 4

Problem of the Day: Global Distances

The Dallas–Fort Worth airport is located at approximately 33°N latitude and 97°W longitude. How far north (to the nearest hundred miles) is it from the equator?

Solution

The location is 33°N latitude, so $\theta = 33°$. Since $r = 3{,}950$ mi, it follows that $C = 2\pi r = 2\pi(3{,}950)$, so that

$$\frac{\theta}{360°} = \frac{s}{C} \qquad \text{Arc length formula}$$

$$\frac{33°}{360°} = \frac{s}{2\pi(3{,}950)} \qquad \text{Substitute known values.}$$

$$\left(\frac{33}{360}\right)2\pi(3{,}950) = s \qquad \text{Multiply both sides by } 2\pi(3{,}950).$$

$$s \approx 2{,}300 \qquad \text{Round to the nearest hundred; the calculator value is 2275.03668.}$$

The Dallas–Fort Worth airport is about 2,300 miles north of the equator. ■

PROBLEM SET 1.5

Essential Concepts

1. **IN YOUR OWN WORDS** Distinguish between equal angles and coterminal angles.

2. **IN YOUR OWN WORDS** Distinguish between equal angles and reference angles.

3. **IN YOUR OWN WORDS** Distinguish between reference angles and coterminal angles.

4. **IN YOUR OWN WORDS** Discuss the concept of arc length. Why not just use a ruler to measure the length of an arc?

Level 1 Drill

Problems 5–16: For each given angle, state:
a. *its quadrant;*
b. *its reference angle;*
c. *a positive coterminal angle less than one revolution;*
d. *a negative coterminal angle less than one revolution.*

5. $135°$ 6. $300°$ 7. $200°$ 8. $80°$

9. $-135°$ 10. $-300°$ 11. $-200°$ 12. $-80°$

13. $450°$ 14. $-800°$ 15. $-750°$ 16. $600°$

17. Sketch the angle $300°$. Classify each of the following angles as equal to, coterminal with, or a reference angle of the sketched angle. If none of these words apply, so state.
    a. $60°$ b. $-60°$ c. $660°$ d. $120°$ e. $-420°$

18. Sketch the angle $-210°$. Classify each of the following angles as equal to, coterminal with, or a reference angle of the sketched angle. If none of these words apply, so state.
    a. $30°$ b. $150°$ c. $-570°$ d. $390°$ e. $210°$

Level 2 Drill

19. **IN YOUR OWN WORDS** Make a verbal statement about the angle θ.
    a. $0° < \theta < 90°$ b. $0° < \theta < 180°$
    c. $-90° < \theta < 90°$ d. $180° < \theta < 270°$

20. **IN YOUR OWN WORDS** Make a verbal statement about the angle θ.
    a. $180° < \theta < 360°$ b. $270° < \theta < 360°$
    c. $0° < 2\theta < 180°$ d. $0° < \frac{1}{2}\theta < 90°$

21. **IN YOUR OWN WORDS** State inequalities that summarize each verbal statement for positive angles that are less than one revolution.
    a. α is in Quadrant I b. β is in Quadrant II
    c. γ is in Quadrant III d. δ is in Quadrant IV

22. **IN YOUR OWN WORDS** State inequalities that summarize each verbal statement. (More than one answer for each is possible.)
    a. θ is in Quadrant I or II (including $90°$).
    b. ω is in Quadrant IV or I (including $0°$).
    c. ϕ is in Quadrant I or II.
    d. λ is in Quadrant IV or I.

Problems 23–30: Find the length of the intercepted arc (to the nearest hundredth) if the central angle θ and radius r are given.

23. $\theta = 360°, r = 3$ in. 24. $\theta = 180°, r = 4$ in.

25. $\theta = 90°, r = 10$ ft 26. $\theta = 45°, r = 7$ ft

27. $\theta = 112°, r = 7.2$ cm 28. $\theta = 1°, r = 1$ m

29. $\theta = 48°, r = 42$ ft 30. $\theta = 20°, r = 8.5$ ft

Level 2 Applications

31. How far (to the nearest hundredth cm) does the tip of an hour hand on a clock move in 3 hours if the hour hand is 2.00 cm long?

32. A 50-cm pendulum on a clock swings through an angle of $100°$. How far (to the nearest cm) does the tip travel in one arc?

33. Through how many revolutions does a pulley with a 20.0-cm diameter turn when 1.00 m of cable is pulled through it without slippage? (See Figure 1.42.) [*Hint:* 1 m = 100 cm; answer to the nearest hundredth revolution.]

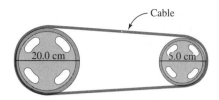

FIGURE 1.42   Pulley with two wheels

34. Answer the question in Problem 33 for a pulley with a 5.0-cm diameter.

35. If Columbia, South Carolina, is located at 34°N latitude and 81°W longitude, what is its approximate distance (to the nearest hundred miles) from Disneyland, which is located at 34°N latitude and 118°W longitude? (See Figure 1.43.) [Assume the radius of the earth is approximately 3,950 mi.]

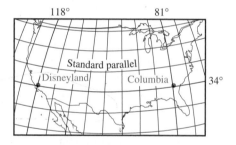

FIGURE 1.43   Distance from Columbia to Disneyland

36. If New York City is at 40°N latitude and 74°W longitude, what is its approximate distance (to the nearest hundred miles) from the equator? [Assume the radius of the earth is 3,950 mi.]

37. Entebbe, Uganda, is located at approximately 33°E longitude, and Stanley Falls in Zaire is located at 25°E longitude. Both these cities lie approximately on the equator. If we know that the radius of the earth is about 6,370 km, what is the distance between the cities to the nearest 10 km?

38. **Historical Question** In about 230 B.C., a mathematician named Eratosthenes estimated the radius of the earth using the following information. (See Figure 1.44.) Syene and Alexandria in Egypt are on the same line of longitude. They are also 800 km apart. At noon on the longest day of the year, when the sun was directly overhead in Syene, Eratosthenes measured the sun to be 7.2° from the

vertical in Alexandria. Because of the distance of the sun from the earth, he assumed that the sun's rays were parallel. Thus, he concluded that the central angle subtending rays from the center of the earth to Syene and Alexandria was also 7.2°. Using this information, find the approximate radius (to the nearest hundred km) of the earth.

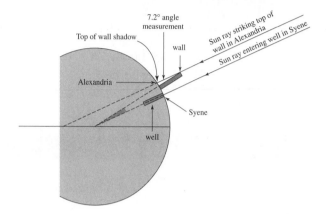

FIGURE 1.44   Estimation for the earth's radius

39. Suppose it is known that the moon subtends an angle of 45.75′ at the center of the earth, as shown in Figure 1.45. It is also known that the center of the moon is 384,417 km from the surface of the earth and the radius of the earth is 6,370 km. What is the diameter of the moon to the nearest 10 km? [*Hint:* For small central angles with large radii, the intercepted arc is approximately equal to its chord.]

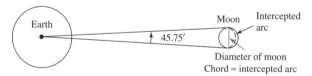

FIGURE 1.45   Determining the diameter of the moon

Level 3 Theory

40. **JOURNAL PROBLEM** [From *The Mathematics Teacher*, September 1993, p. 526, Carol Murray, Algonquin Regional High School, Northboro, MA.] Find a formula for the number of degrees between the hour, h, and minute, m, hands of a clock. For example, if the time is 8:15, then $h = 8$ and $m = 15$;

what we want is a formula (in terms of h and m) that gives the number of degrees between the hour and minute hands. For this example, the formula should give 157.5°.

Level 3
Individual Project

41. Historical Question The division of one revolution (a circle) into 360 equal parts (called *degrees*) is no doubt due to the sexagesimal (base 60) numeration system used by the Babylonians. Several explanations have been put forward to account for

their choice of this number. (For example, see Howard Eves' *In Mathematical Circles.*) One possible explanation is based on the fact that the radius of a circle can be applied exactly six times to its circumference as a chord. Show this with a sketch. Notice that sexagesimal division (that is, division into 60 equal parts) makes this geometric construction easy. It seems natural that each chord should be further divided into 60 equal parts, resulting in the division of the circle into 360 equal parts. Write a short research paper on the history of degree measure for an angle.

1.6 CHAPTER ONE SUMMARY

Objectives

Section 1.1	Angles and Degree Measure
Objective 1.1	Know the definition and notation for angles, including positive and negative angles. Know the Greek letters.
Objective 1.2	Know the Pythagorean theorem and use it to find the length of an unknown side in a right triangle.
Objective 1.3	Draw angles whose measures are given in degrees; know the approximate sizes of angles 0°, 30°, 45°, 60°, 90°, and multiples of these angles.
Objective 1.4	Convert degrees/minutes/seconds to decimal degrees.

Section 1.2	Similar Triangles
Objective 1.5	Know that the sum of the angles of a triangle is 180°, and use it to find the measure of an unknown angle in a triangle.
Objective 1.6	Name corresponding parts of pairs of triangles.
Objective 1.7	Decide whether a pair of given triangles are similar.

Objective 1.8	Given similar triangles, find an unknown length.
Objective 1.9	Solve applied problems using similar triangles.

Section 1.3	Trigonometric Ratios
Objective 1.10	State the right-triangle definition of the trigonometric functions.
Objective 1.11	Evaluate trigonometric functions of cosine, sine, and tangent.
Objective 1.12	Evaluate inverse trigonometric functions of cosine, sine, and tangent.
Objective 1.13	Solve right triangles for all six parts by using the right-triangle definition of cosine, sine, and tangent.
Objective 1.14	Round triangle descriptions to the correct number of significant digits.
Objective 1.15	Solve applied right-triangle problems.

Section 1.4	Right-Triangle Applications
Objective 1.16	Solve applied right-triangle problems involving angles of elevation or depression.

Objective 1.17	Solve applied right-triangle problems involving inaccessible distances.	**Objective 1.20**	Find positive angles coterminal with a given angle and less than one revolution.
Objective 1.18	Solve applied right-triangle problems involving bearing.	**Objective 1.21**	Find the reference angle for a given angle.
Section 1.5	Angles and Arc Length	**Objective 1.22**	Know the arc length formula and be able to apply it.
Objective 1.19	Distinguish among equal, coterminal, and reference angles. Know what it means for an angle to be in standard position.		

Terms

Accuracy [1.3]
Acute angle [1.1]
Adjacent side [1.3]
Alpha (α) [1.1]
Angle [1.1]
Angle of depression [1.4]
Angle of elevation [1.4]
Arc [1.5]
Arc length [1.5]
Bearing [1.4]
Beta (β) [1.1]
Central angle [1.5]
Circumference [1.5]
Complementary angles [1.3]
Corresponding angles [1.2]
Corresponding sides [1.2]

Cosine [1.3]
Coterminal angles [1.5]
Degree [1.1]
Degree mode [1.3]
Delta (δ) [1.1]
Evaluation of a trigonometric function [1.3]
Gamma (γ) [1.1]
Hypotenuse [1.3]
Initial side [1.1]
Inverse cosine [1.3]
Inverse sine [1.3]
Inverse tangent [1.3]
Lambda (λ) [1.1]
Latitude [1.5]
Longitude [1.5]

Minute [1.1]
Mu (μ) [1.1]
Nautical mile [1.5]
Negative angle [1.1]
Obtuse angle [1.1]
Omega (ω) [1.1]
Opposite side [1.3]
Phi (ϕ or φ) [1.1]
Positive angle [1.1]
Precision [1.3]
Principal values of the inverse trig functions [1.3]
Pythagorean theorem [1.1]
Quadrant [1.5]
Quadrantal angle [1.5]

Reference angle [1.5]
Right angle [1.1]
Second [1.1]
Similar triangles [1.2]
Sine [1.3]
Solve a triangle [1.3]
Standard-position angle [1.5]
Straight angle [1.1]
Subtended angle [1.5]
Tangent [1.3]
Terminal side [1.1]
Theta (θ) [1.1]
Trigonometric functions [1.3]
Trigonometric ratios [1.3]
Vertex [1.1]

Sample Test

All of the answers for this sample test are given in the back of the book.

1. Name the following Greek letters.
   a. α **b.** θ **c.** γ **d.** λ **e.** β **f.** δ

2. Draw from memory the angle whose measure is given, and then find the reference angle.
   a. 30°   **b.** 45°   **c.** −60°   **d.** 180°   **e.** 300°

3. Change 50°36′ to decimal degrees correct to the nearest hundredth degree.

4. **a.** State the right-triangle definition of cos θ, sin θ, and tan θ.
   b. Using this definition solve the right triangle where $a = \sqrt{7}$ and $b = 3.0$.

5. Evaluate each function (rounded to four decimal places).
   a. cos 20°   **b.** sin 40°   **c.** tan 15°

6. Round each set of triangle descriptions to the least accurate of the measurements.

a. $a = 59.8$, $b = 77.9$, $c = 111.6$;
$\alpha = 30.72298°$, $\beta = 41.72231°$, $\gamma = 107.5547°$

b. $a = 3.488649$, $b = 0.8917898$, $c = 3.55$;
$\alpha = 79°$, $\beta = 87.3°$, $\gamma = 13.7°$

c. $a = 0.5976337$, $b = 7.7$, $c = 8.0$;
$\alpha = 3.77°$, $\beta = 58°$, $\gamma = 118.226°$

7. The tallest human-built structure in the world is the transmitter tower of KTHI-TV in North Dakota. Find its height if the angle of elevation at 501.0 ft away is 76.35°.

8. Consider Figure 1.46.

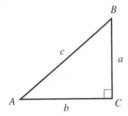

FIGURE 1.46   Standard right triangle

a. What is $\cos \beta$?
b. What is $\sin \beta$?
c. What is $\tan \alpha$?
d. If $a = 5.30$ and $b = 12.2$, solve the triangle.

9. To measure the distance across the Rainbow Bridge in Utah, a surveyor selected two points P and Q on opposite ends of the bridge. From point Q, the surveyor measured 500 ft in the direction N38.4°E to a point R. Point P was then determined to be in the direction S67.5°W from R. What is the distance across the Rainbow Bridge if all the preceding measurements are in the same plane and $\angle PQR$ is a right angle?

10. A curve on a highway is laid out as the arc of a circle of radius 500 m. If the curve subtends a central angle of 18.0°, what is the distance around this section of road? Give the exact answer and an answer rounded to the nearest meter.

Discussion Problems

1. Discuss the measurement of angles. Include measurements such as revolutions and degrees, as well as the possibility of other types of measures.

2. Discuss the concept of arc length. Why not just use a ruler to measure the length of an arc?

3. Write a 500-word paper on the history of trigonometry.

4. Write a computer program for solving right triangles.

5. Discuss why we need to define three trigonometric functions to solve a triangle.

Miscellaneous Problems—Chapter 1

Problems 1–4: Find the arc length (approximate answers correct to the nearest unit).

1. $r = 5$ in.; $\theta = 200°$ **2.** $r = \sqrt{50}$ ft; $\theta = 250°$

3. $r = 6.4$ ft; $\theta = 65°$ **4.** $r = 34$ in.; $\theta = 125°$

Problems 5–8: Determine the quadrant of each angle, and find a positive angle less than one revolution that is coterminal with the given angle.

5. 375°   **6.** −15°   **7.** 815°   **8.** −160°

Problems 9–20: Use a calculator to find the value of the trigonometric functions correct to four decimal places.

9. $\sin 85.2°$ **10.** $\tan 73.1°$

11. $\cos 2°$ **12.** $\sin 60°$

13. $\cos 35°$ **14.** $\sin 18.5°$

15. $\tan 60°$ **16.** $\cos 22°$

17. $-\tan 34.5°$ **18.** $\sin(-55.6°)$

19. $-\sin 55.6°$ **20.** $\sin 0.5°$

Problems 21–32: *Solve the right triangles where γ is a right angle.*

21. $b = 3.6$, $\beta = 63°$ **22.** $a = 42.15$, $\beta = 57.00°$

23. $c = 23.2$, $\beta = 18.8°$ **24.** $c = 2.9$, $\beta = 59°$

25. $a = 2.71$, $c = 7.0$ **26.** $a = 3.014$; $c = 7.8$

27. $b = 3.91$, $c = 7.00$ **28.** $b = 4.484$, $c = 6.8$

29. $c = 8.5$, $\beta = 76°$ **30.** $c = 5.0$, $\beta = 32.00°$

31. $c = 50.05$, $\beta = 23.7°$ **32.** $a = 4.45$, $\beta = 34°$

Applications

33. In Figure 1.41, a nautical mile is defined to be the arc length for a central angle of $1'$ at the center of the earth. If the radius of the earth is approximately 3,950 mi, approximate the number of statute miles in a nautical mile.

34. In Figure 1.1, a right triangle and the Pythagorean theorem were used to construct a length of $\sqrt{2}$. Use the Pythagorean theorem to construct a length of $\sqrt{3}$.

35. Use the spiral of roots shown in Figure 1.47 to find the length marked x.

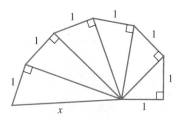

FIGURE 1.47   Spiral of roots

36. Use a spiral of roots (see Problem 35) to draw a segment with length $2\sqrt{3}$.

37. If Chicago is located at 42°N latitude and 88°W longitude, what is its approximate distance (to the nearest hundred miles) from the equator? Assume the radius of the earth is 3,950 mi.

38. Omaha, Nebraska, is located at approximately 97°W longitude, 41°N latitude; Wichita, Kansas, is located at approximately 97°W longitude, 37°N latitude. Notice that these two cities have the same longitude. If we know that the radius of the earth at that longitude is about 6,370 kilometers (km), what is the distance between these cities to the nearest 10 km?

39. Sacramento, California, is approximately 90 miles due east of Santa Rosa, California (see Figure 1.48). How much sooner (to the nearest minute) would a person in Sacramento see the rising sun than a person in Santa Rosa? Assume that the radius of the earth is 3,950 miles and rotates once every 24 hours.

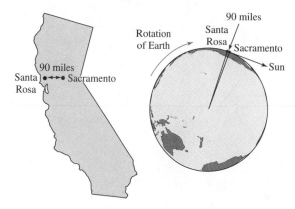

FIGURE 1.48   Time to sunrise

40. When viewing Angel Falls (the world's highest waterfall) from Observation Platform A, located on the same level as the bottom of the waterfall, we calculate the angle of elevation to the top of the waterfall to be 69.30°. From Observation Platform B, which is located on the same level exactly 1,000 ft from the first observation point, we calculate the angle of elevation to the top of the waterfall to be 52.90°. How high is the waterfall?

41. A large advertising sign stands on top of a tall building. (See Figure 1.49.) The angle of elevation from a point on level ground 275 ft from the base of the building to the bottom of the sign is 18.5°, and to the top of the sign is 21.5°. How tall is the sign?

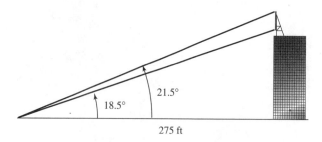

FIGURE 1.49   Determining the height of a sign

CHAPTER TWO

TRIGONOMETRIC FUNCTIONS

PREVIEW

Even though degree measure of angles as introduced in Chapter 1 seems relatively straightforward, for more advanced work in mathematics there is a more convenient and natural measure for angles, called radian measure. We introduce this measure in the first section of this chapter, and then we generalize the right-triangle definition of the cosine, sine, and tangent to be functions of any real number.

In Section 2.2, we give names for all six ratios formed by the sides of a triangle, and we do this in terms of a unit circle. This definition leads naturally to what are called the fundamental trigonometric identities, *which are introduced in Section 2.3. The chapter concludes by finding exact values of certain angles in Section 2.4.*

Solving problems is a practical art, like swimming, or skiing, or playing the piano; you can learn it only by imitation and practice . . . if you wish to learn swimming you will have to go into the water, and if you wish to become a problem solver you have to solve problems.
   George Polya
   Mathematical Discovery, Vol I (New York: John Wiley and Sons, 1962, p. v).

RADIAN MEASURE

PROBLEM OF THE DAY

Suppose a bike wheel with a 24-in. diameter turns completely around five times every second. How fast is this bike traveling?

How would you answer this question?

There is a relationship between the rate at which the bike's wheel is turning, called *angular velocity,* and the speed of the bike, called *linear velocity.* The speed of the bike is related to the circumference of the bike's wheel. Recall the formula that relates distance, rate, and time: $d = rt$, where d is the distance traveled, r is the rate, and t is the time.

Before answering the question for the Problem of the Day, we will introduce a measurement for angles that is more useful than the historical degree unit of measurement. This measurement system is called *radian measure*. It is used in most mathematical and physics formulas because it simplifies many formulas and uses the set of real numbers (decimal system) rather than the sexagesimal (base 60) system of degrees.

Radian Measurement

In calculus and scientific work, a simple method that uses real numbers to measure angles is used. To understand radian measure, draw a circle with any nonzero radius r. Next, measure an arc with length r. Figure 2.1a shows the case in which $r = 1$, and Figure 2.1b shows $r = 2$. Regardless of the choice for r, when the radius of the circle equals the length of the arc, the angle determined by this arc of length r is the same. (It is labeled θ in Figure 2.1.) This angle is used as a basic unit of measurement and is called a **radian.**

Historical Note

Many "forgotten women" astronomers did much of the real work in the elaborate calculations of planetary orbits. Marie Cunitz was the first woman to attempt to correct Kepler's tables of planetary motion. Her work, *Urania Propitia,* displayed a nice example of role reversal—she acknowledged the able assistance of her husband in the preface.

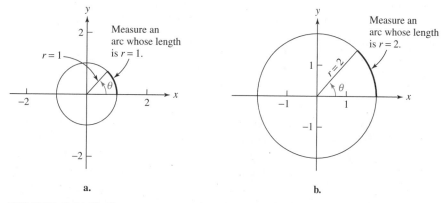

FIGURE 2.1 Radian measure; angles of measure 1. The size of the angle is not dependent on the size of the circle.

> **RADIAN**
>
> An angle that has its vertex at the center of a circle and intercepts an arc on the circle equal in length to the length of the radius of the circle has a measure of **one radian.**

Angle Radian Measure

One revolution	2π
Straight angle $\left(\frac{1}{2}\text{rev}\right)$	π
Right angle $\left(\frac{1}{4}\text{rev}\right)$	$\frac{\pi}{2}$

The circumference of a circle generates an angle of one revolution. Because $C = 2\pi r$, and since the radian is measured in terms of r, an angle of one revolution has a measure of 2π, and a straight angle (one-half revolution) has a measure of π. When measuring angles in radians, we are using *real numbers*. Because radian measure is used so frequently in more advanced work, we agree that radian measure is understood when *no* units of measure are indicated.

> When measuring in degrees in this book, we will always say *degrees* or use the degree symbol, as with $\theta = 5°$. When measuring in radians, we will **omit** the word *radian*. That is, *when no unit is given, radian measure is understood*. For example, $\theta = 5$ means the measure of θ is 5 radians.

A protractor for radian measure is shown in Figure 2.2.

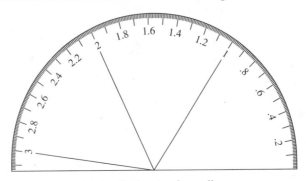

FIGURE 2.2　Protractor for radian measure

Just as with degree measure, you should have the size of 1 radian memorized so that you can draw angles of various sizes from memory with reasonable accuracy. *Practice thinking in terms of radian measure. You should memorize the approximate size of an angle of measure 1 in much the same way you have memorized the approximate size of an angle whose measure is 45°.*

EXAMPLE 1　**Drawing Angles in Radian Measure**

Sketch the angles: **a.** $\dfrac{\pi}{2}$　**b.** $\dfrac{\pi}{3}$　**c.** $\dfrac{\pi}{6}$　**d.** $-\dfrac{5\pi}{6}$　**e.** $-\dfrac{3\pi}{2}$

Solution　**a.** Think of $\dfrac{\pi}{2}$ as $\dfrac{1}{2}(\pi)$.

 A right angle has measure $\dfrac{\pi}{2}$.

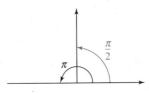

Note:　$\dfrac{\pi}{2} + \dfrac{\pi}{2} = \pi$

b. Think of $\dfrac{\pi}{3}$ as $\dfrac{1}{3}(\pi)$.

Note: $\dfrac{\pi}{3} + \dfrac{\pi}{3} + \dfrac{\pi}{3} = \pi$

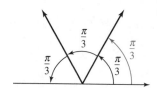

c. Think of $\dfrac{\pi}{6}$ as $\dfrac{1}{2}\left(\dfrac{\pi}{3}\right)$.

Note: $\dfrac{\pi}{6} + \dfrac{\pi}{6} = \dfrac{2\pi}{6} = \dfrac{\pi}{3}$

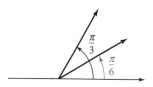

d. Think of $-\dfrac{5\pi}{6}$ as $5\left(\dfrac{\pi}{6}\right)$ in the negative direction.

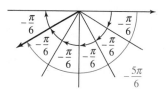

e. Think of $-\dfrac{3\pi}{2}$ as $3\left(\dfrac{\pi}{2}\right)$ in the negative direction.

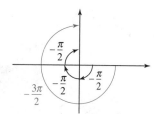

Recall, for a given angle θ, the reference angle θ' is the smallest positive angle that the terminal side of θ makes with the x-axis. In radians, this means that a reference angle must satisfy the given inequalities:

	EXACT VALUES	*APPROXIMATE VALUES*
Reference angles, θ':	$0 < \theta' < \dfrac{\pi}{2}$	$0 < \theta' < 1.57$

EXAMPLE 2

Finding Reference Angles in Radians

For each of the given angles, find the reference angle.

a. $-\dfrac{5\pi}{3}$ **b.** 2.5

Solution   **a.** Draw the angle $-\dfrac{5\pi}{3}$, as shown:

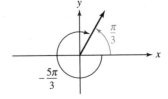

The reference angle is

$$2\pi - \frac{5\pi}{3} = \frac{\pi}{3}$$

b. Draw the angle 2.5 as shown:

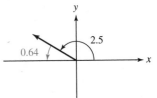

The reference angle is $\pi - 2.5$. This answer can be approximated as 0.64.

Relationship Between Degree and Radian Measures

One revolution is measured by 360° or by 2π radians. Thus

$$\text{Number of revolutions} = \frac{\text{Angle in degrees}}{360}$$

$$\text{Number of revolutions} = \frac{\text{Angle in radians}}{2\pi}$$

Therefore, the following is true.

RELATIONSHIP BETWEEN DEGREE AND RADIAN MEASURE

$$\frac{\text{Angle in degrees}}{360} = \frac{\text{Angle in radians}}{2\pi}$$

Help Window   ? X

To change from *degree measure to radian measure*, let θ be the angle measured in radians, and multiply both sides by 2π to obtain the following formula:

$$\theta = \frac{\pi}{180} \text{(DEGREE MEASURE OF ANGLE)}$$

To change from *radian measure to degree measure*, let θ be the angle measured in degrees, and multiply both sides by 360 to obtain the following formula:

$$\theta = \frac{180}{\pi} \text{(RADIAN MEASURE OF ANGLE)}$$

EXAMPLE 3   **Changing from Degree Measure to Radian Measure**

Change to radians: **a.** 45° **b.** 1° **c.** 123.45°

Solution   In each case, let θ be the angle we seek, measured in radians.

a.
$$\frac{45}{360} = \frac{\theta}{2\pi}$$

$$\frac{45(2\pi)}{360} = \theta$$

$$\frac{\pi}{4} = \theta$$

An alternative method is to remember that π radians is 180°. Because 45° is $\frac{1}{4}$ of 180°, we know that the radian measure is $\frac{\pi}{4}$.

By formula, $\theta = \frac{\pi}{180}(45) = \frac{\pi}{4}$.

The answer $\frac{\pi}{4}$ is called the *exact value* of the radian angle measure. If you use your calculator, θ is approximated as

$$\theta = \frac{\pi}{180}(45) \approx 0.78539816$$

Sometimes students have trouble seeing numbers like π and $\pi/4$ as real numbers. Since the number π is between 3 and 4, $\pi/4$ is a number a little larger than three-quarters.

b.
$$\frac{1}{360} = \frac{\theta}{2\pi}$$ *Substitute 1 for degree measure.*

$$\theta = \frac{\pi}{180}$$ *Exact answer; by formula, $\theta = \frac{\pi}{180}(1) = \frac{\pi}{180}$.*

The approximate value is $\frac{\pi}{180} \approx 0.0174532925$.

c. We find a decimal approximation correct to two decimal places.

$$\frac{123.45}{360} = \frac{\theta}{2\pi}$$ *By formula, $\theta = \frac{\pi}{180}(123.45) \approx 2.15$.*

$$\theta = \frac{123.45}{360}(2\pi)$$ *Notice that this angle is in Quadrant II.*

$$\approx 2.15$$ ∎

You can use the proportion or the formula, but as you can see, the formula is usually easier to use and to remember.

EXAMPLE 4   **Changing Radian Measure to Degree Measure**

Change to degrees: **a.** $\dfrac{\pi}{9}$ **b.** 1 **c.** 4.30

Solution   In each case, let θ be the degree measure of the angle we seek.

a. $\theta = \dfrac{180}{\pi}\left(\dfrac{\pi}{9}\right)$ *Substitute $\dfrac{\pi}{9}$ for radian measure.*

$= 20$

The degree measure of the angle is $20°$.

b. $\theta = \dfrac{180}{\pi}(1)$ *Substitute 1 for radian measure.*

$= \dfrac{180}{\pi}$ *This is the exact answer.*

A calculator approximation is $\dfrac{180°}{\pi} \approx 57.30°$.

c. We find the decimal approximation correct to two decimal places.

$\theta = \dfrac{180}{\pi}(4.30)$ *Substitute 4.30 for radian measure.*

≈ 246.3718519 *Calculator approximation*

To two decimal places, the measure of the angle is $246.37°$. This angle is in Quadrant III.   ■

Calculator Window **?** **X**

Even though the Calculator Windows will not show specific keystrokes, we will note some of the special calculator functions or operations. In each case, you will need to check the owner's manual for your particular calculator. First, make sure your calculator is set on degree mode if you are working in degrees and on radian mode if you are working in radians. On the TI-family of calculators, these choices are found under the MODE key. On the Casio CFX-9850G and 9950G calculators, they are found in the Set UP key of the RUN menu.

   Many calculators have built-in programs that convert angles from degree to radian measure, or from radian to degree measure. However, many calculators can easily be programmed to carry out simple tasks such as this one. Check your owner's manual and see if you can write a simple program to do this.

For the more common measures of angles, it is a good idea to memorize the equivalent degree and radian measures shown in Table 2.1.

TABLE 2.1 Relationship Between Degree and Radian Measure of Angles

Degree	Radians
0°	0
30°	$\frac{\pi}{6} \approx 0.52$
45°	$\frac{\pi}{4} \approx 0.79$
60°	$\frac{\pi}{3} \approx 1.05$
90°	$\frac{\pi}{2} \approx 1.57$
180°	$\pi \approx 3.14$
270°	$\frac{3\pi}{2} \approx 4.71$
360°	$2\pi \approx 6.28$

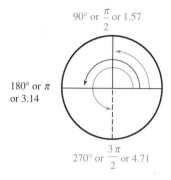

Quadrantal angles

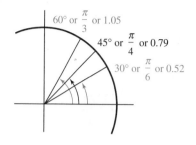

Common first-quadrant angles

$\leftarrow$

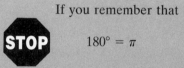

If you remember that

$$180° = \pi$$

you will be able to remember the other measures in this table easily.

Arc Length

Suppose we consider a central angle θ measured in radians with arc length s. We can relate the radian measure of an angle to a circle by finding a simplified arc length formula (which, by the way, is sometimes used to define radian measure).

$$\text{ANGLE IN REVOLUTIONS} = \frac{s}{2\pi r}$$

since one revolution has an arc length (circumference of a circle) of $2\pi r$. Also,

$$\text{ANGLE IN RADIANS} = (\text{ANGLE IN REVOLUTIONS})(2\pi)$$

$$= \frac{s}{2\pi r}(2\pi)$$

$$= \frac{s}{r}$$

This leads us to the **radian arc length formula.**

> ### RADIAN ARC LENGTH FORMULA
>
> The **arc length** subtended by a central angle θ (measured in radians) of a circle of radius r is denoted by s and is found by
>
> $$s = r\theta$$

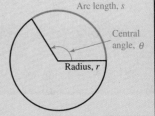

EXAMPLE 5

Finding Arc Length

Find the arc length of a circle of radius 10 cm if the central angle is 2.

Solution

Students often ask the question "2 what?" Do not forget our agreement: Whenever the unit measure is not specified, we mean radian measure. So this means 2 radians. Can you draw this angle from memory?

Use the formula $s = r\theta$:

$$s = r\theta \qquad \text{\textit{Arc length formula}}$$
$$= (10 \text{ cm})(2) \qquad \text{\textit{r = 10 cm; }}\theta = 2$$
$$= 20 \text{ cm}$$

The length of the arc is 20 cm. ∎

Linear and Angular Velocity

We now return to our Problem of the Day. Suppose a bike wheel with a 24-in. diameter turns completely around five times every second. How fast is this bike traveling (to the nearest mi/h)? Now, how would you answer this question?

The speed of the bike is called the *linear velocity,* and to find this velocity, use the distance formula $d = rt$.

Calculate d: The wheel turns 5 revolutions every second, which is $60 \cdot 5$ revolutions every minute, or $60^2 \cdot 5 = 18,000$ revolutions per hour. Furthermore, one revolution is equal to one circumference of the circle ($C = 2\pi r$). The diameter of the wheel is 24 in., so the radius is 12 in.

$$C = 2\pi(12 \text{ in.}) = 24\pi \text{ in.} = 2\pi \text{ ft}$$

Thus, in one hour ($t = 1$), the distance d is found:

$$d = (2\pi \text{ ft})(18,000) = 36,000\pi \text{ ft}$$

Finally, convert this distance to miles by dividing by 5,280 (1 mi = 5,280 ft):

$$d = \frac{36,000\pi}{5,280} \text{ mi}$$

Calculate r:　The rate is the unknown, so we write $d = rt$ as

$$r = \frac{d}{t} = \frac{\dfrac{36,000\pi}{5,280}\,\text{mi}}{1\,\text{h}} \approx 21.4199499\,\text{mi/h}$$

The bike is traveling at about 21 mi/h. *Note:*　You might be used to abbreviating miles per hour as mph, but in scientific work this is usually denoted in fractional form as mi/h.

　　We will formulate the Problem of the Day in more general terms. As noted at the start of this section, there are clearly two ways to calculate velocity. One way is *angular velocity,* which is the rate at which the wheel is turning. This answer is in revolutions per minute (rev/min), radians per second (rad/s), and so on. For this example, the angular velocity is 5 rev/s. The second way to calculate velocity is *linear velocity* (21 mi/h for this example). The linear velocity is given in linear units per unit of time: miles per hour (written mi/h), feet per second (ft/s), kilometers per minute (km/min), and so on.

　　Consider a point P on the wheel of the bike, as shown in Figure 2.3. The measure of the change in $\angle AOP = \theta$ is the *angular velocity,* denoted by ω, which is θ per unit of time. That is, $\omega = \theta/t$. To find the *linear velocity,* begin with the distance formula:

$d = rt$

$s = vt$　　*Since the distance traveled is arc length s, substitute s for d;*
　　　　　　the rate in the distance formula is the velocity v in this application.

$v = \dfrac{s}{t}$　　*Divide both sides by t.*

$v = \dfrac{r\theta}{t}$　　*Arc length formula, $s = r\theta$*

$v = r\omega$　　*Since $\dfrac{\theta}{t} = \omega$*

We summarize these formulas in the following box.

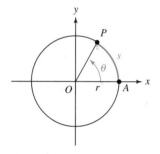

FIGURE 2.3
Angular and linear velocity

**ANGULAR VELOCITY AND
LINEAR VELOCITY**

If P is a point moving with uniform circular motion on a circle of radius r, and a line from the center of the circle through P sweeps out a central angle of θ (measured in radians) in an amount of time t, then the **angular velocity,** denoted by ω, and the **linear velocity,** denoted by v, of the point P are given by these formulas:

$$\omega = \frac{\theta}{t} \quad \text{and} \quad v = r\omega$$

EXAMPLE 6

Hubble Space Telescope

The Hubble Space Telescope (HST) is a joint mission of the National Aeronautics and Space Administration and the European Space Agency. It was launched on April 25, 1990, and was upgraded on February 18, 1997. The reflecting telescope is deployed in low-earth orbit (600 km) with each orbit lasting about 95 minutes. If the radius of the earth is about 6,370 km, what are the angular and linear velocities of the Hubble Space Telescope?

Solution

To find the angular velocity of the HST, we note that it completes

$$\frac{95}{60} = 1.58\overline{3}\,\text{h/orbit} \quad \text{or} \quad \frac{60}{95} \approx 0.63\,\text{orbit/h}$$

or about 15 orbits/day. Since one orbit is the same as one revolution (2π radians), we have

$$\omega = \frac{\theta}{t} = \frac{2\pi}{95/60} = \frac{120\pi}{95} \approx 3.968327562\,\text{rad/h}$$

or about 4 radians per hour. The linear velocity is

$$v = r\omega = (6{,}370 + 600)\left(\frac{120\pi}{95}\right) \approx 27{,}659.24311\,\text{km/h}$$

or about 28,000 kilometers per hour. ∎

National Aeronautics and Space Administration

PROBLEM SET 2.1

Essential Concepts

WHAT IS WRONG, *if anything, with each of the statements in Problems 1–6? Explain.*

1. $30° = \dfrac{\pi}{3}$ **2.** $390° = 30°$

3. $-60° = 60°$ **4.** $-\dfrac{\pi}{4} = \dfrac{7\pi}{4}$

5. A circle with $r = 10$ has a circumference of 10π.

6. The arc length formula is $s = r\theta$, so the length of an arc of a circle of radius 10 with an angle of $10°$ is $s = (10)(10) = 100$.

7. From memory, give the radian measure for each angle whose degree measure is stated. Also sketch the angle.
   a. $30°$ **b.** $90°$ **c.** $45°$ **d.** $60°$ **e.** $180°$

8. From memory, give the radian measure for each angle whose degree measure is stated. Also sketch the angle.
   a. $120°$ **b.** $660°$ **c.** $-135°$ **d.** $-240°$ **e.** $300°$

9. From memory, give the degree measure for each angle whose radian measure is stated. Also sketch the angle.
   a. $\dfrac{\pi}{3}$ **b.** $\dfrac{\pi}{6}$ **c.** $\dfrac{\pi}{2}$ **d.** $\dfrac{\pi}{4}$ **e.** 2π

10. From memory, give the degree measure for each angle whose radian measure is stated. Also sketch the angle.
   a. $\dfrac{5\pi}{6}$ **b.** $\dfrac{5\pi}{3}$ **c.** $-\dfrac{\pi}{4}$ **d.** $-\dfrac{5\pi}{2}$ **e.** $\dfrac{11\pi}{6}$

11. Pick the coin that most closely approximates the given dollar value. For example, $\pi/6$ is a decimal approximately equal to 0.52. This is most closely approximated by a half dollar (Answer E).
A. penny   B. nickel   C. dime   D. quarter
E. Kennedy half-dollar   F. Susan B. Anthony dollar
a. $\pi/3$ **b.** $\pi/12$ **c.** $\pi/180$
d. $\dfrac{\pi}{4} - \dfrac{1}{4}$ **e.** $\dfrac{1}{3}\left(\dfrac{\pi}{10}\right)$ **f.** $\dfrac{1}{50}(\pi)$

12. In degree measure, $50' + 40' = 90'$ or $1°30'$, a very small angle. In radian measure, $50 + 40 = 90$, which is an angle over 14 revolutions.

$$\frac{\pi}{4} + \frac{\pi}{3} = \frac{3\pi}{12} + \frac{4\pi}{12} = \frac{7\pi}{12}$$

Complete the given addition problems.
a. $30' + 50'$ **b.** $30 + 50$ **c.** $5°25' + 8°55'$
d. $\dfrac{\pi}{6} + \dfrac{\pi}{2}$ **e.** $121°16'45'' + 16°55'50''$
f. $2\pi - \dfrac{\pi}{6}$ **g.** $\pi + \dfrac{\pi}{3}$ **h.** $\dfrac{3\pi}{4} + 2\pi$

Level 1 Drill

13. Sketch the angle $2\pi/3$. Classify each of the following angles as equal to, coterminal with, or a reference angle of the sketched angle. If none of these words apply, so state.
a. $\dfrac{\pi}{3}$ **b.** $-\dfrac{\pi}{3}$ **c.** $\dfrac{8\pi}{3}$ **d.** $-\dfrac{4\pi}{3}$ **e.** $\dfrac{4\pi}{6}$

14. Sketch the angle $13\pi/6$. Classify each of the following angles as equal to, coterminal with, or a reference angle of the sketched angle. If none of these words apply, so state.
a. $\dfrac{26\pi}{12}$ **b.** $\dfrac{25\pi}{6}$ **c.** $-\dfrac{11\pi}{6}$ **d.** $\dfrac{\pi}{6}$ **e.** $\dfrac{13\pi}{3}$

15. Sketch the angle $5\pi/6$. Classify each of the following angles as equal to, coterminal with, or a reference angle of the sketched angle. If none of these words apply, so state.
a. $30°$ **b.** $150°$ **c.** $330°$ **d.** $60°$ **e.** $-210°$

Problems 16–19: Sketch each angle and change to radians using exact values.

16. **a.** $270°$ **b.** $480°$ **c.** $40°$
17. **a.** $150°$ **b.** $135°$ **c.** $20°$
18. **a.** $300°$ **b.** $-150°$ **c.** $85°$
19. **a.** $225°$ **b.** $-240°$ **c.** $250°$

Problems 20–23: Sketch each angle and change to radians correct to the nearest hundredth.

20. **a.** $120°$ **b.** $-115°$ **c.** $100°$
21. **a.** $-60°$ **b.** $400°$ **c.** $23.7°$
22. **a.** $350°$ **b.** $525°$ **c.** $-45°$
23. **a.** $38.4°$ **b.** $-210°$ **c.** $-825°$

Problems 24–27: Sketch each angle and change to degrees using exact values.

24. **a.** 5π **b.** $-\dfrac{5\pi}{2}$ **c.** $\dfrac{11\pi}{6}$
25. **a.** $-\dfrac{5\pi}{3}$ **b.** $\dfrac{5\pi}{3}$ **c.** -2π
26. **a.** $\dfrac{2\pi}{3}$ **b.** $\dfrac{4\pi}{3}$ **c.** $\dfrac{11\pi}{3}$
27. **a.** $-\dfrac{\pi}{4}$ **b.** $\dfrac{5\pi}{4}$ **c.** $-\dfrac{11\pi}{4}$

Problems 28–31: Sketch each angle and change to decimal degrees correct to the nearest degree.

28. **a.** $\dfrac{2\pi}{9}$ **b.** $\dfrac{\pi}{2}$ **c.** $\dfrac{5\pi}{3}$
29. **a.** 2 **b.** -3 **c.** 0.5
30. **a.** -0.25 **b.** 0.4 **c.** 7
31. **a.** 12 **b.** -1.5 **c.** 4.712389

Problems 32–33: Find the exact value of the positive angle less than one revolution that is coterminal with each of the given angles.

32. **a.** 3π **b.** $\dfrac{13\pi}{6}$ **c.** $-\pi$
    d. 7 **e.** -7 **f.** $-\dfrac{5\pi}{4}$
33. **a.** $-\dfrac{\pi}{4}$ **b.** $\dfrac{17\pi}{4}$ **c.** $\dfrac{11\pi}{3}$
    d. -2 **e.** 8 **f.** $-\dfrac{\pi}{6}$

Problems 34–37: Find the positive angle less than one revolution correct to four decimal places so that it is coterminal with each of the given angles.

34. **a.** 9 **b.** -5 **c.** $\sqrt{50}$
35. **a.** -6 **b.** 6.2832 **c.** $3\sqrt{5}$
36. **a.** 30 **b.** -3.1416 **c.** $-\dfrac{5\pi}{3}$
37. **a.** -0.7854 **b.** 6.8068 **c.** $\dfrac{9\pi}{4}$

Problems 38–39: *Find the length of the intercepted arc (to the nearest hundredth) if the central angle and radius are given.*

38. a. $\theta = 1, r = 1$ m   **b.** $\theta = \frac{\pi}{3}, r = 4$ in.

39. a. $\theta = 2.34, r = 6$ cm   **b.** $\theta = \frac{2\pi}{3}, r = 15$ cm

Problems 40–43: *Find the angular and linear velocities.*

40. $r = 6$ cm, $\theta = \frac{\pi}{4}, t = 2$ sec

41. $r = 5$ ft, $\theta = 18\pi, t = 5$ min

42. $r = 6,500$ mi, $\theta = 1$ revolution, $t = 10$ hr (round to the nearest mi/h)

43. $r = 93,000,000$ mi, $\theta = 1$ revolution, $t = 365$ days (give the linear velocity in mi/h rounded to three significant figures)

44. An offshore lighthouse has a light that rotates through one rotation every 4 seconds. What is the angular velocity in radians per second?

45. How high will the weight in Figure 2.4 be lifted if the pulley of radius 10 in. is rotated through an angle of 4.5π? Express your answer to the nearest tenth of an inch.

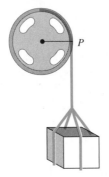

FIGURE 2.4
Pulley lift

46. A flywheel (see Figure 2.5) with radius 15 cm is spinning so that a point on the rim has a linear velocity of 120 cm/s. Find the angular velocity.

FIGURE 2.5 Flywheel

47. The wheel of an automobile is 60 cm in diameter. What is the angular velocity of the wheel in rad/s for this car if it is traveling at 30 km/h? [*Hint:* 60 cm = 0.0006 km; give answer to two significant figures.]

48. A gas gauge (see Figure 2.6) is scaled from empty (E) to full (F) with an arc length of 5 cm with a radius of 2 cm. What is the angle (to the nearest degree) between the E and F readings?

FIGURE 2.6
Gas gauge

49. A speedometer (see Figure 2.7) showing 80 mi/h forms an angle of 2 radians (approximately 115°) with the mark for 0 mi/h. If the radius is 2 cm, find the length of the arc from 0 to 80.

FIGURE 2.7
Speedometer

50. An airplane propeller measures 4 m from tip to tip and rotates at 1,800 rev/min. Find the angular velocity and the linear velocity of a point on the tip of one of the blades. Assume that the airplane itself is not moving.

51. Suppose a bike wheel with a 22-in. diameter is making three complete revolutions every second. Find the linear velocity of the bike in ft/s. (Round to the nearest ft/s.)

52. Find the linear velocity of the bike in Problem 51 in mi/h. (Round to the nearest mi/h.)

53. A simple chain link from the drive sprocket (pedals) to the rear wheel of a bike is shown in Figure 2.8. If the sprocket has radius 4 cm and the drive sprocket has radius 10 cm, what is the number of rotations for the wheel for one rotation of the drive sprocket?

FIGURE 2.8 Drive sprocket bike assembly

54. A belt runs a pulley of radius 40 cm at 50 rev/min. Find the angular velocity of the pulley and the linear velocity (in km/h) of the belt.

55. The wheel of an automobile is 30 in. in diameter. What is the angular velocity of the wheel in rad/s for this car if it is traveling at 30 mi/h?

56. An earth satellite travels in a circular orbit at 32,000 km/h. If the radius of the orbit is 6,770 km, what is the angular velocity (in rad/h)? (Round to two significant digits.)

57. Historical Question   On January 28, 1978, the first federally funded commercial wind generator in the United States was dedicated in Clayton, New Mexico. The rotors span 125 feet each ($r = 125$) and generate enough electricity for about 60 homes. If the rotor rotates at 40 rev/min, find the angular velocity and the linear velocity of a point on the tip of one of the rotors.

Photo Edit © Bonnie Kamin

FIGURE 2.9
Commercial wind generators

58. The biggest windmill in the world was built to generate electricity and is located in Ulfborg, Denmark. Its concrete tower soars 54 meters and is topped by three giant fiberglass propeller blades (see Figure 2.10), each 27 meters long. If the propeller rotates at 30 rev/min, find the angular velocity and the linear velocity of a point on the tip of one of the blades.

FIGURE 2.10
Windmill

59. If the sprocket of the sprocket assembly shown in Figure 2.8 of radius r_1 rotates through an angle of θ_1, what is the corresponding angle of rotation θ_2 for the sprocket of radius r_2?

Level 3
Individual Project

60. Find the angular velocity (in radians per hour) and the linear velocity (rounded to the nearest hundred miles per hour) with respect to the sun for the planets in our solar system. Assume a circular orbit for each planet.

Planet	Mean distance from sun (in millions of miles)	Period of revolution	Equatorial diameter (in miles)	Rotation period
Mercury	36.0	88 days	3,021	59 days
Venus	67.2	224.7 days	7,519	243 days
Earth	92.9	365.2 days	7,926	23 h 56 min
Mars	141.7	687.0 days	4,194	24 h 37 min
Jupiter	483.9	11.86 yr	88,736	9 h 55 min
Saturn	887.1	29.46 yr	74,978	10 h 40 min
Uranus	1,784.0	84 yr	32,193	16 h 48 min
Neptune	2,796.5	165 yr	30,775	16 h 11 min
Pluto	3,666.0	248 yr	1,423	153 h 18 min

SECTION 2.2

TRIGONOMETRIC FUNCTIONS ON A UNIT CIRCLE

PROBLEM OF THE DAY

What are the coordinates of the point of intersection of the terminal side of a standard-position angle θ and a unit circle?

How would you answer this question?

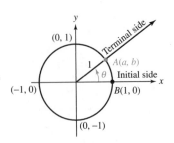

FIGURE 2.11
The unit circle with an angle θ

We will use this question to introduce the trigonometric functions. Consider a relationship between angles and circles. The **unit circle** is the circle centered at the origin with radius one. The equation of the unit circle is

$$x^2 + y^2 = 1$$

Draw a unit circle with an angle θ in standard position, as shown in Figure 2.11. Label the point $(1, 0)$ as B. A second point, A, is the intersection of the terminal side of the angle θ and the unit circle. If the coordinates of A are (a, b), we define two functions of θ:

$$c(\theta) = a \quad \text{and} \quad s(\theta) = b$$

The coordinates of point A, the point of intersection of the terminal side of θ and the unit circle, are found by evaluating c and s. For certain values, this evaluation is easy, as shown in the following example.

EXAMPLE 1

Evaluating c and s Using a Unit Circle

Find: **a.** $s(90°)$ **b.** $c\left(\frac{\pi}{2}\right)$ **c.** $c\left(\frac{3\pi}{2}\right)$ **d.** $s(270°)$ **e.** $c(-3.1416)$

Solution

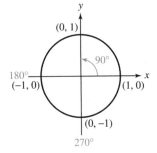

FIGURE 2.12
Example 1

a. $s(90°)$; this is the second component of the ordered pair (a, b), where $A(a, b)$ is the point of intersection of the terminal side of a 90° angle and a unit circle, as shown in Figure 2.12. By inspection, it is 1. Thus, $s(90°) = 1$.

b. $c\left(\frac{\pi}{2}\right) = 0$; for an angle of $\frac{\pi}{2}$ (a right angle), the first component of A is 0.

c. $c\left(\frac{3\pi}{2}\right) = 0$; for an angle of $\frac{3\pi}{2}$, the first component of A is 0.

d. $s(270°) = -1$; for an angle of 270°, the second component of A is -1.

e. $c(-3.1416) = -1$; for an angle of $-3.1416 \approx -\pi$, the first component of A is -1. ∎

We now use the functions we have called c and s to define the trigonometric functions.

Definition of the Trigonometric Functions

The function $c(\theta)$ is called the **cosine function,** and the function $s(\theta)$ is called the **sine function.** These functions, along with four others, make up the **trigonometric functions.**

*This is an important definition. Spend some time learning this before going on. We list the cosine first because cosine always comes first (it is the **first** component); sine is second because it is the second component (also second alphabetically). Note that, with this definition, θ is no longer restricted as it was in Section 1.3.*

This may seem like fine print to you, but spending some time on this help window now will pay dividends to you later in this book.

UNIT CIRCLE DEFINITION OF THE TRIGONOMETRIC FUNCTIONS

Let θ be an angle with vertex at $(0, 0)$ drawn so that the initial side of θ is the positive x-axis, and $A(a, b)$ is the point of intersection of the terminal side of θ with the *unit circle*. Then the six trigonometric functions are defined as follows:

Cosine function: $\cos \theta = a$ (read "cosine of θ is a")

Sine function: $\sin \theta = b$ (read "sine of θ is b")

Tangent function: $\tan \theta = \dfrac{b}{a}, a \neq 0$ (read "tangent of θ is b/a")

Secant function: $\sec \theta = \dfrac{1}{a}, a \neq 0$ (read "secant of θ is $1/a$")

Cosecant function: $\csc \theta = \dfrac{1}{b}, b \neq 0$ (read "cosecant of θ is $1/b$")

Cotangent function: $\cot \theta = \dfrac{a}{b}, b \neq 0$ (read "cotangent of θ is a/b")

Help Window   **?** **X**

Trigonometry is about the study of these six functions, so it is important to remember the unit circle definition. The angle θ is an important part of this definition. The words *cos, sin, tan,* and so on are meaningless without θ. You can speak about the cosine function, or you can speak about cos θ, but you cannot correctly write only *cos*.

Also, take special note of the conditions on the tangent, secant, cosecant, and cotangent functions. These conditions exclude division by 0. For example, $a \neq 0$ means that sec θ and tan θ are not defined when θ is 90°, 270°, or any angle coterminal with either of these angles. If $b \neq 0$, then csc θ and cot θ are not defined when θ is 0°, 180°, or any angle coterminal with these angles. Thus, *cosecant and cotangent are not defined for 0° or 180°*. In this book, we assume all values that cause division by zero are excluded from the domain.

The angle θ in the unit circle definition is called the **argument** of the function. The argument, of course, does not need to be the same as the variable. For example, in $\cos(2\theta + 1)$, we say the *function* is $\cos(2\theta + 1)$, the *argument* is $2\theta + 1$, and the variable is θ.

Evaluation of Trigonometric Functions

We now can answer the Problem of the Day. In many applications, we know the argument and want to find one or more of its trigonometric functions. To help you see the relationship between the angle θ and the function in the definition, consider Example 2.

EXAMPLE 2

Evaluating Trigonometric Functions Using the Definition

Use the definition to evaluate the following:

a. $\cos 110°$ **b.** $\sin 110°$ **c.** $\tan 110°$ **d.** $\sec 110°$

e. $\cos 200°$ **f.** $-\sin 50°$ **g.** $\sin(-50°)$ **h.** $\cos 4$

Solution   We draw a large circle with $r = 1$, as shown in Figure 2.13. Draw an angle where $\theta = 110°$, and estimate the coordinates of $A(a, b)$: $(a, b) \approx (-0.35, 0.95)$.

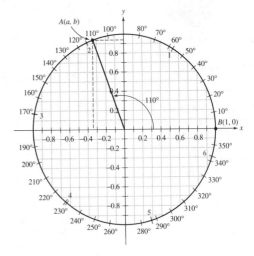

FIGURE 2.13
Approximate values for cosine and sine

a. $\cos 110° \approx -0.35$ (cosine is the first component)

b. $\sin 110° \approx 0.95$ (sine is the second component)

c. $\tan 110° \approx \dfrac{0.95}{-0.35} \approx -2.7$ **d.** $\sec 110° \approx \dfrac{1}{-0.35} \approx -2.9$

e. Look at Figure 2.13, and *think* of an angle $\theta = 200°$. We are looking for the *first* component (remember *cosine is first, sine is second*); $\cos 200° \approx -0.95$.

f. We estimate the *second* component of the point determined by $\theta = 50°$; $\sin 50° \approx 0.75$. Thus, $-\sin 50°$ is the opposite of $\sin 50°$: $-\sin 50° \approx -0.75$.

g. We estimate the *second* component of the point determined by $\theta = -50°$; $\sin(-50°) \approx -0.75$.

h. We estimate the *first* component of the point determined by $\theta = 4$; $\cos 4 \approx -0.66$. ∎

30° (210°)				(329°) 149°
′	Sin	Tan	Cot*	Cos	′
0	.50000	.57735	1.7321	.86603	**60**
1	.50025	.57774	1.7309	.86588	59
2	.50050	.57813	1.7297	.86573	58
3	.50076	.57851	1.7286	.86559	57
4	.50101	.57890	1.7274	.86544	56
5	.50126	.57929	1.7262	.86530	**55**
6	.50151	.57968	1.7251	.86515	54
7	.50176	.58007	1.7239	.86501	53
8	.50201	.58046	1.7228	.86486	52
9	.50227	.58085	1.7216	.86471	51
10	.50252	.58124	1.7205	.86457	**50**
11	.50277	.58162	1.7193	.86442	49
12	.50302	.58201	1.7182	.86427	48
13	.50327	.58240	1.7170	.86413	47
14	.50352	.58279	1.7159	.86398	46
15	.50377	.58318	1.7147	.86384	**45**
16	.50403	.58357	1.7136	.86369	44
17	.50428	.58396	1.7124	.86354	43
18	.50453	.58435	1.7113	.86340	42
19	.50478	.58474	1.7102	.86325	41

Historical Note

For years, the primary method for evaluating trigonometric functions used what is known as a trigonometric table. Typically, the values of the sine, cosine, and tangent in the first quadrant were given. The availability of inexpensive calculators eliminated the need for extensive trigonometric tables.

TABLE 2.2 Signs of the Trigonometric Functions

II		I
Sine and Cosecant positive; others negative		**A**ll positive
Tangent and Cotangent positive; others negative	**C**osine and Secant positive; others negative
III		**IV**

Easy-to-remember form:

A Smart **T**rig **C**lass

II		I
Smart		**A**
Trig		**C**lass
III		**IV**

Notice from Example 2 that some of the trigonometric functions for an angle, 110° for example, are positive and others are negative. In many applications, it is important to know whether a particular function is positive or negative for a particular angle. Consider the point (a, b) in the definition of the trigonometric functions. In Quadrant I, a and b are both positive, so all six trigonometric functions must be positive. In Quadrant II, a is negative and b is positive, so from the definition of trigonometric functions, all are negative except the sine and cosecant. In Quadrant III, a and b are both negative, so all the functions are negative except tangent and cotangent because those are ratios of two negatives. Finally, in Quadrant IV, a is positive and b is negative, so all are negative except cosine and secant. These results are summarized in Table 2.2.

 *Learn the signs of the trigonometric functions in each of the quadrants.*

Reciprocal Identities

Certain relationships among the trigonometric functions enable you to evaluate these functions. These relationships are commonly referred to as the **identities.** Recall from algebra that an **identity** is an open equation (has at least one variable) that is true for all values in the domain.

The first three identities we consider are called the **reciprocal identities.** Remember from arithmetic that 5 and $\frac{1}{5}$ are reciprocals, as are $\frac{2}{3}$ and $\frac{3}{2}$; in general, two numbers are reciprocals if their product is 1.

RECIPROCAL IDENTITIES

1. $\sec \theta = \dfrac{1}{\cos \theta}$ The cosine and secant are reciprocal functions.

2. $\csc \theta = \dfrac{1}{\sin \theta}$ The sine and cosecant are reciprocal functions.

3. $\cot \theta = \dfrac{1}{\tan \theta}$ The tangent and cotangent are reciprocal functions.

STOP

Learn these identities to use your calculator to evaluate the trigonometric functions.

It is easy to prove that these are identities by looking at the definition of secant, cosecant, and cotangent. Also, notice that each of these can be rewritten in other forms. For example, identity 1 can be written as

$$\cos \theta \sec \theta = 1 \quad \text{or} \quad \cos \theta = \frac{1}{\sec \theta}$$

Since the method shown in Example 2 for evaluating the trigonometric functions is not very practical, other procedures are often utilized. Today, we commonly use calculators.

Calculator Window   **?**   **X**

The procedure for evaluating the trigonometric functions is straightforward, but you should be aware of several details:

1. Note the unit of measure used: degree or radian. Calculators have a variety of ways of changing from radian to degree format, so consult your owner's manual to find out how your particular calculator does it. Most, however, simply have a switch (similar to an on/off switch) that sets the calculator in either degree or radian mode. *Remember:* If later in the course, when you are evaluating trigonometric functions, you suddenly start obtaining strange answers and have no idea what you are doing wrong, double-check to make sure your calculator is in the proper mode.

2. With most scientific calculators: | Input angle measure | Input trig function |

With most graphing calculators: | Input trig function | Input angle measure |

3. You must remember which functions are reciprocals, since most calculators do not have sec θ, csc θ, and cot θ keys. These calculators, however, do have a reciprocal key, which is labeled | 1/x | or | x⁻¹ |. For the secant, cosecant, and cotangent functions, evaluate the reciprocal function and then take the reciprocal of the result. *Note:* The key labeled | sin⁻¹ | does NOT represent the reciprocal of sine. Also, | cos⁻¹ | and | tan⁻¹ | do NOT represent reciprocals.

EXAMPLE 3   **Checking Calculator Evaluation of the Trigonometric Functions**

Rework Example 2 using a calculator. Check your calculator answers with the answers shown here.

a. cos 110°   **b.** sin 110°   **c.** tan 110°   **d.** sec 110°

e. cos 200°   **f.** −sin 50°   **g.** sin(−50°)   **h.** cos 4

Solution   **a.** cos 110° ≈ −0.3420201433   **b.** sin 110° ≈ 0.9396926208

c. tan 110° ≈ −2.747477419   **d.** sec 110° ≈ −2.9238044

e. cos 200° ≈ −0.9396926208   **f.** −sin 50° ≈ −0.7660444431

g. sin(−50°) ≈ −0.7660444431   *Do not confuse this with −sin 50°; take the opposite of the angle, then evaluate the function.*

h. cos 4 ≈ −0.6536436209   *If you obtained 0.9975640503, you forgot to change to radian mode.* ∎

PROBLEM SET 2.2

Essential Concepts

1. State the unit circle definition of the trigonometric functions from memory.

Problems 2–5: Find the coordinates of the requested point. Assume that each circle is a unit circle.

2. Point P **3.** Point R

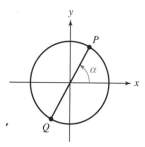
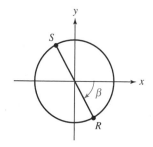

4. Point Q **5.** Point S

6. Name the reciprocal of each trigonometric function.
   a. cosine   **b.** cotangent   **c.** sine
   d. secant   **e.** cosecant   **f.** tangent

WHAT IS WRONG, *if anything, with each of the statements in Problems 7–14? Explain your reasoning.*

7. sin 30° = −0.9880316241

8. $\sin = \dfrac{1}{\sec}$

9. $\cos^{-1}\theta = \dfrac{1}{\csc\theta}$

10. $\sin^{-1}\theta = \dfrac{1}{\sec\theta}$

11. Sine is positive in Quadrant IV.

12. Cosine is positive in Quadrant IV.

13. Both sine and cosine are negative in Quadrant II.

14. sin 2θ = 2 sin θ

Level 1 Drill

Problems 15–18: *Use Figure 2.14 and the definitions of cosine and sine to evaluate the functions given in Problems 15–18 to one decimal place. Check your answers using a calculator.*

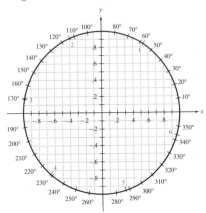

FIGURE 2.14 Using a unit circle definition to evaluate circle functions

15. a. $\cos 50°$ **b.** $\sin 70°$ **c.** $\sin(-150°)$

16. a. $\cos 200°$ **b.** $\cos 250°$ **c.** $\sin(-45°)$

17. a. $\cos 2$ **b.** $\sin 4$ **c.** $\tan(-6)$

18. a. $\sec 3$ **b.** $\cot 5$ **c.** $\csc(-3)$

Problems 19–25: *Use a calculator to evaluate the given functions.*

19. a. $\cos 50°$ **b.** $\sin 20°$ **c.** $\sec 70°$
   d. $\cos(-34°)$ **e.** $\sin(-95°)$ **f.** $\cot 250°$

20. a. $\tan \frac{2\pi}{3}$ **b.** $\cot \frac{\pi}{3}$ **c.** $\csc \frac{\pi}{6}$
   d. $2 \sin 15°$ **e.** $\sin(2 \cdot 15°)$ **f.** $5 \sec 15°$

21. a. $\cos\left(\frac{103°}{2}\right)$ **b.** $\dfrac{\cos 103°}{2}$ **c.** $\dfrac{\cot 103°}{2}$
   d. $\cos\left(-\frac{5\pi}{4}\right)$ **e.** $\sin\left(-\frac{5\pi}{4}\right)$ **f.** $-\cot\left(-\frac{5\pi}{4}\right)$

22. a. $\sin(-2)$ **b.** $-\sin 2$ **c.** $\sec 2$
   d. $\cos(-4)$ **e.** $-\cos(-4)$ **f.** $-\csc(-4)$

23. a. $\dfrac{1}{\csc 2}$ **b.** $\dfrac{1}{\sec 3.5}$ **c.** $\dfrac{1}{\cot 4.5}$

24. a. $\dfrac{1}{\sin 183°}$ **b.** $\dfrac{1}{\cos 215°}$ **c.** $\dfrac{1}{\tan 335°}$

25. a. $\dfrac{\sin 50°}{\cos 50°}$ **b.** $\dfrac{\cos 5}{\sin 5}$ **c.** $\dfrac{\cos 200°}{\sin 200°}$

Problems 26–32: *Tell whether each of the given functions is positive or negative. You should be able to do this without using a calculator.*

26. a. sine, Quadrant I   **b.** cosine, Quadrant II
   c. tangent, Quadrant III   **d.** cotangent, Quadrant IV

27. a. secant, Quadrant I   **b.** cosecant, Quadrant II
   c. sine, Quadrant III   **d.** cosine, Quadrant IV

28. a. $\sin 1$ **b.** $\sin 2$ **c.** $\sin 3$

29. a. $\sin 4$ **b.** $\sin 5$ **c.** $\sin 6$

30. a. $\sin 7$ **b.** $\cos(-1)$ **c.** $\tan 2$

31. a. $\cot 3$ **b.** $\sec 4$ **c.** $\csc 5$

32. a. $\cot 6$ **b.** $\sin(-3)$ **c.** $\cos 10$

Problems 33–37: *State in which quadrant(s) a standard-position angle θ could lie and satisfy the conditions.*

33. a. $\cos \theta = 0.1234$ **b.** $\sin \theta > 0$

34. a. $\sin \theta = -0.85$ **b.** $\cos \theta < 0$

35. a. $\sec \theta = -1.45$ **b.** $\tan \theta < 0$

36. a. $\sin \theta < 0$ and $\tan \theta > 0$
   b. $\sin \theta > 0$ and $\tan \theta < 0$

37. a. $\sin \theta > 0$ and $\cos \theta < 0$
   b. $\cos \theta < 0$ and $\sin \theta < 0$

Level 3 Theory

Problems 38–39: *Prove the fundamental identity using the unit circle definition.*

38. $\csc \theta = \dfrac{1}{\sin \theta}$ **39.** $\sec \theta = \dfrac{1}{\cos \theta}$

40. Fill in the values in the given table.

θ	1	0.5	0.1	0.01	0.001	0.0001
$\sin \theta$						

As θ approaches 0, written $\theta \to 0$, make a conjecture about the values of $\sin \theta$.

41. Fill in the values in the given table.

θ	1	0.5	0.1	0.01	0.001	0.0001
$(\sin \theta)/\theta$						

As θ approaches 0, written $\theta \to 0$, make a conjecture about the values of $\dfrac{\sin \theta}{\theta}$.

42. **PROBLEM FROM CALCULUS** You will learn in calculus that

$$\cos x = 1 - \frac{x^2}{2!} + \frac{x^4}{4!} - \frac{x^6}{6!} + \cdots$$

Use this equation to find cos 1 correct to four decimal places.
[*Hint:* $n! = n(n-1)(n-2) \cdots 3 \cdot 2 \cdot 1$]

Level 3
Individual Projects

Calculator Window   **?**   **X**

43. **HISTORICAL QUESTION** There is evidence (see Eves' *In Mathematical Circles*) that the Babylonians had constructed tables of the trigonometric functions as early as 1900 to 1600 B.C. A tablet known as the Plimpton 322 can be used to find the secant values from 45° to 31°. There are probably lost companion tablets for other functions and other values. The Babylonians used primitive Pythagorean triplets and a right triangle to build their tables. Today, we use computers. Write a computer program that will output a table of values for the trigonometry functions sine, cosine, and tangent for every degree from 0° to 90°.

44. **IN YOUR OWN WORDS Historical Question**
Do some research on primitive Pythagorean triplets (see Problem 43), and write a paper on this topic.

Historical Note

The words *tangent* and *secant* come from the relationship of the ratios to the lengths of the tangent and secant lines drawn on a circle. However, the word *sine* has a more interesting origin. According to Howard Eves in his book *In Mathematical Circles*, Aryabhata called it *jya-adga* (chord half) and abbreviated it *jya*, which the Arabs first wrote as *jiba* but later shortened to *jb*. Later writers saw *jb*, and since *jiba* was meaningless to them, they substituted *jaib* (cove or bay). The Latin equivalent for *jaib* is *sinus*, from which our present word *sine* is derived. Finally, Edmund Gunter (1581–1626) first used the prefix *co* to invent the words *cosine* and *cotangent*. In 1620, Gunter published a seven-place table of common logarithms of the sines and tangents of angle intervals of a minute of an arc. Gunter originally entered the ministry but later decided on astronomy as a career. He was such a poor preacher, Eves states, that Gunter left the ministry in 1619 to the "benefit of both occupations."

SECTION 2.3

FUNDAMENTAL IDENTITIES

PROBLEM OF THE DAY

Graph the path of a projectile with an initial velocity of 30 ft/s and launch angle 45°.

How would you answer this question?

When cannons were introduced in the 13th century, their primary use was to demoralize the enemy. It was much later that they were used for strategic purposes. In fact, cannons existed nearly three centuries before enough was known about the behavior of projectiles to use them with any accuracy. To describe the path of a projectile, or cannonball, we will need to develop some additional trigonometric identities and introduce the idea of *parametric equations.*

In the previous section, we introduced three reciprocal identities:

$$\textbf{1. } \sec \theta = \frac{1}{\cos \theta} \qquad \textbf{2. } \csc \theta = \frac{1}{\sin \theta} \qquad \textbf{3. } \cot \theta = \frac{1}{\tan \theta}$$

These reciprocal identities are necessary to be able to use your calculator to evaluate what we call the **reciprocal trigonometric functions**—namely, secant, cosecant, and cotangent. The building blocks of trigonometry require a knowledge of not only these three reciprocal identities, but five additional identities known as the **fundamental identities.**

Ratio Identities

The next two identities we consider are called the **ratio identities;** they express the tangent and cotangent functions using sine and cosine.

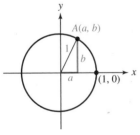

You will need to know these ratio identities to write the tangent and cotangent functions in terms of sines and cosines.

RATIO IDENTITIES

$$\textbf{4. } \tan \theta = \frac{\sin \theta}{\cos \theta} \qquad\qquad \textbf{5. } \cot \theta = \frac{\cos \theta}{\sin \theta}$$

To prove these identities, we use the definition of the trigonometric functions:

$$\tan \theta = \frac{b}{a} \qquad \textit{Definition of tangent}$$

$$= \frac{\sin \theta}{\cos \theta} \qquad \begin{array}{l}\textit{Substitution of values from the}\\ \textit{definition: } \sin \theta = b, \textit{ and } \cos \theta = a\end{array}$$

The ratio identity of the cotangent is proved in exactly the same way.

Pythagorean Identities

The last three identities are called the **Pythagorean identities** because they follow directly from the Pythagorean theorem. Since (a, b) is on a unit circle (see Figure 2.15), the Pythagorean theorem tells us

$$a^2 + b^2 = 1$$

From the definition of sine and cosine (and by substitution), we obtain

$$(\cos \theta)^2 + (\sin \theta)^2 = 1$$

FIGURE 2.15
(a, b) on a unit circle

In trigonometry, we simplify the notation involving the square of a trigonometric function by writing the previous equation as follows:

$$\cos^2\theta + \sin^2\theta = 1$$

PYTHAGOREAN IDENTITIES

6. $\cos^2\theta + \sin^2\theta = 1$ **7.** $1 + \tan^2\theta = \sec^2\theta$ **8.** $\cot^2\theta + 1 = \csc^2\theta$

---

*Memorize the eight identities presented in this section. They are known as the **fundamental identities,** and a great deal of the work we do in trigonometry, and even in more advanced work, is dependent on them.*

To prove identities 7 and 8, we need only divide both sides of $a^2 + b^2 = 1$ by a^2 (for identity 7) or b^2 (for identity 8):

$$a^2 + b^2 = 1 \qquad \textit{Pythagorean theorem}$$

$$\frac{a^2}{a^2} + \frac{b^2}{a^2} = \frac{1}{a^2} \qquad \textit{Divide both sides by } a^2 \ (a \neq 0).$$

$$1 + \left(\frac{b}{a}\right)^2 = \left(\frac{1}{a}\right)^2 \qquad \textit{Properties of exponents}$$

$$1 + (\tan \theta)^2 = (\sec \theta)^2 \qquad \textit{Definition of } \tan \theta \textit{ and } \sec \theta, \textit{ and substitution}$$

$$1 + \tan^2\theta = \sec^2\theta \qquad \textit{Notation change}$$

The proof of identity 8 is left as an exercise.

These Pythagorean identities are frequently written in a different form. For example, we can solve identity 7 for $\tan \theta$:

$$1 + \tan^2\theta = \sec^2\theta \qquad \textit{Identity 7}$$

$$\tan^2\theta = \sec^2\theta - 1 \qquad \textit{Subtract 1 from both sides.}$$

$$\tan \theta = \pm\sqrt{\sec^2\theta - 1} \qquad \textit{Do not forget the} \pm \textit{when you use the square root property from algebra.}$$

EXAMPLE 1

Using Fundamental Identities

Write all the trigonometric functions in terms of $\sin \theta$ by using the eight fundamental identities.

Solution **a.** $\sin \theta = \sin \theta$ *Begin with the easiest function to find.*

b. $\cos^2\theta + \sin^2\theta = 1$ *Identity 6 to find* $\cos \theta$

$$\cos^2\theta = 1 - \sin^2\theta$$

$$\cos \theta = \pm\sqrt{1 - \sin^2\theta}$$

c. $\tan \theta = \dfrac{\sin \theta}{\cos \theta}$ 　　　　　　　 *Identity 4*

$= \dfrac{\sin \theta}{\pm\sqrt{1 - \sin^2\theta}}$ 　　　 *Substitute from part b.**

d. $\cot \theta = \dfrac{1}{\tan \theta}$ 　　　　　　　 *Identity 3*

$= \dfrac{1}{\dfrac{\sin \theta}{\pm\sqrt{1 - \sin^2\theta}}}$ 　　　 *Substitute from part c.*

$= \dfrac{\pm\sqrt{1 - \sin^2\theta}}{\sin \theta}$ 　　　 *Invert and multiply.*

e. $\csc \theta = \dfrac{1}{\sin \theta}$ 　　　　　　　 *Identity 2*

f. $\sec \theta = \dfrac{1}{\cos \theta}$ 　　　　　　　 *Identity 1*

$= \dfrac{1}{\pm\sqrt{1 - \sin^2\theta}}$ 　　　 *Substitute from part b.*   ■

Trigonometric Substitutions

In calculus, functions that contain the following forms ($a > 0$) are often encountered:

$$\sqrt{a^2 - u^2}, \quad \sqrt{u^2 - a^2}, \quad \text{and} \quad \sqrt{u^2 + a^2}$$

In these cases, trigonometry can be used in what is called a *trigonometric substitution*. These substitutions are based on the right-triangle definition of the trigonometric functions. For example, to simplify the expression $\sqrt{a^2 - u^2}$, we can write

$$\sqrt{a^2 - u^2} = \sqrt{a^2\left(1 - \frac{u^2}{a^2}\right)} = a\sqrt{1 - \left(\frac{u}{a}\right)^2}$$

If we consider a triangle with sides of length a and u (as shown in Figure 2.16), we see $\sin \theta = u/a$ so that $u = a \sin \theta$. This means if we make the

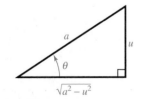

FIGURE 2.16 $u = a \sin \theta$

---

* In algebra, you needed to rationalize the denominator; however, since we work with many reciprocals in trigonometry, we generally relax the requirement that expressions with a radical in a denominator be rationalized.

substitution $u = a \sin \theta$, where θ is an angle in a right triangle, we have

$$
\begin{aligned}
\sqrt{a^2 - u^2} &= \sqrt{a^2 - (a \sin \theta)^2} \\
&= \sqrt{a^2 - a^2 \sin^2\theta} \\
&= \sqrt{a^2(1 - \sin^2\theta)} \\
&= \sqrt{a^2\cos^2\theta} \qquad \text{\textit{Fundamental identity:}} \\
&\phantom{= \sqrt{a^2\cos^2\theta}} \quad \text{\textit{$\cos^2\theta + \sin^2\theta = 1$ written as}} \\
&= a|\cos \theta| \qquad \text{\textit{$\cos^2\theta = 1 - \sin^2\theta$}} \\
&= a \cos \theta \qquad \text{\textit{Because θ is an angle in a right}} \\
& \quad \text{\textit{triangle, it is in Quadrant I.}}
\end{aligned}
$$

Another way of writing this substitution is to note that $\sin \theta = \dfrac{u}{a}$, so that $\theta = \sin^{-1}(u/a)$. These relationships can be shown using a right triangle.

EXAMPLE 2

Trigonometric Substitution

Simplify the expression $\dfrac{du}{u^2\sqrt{4 - u^2}}$ by making the trigonometric substitution $u = 2 \sin \theta$, $-\dfrac{\pi}{2} < \theta < \dfrac{\pi}{2}$, for an angle θ in a right triangle, and where du is a single variable representing $2 \cos \theta$. Leave your answer as a function of u.

Solution

$$
\begin{aligned}
\frac{du}{u^2\sqrt{4 - u^2}} &= \frac{2 \cos \theta}{4 \sin^2\theta\sqrt{4 - 4 \sin^2\theta}} \qquad \text{\textit{Substitute.}} \\[2mm]
&= \frac{2 \cos \theta}{4 \sin^2\theta\sqrt{4(1 - \sin^2\theta)}} \qquad \text{\textit{Factor.}} \\[2mm]
&= \frac{2 \cos \theta}{4 \sin^2\theta\sqrt{4 \cos^2\theta}} \qquad \text{\textit{Fundamental identity}} \\[2mm]
&= \frac{2 \cos \theta}{4 \sin^2\theta(2 \cos \theta)} \qquad \text{\textit{$\sqrt{4 \cos^2\theta} = 2|\cos \theta| = 2 \cos \theta$}} \\
&\phantom{= \frac{2 \cos \theta}{4 \sin^2\theta(2 \cos \theta)}} \quad \text{\textit{since cosine is positive when}} \\
&\phantom{= \frac{2 \cos \theta}{4 \sin^2\theta(2 \cos \theta)}} \quad \text{\textit{$-\pi/2 < \theta < \pi/2$}} \\[2mm]
&= \frac{1}{4 \sin^2\theta} \qquad \text{\textit{Simplify.}} \\[2mm]
&= \frac{1}{4}\csc^2\theta \qquad \text{\textit{Fundamental identity}} \\[2mm]
&= \frac{1}{4}\left(\frac{2}{u}\right)^2 \qquad \text{\textit{$\csc \theta = \dfrac{2}{u}$ from Figure 2.16}} \\[2mm]
&= \frac{1}{u^2}
\end{aligned}
$$

∎

Trigonometric substitutions are summarized in Table 2.3 (page 70).

TABLE 2.3

Algebraic expression	Trigonometric substitution	Reference triangle
$\sqrt{a^2 - u^2}$	$u = a \sin \theta$	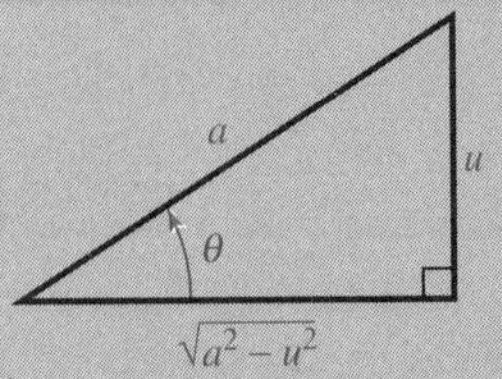
$\sqrt{u^2 - a^2}$	$u = a \sec \theta$	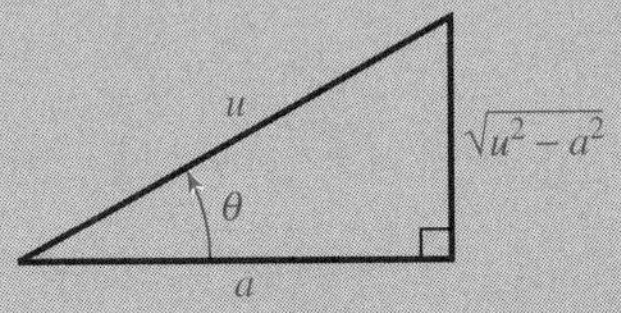
$\sqrt{a^2 + u^2}$	$u = a \tan \theta$	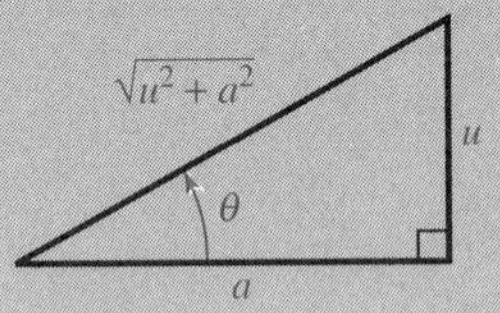

EXAMPLE 3

Finding the Trigonometric Functions Using a Reference Triangle

Using the reference triangle for the substitution $u = a \tan \theta$, write all six trigonometric functions.

Solution   Use the reference triangle to find:

$$\cos \theta = \frac{a}{\sqrt{u^2 + a^2}}; \quad \sin \theta = \frac{u}{\sqrt{u^2 + a^2}}; \quad \tan \theta = \frac{u}{a}$$

The reciprocal functions are

$$\sec \theta = \frac{\sqrt{u^2 + a^2}}{a}; \quad \csc \theta = \frac{\sqrt{u^2 + a^2}}{u}; \quad \cot \theta = \frac{a}{u} \quad \blacksquare$$

Graphing with a Trigonometric Parameter

We assume you are familiar with graphing on a coordinate system from your previous work in mathematics. For example, you have probably done graphs similar to those shown in Figure 2.17.

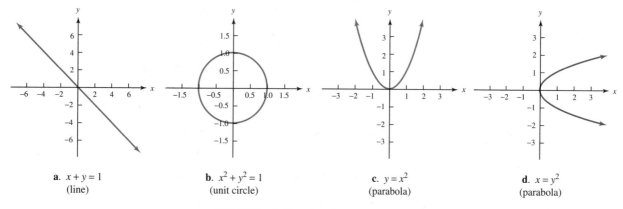

a. $x + y = 1$
(line)

b. $x^2 + y^2 = 1$
(unit circle)

c. $y = x^2$
(parabola)

d. $x = y^2$
(parabola)

FIGURE 2.17   Graphs from algebra

Sometimes we plot points (x, y) by defining both x and y in terms of another variable, say θ. When we do this, we call θ a **parameter** and the equations that define x and y **parametric equations.**

EXAMPLE 4   **Graphing with a Unit Circle Parameter**

Plot the curve represented by the parametric equations

$$x = \cos \theta, \qquad y = \sin \theta$$

Solution   The parameter is θ, and we can generate a table of values by using a calculator.

θ	0°	15°	30°	45°	60°	75°	90°	$\cdots$	120°	$\cdots$
x	1.00	0.97	0.87	0.71	0.50	0.26	0.00	$\cdots$	-0.50	$\cdots$
y	0.00	0.26	0.50	0.71	0.87	0.97	1.00	$\cdots$	0.87	$\cdots$

These points are plotted in Figure 2.18. If the plotted points are connected, you can see that the curve is a circle with center at the origin and radius 1. We can also show this by eliminating the parameter.

$$x^2 + y^2 = (\cos \theta)^2 + (\sin \theta)^2 \quad \text{\textit{Substitute} } x = \cos\theta \text{ \textit{and} } y = \sin\theta.$$
$$= \cos^2\theta + \sin^2\theta \quad \text{\textit{Notational change}}$$
$$= 1 \quad \text{\textit{Pythagorean identity}}$$

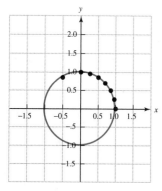

FIGURE 2.18
Graph of $x = \cos \theta$, $y = \sin \theta$; this is the unit circle.

We recognize the equation

$$x^2 + y^2 = 1$$

as the equation of the unit circle; center at $(0, 0)$ and radius 1. We can use these unit circle parameters for circles whose radius is not 1, say, r. Let $x = r \cos \theta$ and $y = r \sin \theta$, so that

$$x^2 + y^2 = (r \cos \theta)^2 + (r \sin \theta)^2 = r^2 \cos^2\theta + r^2 \sin^2\theta = r^2$$

which is a circle with center $(0, 0)$ and radius r.   ■

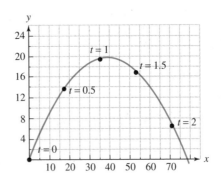

FIGURE 2.19
Path of a projectile

EXAMPLE 5

Problem of the Day: Modeling Projectile Motion

Using Newton's laws of motion, as well as concepts from calculus, it can be shown that if air resistance is neglected, the path of a projectile is given by the parametric equations

$$x = (v_0 \cos \theta)t, \qquad y = h_0 + (v_0 \sin \theta)t - 16t^2$$

where h_0 is the initial height off the ground, and v_0 is the initial velocity of the projectile in the direction of θ with the horizontal ($0° \leq \theta \leq 180°$), as shown in Figure 2.19. The variables x and y represent the horizontal and vertical distances, respectively, measured in feet, and the parameter t represents the time in seconds after firing the projectile.

Graph the path of a projectile with initial velocity $v_0 = 50$ ft/sec and angle $\theta = 45°$. Also, estimate how far downrange the projectile will hit, and the time until impact.

Solution   We see $h_0 = 0$, $v_0 = 50$, so that the desired equations are

$$x = (50 \cos 45°)t, \qquad y = (50 \sin 45°)t - 16t^2$$

We can set up a table of values or use a graphing calculator to show the values correct to the nearest tenth.

t	0	0.5	1	1.5	2	2.5
x	0	17.7	35.4	53.0	70.7	88.4
y	0	13.7	19.4	17.0	6.7	−11.6

Sketch the curve:

If we wish to estimate how far downrange the projectile will hit or when it will impact, we can estimate the point of intersection of the curve with the x-axis [or solve $y = 0$ for t, and find $x(t)$]. We obtain the following values:

t	2.1	2.2	2.3
x	74.2	77.8	81.3
y	3.7	0.34	-3.3

We see that the time until impact is between 2.2 and 2.3 seconds, and the point of impact is between 77.8 and 81.3 ft downrange. ∎

Calculator Window ? X

You should look in Appendix A for information about graphing calculators. For Example 5, you will need to change to PARAMETRIC mode. On the calculator, the formulas for x and y are input as X_{1T} and Y_{1T}, as shown at the right.

After obtaining the graph, you can use the TRACE key. This calculator function lights up the **cursor,** which travels along the curve giving you the values of x, y, and T.

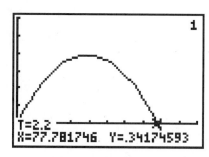

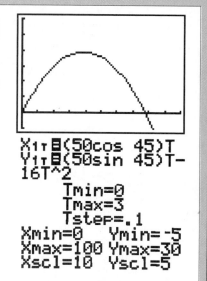

Notice at the bottom we can see that, when $T = 2.2$, the calculator approximates the coordinates of x and y as (77.781746, .34174593). This means that the projectile impacts after about 2.2 seconds and it travels downrange approximately 78 ft.

PROBLEM SET 2.3

Essential Concepts

1. State the reciprocal identities from memory.

2. State the ratio identities from memory.

3. State the Pythagorean identities from memory.

4. Solve $\cos^2\theta + \sin^2\theta = 1$ for $\cos\theta$.

5. Solve $\cos^2\theta + \sin^2\theta = 1$ for $\sin\theta$.

6. Solve $1 + \tan^2\theta = \sec^2\theta$ for $\sec\theta$.

7. Solve $\cot^2\theta + 1 = \csc^2\theta$ for $\cot\theta$.

8. If $\sin\theta = \dfrac{u}{a}$, write the other trigonometric functions in terms of a and u.

9. If $\cos \theta = \dfrac{a}{u}$, write the other trigonometric functions in terms of a and u.

Level 1 Drill

Problems 10–27: *First use a calculator to evaluate each expression, and then evaluate the expression using one of the fundamental identities before using your calculator.*

10. $\dfrac{1}{\csc 135°}$ **11.** $\dfrac{1}{\sec 352°}$ **12.** $\dfrac{1}{\cot 1.4}$ **13.** $\dfrac{1}{\sin 2.6}$

14. $\dfrac{\cos 4.2}{\sin 4.2}$ **15.** $\dfrac{\sin 0.25}{\cos 0.25}$ **16.** $\dfrac{\sin 28°}{\cos 28°}$ **17.** $\dfrac{\cos 128°}{\sin 128°}$

18. $1 - \cos^2 28°$ **19.** $1 - \sin^2 0.5$

20. $-\sqrt{1 - \cos^2 190°}$ **21.** $-\sqrt{1 - \sin^2 100°}$

22. $-\sqrt{1 - \sin^2 190°}$ **23.** $-\sqrt{\sec^2 135° - 1}$

24. $\tan \dfrac{\pi}{3} \cos \dfrac{\pi}{3}$ **25.** $\tan \dfrac{5\pi}{6} \cos \dfrac{5\pi}{6}$

26. $\sin \dfrac{11\pi}{6} \cot \dfrac{11\pi}{6}$ **27.** $\sin\left(-\dfrac{2\pi}{3}\right) \cot\left(-\dfrac{2\pi}{3}\right)$

Level 2 Drill

Problems 28–32: *Write all the trigonometric functions in terms of the given functions. You do not need to rationalize denominators.*

28. $\cos \theta$ **29.** $\tan \theta$ **30.** $\cot \theta$ **31.** $\sec \theta$ **32.** $\csc \theta$

Problems 33–36: *Use the fundamental identities to find the other trigonometric functions of θ using the given information. State the answers correct to the nearest hundredth.*

33. $\cos \theta = \dfrac{5}{13}; \tan \theta < 0$ **34.** $\tan \theta = \dfrac{5}{12}; \sin \theta > 0$

35. $\sin \theta = 0.65; \sec \theta > 0$ **36.** $\cot \theta = 1.25; \cos \theta < 0$

Problems 37–40: PROBLEMS FROM CALCULUS
Make the appropriate trigonometric substitutions to simplify each algebraic expression. Leave your answer as a function of u. [These problems are from *Calculus, 4th edition* by Howard Anton (New York: John Wiley & Sons, Inc., 1992, pp. 576–581).]

37. $\dfrac{du}{\sqrt{u^2 + a^2}}$
    where du is a single variable representing $a \sec^2 \theta$.

38. $\dfrac{\sqrt{u^2 - 25}}{u} \, du$
    where du is a single variable representing $5 \sec \theta \tan \theta$.

39. $\dfrac{u^2 \, du}{\sqrt{u^2 + 5}}$
    where du is a single variable representing $\sqrt{5} \sec^2 \theta$.

40. $\dfrac{du}{u^2 \sqrt{u^2 - 16}}$
    where du is a single variable representing $4 \sec \theta \tan \theta$.

Problems 41–47: *Sketch the curves by plotting points.*

41. $x = \cos \theta, y = \sin \theta, 0° \leq \theta \leq 180°$

42. $x = 10 \cos \theta, y = 10 \sin \theta, 90° \leq \theta \leq 270°$

43. $x = 8 \cos \theta, y = 8 \sin \theta, 45° \leq \theta \leq 135°$

44. $x = 4 \cos \theta, y = 3 \sin \theta, 0° \leq \theta \leq 360°$

45. $x = 5 \cos \theta, y = 2 \sin \theta, 0° \leq \theta \leq 360°$

46. $x = 2 - \sin \theta, y = 1 - \cos \theta, 0° \leq \theta \leq 360°$

47. $x = 1 + \sin \theta, y = 1 - \cos \theta, 0° \leq \theta \leq 360°$

Level 2 Applications

48. IN YOUR OWN WORDS Graph the path of the projectile in Example 5 for an angle of 60° instead of 45°. How does increasing the size of this angle affect the point of impact as well as the time of flight?

49. IN YOUR OWN WORDS Graph the path of the projectile in Example 5 for an angle of 30° instead of 45°. How does decreasing the size of this angle affect the point of impact as well as the time of flight?

50. IN YOUR OWN WORDS Graph the path of the projectile in Example 5 for an angle of 72° instead of 45°. How does increasing the size of this angle affect the point of impact as well as the time of flight?

51. A cannon is fired with an initial speed of 300 ft/sec at an angle of 42° to the horizontal. Neglecting air resistance, sketch the path of the projectile. What is the time of impact (to the nearest tenth of a second)?

52. The shortest distance from home plate to an outfield fence at the old Fenway Park in Boston is 302 ft. If a batter hits a baseball 3 ft above the ground with an initial velocity of 100 ft/sec ($\approx$ 68 mi/h) and an angle of 45° with respect to the ground, will the baseball clear a 10-ft high fence located 302 ft from home plate?

Level 3 Theory

Problems 53–55: *Prove the fundamental identity using the unit circle definition.*

53. $\cot \theta = \dfrac{\cos \theta}{\sin \theta}$ **54.** $1 + \tan^2\theta = \sec^2\theta$

55. $\cot^2\theta + 1 = \csc^2\theta$

SECTION 2.4

TRIGONOMETRIC FUNCTIONS OF ANY ANGLE

PROBLEM OF THE DAY

Let $P(x, y)$ be any point on the terminal side of an angle θ, other than the origin. What are the coordinates of P in terms of θ?

How would you answer this question?

It is not always convenient to work with the unit circle definition of the trigonometric functions developed in the last section. In this section, we use a result from geometry concerning similar triangles to generalize the unit circle definition. This ratio definition of the trigonometric functions will answer the Problem of the Day.

Ratio Definition of the Trigonometric Functions

Suppose you want to find the trigonometric functions of an angle whose terminal side passes through some known point, (3, 4). To apply the unit circle definition, we need to find the point (a, b), as shown in Figure 2.20. Let (3, 4) be denoted by P and (a, b) by A. Let B be the point $(a, 0)$ and Q be the point (3, 0): $|\overline{OA}| = 1$, $|\overline{OP}| = \sqrt{3^2 + 4^2} = 5$. Now consider $\triangle AOB$ and $\triangle POQ$. Recall from Section 1.2 that two triangles are *similar* if two angles of one are congruent to two angles of the other. For these triangles, $\angle OBA$ is congruent to $\angle OQP$ because both are right angles, and $\angle O$ is congruent to $\angle O$ because equal angles are also congruent. Thus, these triangles are similar, which is denoted by

$$\triangle AOB \sim \triangle POQ$$

The important property of similar triangles is that corresponding parts of similar triangles are proportional. Thus,

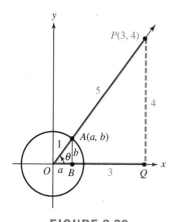

FIGURE 2.20
Finding (a, b) when a point on the terminal side is known

$$a = \frac{a}{1} = \frac{3}{5} \quad \text{and} \quad b = \frac{b}{1} = \frac{4}{5}$$

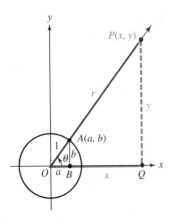

FIGURE 2.21
Similar triangles used to obtain the ratio definition of the trigonometric functions

Once again, we present a fundamental definition for the study of trigonometry. You should memorize this definition.

Thus, $\cos \theta = \dfrac{3}{5}$, $\sin \theta = \dfrac{4}{5}$, and $\tan \theta = \dfrac{4/5}{3/5} = \dfrac{4}{3}$. The reciprocals are $\sec \theta = \dfrac{5}{3}$, $\csc \theta = \dfrac{5}{4}$, and $\cot \theta = \dfrac{3}{4}$.

If we carry out these steps for a point $P(x, y)$ instead of $(3, 4)$, as shown in Figure 2.21, it is still true that

$$\triangle AOB \sim \triangle POQ$$

Let r be the distance from O to P. That is, let $r = \sqrt{x^2 + y^2}$. Then,

$$a = \frac{a}{1} = \frac{x}{r} \qquad\qquad \frac{1}{a} = \frac{r}{x}$$

$$b = \frac{b}{1} = \frac{y}{r} \qquad\qquad \frac{1}{b} = \frac{r}{y}$$

$$\frac{b}{a} = \frac{y/r}{x/r} = \frac{y}{x} \qquad\qquad \frac{a}{b} = \frac{x}{y}$$

These ratios lead to an alternative definition of the trigonometric functions that allows you to choose *any* point (x, y). In practice, this definition is the more frequently used.

RATIO DEFINITION OF THE TRIGONOMETRIC FUNCTIONS

Let θ be any angle in standard position and let $P(x, y)$ be any point on the terminal side of the angle at a distance of r from the origin ($r \neq 0$). Then

$$\cos \theta = \frac{x}{r} \qquad\qquad \sin \theta = \frac{y}{r} \qquad\qquad \tan \theta = \frac{y}{x} \quad (x \neq 0)$$

$$\sec \theta = \frac{r}{x} \quad (x \neq 0) \qquad \csc \theta = \frac{r}{y} \quad (y \neq 0) \qquad \cot \theta = \frac{x}{y} \quad (y \neq 0)$$

Help Window   **?** **X**

This **ratio definition of the trigonometric functions** should be memorized because a great deal of your success with the trigonometric functions depends on a quick recall of this definition. Note from the Pythagorean theorem that r is determined by x and y:

$$r^2 = x^2 + y^2 \qquad \text{*Pythagorean theorem*}$$
$$r = \pm\sqrt{x^2 + y^2} \qquad \text{*Square root property*}$$
$$r = \sqrt{x^2 + y^2} \qquad \text{*Positive value since r is a distance*}$$

Note that since r is a distance, it is always positive; on the other hand, (x, y) is a point, so x and y could be positive, negative, or zero. This is why there are exclusions for certain parts of the definition.

EXAMPLE 1   **Evaluating Trigonometric Functions Given a Point on the Terminal Side**

Find the values of the six trigonometric functions for an angle θ in standard position with the terminal side passing through $(-5, -2)$.

Solution $x = -5$, $y = -2$, and $r = \sqrt{(-5)^2 + (-2)^2} = \sqrt{29}$. Thus

$$\cos \theta = \frac{-5}{\sqrt{29}} \qquad \sec \theta = \frac{-\sqrt{29}}{5}$$

$$\sin \theta = \frac{-2}{\sqrt{29}} \qquad \csc \theta = \frac{-\sqrt{29}}{2}$$

$$\tan \theta = \frac{-2}{-5} = \frac{2}{5} \qquad \cot \theta = \frac{5}{2}$$

■

There are certain angles for which we can find the **exact value** (as opposed to calculator approximations) of the trigonometric functions.

EXAMPLE 2   **Exact Values of Trigonometric Functions of $\frac{\pi}{4} = 45°$**

Find the exact values of the trigonometric functions of $\frac{\pi}{4}$ (see Figure 2.22).

Solution $$\cos \frac{\pi}{4} = \frac{x}{r} \qquad\qquad\quad \textit{From the ratio definition (this is true for any angle)}$$

$$= \frac{x}{\sqrt{x^2 + y^2}} \qquad r = \sqrt{x^2 + y^2} \textit{ for any angle}$$

$$= \frac{x}{\sqrt{x^2 + x^2}} \qquad \textit{If } \theta = \frac{\pi}{4}, \textit{ then } \theta \textit{ bisects Quadrant I so that } x = y.$$

$$= \frac{x}{\sqrt{2x^2}}$$

$$= \frac{x}{x\sqrt{2}} \qquad\qquad \sqrt{x^2} = |x| = x \textit{ since } x \textit{ is positive in Quadrant I.}$$

$$= \frac{1}{\sqrt{2}} \quad \text{or} \quad \tfrac{1}{2}\sqrt{2}$$

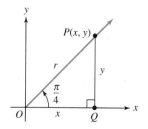

FIGURE 2.22 $\theta = \frac{\pi}{4}$

Similarly (remember $y = x$),

$$\sin \frac{\pi}{4} = \frac{y}{r} = \frac{y}{\sqrt{x^2 + y^2}} = \frac{x}{\sqrt{x^2 + x^2}} = \frac{x}{\sqrt{2x^2}} = \frac{1}{\sqrt{2}} \quad \text{or} \quad \tfrac{1}{2}\sqrt{2}$$

By the ratio identity, we find

$$\tan \frac{\pi}{4} = \frac{\sin \frac{\pi}{4}}{\cos \frac{\pi}{4}} = \frac{\tfrac{1}{2}\sqrt{2}}{\tfrac{1}{2}\sqrt{2}} = 1$$

Finally, reciprocals give us the reciprocal functions:

$$\sec \tfrac{\pi}{4} = \sqrt{2}, \quad \csc \tfrac{\pi}{4} = \sqrt{2}, \quad \text{and} \quad \cot \tfrac{\pi}{4} = 1 \qquad \blacksquare$$

EXAMPLE 3   **Exact Values of Trigonometric Functions of $\frac{\pi}{6} = 30°$**

Evaluate the trigonometric functions of $30°$.

Solution   Consider not only the standard-position angle $30°$, but also the standard position angle $-30°$. Choose $P_1(x, y)$ and $P_2(x, -y)$, respectively, on the terminal sides, as shown in Figure 2.23.

   Since $\angle OQP_1$ and $\angle OQP_2$ are right angles, $\angle OP_1Q = 60°$ and $\angle OP_2Q = 60°$. Thus, $\triangle OP_1P_2$ is an equiangular triangle. (That is, all angles measure $60°$.) From geometry, we know that an equiangular triangle has sides the same length. Thus, $2y = r$. Notice the following relationship between x and y:

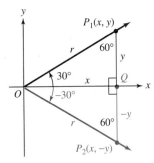

FIGURE 2.23 $\theta = \frac{\pi}{6}$

$$r^2 = x^2 + y^2$$
$$(2y)^2 = x^2 + y^2 \quad \text{\textit{Since }} 2y = r$$
$$3y^2 = x^2$$
$$\sqrt{3}\,|y| = |x|$$

Since x and y are both positive when $\theta = 30°$, we see $x = \sqrt{3}y$.

$$\cos 30° = \frac{x}{r} = \frac{\sqrt{3}y}{2y} = \frac{\sqrt{3}}{2} \qquad\qquad \sec 30° = \frac{2}{\sqrt{3}} \quad \text{or} \quad \tfrac{2}{3}\sqrt{3}$$

$$\sin 30° = \frac{y}{r} = \frac{y}{2y} = \frac{1}{2} \qquad\qquad\qquad \csc 30° = 2$$

$$\tan 30° = \frac{y}{x} = \frac{y}{\sqrt{3}y} = \frac{1}{\sqrt{3}} \quad \text{or} \quad \tfrac{1}{3}\sqrt{3} \qquad \cot 30° = \sqrt{3} \qquad \blacksquare$$

EXAMPLE 4   **Exact Values of Trigonometric Functions of $\frac{\pi}{3} = 60°$**

Find the exact values for the trigonometric functions of $\frac{\pi}{3}$.

Solution   For an angle of $60°$ instead of $30°$, it can be shown that $r = 2x$, so that

$$r^2 = x^2 + y^2 \quad \text{\textit{Pythagorean theorem}}$$
$$(2x)^2 = x^2 + y^2 \quad \text{\textit{Since }} r = 2x$$
$$3x^2 = y^2 \quad\quad \text{\textit{Subtract }} x^2 \text{ \textit{from both sides.}}$$
$$\sqrt{3}\,|x| = |y|$$

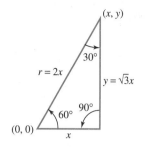

FIGURE 2.24 $\frac{\pi}{3} = 60°$ angle

For 60° (see Figure 2.24), x and y are both positive, so $y = \sqrt{3}\,x$.

$$\cos 60° = \frac{x}{r} = \frac{x}{2x} = \frac{1}{2} \qquad\qquad \sec 60° = 2$$

$$\sin 60° = \frac{y}{r} = \frac{\sqrt{3}\,x}{2x} = \frac{\sqrt{3}}{2} \qquad \csc 60° = \frac{2}{\sqrt{3}} = \tfrac{2}{3}\sqrt{3}$$

$$\tan 60° = \frac{y}{x} = \frac{\sqrt{3}\,x}{x} = \sqrt{3} \qquad \cot 60° = \frac{1}{\sqrt{3}} = \tfrac{1}{3}\sqrt{3}$$

∎

The results of Examples 2–4 are summarized in Table 2.4. You should now be able to verify all the entries in this table.

STOP

You should spend some time memorizing this table. These exact values are used throughout this book and will also be used in your future work in mathematics. These exact values can be represented on a unit circle as shown below:

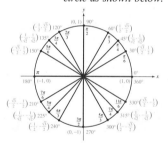

TABLE 2.4 Table of Exact Values

Angle θ	0	$\frac{\pi}{6}$	$\frac{\pi}{4}$	$\frac{\pi}{3}$	$\frac{\pi}{2}$	π	$\frac{3\pi}{2}$
$\cos\theta$	1	$\frac{\sqrt{3}}{2}$	$\frac{\sqrt{2}}{2}$	$\frac{1}{2}$	0	−1	0
$\sin\theta$	0	$\frac{1}{2}$	$\frac{\sqrt{2}}{2}$	$\frac{\sqrt{3}}{2}$	1	0	−1
$\tan\theta$	0	$\frac{\sqrt{3}}{3}$	1	$\sqrt{3}$	undef.	0	undef.
$\sec\theta$	1	$\frac{2}{\sqrt{3}}$	$\frac{2}{\sqrt{2}}$	2	undef.	−1	undef.
$\csc\theta$	undef.	2	$\frac{2}{\sqrt{2}}$	$\frac{2}{\sqrt{3}}$	1	undef.	−1
$\cot\theta$	undef.	$\sqrt{3}$	1	$\frac{1}{\sqrt{3}}$	0	undef.	0

You can find exact values of the trigonometric functions that are multiples of those in Table 2.4 by using the idea of a reference angle and the **reduction principle:**

REDUCTION PRINCIPLE

If t represents any of the six trigonometric functions, then

$$t(\theta) = \pm t(\theta')$$

where θ' is the reference angle of θ and the plus or minus sign depends on the quadrant of the terminal side of the angle θ.

EXAMPLE 5

Exact Values by Using the Reduction Principle

Evaluate
a. $\cos 135°$ **b.** $\tan 210°$ **c.** $\sin\left(-\frac{7\pi}{6}\right)$ **d.** $\tan \frac{5\pi}{3}$ **e.** $\csc\left(-\frac{5\pi}{3}\right)$

Solution

a. $\cos 135° = -\cos \underbrace{45°}_{\substack{\uparrow \\ \text{reference angle}}} = -\frac{\sqrt{2}}{2}$

(*Quadrant* II, *so cosine is negative*)

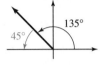

b. $\tan 210° = \underset{\uparrow}{+}\tan 30° = \frac{\sqrt{3}}{3}$

(*Quadrant* III, *so tangent is positive*)

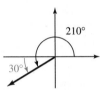

c. $\sin\left(-\frac{7\pi}{6}\right) = \underset{\uparrow}{+}\sin \frac{\pi}{6} = \frac{1}{2}$

(*Quadrant* II; *sine positive*)

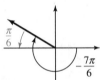

d. $\tan \frac{5\pi}{3} = \underset{\uparrow}{-}\tan \frac{\pi}{3} = -\sqrt{3}$

(*Quadrant* IV; *tangent negative*)

e. $\csc\left(-\frac{5\pi}{3}\right) = +\csc \frac{\pi}{3} = \frac{2}{\sqrt{3}}$ or $\frac{2}{3}\sqrt{3}$ (*Quadrant* I)

PROBLEM SET 2.4

Essential Concepts

1. **IN YOUR OWN WORDS** State the ratio definition of the trigonometric functions from memory.

2. **IN YOUR OWN WORDS** State the reduction principle, and then explain what it is saying. Include some examples.

Problems **3–9:** *Find the exact value.*

3. **a.** $\tan \frac{\pi}{4}$ **b.** $\cos 0$ **c.** $\sin 60°$
   d. $\cos 30°$ **e.** $\cos 270°$ **f.** $\tan \frac{\pi}{6}$

4. **a.** $\tan 180°$ **b.** $\sin 45°$ **c.** $\sin \pi$
   d. $\sin \frac{\pi}{2}$ **e.** $\tan 0$ **f.** $\cos \frac{\pi}{4}$

5. **a.** $\cos 60°$ **b.** $\sin \frac{\pi}{6}$ **c.** $\sin 0°$
   d. $\cos \pi$ **e.** $\tan \frac{\pi}{3}$ **f.** $\cos \frac{\pi}{2}$

6. **a.** $\sec \frac{\pi}{6}$ **b.** $\csc 0$ **c.** $\sec \frac{\pi}{4}$
   d. $\sec 0°$ **e.** $\cot \frac{3\pi}{2}$ **f.** $\cot \pi$

7. **a.** $\sec \frac{\pi}{3}$ **b.** $\cot \frac{\pi}{6}$ **c.** $\cot 90°$
   d. $\csc 60°$ **e.** $\cot \frac{\pi}{3}$ **f.** $\cot 0$

8. **a.** $\csc 30°$ **b.** $\csc \pi$ **c.** $\cot \frac{3\pi}{2}$
   d. $\csc 90°$ **e.** $\sec \frac{3\pi}{2}$ **f.** $\sec 90°$

9. **a.** $\tan 90°$ **b.** $\sin \frac{3\pi}{2}$ **c.** $\tan \frac{3\pi}{2}$
   d. $\cot \frac{\pi}{4}$ **e.** $\csc \frac{\pi}{4}$ **f.** $\sec 180°$

Level 1 Drill

Problems 10–13: *Use the reduction principle to give the exact values in simplified form.*

10. a. $\sin \frac{5\pi}{6}$ **b.** $\csc\left(-\frac{3\pi}{2}\right)$ **c.** $\cos(-300°)$

d. $\sin 390°$ **e.** $\sin \frac{17\pi}{4}$ **f.** $\cos(-6\pi)$

11. a. $\cos \frac{9\pi}{2}$ **b.** $\cos 495°$ **c.** $\sin(-765°)$

d. $\cos 300°$ **e.** $\sin 120°$ **f.** $\tan 120°$

12. a. $\cot 240°$ **b.** $\sec 120°$ **c.** $\tan 135°$

d. $\csc \frac{3\pi}{2}$ **e.** $\sec(-420°)$ **f.** $\csc(-6\pi)$

13. a. $\sin(-390°)$ **b.** $\tan \frac{11\pi}{4}$ **c.** $\cos\left(-\frac{17\pi}{6}\right)$

Problems 14–19: *Find the values of the six trigonometric functions for angle θ in standard position whose terminal side passes through the given points. Draw a sketch showing θ and the reference angle.*

14. $(3, 4)$ **15.** $(3, -4)$ **16.** $(-5, -12)$

17. $(-5, 12)$ **18.** $(-6, 1)$ **19.** $(-2, -3)$

WHAT IS WRONG, *if anything, with each of the statements? That is, decide whether each statement in Problems 20–35 is true or false by choosing various values for θ. If it is false, give a counterexample (that is, an example that shows it is not true).*

20. $\sin \theta \csc \theta = 1$ **21.** $\cos \theta \csc \theta = 1$

22. $\sin^2\theta + \cos^2\theta = 1$ **23.** $(\sin \theta + \cos \theta)^2 = 1$

24. $\tan^2\theta - \sec^2\theta = 1$ **25.** $\cos \theta = \dfrac{\sin \theta}{\tan \theta}$

26. $\cos(-\theta) = \cos \theta$ **27.** $\sin(-\theta) = \sin \theta$

28. $\dfrac{\cos 2\theta}{2} = \cos \theta$ **29.** $\dfrac{\sin 5\theta}{5} = \sin \theta$

30. $2 \sin \theta \cos \theta = \sin 2\theta$ **31.** $\sin \frac{1}{2}\theta = \frac{1}{2} \sin \theta$

32. $\sin \frac{1}{2}\theta = \sqrt{\dfrac{1 - \cos \theta}{2}}$

33. $\sin \frac{1}{2}\theta = -\sqrt{\dfrac{1 - \cos \theta}{2}}$

34. $\sin \theta = \sqrt{1 - \cos^2\theta}$

35. $\cos \theta = -\sqrt{1 - \sin^2\theta}$

Level 2 Drill

36. Verify the entries in Table 2.4 for the angle $\pi/3$.

37. Find $\cos \frac{5\pi}{4}$ by using the procedure illustrated in Example 2.

38. Find $\cos 135°$ by choosing an arbitrary point (x, y) on the terminal side of $135°$ and applying the ratio definition of the trigonometric functions.

39. Find $\sin\left(-\frac{\pi}{4}\right)$ by choosing an arbitrary point (x, y) on the terminal side of $-\frac{\pi}{4}$ and applying the ratio definition of the trigonometric functions.

40. Find $\sin 210°$ by choosing an arbitrary point (x, y) on the terminal side of $210°$ and applying the ratio definition of the trigonometric functions.

41. Find $\cos 210°$ by choosing an arbitrary point (x, y) on the terminal side of $210°$ and applying the ratio definition of the trigonometric functions.

Problems 42–54: *Substitute the exact values for the trigonometric functions in the given expressions and simplify.*

42. a. $\sin 30° + \cos 0°$ **b.** $\sin \frac{\pi}{2} + 3 \cos \frac{\pi}{2}$

43. a. $2 \cos \frac{\pi}{2}$ **b.** $\cos \frac{2\pi}{2}$

44. a. $\sin \frac{2\pi}{4}$ **b.** $2 \sin \frac{\pi}{4}$

45. a. $\sin^2 60°$ **b.** $\cos^2 \frac{\pi}{4}$

46. a. $\sin^2 \frac{\pi}{6} + \cos^2 \frac{\pi}{6}$ **b.** $\sin^2 \frac{\pi}{2} + \cos^2 \frac{\pi}{2}$

47. a. $\sin^2 \frac{\pi}{3} + \cos^2 \frac{\pi}{3}$ **b.** $\sin^2 \frac{\pi}{6} + \cos^2 \frac{\pi}{3}$

48. a. $\sin \frac{\pi}{6} \csc \frac{\pi}{6}$ **b.** $\csc \frac{\pi}{2} \sin \frac{\pi}{2}$

49. a. $\cos\left(\frac{\pi}{4} - \frac{\pi}{2}\right)$ **b.** $\cos \frac{\pi}{4} - \cos \frac{\pi}{2}$

50. a. $\tan(2 \cdot 30°)$ **b.** $2 \tan 30°$

51. a. $\csc\left(\frac{1}{2} \cdot 60°\right)$ **b.** $\dfrac{\csc 60°}{2}$

52. a. $\cos\left(\frac{1}{2} \cdot 60°\right)$ **b.** $\sqrt{\dfrac{1 + \cos 60°}{2}}$

53. a. $\tan(2 \cdot 60°)$ **b.** $\dfrac{2 \tan 60°}{1 - \tan^2 60°}$

54. a. $\cos\left(\frac{\pi}{2} - \frac{\pi}{6}\right)$ **b.** $\cos \frac{\pi}{2} \cos \frac{\pi}{6} + \sin \frac{\pi}{2} \sin \frac{\pi}{6}$

55. It can be shown that the exact value for $\sin 3°$ is

$$\frac{1}{16}[\sqrt{30} + \sqrt{10} - \sqrt{6} - \sqrt{2}]$$
$$-\frac{1}{8}\sqrt{20 + 4\sqrt{5} - 2\sqrt{15} - 10\sqrt{3}}$$

Use your calculator to verify this exact value.

56. It can be shown that

$$\sin 15° = \frac{1}{\sqrt{6} + \sqrt{2}} \quad \text{and} \quad \cos 15° = \frac{2 + \sqrt{3}}{\sqrt{6} + \sqrt{2}}$$

a. Simplify each of these exact values.
b. Verify, using the exact values, that

$$\cos^2 15° + \sin^2 15° = 1$$

Level 2 Applications

Problems 57–61: *A useful trigonometric formulation of slope involves the **angle of inclination**.*

> The **angle of inclination** of a line ℓ is the angle ϕ $(0 \leq \phi < \pi)$ between line ℓ and the positive x-axis. Then the **slope** of line ℓ with inclination ϕ is $m = \tan \phi$.

57. Draw the line passing through $(1, 2)$ with an angle of $60°$. What is the slope of this line? Graph this line.

58. Recall from algebra that the point–slope form of the equation of a line passing through the point (x_1, y_1) with slope m is

$$y - y_1 = m(x - x_1)$$

Find the equation of the line in Problem 57.

59. Find the equations of the lines passing through the given points with the given angles of inclination. Write your answers in the form $y = m(x - h) + k$.
a. $30°$; $(2, 3)$ **b.** $120°$; $(1, 4)$
c. $45°$; $(9, -5)$ **d.** $135°$; $(-3, -8)$

60. Find the approximate equation of the lines passing through each point with the given angle of inclination. Write your equation in the form $y = mx + b$, and round all numbers and coefficients to two decimal places.
a. $23°$; $(0.81, 0.53)$ **b.** $83°$; $(-1.45, 0.61)$
c. $138°$; $(1, 8)$ **d.** $175°$; $(-3, -2)$

61. In terms of the angle of inclination ϕ, describe the following lines:
a. A line that is *rising*   **b.** A line that is *falling*
c. A horizontal line   **d.** A vertical line

62. Use the definition of tangent to show that the definition of slope based on angle of inclination is the same as the following definition of slope from algebra: *The slope of the line passing through the points $P_1(x_1, y_1)$ and $P_2(x_2, y_2)$ is*

$$m = \frac{y_2 - y_1}{x_2 - x_1}$$

63. The position of a piston, connecting rod, and crankshaft in an internal combustion engine are related as shown in Figure 2.25.

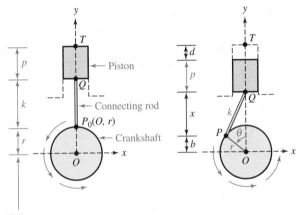

All these distances are constants.

FIGURE 2.25   Piston in an internal combustion engine

It can be shown that

$$d = k + r - r\cos(2\pi vt) - \sqrt{k^2 - r^2 \sin^2(2\pi vt)}$$

where d = depth of piston stroke
$\quad k$ = length of connecting rod
$\quad r$ = radius of the crankshaft
$\quad v$ = velocity of the crankshaft in rev/min
$\quad t$ = time in minutes

Find the position of a piston after 0.06 sec and if the connecting rod is 15 cm and the radius of the crankshaft is 5 cm. Assume that the crankshaft is turning at 600 rev/min.

64. Find the position of a piston after $\frac{1}{2}$ second if the crankshaft (see Problem 63) is turning at 60 rev/min and has radius 4 in. Assume that the connecting rod is 8 inches long.

Level 3 Theory

65. Let A and B be any two points on a unit circle where A is a point on the terminal side of an angle α and B is a point on the terminal side of an angle β. Find the slope of the line passing through A and B.

66. Let $P(x, y)$ be any point in the plane. Show that $P(r \cos \theta, r \sin \theta)$ is a representation for P, where θ is the standard-position angle formed by drawing ray $\overrightarrow{OP}$.

67. a. If θ is in Quadrant I, then $\theta + \pi$ is in Quadrant III with a reference angle θ. Use this fact and the reduction principle to show that

$$\sin(\theta + \pi) = -\sin \theta$$

    b. Show that $\sin(\theta + \pi) = -\sin \theta$ if θ is in Quadrant II.

    c. Show that $\sin(\theta + \pi) = -\sin \theta$ if θ is in Quadrant III.

    d. Show that $\sin(\theta + \pi) = -\sin \theta$ if θ is in Quadrant IV.

68. Show that $\cos(\theta + \pi) = -\cos \theta$ for any angle θ.

Level 3 Individual Projects

69. Historical Question In this book, we measure angles in degrees and in radians. Another measurement, called a *grad*, was proposed in France and is used in Europe. A grad, also called a *new degree*, is found on many calculators and is defined as a unit of measurement that divides one revolution into 400 equal parts. These new degrees are denoted by 1^g. Notice that 100^g is a right angle and 200^g is a straight angle. Write a paper discussing converting grads to degrees and radians and also degrees and radians to grads. Illustrate with examples, including $\cos 50^g$, $\sin 150^g$, and $\tan 261^g$.

70. Figure 2.26 shows the *gradient* of a road, which uses the tangent function. For example, a gradient of 8% means a difference in height of 8 m over a map distance of 100 m.

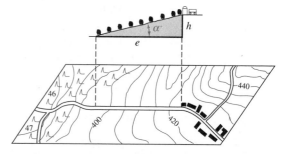

FIGURE 2.26 Gradient (From *The VNR Concise Encyclopedia of Mathematics,* W. Gellert, H. Kustner, M. Hellwich, and H. Kastner, ed., p. 221. © VEB Bibliographisches Institut Leipzig, 1975.)

Write a paper about road or topological gradients and how they are determined.

2.5 CHAPTER TWO SUMMARY

Objectives

Section 2.1   Radian Measure

Objective 2.1   Be familiar with the radian measure of an angle and be able to approximate the angle associated with a given radian measure without using any measuring devices.

Objective 2.2   Find the reference angle for a given angle.

Objective 2.3   Change from radian measure to degree measure and from degree measure to radian measure; know the commonly used radian and degree measure equivalences.

Objective 2.4   Know the radian arc length formula and be able to apply it.

Objective 2.5   Know the formula for the area of a sector and be able to apply it.

Objective 2.6	Know the difference between angular and linear velocity, and be able to compute them.	**Objective 2.12**	Graph with a unit circle using a parameter.
		Objective 2.13	Use trigonometric functions to model projectile motion.
Section 2.2	Trigonometric Functions on a Unit Circle	Section 2.4	Trigonometric Functions of Any Angle
Objective 2.7	Know the unit circle definition of the trigonometric functions.	**Objective 2.14**	Know the ratio definition of the trigonometric functions.
Objective 2.8	Evaluate trigonometric functions using the definition and using a calculator.	**Objective 2.15**	Use the definition of the trigonometric functions to approximate their values for a given angle or for an angle passing through a given point.
Objective 2.9	State and prove the eight fundamental identities.		
Objective 2.10	Know the signs of the six trigonometric functions in each of the four quadrants. Name the function(s) that is (are) positive in the given quadrant.	**Objective 2.16**	Know and be able to derive the table of exact values.
		Objective 2.17	Use the reduction principle, along with the table of exact values, to evaluate certain trigonometric functions.
Section 2.3	Fundamental Identities		
Objective 2.11	Use the fundamental identities to find the other trigonometric functions if you are given the value of one function and want to find the others.		

Terms

Angular velocity [2.1]
Arc length formula [2.1]
Argument [2.2]
Cosecant [2.2]
Cosine [2.2]
Cotangent [2.2]
Exact values [2.4]
Fundamental identity [2.3]

Identity [2.2]
Linear velocity [2.1]
Parameter [2.3]
Parametric equation [2.3]
Pythagorean identities [2.3]
Radian [2.1]

Ratio definition of the trigonometric functions [2.4]
Ratio identities [2.2]
Reciprocal functions [2.3]
Reciprocal identities [2.2]
Reduction principle [2.4]
Secant [2.2]

Sine [2.2]
Tangent [2.2]
Trigonometric functions [2.2]
Unit circle [1.3]
Unit circle definition of the trigonometric functions [1.3]

Sample Test

All of the answers of this sample test are given in the back of the book.

1. State the ratio definition of the trigonometric functions.

2. Change 500° to radians correct to the nearest hundredth.

3. Change -3.8 to degrees correct to the nearest degree.

4. Find the values of the trigonometric functions for an angle α whose terminal side passes through $(\sqrt{7}, 3)$.

5. Evaluate each function (rounded to four decimal places).
  a. $\cos 2$ **b.** $\sin 4$ **c.** $\tan 0.85$
  d. $\csc(-1.5)$ **e.** $\sec(-8.5)$

6. If α is an angle in standard position, what are the coordinates of P_α, which is the point of intersection of the terminal side of α and the unit circle?

7. Write each as a function of an acute angle.
  a. $\cos \frac{5\pi}{3}$ **b.** $\sin\left(-\frac{3\pi}{4}\right)$ **c.** $\tan\left(-\frac{7\pi}{4}\right)$
  d. $\cot \frac{5\pi}{4}$ **e.** $\csc 2$ **f.** $\sec 4$

8. a. State and prove one of the reciprocal identities.
  b. State and prove one of the ratio identities.
  c. State and prove one of the Pythagorean identities.

9. Complete the table of exact values.

degrees:	$-45°$	a.	b.	$210°$	$90°$
radians:	c.	π	$-\frac{4\pi}{3}$	d.	e.
sine:	f.	g.	h.	i.	j.
cosine:	k.	$\ell.$	m.	n.	o.
tangent:	p.	q.	r.	s.	t.

10. If a 10.0-ft water wheel (see Figure 2.27) is rotating at a speed of 5.0 revolutions per minute, what is the speed of the current (in miles per hour, rounded to the nearest tenth)?

FIGURE 2.27 Waterwheel

Discussion Problems

1. Discuss the measurement of angles. Include measurements such as revolutions, degrees, radians, as well as the possibility of other types of measures.

2. Discuss measuring arc length using degree measure and radian measure. Which method is easier to apply? Why?

3. Discuss the evaluation of trigonometric functions. Why do we sometimes obtain approximate values and sometimes exact values? Why derive any exact values?

4. Explain the differences and similarities between 10-speed and 18-speed derailleurs on a bicycle. Include a discussion of both linear and angular speeds.

Miscellaneous Problems—Chapters 1–2

1. If α is an angle in standard position, what are the coordinates of the point of intersection of the terminal side of α and a circle with radius 2?

2. If $23°$ is the measure of an angle α in standard position, what are the coordinates of the point of intersection of the terminal side of α and a unit circle?

Problems **3–6:** *Change the degree measures to radian measures (correct to the nearest hundredth).*

  3. $150°$ **4.** $-30°$ **5.** $-100°$ **6.** $24.5°$

Problems **7–10:** *Change the radian measures to degree measures (correct to the nearest degree).*

  7. $\frac{2\pi}{3}$ **8.** $-\frac{5\pi}{4}$ **9.** -2.5 **10.** 0.4589

Problems **11–14:** *Find the arc length* (*approximate answers correct to the nearest unit*).

11. $r = 5$ in.; $\theta = 6$ **12.** $r = \sqrt{50}$ ft; $\theta = \frac{\pi}{6}$

13. $r = 6.4$ ft; $\theta = 2.5$ **14.** $r = 34$ in.; $\theta = 1$

Problems **15–18:** *Determine the quadrant of each angle, and find a positive angle less than one revolution that is coterminal with the given angle.*

15. 9   **16.** -2 **17.** $-\frac{2\pi}{3}$ **18.** $-\frac{5\pi}{4}$

Problems **19–28:** *Use a calculator to find the value of the trigonometric functions correct to four decimal places.*

19. $\sin 8.3$ **20.** $\cos 0.64$

21. $\cos(-0.42)$ **22.** $\sin(-6.2)$

23. $-\sin 6.2$ **24.** $-\cos 0.42$

25. $\cot 6$ **26.** $\sec 0.5$

27. $\csc \frac{\pi}{7}$ **28.** $\csc\left(-\frac{3\pi}{8}\right)$

Problems **29–32:** *Use the definition of the trigonometric functions to find the values for an angle θ whose terminal side passes through the given point.*

29. $(4, 5)$ **30.** $(-2, 3)$ **31.** $(5, \sqrt{11})$ **32.** $(\sqrt{55}, 3)$

Problems **33–47:** *Find the exact value of the following trigonometric functions.*

33. $\sin \frac{2\pi}{3}$ **34.** $\tan \frac{5\pi}{4}$ **35.** $\cos 240°$

36. $\cos\left(-\frac{5\pi}{3}\right)$ **37.** $\sec \frac{5\pi}{6}$ **38.** $\sin 300°$

39. $\tan \frac{\pi}{3}$ **40.** $\cot \frac{5\pi}{3}$ **41.** $\tan 120°$

42. $\sec 0$ **43.** $\csc 2\pi$ **44.** $\tan \pi$

45. $\cot(-30°)$ **46.** $\tan(-135°)$ **47.** $\sec \frac{3\pi}{4}$

Applications

48. A simple chain link from the drive sprocket (pedals) to the rear wheel of a bike is shown in Figure 2.28.

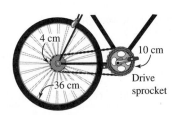

FIGURE 2.28   Bike drive-sprocket assembly

If the sprocket has radius 4 cm and the drive sprocket has radius 10 cm, what is the angle of rotation for the wheel for one rotation of the drive sprocket? What is the speed of a bike (in km/h) when the drive sprocket is rotating at 250 revolutions per minute and the radius of the wheel is 36 cm?

49. Suppose a belt drives two wheels, of radii 3 in. and 9 in., respectively, as shown in Figure 2.29. Assume the larger wheel is rotating at 100 rev/min.

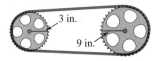

FIGURE 2.29   Two belt-driven gears

a. What is the angular speed of the larger wheel in radians/min?

b. What is the linear speed of the larger wheel (in mi/h, rounded to two decimal places)?

c. What is the angular speed of the smaller wheel (in rev/min, rounded to the nearest unit)?

50. How fast would you need to travel in an airplane at 31,680 ft above the surface of the earth to keep up with the sun? Assume that the earth has radius 3,950 miles and rotates once every 24 hours.

Group Research Projects

Working in small groups is typical of most work environments, and this book seeks to develop skills with group activities. At the end of each chapter we present a list of suggested projects, and even though they could be done as individual projects, we suggest that these projects be done in groups of three or four students.

1. Write a paper considering the pros and cons of measuring in the degree system of angle measurement and also in the radian system of measurement. Reach a consensus as to the better method of measurement and give a rationale for your conclusion.

2. Suppose a bullet is fired into the air at an angle of 45°. How long does it take (rounded to the nearest tenth of a second) for the bullet to reach a height of 1,000 ft?

   First hypothesis: Assume that the bullet travels in a straight line and neglect the forces of gravity. Sketch the path of the projectile, and answer the question.

   Second hypothesis: Assume the following parametric equations for the path of a projectile, which do not neglect the forces of gravity:

   $$x = (v_0 \cos \theta)t, \quad y = (v_0 \sin \theta)t - 16t^2$$

   where v_0 is the initial velocity of the projectile in the direction of θ with the horizontal. Sketch the path of the projectile and answer the question.

   When will the bullet hit the ground (rounded to the nearest tenth of a second) for each of these two hypotheses?

3. A basketball rim is in a plane 10 feet above the floor of the gymnasium, and the front of the rim is 13 feet (horizontally) from the free throw line. The three crucial parameters of the shot are initial height, launch angle, and initial velocity of the basketball as the shot is released.*

Suppose the ball is launched with an initial velocity of 24 ft/s at an initial height of 6 ft. Also, according to the official regulations, an official basketball rim has an inside diameter of 18 in. and the basketball is from 29.5 to 30.5 inches in circumference. Therefore, suppose for this problem that, since the inside of the front of the rim is 13 ft from the foul line, any shot that places the ball between approximately 13.4 ft and 14.1 ft from the foul line when the ball is approximately 10 ft above the floor is a swish. Assume that the error of the shooter is not to the right or to the left. Find the parametric equations for a winning shot. That is, find an angle (in degrees) that is necessary to make a basket.

Write a program that will simulate the situation. Indicate the software, program, or calculator you are using. For example, the following graphs were done using a program on a TI-92 calculator.

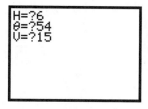 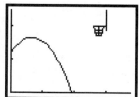

Now hold the angle constant and consider different velocities. Write a paper on your results.

4. Repeat Problem 3, except this time do not neglect air resistance. The equations in the text (Example 5 in Section 2.3) assume that air resistance is neglected. Do some research to find appropriate equations that account for factors such as elevation above sea level, air temperature, and relative humidity.

---

* From "A Mathematical Look at a Free Throw Using Technology," by Charles Vonder Embse and Arne Engebretsen, *The Mathematics Teacher*, December 1996, pp. 774–779.

CHAPTER THREE

GRAPHS OF TRIGONOMETRIC FUNCTIONS

PREVIEW

In this chapter, we introduce the concept of graphing the trigonometric functions. You can understand these functions more easily if you form a mental image or "picture" of them. In the first section, a quick, simple method called framing *is discussed and is then used in the second section to graph general cosine, sine, and tangent functions easily and efficiently in one step.*

Throughout this book, we will avail ourselves of computer and calculator technology to "see" many relationships more easily, but that usage does not reduce our need to understand the nature and properties of trigonometric graphs.

In the third section, we use the general cosine, sine, and tangent curves to graph sums or products of the general trigonometric functions. We conclude this section by considering trigonometric systems by graphs and some applications of waves and motion.

In the last section of this chapter, we use inverse trigonometric functions to discuss the idea of reversing the process of finding a trigonometric value. You will use the skills you learn here later in this book to solve both trigonometric equations and triangles (Chapters 4 and 5).

Trigonometry contains the science of continually undulating magnitude: meaning magnitude which becomes alternately greater and less, without any termination to succession of increase and decrease. . . .

Augustus De Morgan, 1849

SECTION 3.1

GRAPHS OF THE STANDARD TRIGONOMETRIC FUNCTIONS

PROBLEM OF THE DAY

In a classic study,* Lotka considered the interdependence of two species, the first of which serves as food for the other. If we consider the populations of foxes and rabbits in a certain region, we might find that their populations go up and down in cycles (but out of phase with one another). Suppose it was found that the population of the rabbits changes according to the formula

$$y = 400 + 100 \cos \frac{\pi x}{4}$$

where y is the number of rabbits after time x (measured in months). What is the graph representing the rabbit population?

How would you answer this question?

x	y
0	500
1	470.7107
2	400
3	329.2893
4	300
5	329.2893
6	400
7	470.7107
8	500
9	470.7107
10	400
11	329.2893
12	300

We will consider the Problem of the Day by plotting points. A table of values is shown in the margin. Remember, x and y are both real numbers; x measures the number of months and y measures the number of rabbits. For example, if $x = 1$, then

$$y = 400 + 100 \cos \frac{\pi x}{4} = 400 + 100 \cos \frac{\pi(1)}{4} \approx 470.7106781$$

Note: If you obtained 499.990605, you need to change from degree mode to radian mode. Remember, if units are not specified, radians are understood.

Graph these points on a Cartesian coordinate system, giving careful attention to the units on the x- and y-axes (it is not practical to choose the same units on each axis). The points are plotted in Figure 3.1a (page 90) and connected in Figure 3.1b.

The points plotted for the Problem of the Day illustrate some characteristic properties of the graphs of the trigonometric functions. The graphs of the cosine and sine functions look like waves. We will consider various properties associated with such graphs. The *amplitude* is half the distance between the greatest and the least values of the second components of the ordered pairs of a periodic curve. The *period* can be measured from any point on the graph to the next point horizontally where the cycle repeats. The *frequency* tells how many cycles occur in a given interval and is defined to be the reciprocal of the period. For example, if the horizontal axis is a function of time, then the frequency might be expressed in *cycles per year,* as in the case of rabbit populations, or in *cycles per second,* as in the case of radio waves.

---

* From *Elements of Mathematical Biology* by Alfred Lotka, New York: Dover Publications, 1956.

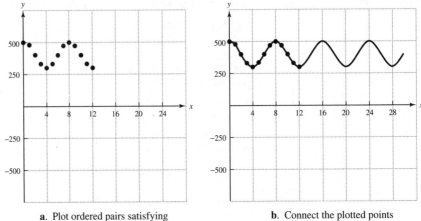

a. Plot ordered pairs satisfying the given equation.

b. Connect the plotted points with a smooth curve.

FIGURE 3.1 Graph of $y = 400 + 100 \cos \dfrac{\pi x}{4}$

EXAMPLE 1 **Finding Amplitude, Period, and Frequency**

Given the calculator graph shown in Figure 3.2, find the period, amplitude, and frequency.

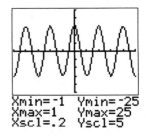

FIGURE 3.2
Computer graph of a periodic function

Solution Calculator graphs do not show the scale on the screen, nor do they label the axes. See Appendix A for the notation used in this book.

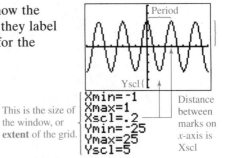

We note here that the x-scale is 0.2 unit per tic mark. The period is measured from any point to the next point horizontally where the cycle repeats. In this case, the period is 0.4 unit.

The y-scale is 5 units per tic mark, so each tic mark represents 5 units, and the amplitude is the maximum height of the wave. In this example, it is 15 units.

The frequency is the reciprocal of the period, so in this example, the frequency is $1/0.4 = 2.5$ or $2\frac{1}{2}$ cycles per unit.   ■

We now turn to graphing the standard trigonometric functions, by plotting points as we did with the Problem of the Day. The process will then be generalized so we can graph the functions without too many calculations concerning points.

To plot points, we need to relate the trigonometric functions to the Cartesian coordinate system. For example, we may wish to write $y = \tan x$, but in so doing, we change the meaning of x and y as they were used in the last chapter. Recall from the *definition* of the trigonometric functions that

$$\tan \theta = \frac{y}{x}$$

where (x, y) represents a point on the terminal side of angle θ. *Now,* when we write

$$y = \tan x$$

x represents the *angle,* and y is the value of the tangent function at x. The **domain** of a function defined by $y = f(x)$ is the set of all values for which x is defined, and the **range** is the set of all resulting y-values.

Graph of Standard Sine Function

To graph $y = \sin x$, begin by plotting familiar (exact) values for the sine function:

x = real number	0	$\frac{\pi}{2}$	π	$\frac{3\pi}{2}$	2π
$y = \sin x$	0	1	0	-1	0

Exact values are used here, but a calculator could be used to generate other values. If graphing with a calculator or computer, you will need to specify a "window" or "frame" of values. If graphing with a pencil and paper, you may find it convenient to choose 12 intervals on the x-axis for π units, and 10 intervals on the y-axis for 1 unit. The smooth curve that connects these points is called the **sine curve** and is shown in Figure 3.3 (page 92).

x	y
-1	-0.84
0	0.00
1	0.84
2	0.91
3	0.14
4	-0.76
5	-0.96
6	-0.28
7	0.66

Note: sine is positive in Quadrants I $\left(0 < x < \frac{\pi}{2}\right)$ and II $\left(\frac{\pi}{2} < x < \pi\right)$, and negative in Quadrants III $\left(\pi < x < \frac{3\pi}{2}\right)$ and IV $\left(\frac{3\pi}{2} < x < 2\pi\right)$.

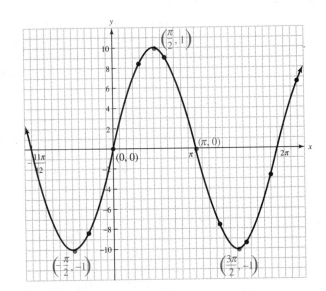

FIGURE 3.3   Graph of $y = \sin x$ by plotting points (see table in margin)

The domain for x is all real numbers, so what about values other than $0 \le x < 2\pi$? Using the reduction principle (which was discussed on page 79), we know that

$$\sin x = \sin(x + 2\pi) = \sin(x - 2\pi) = \sin(x + 4\pi) = \cdots$$

More generally,

$$\sin(\theta + 2n\pi) = \sin \theta$$

for any integer n. In other words, the values of the sine function repeat themselves after 2π. We describe this by saying that the sine function is **periodic with period 2π**. The sine curve is shown in Figure 3.4. Notice that even though the domain of the sine function is all real numbers, the range is restricted to values between -1 and 1 (inclusive).

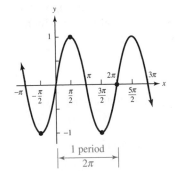

FIGURE 3.4
Graph of $y = \sin x$
amplitude: 1
period: 2π
domain: all real x,
         written $\mathbb{R}$
range: $-1 \le y \le 1$

Framing a Sine Curve

For the period labeled in Figure 3.4, the sine curve starts at $(0, 0)$, goes *up* to $\left(\frac{\pi}{2}, 1\right)$ and then *down to* $\left(\frac{3\pi}{2}, -1\right)$ passing through $(\pi, 0)$, and then goes back up to $(2\pi, 0)$, which completes one period. This graph shows that the range of the sine function is $-1 \le y \le 1$. We summarize a technique for sketching the sine curve (see Figure 3.5) called **framing the curve**.

FRAMING A SINE CURVE

The standard sine function

$$y = \sin x$$

has domain $\mathbb{R}$ (all real numbers) and range $-1 \leq y \leq 1$, and is periodic with period 2π. One period of this curve can be sketched by framing, as follows:

1. *Start* at the origin $(0, 0)$.
2. *Height* of this frame is two units: one unit up and one unit down from the starting point $(-1 \leq y \leq 1)$.
3. *Length* of this frame is 2π units (about 6.28) from the starting point (the period is 2π).
4. The curve is now framed. Plot *five* critical points within the frame:
   a. Endpoints (along axis)
   b. Midpoint (along axis)
   c. Quarterpoints (up first, then down)
5. Draw the curve through the critical points, remembering the shape of the sine curve.

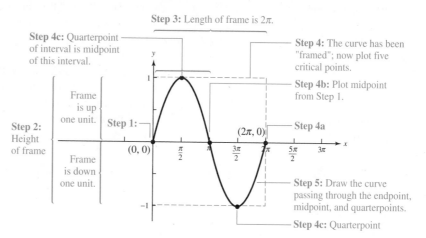

FIGURE 3.5 Procedure for framing the sine curve

EXAMPLE 2

Drawing One Period of a Standard Sine Curve

Draw one period of a standard sine curve using $(3, 1)$ as the starting point for building a frame.

Solution

Plot the point $(3, 1)$ as the starting point. Draw the standard frame as shown in Figure 3.6 on page 94.

*Remember, when framing a curve, first plot the starting point, and then count squares on **your graphing paper** to build the frame.*

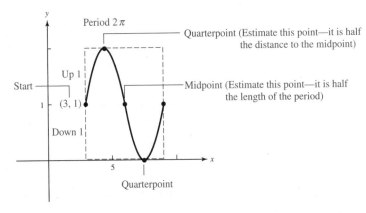

FIGURE 3.6   Graph of one period of a sine curve by framing   ■

Calculator Window **?** **X**

Curve-sketching by plotting points can be done efficiently on a calculator. Many people believe that it is not necessary to *think* when they use a calculator, that the "machine" will do all the work. However, this is not true; with a calculator, your attention is simply focused on different matters. For example, in this section we are concerned with graphing $y = \sin x$. The first task is to decide on the domain and the range. If you know what the curve looks like, then it is easy to make an intelligent choice about the domain and range. However, if you do not know anything about the curve, you may have to make several attempts at setting the domain and range before you obtain a satisfactory graph. Without the work of this section, graphing a sine curve on a calculator would be very difficult. On almost all graphing calculators, the scale is fixed automatically by pressing ZOOM and choosing the *trigonometric* scale. Notice the "strange" scale for the *x*-axis is an approximation for $\pi/2$ and the Xmin and Xmax values are automatically set by choosing the trigonometric scale.

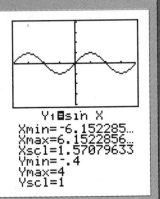

EXAMPLE 3   **Sketching a Curve by Plotting Points**

Graph $f(x) = \dfrac{\sin x}{x}$.

Solution   We note that f is not defined when $x = 0$. Set up a table of values, and plot the corresponding points on a coordinate system, as shown in Figure 3.7. Draw an open circle at $x = 0$, since the graph does not exist at that point.

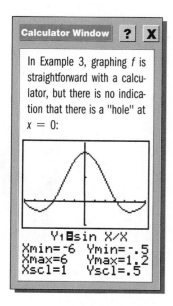

x	y
-6	-0.047
-5	-0.192
-4	-0.189
-3	0.047
-2	0.455
-1	0.841
1	0.841
2	0.455
3	0.047
4	-0.189
5	-0.192
6	-0.047

FIGURE 3.7 Graph of $y = \dfrac{\sin x}{x}$

A calculator graph for this function is shown in the margin. ■

Figure 3.7 shows that portion of f for $-2\pi \le x \le 2\pi$. In the problem set, you are asked to graph this function for other possible domains (that is, for other values of x).

Graph of Standard Cosine Function

We can graph the cosine curve by plotting points, as we did with the sine curve. The details of plotting points are left as an exercise. The cosine curve $y = \cos x$ is "framed" as described in Figure 3.8.

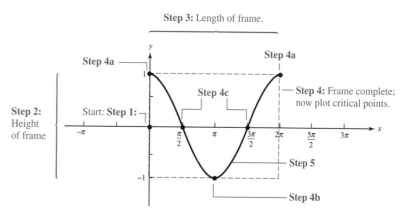

FIGURE 3.8 Procedure for framing a cosine curve

Notice that the only difference in framing a cosine and a sine curve is in Step 4; the procedures for building the frame are identical.

FRAMING A COSINE CURVE

The standard cosine function

$$y = \cos x$$

has domain $\mathbb{R}$ and range $-1 \leq y \leq 1$, and is periodic with period 2π. One period of this curve can be sketched by framing, as follows:

1. *Start* at the origin $(0, 0)$.
2. *Height* of this frame is two units: one unit up and one unit down from the starting point $(-1 \leq y \leq 1)$.
3. *Length* of this frame is 2π units (about 6.28) from the starting point (the period is 2π).
4. The curve is now framed. Plot *five* critical points within the frame:
   a. Endpoints (at the top corners of the frame)
   b. Midpoint (at the bottom of the frame)
   c. Quarterpoints (along the axis)
5. Draw the curve through the critical points, remembering the shape of the cosine curve.

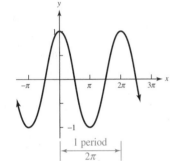

FIGURE 3.9
Graph of $y = \cos x$
amplitude: 1
period: 2π
domain: $\mathbb{R}$
range: $-1 \leq y \leq 1$

Since values for x greater than 2π or less than 0 are coterminal with those already considered, we see that **the period of the cosine function is 2π.** The cosine curve is shown in Figure 3.9. Notice that the domain and range of the cosine function are the same as they are for the sine function.

Graph of Standard Tangent Function

By setting up a table of values and plotting points (the details are left as an exercise), we notice that $y = \tan x$ does not exist at $\frac{\pi}{2}, \frac{3\pi}{2}$, or $\frac{\pi}{2} + n\pi$ for any integer n. The lines $x = \frac{\pi}{2}, x = \frac{3\pi}{2}, \ldots, x = \frac{\pi}{2} + n\pi$ for which the tangent is not defined are *vertical asymptotes*. The procedure is summarized in Figure 3.10.

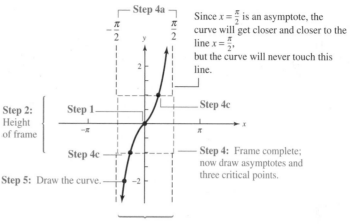

FIGURE 3.10 Procedure for framing a tangent curve

The standard tangent curve has period π.

FRAMING A TANGENT CURVE

The standard tangent function

$$y = \tan x$$

has domain all real numbers where $x \neq \frac{\pi}{2} + n\pi$ (for any integer n) and range $\mathbb{R}$. The standard tangent function is periodic with period π. One period of this curve can be sketched by framing, as follows:

1. *Start* at the origin $(0, 0)$; for the tangent curve this is the *center* of the frame.
2. *Height* of this frame is two units: one unit up and one unit down from the starting point.
3. *Length* of this frame is π units (about 3.14) and is drawn so that it is $\frac{\pi}{2}$ (about 1.57) units on each side of the starting point.
4. The curve is now framed. Draw the asymptotes and plot three critical points within the frame:
   a. Extend the vertical sides of the frame; these are the asymptotes
   b. Midpoint (this is the starting point)
   c. Quarterpoints (down first, then up to help you remember this, we note as you read from left to right the curve is increasing—up/up and away)
5. Draw the curve through the critical points, using the asymptotes as guides and remembering the shape of the tangent curve.

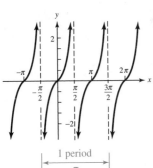

FIGURE 3.11

Graph of $y = \tan x$

period: π

domain: all real x,
                $x \neq \pi/2 + n\pi$
                (n any integer)

range: $\mathbb{R}$

The tangent curve is indicated in Figure 3.11. Even though the curve repeats for values of x greater than 2π or less than 0, it also repeats after it has passed through an interval with length π. For this reason, $\tan(\theta + n\pi) = \tan\theta$ for any integer n, and we see that **the tangent function has a period of π.** The domain of the tangent function is restricted so that multiples of π added to $\frac{\pi}{2}$ are excluded, because the tangent is not defined for these values. The range, on the other hand, is unrestricted; it is the set of all real numbers.

Graphs of Reciprocal Functions

The graphs of the other three trigonometric functions could be done in the same fashion as sine, cosine, and tangent. Instead, however, we make use of the reciprocal relationships and graph them as shown in the following example.

EXAMPLE 4

Graph of the Standard Secant Function Using Reciprocals

Sketch $y = \sec x$ by sketching the reciprocal of $y = \cos x$.

Solution   Begin by sketching $y = \cos x$ (black dashed curve in Figure 3.12). Wherever $\cos x = 0$, $\sec x$ is undefined; draw asymptotes at these points. Now, plot points by finding the reciprocals of the ordinates of points previously plotted. When $y = \cos x = \frac{1}{2}$, for example, the reciprocal is

$$y = \sec x = \frac{1}{\cos x} = \frac{1}{\frac{1}{2}} = 2$$

The completed graph is shown in Figure 3.12.

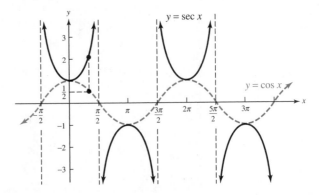

FIGURE 3.12   Graph of $y = \sec x$

The graphs of the other reciprocal trigonometric functions are shown in Figure 3.13.

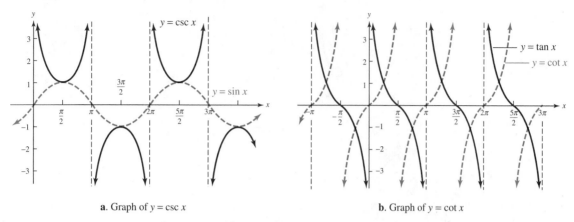

a. Graph of $y = \csc x$ **b.** Graph of $y = \cot x$

FIGURE 3.13   Graphs of the reciprocal functions

Remember the shapes of the standard trigonometric curves, as summarized in Table 3.1.

TABLE 3.1 Directory of Curves (Let n = an integer.)

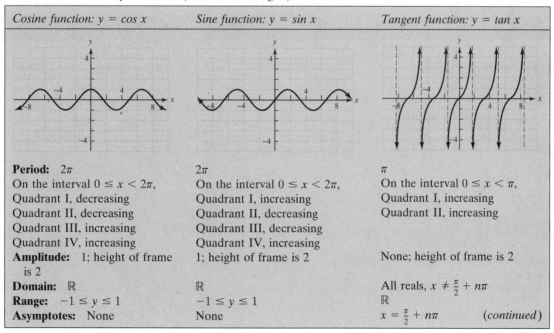

Cosine function: $y = \cos x$	*Sine function:* $y = \sin x$	*Tangent function:* $y = \tan x$
Period: 2π	2π	π
On the interval $0 \leq x < 2\pi$,	On the interval $0 \leq x < 2\pi$,	On the interval $0 \leq x < \pi$,
Quadrant I, decreasing	Quadrant I, increasing	Quadrant I, increasing
Quadrant II, decreasing	Quadrant II, decreasing	Quadrant II, increasing
Quadrant III, increasing	Quadrant III, decreasing	
Quadrant IV, increasing	Quadrant IV, increasing	
Amplitude: 1; height of frame is 2	1; height of frame is 2	None; height of frame is 2
Domain: $\mathbb{R}$	$\mathbb{R}$	All reals, $x \neq \frac{\pi}{2} + n\pi$
Range: $-1 \leq y \leq 1$	$-1 \leq y \leq 1$	$\mathbb{R}$
Asymptotes: None	None	$x = \frac{\pi}{2} + n\pi$ (*continued*)

TABLE 3.1 *Continued*

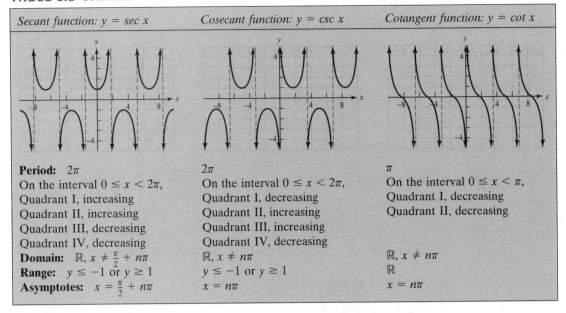

Secant function: $y = \sec x$	Cosecant function: $y = \csc x$	Cotangent function: $y = \cot x$
Period: 2π	2π	π
On the interval $0 \le x < 2\pi$,	On the interval $0 \le x < 2\pi$,	On the interval $0 \le x < \pi$,
Quadrant I, increasing	Quadrant I, decreasing	Quadrant I, decreasing
Quadrant II, increasing	Quadrant II, increasing	Quadrant II, decreasing
Quadrant III, decreasing	Quadrant III, increasing	
Quadrant IV, decreasing	Quadrant IV, decreasing	
Domain: $\mathbb{R}, x \ne \frac{\pi}{2} + n\pi$	$\mathbb{R}, x \ne n\pi$	$\mathbb{R}, x \ne n\pi$
Range: $y \le -1$ or $y \ge 1$	$y \le -1$ or $y \ge 1$	$\mathbb{R}$
Asymptotes: $x = \frac{\pi}{2} + n\pi$	$x = n\pi$	$x = n\pi$

PROBLEM SET 3.1

Essential Concepts

1. What are the amplitude and period of the standard cosine, sine, and tangent curves?

Problems 2–7: Draw a quick sketch of each curve from memory.

2. $y = \cos x$ **3.** $y = \sin x$ **4.** $y = \tan x$

5. $y = \sec x$ **6.** $y = \csc x$ **7.** $y = \cot x$

Level 1 Drill

8. Complete the following table of values for $y = \cos x$.

x = angle: $\frac{2\pi}{3}$ $\frac{3\pi}{4}$ $\frac{5\pi}{6}$ $\frac{7\pi}{6}$ $\frac{5\pi}{4}$ $\frac{4\pi}{3}$ $\frac{7\pi}{4}$ $\frac{11\pi}{6}$
Quadrant:
$y = \cos x$:
y approx.:

Use this table, along with other values if necessary, to plot $y = \cos x$.

9. Complete the following table of values for $y = \tan x$.

x = angle: $\frac{2\pi}{3}$ $\frac{3\pi}{4}$ $\frac{5\pi}{6}$ $\frac{7\pi}{6}$ $\frac{5\pi}{4}$ $\frac{4\pi}{3}$ $\frac{7\pi}{4}$ $\frac{11\pi}{6}$
Quadrant:
$y = \tan x$:
y approx.:

Use this table, along with other values if necessary, to plot $y = \tan x$.

Problems 10–13: *State the amplitude and period.*

10.

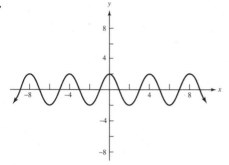

11.

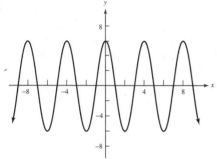

12.

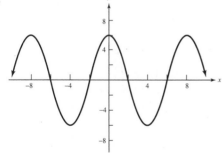

13.

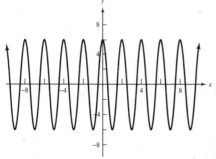

Problems 14–17: *State the amplitude, period, and frequency for the given graphs.*

14.

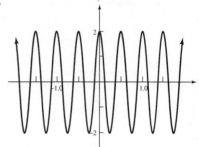

15.

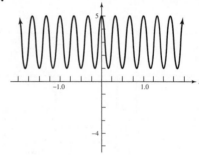

16.

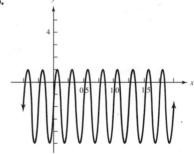

17.

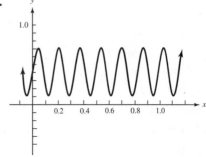

WHAT IS WRONG, *if anything, with each of the calculator graphs in Problems 18–21? Explain your reasoning.*

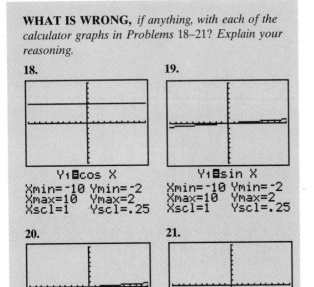

18.

Y₁=cos X
Xmin=-10 Ymin=-2
Xmax=10   Ymax=2
Xscl=1   Yscl=.25

19.

Y₁=sin X
Xmin=-10 Ymin=-2
Xmax=10   Ymax=2
Xscl=1   Yscl=.25

20.

Y₁=tan X
Xmin=-10 Ymin=-2
Xmax=10   Ymax=2
Xscl=1   Yscl=.25

21.

Y₁=1/sin X
Xmin=-10 Ymin=-2
Xmax=10   Ymax=2
Xscl=1   Yscl=.25

Problems 22–35: *Plot the given point; then use the framing procedure along with the given point to draw one period of the requested curve.*

22. $(\pi, 1)$; cosine curve

23. $(\pi, 1)$; sine curve

24. $(\pi, 1)$; tangent curve

25. $\left(-\frac{\pi}{2}, 2\right)$; cosine curve

26. $\left(-\frac{\pi}{2}, 2\right)$; sine curve

27. $\left(-\frac{\pi}{2}, 2\right)$; tangent curve

28. $\left(-\frac{\pi}{4}, -2\right)$; cosine curve

29. $\left(-\frac{\pi}{4}, -2\right)$; sine curve

30. $\left(-\frac{\pi}{4}, -2\right)$; tangent curve

31. $\left(-\frac{\pi}{4}, 2\right)$; tangent curve

32. $(-1, -2)$; sine curve

33. $(-1, -2)$; cosine curve

34. $(-1, -2)$; tangent curve

35. $(1, 0.5)$; sine curve

Problems 36–41: *The vertical asymptotes for the graph of a trigonometric function occur where the function is not defined. Find the equations for the vertical asymptotes of the given trigonometric functions. If the curve has no vertical asymptotes, so state.*

36. $y = \cos x$ **37.** $y = \sin x$

38. $y = \tan x$ **39.** $y = \sec x$

40. $y = \csc x$ **41.** $y = \cot x$

Level 2 Drill

42. Complete a table of values to plot $y = \sec x$ for $0 \le x \le 2\pi$. For these values of x, the secant is not defined for $x = \pi/2$ and $x = 3\pi/2$.

43. Complete a table of values to plot $y = \csc x$ for $0 < x < 2\pi$. For these values of x, the cosecant is not defined for $x = \pi$.

44. Complete a table of values to plot $y = \cot x$ for $0 < x < 2\pi$. For these values of x, the cotangent is not defined for $x = \pi$.

Problems 45–48: *Sketch the curves by plotting points or by using a graphing calculator.*

45. a. $y = |\sin x|$ **b.** $y = \sin|x|$

46. a. $y = |\cos x|$ **b.** $y = \cos|x|$

47. a. $y = |\tan x|$ **b.** $y = \tan|x|$

48. a. $y = |\sec x|$ **b.** $y = \sec|x|$

Problems 49–56: *Plot points (with or without a graphing calculator) to sketch the graphs for the indicated values. Pay particular attention to values for which the denominator is 0.*

49. $f(x) = \dfrac{\sin x}{x}$ for $\pi \le x \le 3\pi$

50. $f(x) = \dfrac{\sin x}{x}$ for $-3\pi \le x < 0$

51. $y = \dfrac{\sin x}{x}$ for $-5 \le x \le 5$;
undefined for $x = 0$

52. $y = \dfrac{\sin 5x}{2x}$ for $-4 \le x \le 4$;
undefined for $x = 0$

53. $y = \dfrac{\cos x - 1}{x}$ for $-\pi \le x \le \pi$;
undefined for $x = 0$

54. $y = \dfrac{\sin x}{1 + x^2}$ for $-1 \le x \le 1$

55. $y = \dfrac{x}{\cos x}$ for $-\pi \le x \le \pi$;

undefined for $x = \pm \dfrac{\pi}{2}$

56. $y = \dfrac{x}{\sin x}$ for $-2\pi < x < 2\pi$;

undefined for $x = 0, \pm\pi$

57. Sketch a figure following these instructions: First, draw a Cartesian coordinate system (be sure to label the x- and y-axes); label the origin as point O. Draw a unit circle with center at $C(-1, 0)$. Draw another coordinate axes with origin at the point C and label these axes x' and y'. Let θ be an angle drawn with the vertex at C, the initial side be the positive x-axis, and the point $P(x', y')$ be the intersection of the terminal side of this angle and the unit circle. Let $|\overline{PQ}|$ be the perpendicular drawn to the x'-axis and $|\overline{PR}|$ be the perpendicular drawn to the y'-axis. Finally, let S be the intersection of the line determined by the terminal side and the y-axis.
a. Show $\sin \theta = |\overline{PQ}|$.
b. Show $\cos \theta = |\overline{PR}|$.
c. Show $\tan \theta = |\overline{SO}|$.

Level 2 Applications

58. The current I (in amperes) in a certain circuit (for some convenient unit of time) generates the following set of data points:

Time	Height	Time	Height
0	−60.0000	10	30.00000
1	−58.6889	11	40.14784
2	−54.8127	12	48.54102
3	−48.5410	13	54.81273
4	−40.1478	14	58.68886
5	−30.0000	15	60.00000
6	−18.5410	16	58.68886
7	−6.27171	17	54.81273
8	6.27171	18	48.54102
9	18.54102	19	40.14784
		20	30.00000

Plot the data points and draw a smooth curve passing through these points. Let x represent the time and y the current.

59. Suppose a point P on a water wheel with a 30-ft radius is d units from the water, as shown in Figure 3.14.

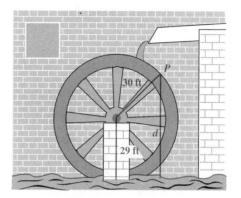

FIGURE 3.14   Water wheel

If the water wheel turns at 6 rev/min, the height of the point P above the water is given by the following set of data points:

Time	Height	Time	Height
0	−1.000	10	−1.000
1	4.729	11	4.729
2	19.729	12	19.729
3	38.271	13	38.270
4	53.271	14	53.270
5	59.000	15	59.000
6	53.271	16	53.271
7	38.271	17	38.271
8	19.729	18	19.730
9	4.729	19	4.730
		20	−1.000

Plot the data points and draw a smooth curve passing through these points. Let x be the time and y the current.

60. IN YOUR OWN WORDS Take a piece of paper and wrap it around a candle, as shown in Figure 3.15 (page 104). Make a perpendicular cut at A and a slanted cut at B as shown. Unroll the paper. Describe what you think the edges A and B will look like. Then perform the experiment to see whether your guess was correct. Describe what you found.

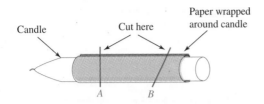

FIGURE 3.15 Candle experiment

Computer Window **?** **X**

61. There are many computer graphics programs available for both IBM and MacIntosh formats. For example, CONVERGE, which is available on IBM format from JEMware (phone 808-523-9911), graphs trigonometric curves. Obtain a similar graphics program and explore graphing trigonometric functions; then write a short paper describing the program you are using.

Level 3 Theory

62. In Problem 57, you discovered that $\sin \theta = |PQ|$. Use this fact to sketch $y = \sin \theta$. For example, when $\theta = \frac{\pi}{4}$, draw this angle, and then measure $|PQ|$ and *plot this height* at the location marked by $\theta = \frac{\pi}{4}$ on the θ-axis, as shown in Figure 3.16. As P makes one revolution, it is easy to quickly plot the points on the curve $y = \sin \theta$. It is also easy to note the relationship between the unit circle definition of sine and what we call the *graph of the sine curve*.

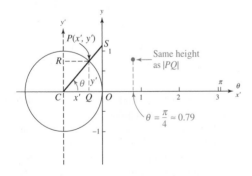

FIGURE 3.16 Unit circle graph of $y = \sin \theta$

63. In Problem 57, you discovered that $\tan \theta = |SO|$. Use this fact to sketch $y = \tan \theta$. For example, when $\theta = \frac{\pi}{4}$, draw this angle as shown in Figure 3.17 and then measure $|SO|$ and *plot this height* at $\theta = \frac{\pi}{4}$ on the θ-axis as shown on the right. As P makes one revolution, notice that $|SO|$ does not exist at $\theta = \frac{\pi}{2}$ and $\frac{3\pi}{2}$.

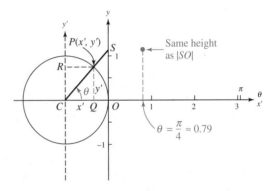

FIGURE 3.17 Unit circle graph of $y = \tan \theta$

64. In Problem 57, you discovered that $\cos \theta = |PR|$. Use this fact to sketch $y = \cos \theta$. To do this, rotate the unit circled described in Problem 57 by 90°, as shown in Figure 3.18. For example, when $\theta = \frac{\pi}{4}$, draw this angle, and then measure $|PR|$ and plot this height at $\theta = \frac{\pi}{4}$ on the θ-axis. As P makes one revolution, it is easy to quickly plot the points on the curve $y = \cos \theta$.

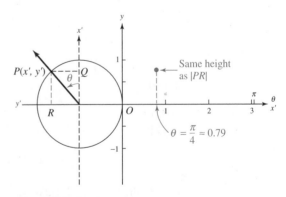

FIGURE 3.18 Unit circle graph of $y = \cos \theta$

**Level 3
Individual Project**

65. Cyclic or periodic phenomena can be seen in all sorts of plants and animals. For example, bean leaves turn limp at night, but stand erect during the day. The graph of a rat's activity falls into a lull during the day, but jumps to a plateau at night. Finally, Figure 3.19 shows the periodic behavior of the housefly, which emerges from its pupal case in the morning and then matures during the day. Do some research on periodic phenomena, and write a report on your investigation.

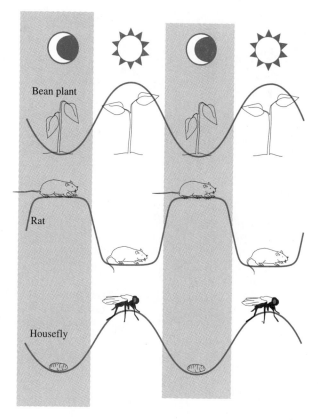

Bean plant

Rat

Housefly

FIGURE 3.19
Many common phenomena demonstrate periodicity.

SECTION 3.2

GENERAL COSINE, SINE, AND TANGENT CURVES

PROBLEM OF THE DAY

The flywheel on a lawn mower has 16 evenly spaced cooling fins. If we rotate the engine through two complete revolutions, we can generate 32 data points, with the first coordinate representing time and the second representing the depth of the piston. What is an equation for the curve generated by the data for the lawn mower engine shown in Table 3.2 (page 106)?

How would you answer this question?

TABLE 3.2 Data Points for Lawn Mower Engine

x	y	x	y
0	0	16	0
1	0.2	17	0.2
2	0.8	18	0.8
3	1.7	19	1.7
4	2.5	20	2.5
5	3.4	21	3.4
6	4.0	22	4.0
7	4.4	23	4.4
8	4.5	24	4.5
9	4.4	25	4.4
10	4.0	26	4.0
11	3.5	27	3.5
12	2.2	28	2.2
13	1.7	29	1.7
14	0.9	30	0.9
15	0.2	31	0.2
		32	0

First we plot the data points.

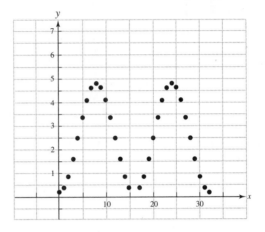

We recognize this as a sine curve (or possibly a cosine curve). The graphs of $y = \cos x$ and $y = \sin x$ have the same shape, except one has been shifted to the right, as shown in Figure 3.20.

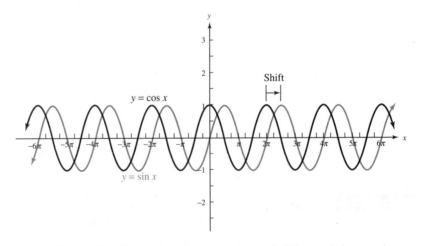

FIGURE 3.20   Comparison between standard cosine and sine graphs

In this section, we will discuss variations of the standard cosine, sine, and tangent graphs. These variations will not only include shifting, as shown in Figure 3.20, but will also include variations in amplitude and period. Since the cosine and sine graphs are identical (except for a possible shift), when we are confronted with a curve such as the one given in the Problem of the Day, we call it a **sinusoidal** curve. That is, the graphs of both the cosine and sine functions are referred to as *sinusoidal* curves.

We begin our discussion by considering a **phase shift** (shown in Figure 3.20), which means that a given curve is shifted to the left or the right. At the same time, we will also consider shifting the curve up or down. We will call these horizontal and vertical shifts **translations.**

Translations and Phase Shift

Once we are given a particular function, it is possible to shift the graph of that function to other locations. For example, let $y = f(x)$ be the function shown in Figure 3.21. It is possible to shift the entire curve up, down, right, or left, as shown in Figure 3.22.

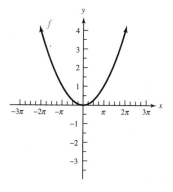

FIGURE 3.21
Graph of $y = f(x)$

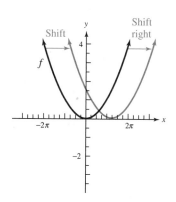

a. Graph of f shifted right

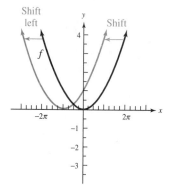

b. Graph of f shifted left

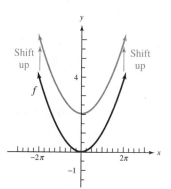

c. Graph of f shifted up

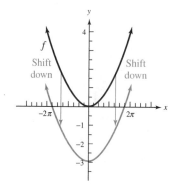

d. Graph of f shifted down

FIGURE 3.22 Shifting the graph of $y = f(x)$

Instead of shifting the curve relative to fixed axes, consider the effect of shifting the axes. If the coordinate axes are shifted up k units, the origin of this new coordinate system corresponds to the point $(0, k)$ on the old coordinate system. If the axes are shifted h units to the right, the origin corresponds

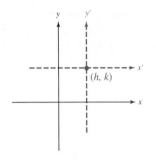

FIGURE 3.23

Shifting the origin to point (h, k)

to the point $(h, 0)$ of the old system. A horizontal shift of h units followed by a vertical shift of k units shifts the new coordinate axes so that the origin corresponds to a point (h, k) on the old axes.

Suppose a *new* coordinate system with origin at (h, k) is drawn and the new axes are labeled x' and y', as shown in Figure 3.23. Every point on a given curve can now be labeled in two ways, as shown in Figure 3.24:

1. As (x, y) measured from the old origin
2. As (x', y') measured from the new origin

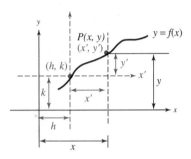

FIGURE 3.24   Comparison of coordinate axes

To find the relationship between (x, y) and (x', y'), consider the graph shown in Figure 3.24.

$$x = x' + h \quad \text{or} \quad x' = x - h$$
$$y = y' + k \quad \text{or} \quad y' = y - k$$

This relationship tells us that if $y - k = f(x - h)$, the graph of this function is the same as the graph $y' = f(x')$, where point (x', y') is measured from the new origin located at (h, k). This fact can greatly simplify the work of graphing, since $y' = f(x')$ is usually easier to graph than $y - k = f(x - h)$.

EXAMPLE 1   **Finding a Translation Point, (h, k)**

Find (h, k) for each given equation.

a. $y - 5 = f(x - 7)$ **b.** $y + 6 = f(x - \pi)$
c. $y - 15 = f\left(x + \frac{1}{3}\right)$ **d.** $x + 5 = (y - 2)^2$
e. $y = \cos(x + 5)$ **f.** $y = \tan\left(x - \frac{\pi}{6}\right) + 4$

Solution   **a.** By inspection, $h = 7$ (it is the number subtracted from x) and $k = 5$, so $(h, k) = (7, 5)$.
b. By inspection, $(h, k) = (\pi, -6)$.   *Did you notice that*
$y + 6 = y - (-6)$, *so that* $k = -6$?

c. By inspection, $(h, k) = \left(-\frac{1}{3}, 15\right)$.

d. Notice that this is a particular equation, instead of a general equation. By inspection, $(h, k) = (-5, 2)$.

e. By inspection, $(h, k) = (-5, 0)$.

f. Write the equation as $y - 4 = \tan\left(x - \frac{\pi}{6}\right)$ to find $(h, k) = \left(\frac{\pi}{6}, 4\right)$.   ■

In the case of the graph of the trigonometric functions, we will simply determine (h, k) and then build a frame at (h, k) rather than at the origin.

EXAMPLE 2   **Graphing One Period of a Translated Sine Curve**

Graph one period of $y = \sin\left(x + \frac{\pi}{2}\right)$.

Solution   First, $(h, k) = \left(-\frac{\pi}{2}, 0\right)$. This is the starting point of the frame. *From this point,* draw the frame as shown in Figure 3.25a. Plot the five critical points for a sine curve (up/down), and then use the frame to draw the curve as shown in Figure 3.25b.

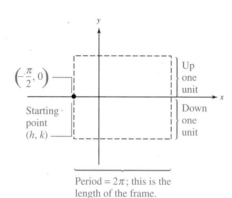

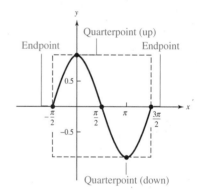

a. Drawing the frame. This step is the same for both sine and cosine curves.

b. Plot critical points and draw one period of the curve.

FIGURE 3.25   Graph of one period of $y = \sin\left(x + \frac{\pi}{2}\right)$ ■

Notice from Figure 3.25 that the graph of $y = \sin\left(x + \frac{\pi}{2}\right)$ is the same as the graph of $y = \cos x$. Thus

$$\sin\left(x + \frac{\pi}{2}\right) = \cos x$$

This confirms the definition of sinusoidal curves.

EXAMPLE 3   **Graphing One Period of a Translated Cosine Function**

Graph one period of $y - 2 = \cos\left(x - \frac{\pi}{6}\right)$.

Solution   Notice that $(h, k) = \left(\frac{\pi}{6}, 2\right)$. Frame the curve, plot the five critical points for a cosine curve (down/up), and connect the points as shown in Figure 3.26.

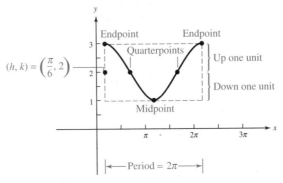

FIGURE 3.26   Graph of one period of $y - 2 = \cos\left(x - \frac{\pi}{6}\right)$

EXAMPLE 4

Graphing One Period of a Translated Tangent Function

Sketch one period of $y + 3 = \tan\left(x + \frac{\pi}{3}\right)$.

Solution   Notice that $(h, k) = \left(-\frac{\pi}{3}, -3\right)$, and remember that this is the center of the frame and that the period of the tangent is π, as shown in Figure 3.27. Plot the midpoint and quarterpoints for the frame of a tangent curve (up/up and away).

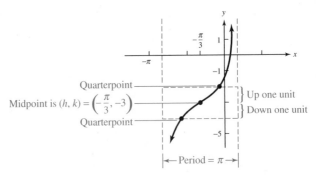

FIGURE 3.27   Graph of one period of $y + 3 = \tan\left(x + \frac{\pi}{3}\right)$

Changes in Amplitude and Period

We will now discuss two additional changes for the function defined by $y = f(x)$. The first, $y = af(x)$, changes the scale on the y-axis; the second, $y = f(bx)$, changes the scale on the x-axis.

For a function $y = af(x)$, it is clear that the y-value is a times the corresponding value of $f(x)$, which means that $f(x)$ is stretched or shrunk in the y-direction by the factor of a. For example, if $y = f(x) = \cos x$, then $y = 3f(x) = 3 \cos x$ is the graph of $\cos x$ that has been stretched so that the high point is at 3 units and the low point is at -3 units (see Figure 3.28). In

general, given

$$y = af(x)$$

where f represents a trigonometric function, $2|a|$ gives the height of the frame for f. For sine and cosine curves, $|a|$ is called the **amplitude** of the function. When $a = 1$, the amplitude is 1, so $y = \sin x$ and $y = \cos x$ are said to have amplitude 1.

EXAMPLE 5   **Graphing a Trigonometric Curve with an Arbitrary Amplitude**

Graph $y = 3 \cos x$.

Solution   From the starting point $(0, 0)$, draw a frame with amplitude $a = 3$ and period 2π, as shown in Figure 3.28. *After* drawing the frame, plot the five critical points for a cosine curve (down/up) using the frame as a guide.

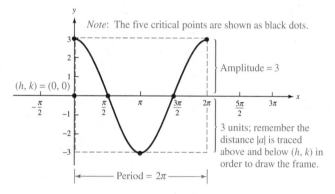

FIGURE 3.28   Graph of one period of $y = 3 \cos x$ ■

For a function $y = f(bx)$, $b > 0$, b affects the scale on the x-axis. Recall that $y = \sin x$ has a period of 2π ($f(x) = \sin x$, so $b = 1$). A function $y = \sin 2x$ ($f(x) = \sin x$ and $f(2x) = \sin 2x$) must complete one period as $2x$ varies from 0 to 2π. This means that one period is completed as x varies from 0 to π. (For each value of x, the result is doubled *before* we find the sine of that number.) In general, the period of $y = \sin bx$ is $2\pi/b$, and the period of $y = \cos bx$ is $2\pi/b$. Since the period of $\tan x$ is π, however, $y = \tan bx$ has a period of π/b. Therefore, when framing the curve, use $2\pi/b$ for the period of the sine and cosine, and π/b for period of the tangent.

Note the periods for standard cosine, sine, and tangent.

EXAMPLE 6   **Graphing a Trigonometric Curve with an Arbitrary Period**

Graph one period of $y = \sin 2x$.

Solution   The period is $p = \dfrac{2\pi}{2} = \pi$; thus, the endpoints of the frame are $(0, 0)$ and $(\pi, 0)$, as shown in Figure 3.29 on page 112. Plot the five critical points for a sine curve (up/down).

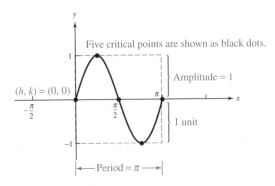

FIGURE 3.29 Graph of one period of $y = \sin 2x$ ■

Graphs of General Trigonometric Curves

Summarizing the preceding results, we have the *general* cosine, sine, and tangent curves.

STOP

The ability to quickly and easily sketch these general trigonometric curves is an important one which you will use frequently in your work in mathematics.

GENERAL FORM OF COSINE, SINE, AND TANGENT CURVES

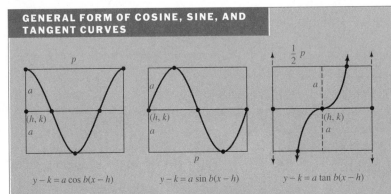

$$y - k = a \cos b(x - h) \qquad y - k = a \sin b(x - h) \qquad y - k = a \tan b(x - h)$$

To sketch a general-form trigonometric curve,

1. Algebraically, put the equation in one of the general forms shown above.
2. Identify (by inspection) the following values:
   (h, k), a, and b. Then calculate p; $p = \frac{2\pi}{b}$ for cosine and sine and $p = \frac{\pi}{b}$ for tangent.
3. Draw the frame.
   a. Plot (h, k); this is the starting point.
   b. Draw **amplitude,** $|a|$; the *height of the frame* is $2|a|$: up a units from (h, k) and down a units from (h, k).*
   c. Draw **period,** p; the *length of the frame* is p. Remember $p = 2\pi/b$ for cosine and sine, and $p = \pi/b$ for tangent.
4. Locate the critical values using the frame as a guide, and then sketch the appropriate graphs. **You do not need to know the coordinates of these critical values.**

* For now, assume $a > 0$; we will consider $a < 0$ in Section 4.3.

EXAMPLE 7

Graphing a General Sine Curve

Graph $y + 1 = 2 \sin \frac{2}{3}\left(x - \frac{\pi}{2}\right)$.

Solution

$$(h, k) = \left(\frac{\pi}{2}, -1\right), a = 2, b = \frac{2}{3}, \text{so } p = \frac{2\pi}{\frac{2}{3}} = 3\pi$$

Now plot (h, k) and frame the curve. Then plot the five critical points (two endpoints, the midpoint, and two quarterpoints). Finally, after sketching one period, draw the other periods as shown in Figure 3.30.

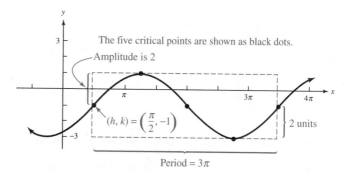

FIGURE 3.30   Graph of $y + 1 = 2 \sin \frac{2}{3}\left(x - \frac{\pi}{2}\right)$

Graph $y + 1 = 2 \sin \frac{2}{3}\left(x - \frac{\pi}{2}\right)$.

If you are using a graphing calculator, you may get a display like the one shown here. Input varies, depending on the model. You must enclose the fraction within parentheses, and it is also necessary to solve the general-form equation for y.

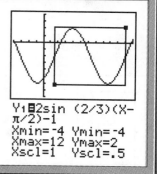

```
Y₁⊟2sin (2/3)(X-
π/2)-1
Xmin=-4 Ymin=-4
Xmax=12 Ymax=2
Xscl=1  Yscl=.5
```

EXAMPLE 8

Graphing a General Cosine Curve

Graph $y = 3 \cos\left(2x + \frac{\pi}{2}\right) - 2$.

Solution

YIELD

Pay attention to the step where the coefficient is factored out.

Rewrite in general form to obtain $y + 2 = 3 \cos 2\left(x + \frac{\pi}{4}\right)$. *Notice that you must factor out the coefficient of x in the argument of cosine.* Note $(h, k) = \left(-\frac{\pi}{4}, -2\right)$; $a = 3$, $b = 2$, and $p = 2\pi/2 = \pi$. Plot (h, k), and frame the cosine curve (down/up) as shown in Figure 3.31 (page 114).

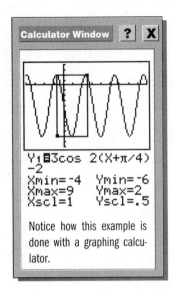

Y₁■3cos 2(X+π/4)
-2
Xmin=-4   Ymin=-6
Xmax=9   Ymax=2
Xscl=1   Yscl=.5

Notice how this example is done with a graphing calculator.

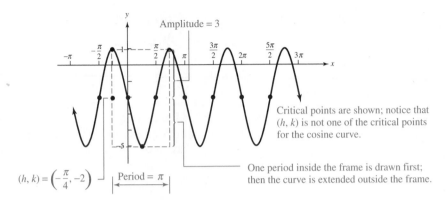

Amplitude = 3

Critical points are shown; notice that (h, k) is not one of the critical points for the cosine curve.

One period inside the frame is drawn first; then the curve is extended outside the frame.

$(h, k) = \left(-\frac{\pi}{4}, -2\right)$ Period $= \pi$

FIGURE 3.31   Graph of $y = 3\cos\left(2x + \frac{\pi}{2}\right) - 2$

EXAMPLE 9

Graphing a General Tangent Curve

Graph $y - 2 = 3\tan\frac{1}{2}\left(x - \frac{\pi}{3}\right)$.

Solution $(h, k) = \left(\frac{\pi}{3}, 2\right)$, $a = 3$, $b = \frac{1}{2}$, so $p = \dfrac{\pi}{\frac{1}{2}} = 2\pi$. Plot (h, k) and frame the

curve, as shown in Figure 3.32. Plot the midpoint and the quarterpoints for a tangent curve (up/up and away).

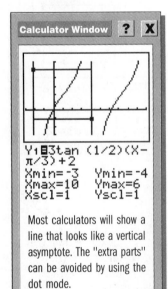

Y₁■3tan (1/2)(X-
π/3)+2
Xmin=-3   Ymin=-4
Xmax=10   Ymax=6
Xscl=1   Yscl=1

Most calculators will show a line that looks like a vertical asymptote. The "extra parts" can be avoided by using the dot mode.

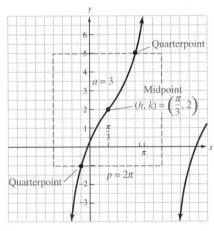

Quarterpoint

$a = 3$

Midpoint

$(h, k) = \left(\dfrac{\pi}{3}, 2\right)$

$\dfrac{\pi}{3}$

π

$p = 2\pi$

Quarterpoint

FIGURE 3.32   Graph of $y - 2 = 3\tan\frac{1}{2}\left(x - \frac{\pi}{3}\right)$

If you are using a calculator (see marginal comment), you will need to solve the general form for y before inputting the function.

EXAMPLE 10   **Graphing a Curve from Given Data Points**

Let us reconsider the Problem of the Day.

The flywheel on a lawn mower has 16 evenly spaced cooling fins. If we rotate the engine through two complete revolutions, we can generate 32 data points, with the first coordinate representing time and the second representing the depth of the piston. Determine an equation of the curve generated by the data as shown in Table 3.3.

Solution   As previously noted, we plot the data points, as shown in Figure 3.33, and note that the points are sinusoidal.

TABLE 3.3 Data Points for Lawn Mower Engine

x	y	x	y
0	0	16	0
1	0.2	17	0.2
2	0.8	18	0.8
3	1.7	19	1.7
4	2.5	20	2.5
5	3.4	21	3.4
6	4.0	22	4.0
7	4.4	23	4.4
8	4.5	24	4.5
9	4.4	25	4.4
10	4.0	26	4.0
11	3.5	27	3.5
12	2.2	28	2.2
13	1.7	29	1.7
14	0.9	30	0.9
15	0.2	31	0.2
		32	0

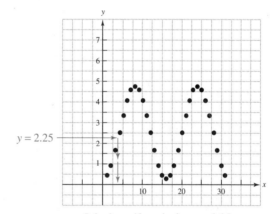

$y = 2.25$

It looks as if $x = 4$ when $y = 2.25$.

FIGURE 3.33   Data points for the Problem of the Day

The graph seems to have a period of 16 and an amplitude of 2.25. This means that for the general sine curve

$$y - k = a \sin b(x - h)$$

Thus, $a = 2.25 = \frac{9}{4}$; period $p = \dfrac{2\pi}{b} = 16$ so that $b = \frac{\pi}{8}$. Finally, we need to find an appropriate starting point, (h, k). There are many points we might choose, but we select a y-value halfway between the maximum and minimum, that is, $y = 2.25$. Next, we approximate the intersection of the line $y = 2.25 = \frac{9}{4}$ and the graph: it appears to be $(4, 2.25)$.

$$y - \tfrac{9}{4} = \tfrac{9}{4}\sin\tfrac{\pi}{8}(x - 4)$$

The graph is shown in Figure 3.34.

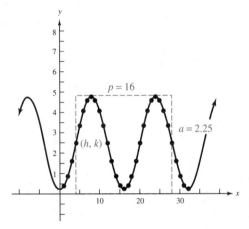

FIGURE 3.34   Graph of $y - \frac{9}{4} = \frac{9}{4} \sin \frac{\pi}{8}(x - 4)$ ■

EXAMPLE 11

Solution

Sound Frequency

A tuning fork is a small steel instrument with two prongs that, when struck, vibrates at a particular frequency so that it produces a certain fixed tone in perfect pitch. It is used as a guide for tuning instruments or for testing hearing. A middle C tuning fork vibrates at 264 Hz, with an amplitude of 0.02 cm.* Write an equation and graph this sound wave for a middle C.

We model this as a sinusoidal curve. This means we are looking for either a sine or a cosine equation of the form $y = a \cos bx$ or $y = a \sin bx$, where y is the amount of vibration. If we assume that x is the time, then a is the amplitude, and the period $p = 2\pi/b$. Also, if $x = 0$, then the time is 0 and there is no vibration, so we select, as a model, the equation $y = a \sin bx$, since $y = 0$ for $x = 0$.

a is the amplitude, so $a = 0.02$. The period is the reciprocal of the frequency, so $p = \dfrac{1}{264}$ and

$$b = \frac{2\pi}{p} = \frac{2\pi}{\frac{1}{264}} = 528\pi$$

The desired equation is

$$y = 0.02 \sin 528\pi x$$

---

* Hz is the abbreviation for *hertz*, a unit of measurement that means "cycles per unit of time." It is named after Heinrich Hertz (1857–1894) who discovered radio waves. An amplitude of 0.02 cm means that the prong vibrates to the right and left a maximum of 0.02 cm (where we assume, as usual, that right is positive and left is negative).

The graph is shown in Figure 3.35. You will need to pay attention to the scale when you graph this function.

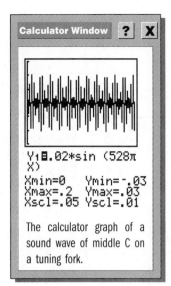

Calculator Window ? X

Y₁**B**.02*sin (528π X)
Xmin=0   Ymin=⁻.03
Xmax=.2   Ymax=.03
Xscl=.05 Yscl=.01

The calculator graph of a sound wave of middle C on a tuning fork.

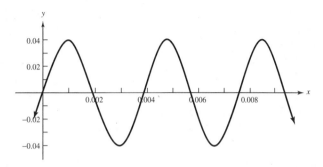

FIGURE 3.35   Graph of sound waves for a middle C tuning fork

Even though the graph shown in Figure 3.35 accurately represents the sound wave, it does not look like pictures of sound waves that you may have seen before. A calculator graph (shown in the margin—note the different scale) shows a more realistic picture of the middle C tuning fork. It would be very difficult to draw this graph without a computer or a graphing calculator. The curve is "bunched up" because the curve is going through so many cycles in a short period. Note that Xmax = 0.2 s, which means that the curve is going through $0.2 \div (1/264) = 52.8$ cycles/s. In Figure 3.35, we let the maximum value of x be 0.01, which produces 1 cycle/s. This produces the more familiar-looking sine curve. ∎

PROBLEM SET 3.2

Essential Concepts

1. What is the difference between the standard and general form of the equations of the trigonometric curves?

2. What is a frame? What are the height, length, and starting point of the frames for the general form of the cosine, sine, and tangent curves?

3. **IN YOUR OWN WORDS** What is the general cosine form? Explain the meaning of each variable you use.

4. **IN YOUR OWN WORDS** What is the general sine form? Explain the meaning of each variable you use.

5. **IN YOUR OWN WORDS** What is the general tangent form? Explain the meaning of each variable you use.

6. **IN YOUR OWN WORDS** Outline a procedure for graphing a general-form trigonometric equation.

Level 1 Drill

Problems 7–18: Graph one period of each function.

7. **a.** $y = \sin(x - 1)$ **b.** $y = \sin x - 1$
   c. IN YOUR OWN WORDS Explain how the graphs in parts **a** and **b** differ.

8. a. $y = \cos\left(x - \frac{\pi}{3}\right)$ **b.** $y = \cos x - \frac{\pi}{3}$
c. IN YOUR OWN WORDS Explain how the graphs in parts **a** and **b** differ.

9. a. $y = 3 \sin x$ **b.** $y = \sin 3x$
c. IN YOUR OWN WORDS Explain how the graphs in parts **a** and **b** differ.

10. a. $y = 2 \cos x$ **b.** $y = \cos 2x$
c. IN YOUR OWN WORDS Explain how the graphs in parts **a** and **b** differ.

11. a. $y = \frac{1}{2} \sin x$ **b.** $y = \sin \frac{1}{2} x$
c. IN YOUR OWN WORDS Explain how the graphs in parts **a** and **b** differ.

12. a. $y = 4 \tan x$ **b.** $y = \tan 4x$
c. IN YOUR OWN WORDS Explain how the graphs in parts **a** and **b** differ.

13. $y - 2 = \sin\left(x - \frac{\pi}{2}\right)$ **14.** $y + 1 = \cos\left(x + \frac{\pi}{3}\right)$

15. $y - 3 = \tan\left(x + \frac{\pi}{6}\right)$ **16.** $y - 1 = \cos 2\left(x - \frac{\pi}{4}\right)$

17. $y + 2 = \sin 3\left(x + \frac{\pi}{6}\right)$ **18.** $y - 1 = \tan 2\left(x - \frac{\pi}{4}\right)$

Problems 19–24: *State the period and amplitude for each graph. Also give a possible equation for the given (h, k) for each of these calculator graphs.*

19.

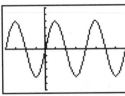

```
Xmin=-6.283185...
Xmax=12.566370...
Xscl=1.5707963...
Ymin=-6
Ymax=6
Yscl=1
```
$(h, k) = (0, 0)$

20.

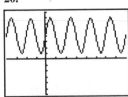

```
Xmin=-6.283185...
Xmax=12.566370...
Xscl=1.5707963...
Ymin=-6
Ymax=8
Yscl=1
```
$(h, k) = (0, 4)$

21.

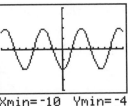

```
Xmin=-10   Ymin=-4
Xmax=10    Ymax=4
Xscl=1     Yscl=1
```
$(h, k) = (\pi, 0)$

22.

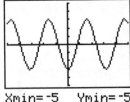

```
Xmin=-5    Ymin=-5
Xmax=5     Ymax=5
Xscl=1     Yscl=1
```
$(h, k) = \left(\frac{\pi}{2}, 0\right)$

23.

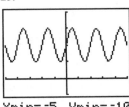

```
Xmin=-5    Ymin=-10
Xmax=5     Ymax=40
Xscl=1     Yscl=10
```
$(h, k) = (0, 20)$

24.

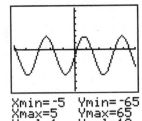

```
Xmin=-5    Ymin=-65
Xmax=5     Ymax=65
Xscl=1     Yscl=10
```
$(h, k) = (-3, -10)$

Level 2 Drill

Problems 25–46: *Specify the period and amplitude for each graph; then graph each curve.*

25. $y = \frac{1}{2} \cos\left(x + \frac{\pi}{6}\right)$ **26.** $y = 2 \sin\left(x - \frac{\pi}{4}\right)$

27. $y = 2 \sin 2\pi x$ **28.** $y = 3 \cos 3\pi x$

29. $y = 4 \tan\left(\frac{\pi x}{5}\right)$ **30.** $y = \frac{1}{4} \tan\left(\frac{\pi x}{3}\right)$

31. $y = \sin(4x + \pi)$ **32.** $y = \sin(3x + \pi)$

33. $y = \tan\left(2x - \frac{\pi}{2}\right)$ **34.** $y = \tan\left(\frac{x}{2} + \frac{\pi}{3}\right)$

35. $y = 2 \cos(3x + 2\pi) - 2$

36. $y = 4 \sin\left(\frac{1}{2}x + 2\right) - 1$

37. $y = \sqrt{2} \cos(x - \sqrt{2}) - 1$

38. $y = \sqrt{3} \sin\left(\frac{1}{3}x - \sqrt{\frac{1}{3}}\right)$

39. $y = 2 \sec x$ **40.** $y = \csc 2x$

41. $y = \frac{1}{2} \cot x$ **42.** $y = \cot \frac{1}{2} x$

43. $y = \csc 2x + 1$ **44.** $y = \csc(2x + 1)$

45. $y = 2 \sec\left(x + \frac{\pi}{3}\right)$ **46.** $y - 2 = 2 \sec\left(x + \frac{2\pi}{3}\right)$

Level 2 Applications

47. The pumping action of the heart consists of two phases: the systolic phase, in which blood rushes from the left ventricle into the aorta, and the diastolic phase, during which the heart muscle relaxes. The length of time between peaks is called the period of the pulse, and the pulse rate is the number of pulse beats in one minute. The graph of the blood pressure as a function of time is shown in Figure 3.36.

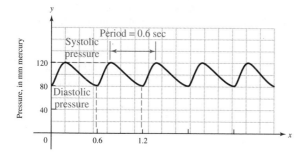

FIGURE 3.36   Blood pressure

a. What is the amplitude?
b. What is the pulse rate?

48. PROBLEM FROM CALCULUS In *Calculus for the Life Sciences* by Rodolfo De Sapio, the motion of a human arm is analyzed. Consider a rhythmically moving arm as shown in Figure 3.37. The upper arm

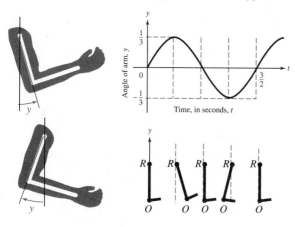

FIGURE 3.37   Arm motion

RO rotates back and forth about the point *R*, and the position of the arm is measured by the angle *y*. The graph shows the relationship between the angle and the time *t*, in seconds. Write an equation for this graph.

49. An E above middle C tuning fork vibrates at 330 Hz, with an amplitude of 0.02 cm. Write an equation and graph this sound wave for an E above middle C.

50. A G above middle C tuning fork vibrates at 396 Hz, with an amplitude of 0.04 cm. Write an equation and graph this sound wave for a G above middle C.

51. In the previous section, we considered the following data for an electrical current *I* (in amperes) in a certain circuit (for some convenient unit of time), which generates the following set of data points:

Time	Height	Time	Height
0	−60.0000	10	30.00000
1	−58.6889	11	40.14784
2	−54.8127	12	48.54102
3	−48.5410	13	54.81273
4	−40.1478	14	58.68886
5	−30.0000	15	60.00000
6	−18.5410	16	58.68886
7	−6.27171	17	54.81273
8	6.27171	18	48.54102
9	18.54102	19	40.14784
		20	30.00000

Determine a possible equation of the curve generated by this circuit.

52. In the previous section, we considered the following data for a point *P* on a water wheel with a 30-ft radius that is *d* units from the water. If the water wheel turns at 6 rev/min, the height of the point *P* above the water is given by the following set of data points:

Time	Height	Time	Height
0	−1.000	10	−1.000
1	4.729	11	4.729
2	19.729	12	19.729
3	38.271	13	38.270
4	53.271	14	53.270
5	59.000	15	59.000
6	53.271	16	53.271
7	38.271	17	38.271
8	19.729	18	19.730
9	4.729	19	4.730
		20	−1.000

Determine a possible equation of the curve generated by this water wheel.

53. A spring at rest is shown in Figure 3.38. If the spring is pulled to position -3 and released, the spring will contract and stretch again for the first two seconds according to the table of values. Graph these data points and then write an appropriate equation to model these data.

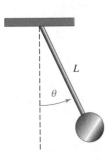

Time	Position	Time	Position
0	-3.0	1.0	-1.5
0.1	-2.6	1.1	-2.6
0.2	-1.5	1.2	-3.0
0.3	0.0	1.3	-2.6
0.4	1.5	1.4	-1.5
0.5	2.6	1.5	0.0
0.6	3	1.6	1.5
0.7	2.6	1.7	2.6
0.8	1.5	1.8	3.0
0.9	0	1.9	2.6
		2.0	1.5

FIGURE 3.38   Spring at rest

54. A pendulum consisting of a 9-in. rod with a weight attached at one end is shown in Figure 3.39. If the string is stretched to 60° and released, the angle through which the pendulum swings can be measured for the first 20 sec after release. We measure swing to the right as a positive angle and swing to the left as a negative angle. After finding values for $t = 0$ to 10 in 1-second intervals, we obtain the following additional measurements (specifically chosen to help you answer this question). Graph these data points and then write an appropriate equation to model these data.

Time	Angle	Time	Angle
0	1.05	0.83	0.01
1	-0.32	1.67	-1.05
2	-0.85	2.50	0.00
3	0.85	3.33	1.05
4	0.32		
5	-1.05		
6	0.33		
7	0.84		
8	-0.85		
9	-0.32		
10	1.05		

FIGURE 3.39   Motion of a pendulum

55. IN YOUR OWN WORDS Problem 54 is an important application from physics. If $t = 0$ for initial angle $\theta_0 = \frac{\pi}{3}$ (that is, 60°), the table value is 1.047. This is an approximate value. Explain where you think the number 1.047 comes from. In physics, the equation for pendulum motion (which is derived in calculus) is

$$\theta = \theta_0 \cos(t\sqrt{g/L})$$

where θ is the angle, θ_0 is the initial angle, t is the time in seconds, g is a constant due to gravity, and L is the length of the pendulum in inches. Do you think the period depends on the length?

56. In physics, it is shown that the velocity for the pendulum in Problem 54 is given by the equation

$$V = -\frac{\pi}{3}\sqrt{\frac{32}{0.75}}\sin\sqrt{\frac{32}{0.75}}\,t$$

where V is in ft/s. Graph this function for two complete cycles. If you have a calculator, draw a graph representing the first 20 sec.

57. The distance that a certain satellite is north or south of the equator is given by

$$y = 3{,}000\cos\left(\frac{\pi}{60}t + \frac{\pi}{5}\right)$$

where t is the number of minutes that have elapsed since lift-off, and y is measured in miles.
a. Graph the equation for $0 \le t \le 120$.
b. What is the greatest distance north of the equator reached by the satellite?
c. How long does it take to complete one period?

58. PROBLEM FROM CALCULUS It is shown in calculus that for small values of x (called a *neighborhood* of the origin in calculus), the graph of $y = \sin x$ and the line $y = x$ (where x is measured in radians) are almost the same. This means, for example, that

$$\sin 0.05 \approx 0.05$$

Illustrate this fact graphically.

Level 3 Theory

Calculator Window **?** **X**

59. IN YOUR OWN WORDS The graph of the sound wave equation

$$y = 0.04 \sin(880\pi x)$$

is shown in Figure 3.40.

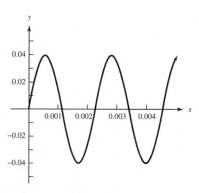

FIGURE 3.40 Graph of $y = 0.04 \sin(880\pi x)$

If we draw this with a graphing calculator using $0 \le x \le 0.1$, we obtain the following graph:

```
Y₁≡.04sin (880πX
)
Xmin=0  Ymin=-.05
Xmax=.1 Ymax=.05
Xscl=.01 Yscl=.01
```

Notice that these graphs look very different. Explain why you think these graphs look different.

60. In this problem, all graphs are of the curve

$$y = 0.04 \sin(880\pi x)$$

(See Problem 59.) We have changed only the x-scale. Use a graphing calculator and match the equation to the graph.

a. $0 \le x \le 3.1$ A.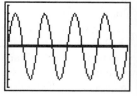

b. $0 \le x \le 1$

c. $0 \le x \le 2$

d. $0 \le x \le 2.9$

e. $0 \le x \le 3$ B.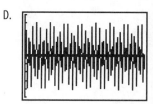

C.

D.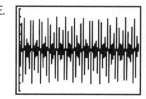

E.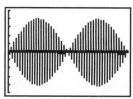

61. IN YOUR OWN WORDS Comment on the graphs from Problem 59. Why do you think they all look so different?

62. Draw the graph for the equation in Problem 59 for $0 \le x \le 50$. Comment on the result in light of the graphs shown in Problem 60.

Level 3
Individual Projects

63. You have no doubt heard a radio station proclaim its identity—"Tune to 97.7 on the FM dial." The letters FM refer to frequency modulation, and 97.7 means that the station broadcasts at 97.7 million Hz. The letters AM on a radio dial refer to amplitude modulation. Write a paper on radio waves and broadcast frequencies. Explain what is meant by frequency modulation (FM) and amplitude modulation (AM).

64. The times of sunrise and sunset for various latitudes are shown in Table 3.4.
   a. Plot the sunrise and sunset for the listed city closest to your home.
   b. Plot the length of daylight for the listed city closest to your home.
   c. **IN YOUR OWN WORDS** Are either of these plotted curves sinusoidal? Discuss.

TABLE 3.4

Date	Time of Sunrise						Time of Sunset					
	20°N LATITUDE (HAWAII)	30°N LATITUDE (NEW ORLEANS)	35°N LATITUDE (ALBUQUERQUE)	40°N LATITUDE (PHILADELPHIA)	45°N LATITUDE (MINNEAPOLIS)	60°N LATITUDE (ALASKA)	20°N LATITUDE (HAWAII)	30°N LATITUDE (NEW ORLEANS)	35°N LATITUDE (ALBUQUERQUE)	40°N LATITUDE (PHILADELPHIA)	45°N LATITUDE (MINNEAPOLIS)	60°N LATITUDE (ALASKA)
	h m	h m	h m	h m	h m	h m	h m	h m	h m	h m	h m	h m
Jan. 1	6 35	6 56	7 08	7 22	7 38	9 03	17 31	17 10	1658	16 44	16 28	15 03
Jan. 15	6 38	6 57	7 08	7 20	7 35	8 48	17 41	17 22	17 12	16 59	16 44	15 31
Jan. 30	6 36	6 52	7 01	7 11	7 23	8 19	17 51	17 35	17 27	17 17	17 05	16 09
Feb. 14	6 30	6 41	6 48	6 55	7 03	7 42	17 59	17 48	17 42	17 34	17 26	16 48
Mar. 1	6 20	6 26	6 30	6 34	6 39	6 59	18 05	17 59	17 56	17 52	17 47	17 27
Mar. 16	6 08	6 09	6 10	6 11	6 11	6 15	18 10	18 09	18 08	18 08	18 07	18 04
Mar. 31	5 55	5 51	5 49	5 46	5 43	5 29	18 14	18 18	18 20	18 23	18 26	18 41
Apr. 15	5 42	5 34	5 28	5 23	5 16	4 44	18 18	18 27	18 32	18 38	18 45	19 18
Apr. 30	5 32	5 18	5 11	5 02	4 51	4 01	18 23	18 37	18 44	18 53	19 04	19 55
May 15	5 24	5 07	4 57	4 45	4 31	3 23	18 29	18 46	18 56	19 08	19 22	20 31
May 30	5 20	5 00	4 48	4 34	4 18	2 53	18 35	18 55	19 07	19 21	19 38	21 04
June 14	5 20	4 58	4 45	4 30	4 13	2 37	18 40	19 02	19 15	19 30	19 48	21 24
June 29	5 23	5 02	4 49	4 34	4 16	2 40	18 43	19 05	19 18	19 33	19 51	21 26
July 14	5 29	5 08	4 56	4 43	4 26	3 01	18 43	19 03	19 15	19 29	19 45	21 09
July 29	5 34	5 17	5 07	4 55	4 41	3 33	18 39	18 56	19 06	19 17	19 31	20 38
Aug. 13	5 39	5 26	5 18	5 09	4 59	4 09	18 30	18 43	18 51	19 00	19 10	19 59
Aug. 28	5 43	5 35	5 29	5 24	5 17	4 45	18 19	18 27	18 33	18 38	18 45	19 16
Sept. 12	5 47	5 43	5 40	5 38	5 35	5 20	18 06	18 10	18 12	18 14	18 17	18 31
Sept. 27	5 50	5 51	5 51	5 52	5 53	5 55	17 52	17 51	17 50	17 49	17 49	17 45
Oct. 12	5 54	6 00	6 03	6 07	6 11	6 31	17 39	17 33	17 29	17 26	17 21	17 01
Oct. 22	5 57	6 06	6 12	6 18	6 25	6 56	17 32	17 22	17 17	17 11	17 04	16 32
Nov. 6	6 04	6 18	6 26	6 35	6 45	7 35	17 23	17 10	17 02	16 52	16 42	15 52
Nov. 21	6 12	6 30	6 40	6 52	7 05	8 12	17 19	17 02	16 51	16 40	16 26	15 19
Dec. 6	6 22	6 42	6 54	7 07	7 23	8 44	17 20	17 00	16 48	16 35	16 19	14 58
Dec. 21	6 30	6 52	7 04	7 18	7 35	9 02	17 26	17 05	16 52	16 38	16 21	14 54

The data in this table are courtesy of the U.S. Naval Observatory. This table of times of sunrise and sunset may be used in any year of the 20th century with an error not exceeding two minutes and generally less than one minute. The times are fairly accurate anywhere in the vicinity of the stated latitude, with an error of less than one minute for each 9 miles from the given latitude.

SECTION 3.3

TRIGONOMETRIC GRAPHS

PROBLEM OF THE DAY

Show that $y = \sin \theta + \sqrt{3} \cos \theta$ is identical to $y = 2 \sin\left(\theta + \frac{\pi}{3}\right)$.

How would you show that these equations are identical?

What does it mean for equations to be identical? We need to prove that

$$\sin \theta + \sqrt{3} \cos \theta = 2 \sin\left(\theta + \frac{\pi}{3}\right)$$

is a true equation for all values of θ. If $\theta = \frac{\pi}{6}$, then

$$\sin \frac{\pi}{6} + \sqrt{3} \cos \frac{\pi}{6} = 2 \sin\left(\frac{\pi}{6} + \frac{\pi}{3}\right)$$

$$\frac{1}{2} + \sqrt{3}\frac{\sqrt{3}}{2} = 2 \sin \frac{\pi}{2} \qquad \textit{Exact values; also } \frac{\pi}{6} + \frac{\pi}{3} = \frac{3\pi}{6} = \frac{\pi}{2}$$

$$\frac{1}{2} + \frac{3}{2} = 2(1) \qquad \textit{Exact value}$$

$$2 = 2 \qquad \textit{True equation}$$

We have now shown the equation is true for one value—namely, $\frac{\pi}{6}$; we need to show that it is true for all values. If we use a calculator, we might graph

$$y_1 = \sin x + \sqrt{3} \cos x$$

Remember, use x for the independent variable when you use a calculator; also use y_1 and y_2 because you are graphing two different functions.

$$y_2 = 2 \sin\left(x + \frac{\pi}{3}\right)$$

Now compare their graphs (shown in Figure 3.41). They appear to be the same for all values of x, so we are led to conclude that $y_1 = y_2$. Keep in mind that these graphs may be off by "only a hair" so that a visual inspection will not reveal their differences. You might also want to create and check a table of values.

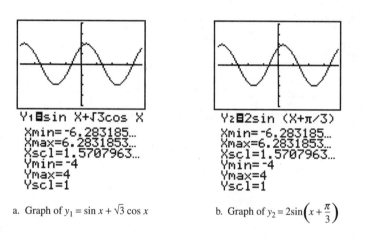

a. Graph of $y_1 = \sin x + \sqrt{3} \cos x$ b. Graph of $y_2 = 2\sin\left(x + \dfrac{\pi}{3}\right)$

FIGURE 3.41 Compare graphs. If you graph both of these equations at once, you will see only one graph because these graphs are identical.

In this section, we discuss a method for graphing equations like y_1 called **adding the ordinates.** We also briefly consider Fourier graphs as well as other trigonometric graphs. Finally, we will find the intersection points of two curves using the graphing method for solving a system of equations.

Adding Ordinates

We can graph the sum of two curves by adding the ordinates (the y-values) at various x-value locations. That is, we add second components, point by point, until we have plotted enough points to see the shape of the curve that is the sum, as shown in Figure 3.42 on page 124.

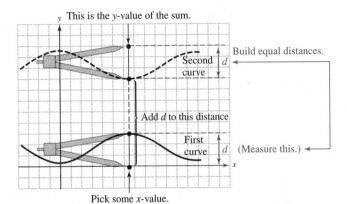

FIGURE 3.42   Adding ordinates (that is, adding second components)

We illustrate this method by graphing the sum function of the Problem of the Day.

EXAMPLE 1   **Graphing by Adding Ordinates**

Graph $y = \sin \theta + \sqrt{3} \cos \theta$.

Solution   Begin by graphing $y_1 = \sin \theta$, as shown in Figure 3.43.

$$(h, k) = (0, 0); \quad a = 1, \quad b = 1, \quad p = \frac{2\pi}{1} = 2\pi$$

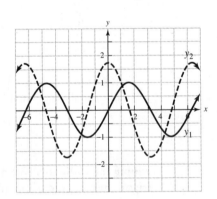

FIGURE 3.43   Individual graphs of functions to be added

Next, plot $y_2 = \sqrt{3} \cos \theta$, as shown in Figure 3.43.

$$(h, k) = (0, 0); \quad a = \sqrt{3}, \quad b = 1, \quad p = \frac{2\pi}{1} = 2\pi$$

Now plot points (which represent the sum of the graphs to be added) one by one. This process is shown in Figure 3.44.

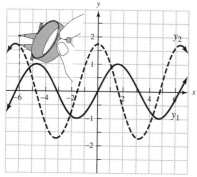
a. Begin by plotting a sum point.

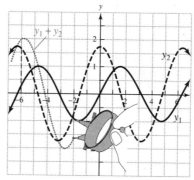
b. Continue by plotting additional points.

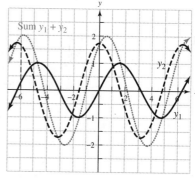
c. Connect the sum points.

FIGURE 3.44 Steps in graphing by adding ordinates

The completed graph is shown by the dots in Figure 3.44c.

EXAMPLE 2 **Graphing a Sum**

Graph $y = 2 \cos 3x + 3 \sin 2x$.

Solution We graph the sum by adding ordinates. Let $y = y_1 + y_2$, where

$$y_1 = 2 \cos 3x \qquad \left(a = 2, b = 3, p = \frac{2\pi}{3} \right)$$

$$y_2 = 3 \sin 2x \qquad (a = 3, b = 2, p = \pi)$$

This graph is shown in Figure 3.45a. Plot the sum points (one by one, as shown in Figure 3.45b) to obtain the curve shown in color in Figure 3.45c.

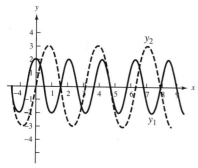
a. Graphs of y_1 and y_2

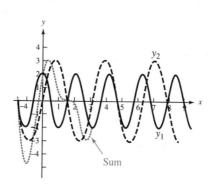
b. Plot sum points one by one.

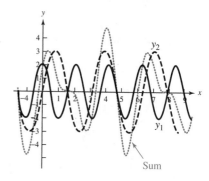
c. Draw sum curve.

FIGURE 3.45 Graphing by adding ordinates

While easy to understand, the method of graphing by adding ordinates is often "messy" because there is so much going on in the graph (as can be

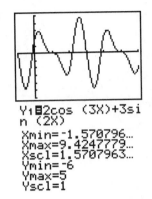

FIGURE 3.46

Calculator graph of
$y = 2 \cos 3x + 3 \sin 2x$

seen in Figure 3.45c). Technology plays a big role in graphing more complicated curves, and even though you may not presently have a graphing calculator, you should take time to look at the calculator output shown in Figure 3.46.

First, notice the input line. Most calculators wrap the text and symbols from one line to the next. This may be in the middle of a word, as you can see in Figure 3.46. The domain (permissible *x*-values) in the output shows decimal approximations, but we can see that the intended domain is

Domain: $-\frac{\pi}{2} \le x \le 3\pi$

Range: $-6 \le y \le 5$

Note that the domain and range used are *not* the domain and range for sum function, but rather indicate the domain and range for the calculator window shown. These values can easily be changed. ■

Fourier Series—An Example

Joseph Fourier (1768–1830) used combinations of trigonometric functions to represent certain types of periodic functions. For example, the sawtooth function illustrated in Figure 3.47 is a periodic function. Fourier noticed that such a function can be approximated by a trigonometric function, shown here for $y = \sin x$.

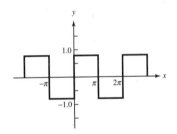

FIGURE 3.47

Sawtooth function

First approximation:

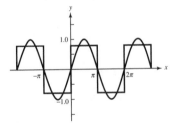

Even though this is a good approximation, it can be improved by considering sums of trigonometric functions:

Second approximation:

$y = \sin x + \dfrac{\sin 3x}{3}$

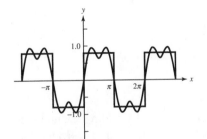

Third approximation:

$y = \sin x + \dfrac{\sin 3x}{3} + \dfrac{\sin 5x}{5}$

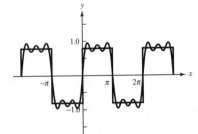

Fourth approximation:

$y = \sin x + \dfrac{\sin 3x}{3} + \dfrac{\sin 5x}{5} + \dfrac{\sin 7x}{7}$

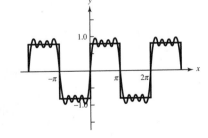

In mathematics, the sum of the terms of a sequence is called a *series,* and in 1822, Fourier published a book that investigated series such as

$$y = \sin x + \frac{\sin 3x}{3} + \frac{\sin 5x}{5} + \frac{\sin 7x}{7} + \frac{\sin 9x}{9} + \cdots$$

to approximate periodic functions. Even though this series, called a *Fourier series,* is beyond the scope of this course, we can consider the first few terms.

In Example 11 in Section 3.2, we considered the sound wave for a given frequency. If a sound, or musical note, is produced by a vibration of frequency f and amplitude a, then the equation has the form

$$y = b + a \cos(2\pi fx) \quad \text{or} \quad y = b + a \sin(2\pi fx)$$

Sound quality is produced by adding *overtones* for musical notes. For example, the Fourier series approximation to the first function is

$$y = b + \underbrace{\cos(2\pi fx)}_{\textit{primary note}} + \underbrace{\frac{\cos 2(2\pi fx)}{2}}_{\textit{1st overtone}} + \underbrace{\frac{\cos 3(2\pi fx)}{3}}_{\textit{2nd overtone}} + \cdots + \underbrace{\frac{\cos n(2\pi fx)}{n}}_{\textit{nth overtone}} + \cdots$$

An application is discussed in the following example.

EXAMPLE 3   **Graphing a Sound Wave**

In "The Computer as a Musical Instrument" by Max Mathews and John Pierce (*Scientific American,* Feb. 1987, pp. 126–133), the authors discuss waveforms shown in Figure 3.48.

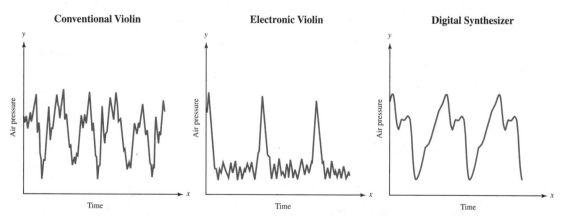

Waveforms show the way a particular sound causes the ambient air pressure to fluctuate. The sound from a conventional violin (left) is characterized by a complex periodic waveform. An electronic violin (middle) converts the motion of a bowed metal string into an electrical signal, which is then filtered to yield a simple violinlike waveform. A popular digital sound synthesizer (right) can mimic the waveform of an actual violin sound more closely than the electronic violin can.

FIGURE 3.48   Sound wave of a violin

Simulate the sound wave with equation

$$y = \sin 400\pi x + \frac{\sin 800\pi x}{2} + 2$$

using a graphing calculator. The first term, $\sin 400\pi x$, gives the musical *tone* and the next term, $\sin(800\pi x)/2$, is called the *first overtone*. If there were additional terms, they would be called the *second, third, . . .* overtones. The constant term, 2, simply moves the entire graph up two units.

Solution   This problem is best done with a graphing calculator. The graph is shown in the margin. If you do not have a graphing calculator, then graph this curve by adding ordinates. Begin by graphing

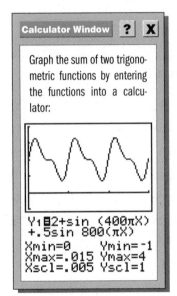

Graph the sum of two trigonometric functions by entering the functions into a calculator:

Y₁⊟2+sin (400πX)
+.5sin 800(πX)
Xmin=0   Ymin=⁻1
Xmax=.015 Ymax=4
Xscl=.005 Yscl=1

$$y_1 = 2 + \sin 400\pi x; \qquad (h, k) = (0, 2); \qquad a = 1, \quad b = 400\pi, \quad p = \frac{2\pi}{400\pi} = \frac{1}{200}$$

$$y_2 = \frac{\sin 800\pi x}{2}; \qquad (h, k) = (0, 0); \qquad a = \tfrac{1}{2}, \quad b = 800\pi, \quad p = \frac{2\pi}{800\pi} = \frac{1}{400}$$

These graphs are shown in Figure 3.49.

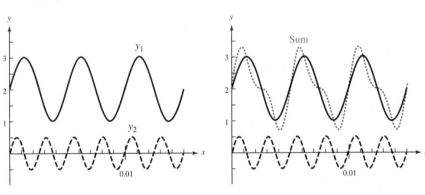

a. Graphs of curves to be added   b. Find the sum by adding ordinates.

FIGURE 3.49   Graph of $y = 2 + \sin 400\pi x + \dfrac{\sin 800\pi x}{2}$ ■

Notice that the solution does not exactly reproduce the wave form for the digital synthesizer. However, with each term of the Fourier series we add, the closer we approach the actual sound wave.

Today, we use Fourier series to study electrical circuits, sound waves, and heat flow, to name just a few applications.

Water Waves

Wave motion is one of the most common examples of periodic motion. There are water waves, sound waves, and radio waves, as well as many other electromagnetic waves. Sometimes waves are classified by considering how the motions of the particles of matter relate to the direction of the waves themselves.

This classification system yields waves called **transverse waves** and others called **longitudinal waves.** In a transverse wave, the particles vibrate at right angles to the direction in which the wave itself is propagated. In a longitudinal wave, the particles vibrate in the same direction as that in which the wave is propagated. Both types of waves are shown in Figure 3.50.

EXAMPLES
Sound waves are longitudinal. Transverse waves can be seen by tying one end of a rope to a fence post and moving the other end up and down.

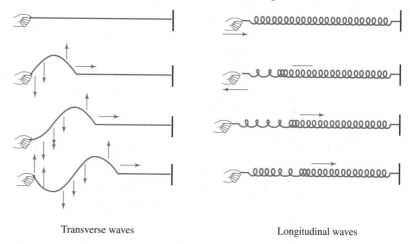

Transverse waves   Longitudinal waves

FIGURE 3.50   Transverse and longitudinal waves

If we look at water waves, we find that the wave action is a combination of both transverse and longitudinal waves. The water molecules move back and forth and, at the same time, up and down. The path of the molecules makes a circle or an ellipse (see Figure 3.51). In their simplest form, water waves can be shown to be sine curves represented by an equation of the form

$$y = a \sin \frac{2\pi}{\lambda} (x - \lambda nt)$$

where

a = *amplitude of the wave*

n = *frequency* (defined as the number of waves passing any given point per second); it is customary to express frequency in vib/s or in cycles/s

λ = *wavelength* (representing the distance between two adjacent points in the wave that have the same phase)

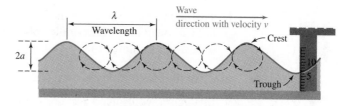

FIGURE 3.51   Water waves

$t = time$

$x = distance\ from\ the\ source$

If p is the *period* of the wave (in seconds), then $p = 1/n$ and it can be shown that the *phase velocity* (the speed of the wave) is

$$v = \sqrt{\frac{g\lambda}{2\pi}} \qquad \text{or} \qquad v = \frac{\lambda}{p} = \lambda n \qquad \begin{array}{l}\text{in ft/s, where } g = 32 \text{ ft/s}^2 \text{ and it can be}\\ \text{shown that } g = 2\pi\lambda/p^2\end{array}$$

Thus, $v \approx \sqrt{5.09\lambda}$, so $\sqrt{5.09\lambda} = \dfrac{\lambda}{p}$, which implies (see Problem 64) that $\lambda \approx 5.09p^2$. In terms of v, the **equation of the wave** is

$$y = a \sin \frac{2\pi}{\lambda}(x - vt)$$

EXAMPLE 4

Finding an Equation for Ocean Waves

Suppose you are watching waves pass a pier piling, and you count 20 waves per minute, each with an amplitude of 2 ft. Find the wavelength (in feet) and phase velocity (in mi/h), then write the equation for one of these waves.

Solution

Since $a = 2$, 20 waves/min is $\frac{1}{3}$ wave/sec, so $n = \frac{1}{3}$ and $p = 3$. Then, the wavelength in feet is

$$\lambda \approx 5.09p^2 \approx 45.81$$

The phase velocity is

$$\begin{aligned} v &\approx \frac{\lambda}{p} = \frac{45.81}{3} = 15.27 \text{ ft/s} \\ &= 15.27(60^2) \text{ ft/h} && \textit{One hour} = 60^2 \, sec \\ &= \frac{15.27(60)^2}{5,280} \text{ mi/h} && \textit{One mile} = 5,280\,ft \\ &\approx 10.41 \text{ mi/h} && \textit{Simplify.} \end{aligned}$$

The equation for the wave is

$$y = a \sin \frac{2\pi}{\lambda}(x - \lambda n t)$$

$$= 2 \sin \frac{2\pi}{45.81}(x - 15.27t)$$

The equation of the wave in Example 4 has three variables, x, y, and t. If we hold t constant (for example, photograph the waves when $t = 5$), we can obtain the graph of the wave shown in Figure 3.52.

Trigonometric Systems of Equations

In algebra, you solved **systems of equations** in which you found replacements for the variables that make all of the equations in the system true *at the same*

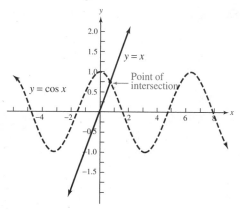

FIGURE 3.52 Wave equation

time. These are sometimes called **simultaneous solutions.** We indicate a system of equations by using a large brace, as illustrated in Example 5. In this section, we solve systems of equations graphically.

EXAMPLE 5 **Solving a Trigonometric System**

Solve $\begin{cases} y = x \\ y = \cos x \end{cases}$

Solution We wish to find all values (x, y) that make both equations true at the same time. We solve this system by graphing $y = x$, and $y = \cos x$ on the same coordinate axes, as shown in Figure 3.53.

FIGURE 3.53 Graphs of $y = x$ and $y = \cos x$ showing the point of intersection of the two curves

We estimate the point of intersection to be $(0.7, 0.7)$. ■

The Calculator Window on page 132 shows how you would solve the system given in Example 5 with a graphing calculator.

A system of equations may have any number of solutions (from zero to infinitely many). Example 6 illustrates a few possibilities.

Calculator Window **?** **X**

To solve a system of equations using a graphing calculator, begin by graphing the two curves:

$$y_1 = x$$

$$y_2 = \cos x$$

Since the scale is not shown on calculator screens, it would be difficult to estimate points of intersection if it were not for a built-in function called *trace*. When you press $\boxed{\text{TRACE}}$, the cursor will appear on one of the graphed curves, and the coordinates of the cursor will appear at the bottom of the screen:

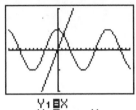

```
Y₁▤X
Y₂▤cos X
Xmin=-6    Ymin=-2
Xmax=9     Ymax=2
Xscl=.5    Yscl=1
```

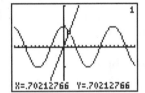

X=.70212766 Y=.70212766

On some calculator models, the numeral at the top tells you which curve the cursor will follow; the 1 here indicates the cursor will trace along the curve whose equation was entered at the y_1 prompt, namely $y = x$.

The trace values shown on your calculator may vary depending on the model and on the window set. We find the estimated solution (0.7, 0.7). Some calculators have an ISECT or INTERSECTION command. If you use such a calculator, you can find the intersection point with greater accuracy. For this example, we find a more accurate (but still not exact) intersection point, (0.7390851, 0.7390851).

EXAMPLE 6 **Estimating Solutions to Systems of Equations**

State the number of solutions and estimate the *x*-value of the solutions.

a. $\begin{cases} y = \frac{1}{5}x \\ y = \cos x \end{cases}$ **b.** $\begin{cases} y = \frac{1}{2} \\ y = \cos x \end{cases}$ **c.** $\begin{cases} y = \frac{3}{2} \\ y = \sin x \end{cases}$

Solution **a.** Graph $y_1 = \frac{1}{5}x$ and $y_2 = \cos x$ as shown in Figure 3.54. Note the use of y_1 and y_2 to indicate how you might enter these equations into a calculator.

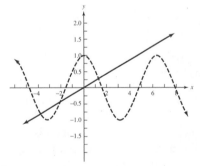

It looks like there are three solutions. We estimate these solutions to be $x = -4, -2,$ and 1.5.

FIGURE 3.54 Graphs of $y = \frac{1}{5}x$ and $y = \cos x$

If you have a graphing calculator, you can obtain a better estimate:

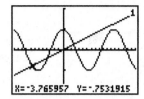 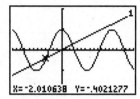 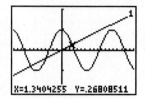

The solutions appear to be at $x = -3.8$, -2.0, and 1.3.

b. Graph $y_1 = \frac{1}{2}$ and $y_2 = \cos x$ as shown in Figure 3.55.

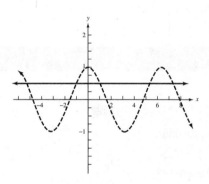

FIGURE 3.55
Graphs of $y = \frac{1}{2}$ and $y = \cos x$

There are infinitely many solutions. We estimate the solutions for

$$0 \le x \le 2\pi$$

to be $x = 1$ and $x = 5$. Since we know that the period of cosine is 2π, we add multiples (both positive and negative) to these estimated solutions:

$$x = 1 \pm 2\pi, \quad x = 1 \pm 4\pi, \quad \ldots$$
$$x = 5 \pm 2\pi, \quad x = 5 \pm 4\pi, \quad \ldots$$

Other values (for an integer k): $x = 1 + 2\pi k, 5 + 2\pi k$

A graphing calculator gives a better estimate:

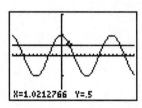 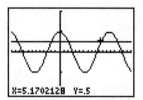

The approximate solutions are $x = 1.0 \pm 6.28k$, $5.2 \pm 6.28k$. Using our knowledge of the exact values for cosine, we know $\cos \frac{\pi}{3} = \frac{1}{2}$, so a complete solution is

$$x = \frac{\pi}{3} \pm 2\pi k, \quad x = \frac{5\pi}{3} \pm 2\pi k$$

We will discuss exact solutions for this system in greater detail in Chapter 4.

c. Graph $y = \frac{3}{2}$ and $y = \sin x$ as shown in Figure 3.56.

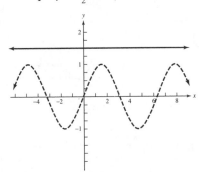

We see from the graph that there is no point of intersection, so we say the solution set is empty or that the system is **inconsistent.**

FIGURE 3.56
Graphs of $y = \frac{3}{2}$ and $y = \sin x$

PROBLEM SET 3.3

Essential Concepts

1. IN YOUR OWN WORDS Describe the procedure called adding ordinates.

2. What does it mean to solve a system of equations?

Level 1 Drill

Problems 3–14: *Solve each system of equations to estimate the x-value of the solutions in the domain* $0 \le x \le 2\pi$.

3. $\begin{cases} y = \frac{1}{4}x \\ y = \sin x \end{cases}$ **4.** $\begin{cases} y = \frac{1}{2}x \\ y = \sin x \end{cases}$

5. $\begin{cases} y = x - 1 \\ y = \tan x \end{cases}$ **6.** $\begin{cases} y = 2x \\ y = \tan x \end{cases}$

7. $\begin{cases} y = \frac{1}{2} \\ y = \cos x \end{cases}$ **8.** $\begin{cases} y = \frac{\sqrt{2}}{2} \\ y = \cos x \end{cases}$

9. $\begin{cases} y = 2x - 1 \\ y = \cos x \end{cases}$ **10.** $\begin{cases} y = -\frac{1}{2}x + 1 \\ y = \cos x \end{cases}$

11. $\begin{cases} y = \frac{\sqrt{2}}{2} \\ y = \sin x \end{cases}$ **12.** $\begin{cases} y = \frac{\sqrt{2}}{2} \\ y = \tan x \end{cases}$

13. $\begin{cases} y = \frac{\sqrt{3}}{3} \\ y = \tan x \end{cases}$ **14.** $\begin{cases} y = \sqrt{3} \\ y = \tan x \end{cases}$

Level 2 Drill

Problems 15–30: *Graph the functions by adding ordinates.*

15. $y = \frac{x}{2} + \sin x$ **16.** $y = \frac{x}{3} + \sin x$

17. $y = \frac{x}{3} + \cos x$ **18.** $y = \frac{2x}{3} + \cos x$

19. $y = \frac{2x}{3} + \sin x$ **20.** $y = x + \sin x$

21. $y = x + \cos x$ **22.** $y = x + \cos 2x$

23. $y = 2x + \cos 2x$ **24.** $y = 3x + \sin 3x$

25. $y = \sin x + 2 \cos x$ **26.** $y = \sin 2x + \cos 2x$

27. $y = 2 \sin \frac{x}{2} + \cos x$ **28.** $y = 2 \sin x + \cos \frac{x}{2}$

29. $y = 2 \sin x + \cos x$ **30.** $y = 3 \sin \frac{x}{3} + \cos \frac{x}{3}$

31. Show that $y_1 = \sin \theta + \cos \theta$ and $y_2 = \sqrt{2} \sin\left(\theta + \frac{\pi}{4}\right)$ are identical.

32. Show that $y_1 = \dfrac{\sqrt{3}}{2} \sin \theta + \dfrac{1}{2} \cos \theta$ and $y_2 = \sin\left(\theta + \frac{\pi}{6}\right)$ are identical.

Level 2 Applications

33. The sales of rototillers are very seasonal. A new business can describe its sales by

$$y = 0.02x + \left| \sin \frac{x\pi}{6} \right|$$

where x is the number of months since going into business on January 1, 1996. Plot the sales graph for the first three years.

34. The pattern of CO_2 in the atmosphere can be illustrated by a graph whose equation is

$$y = x + \sin x$$

Graph this curve for $0 \le x \le 20$.

35. The sun/moon tide curves are shown for a new moon.* During a new moon, the sun and moon tidal bulges are centered at the same longitude, so their effects are added to produce maximum high tides and minimum low tides. This produces maximum tidal range.

NEW MOON (Spring tide)

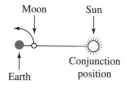

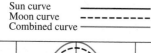

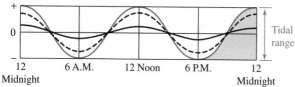

Write possible equations for the sun and moon curves, and also for the combined curve. Assume the tidal range is 10 ft and the period is 12.

* From *Introductory Oceanography*, 5th ed., by H. V. Thurman, p. 253. Reprinted with permission of Merrill, an imprint of Macmillan Publishing Company. Copyright © 1988 Merrill Publishing Company, Columbus, Ohio.

36. The sun/moon tide curves are shown for the third quarter or neap tide. During the neap tide, the sun and moon tidal bulges are at right angles to each other. The sun tide thus reduces the effect of the moon tide.

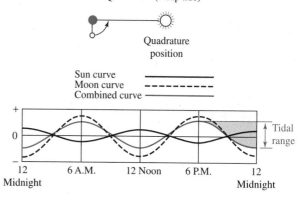

Write possible equations for the sun and moon curves, and also for the combined curve. Assume the tidal range is 4 ft and the period is 12.

37. Two sound waves are said to be *destructive* if the amplitude of the sum of the waves is zero at every point.

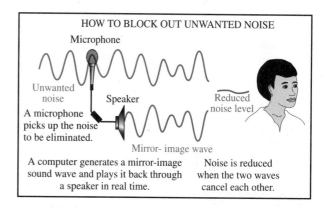

Show that $y = \sin x$ and $y = \cos\left(x + \frac{\pi}{2}\right)$ combine to form destructive interference.

38. Graph the sound wave (and first overtone) given by

$$y = 0.04 \sin 200\pi x + 0.06 \sin 400\pi x$$

for $0 \le x \le 0.01$.

39. Graph the sound wave for a frequency of 294 Hz. [Use a cosine function and assume $a = 1$ and $(h, k) = (0, 2)$.]

40. Graph the sound wave for a frequency of 294 Hz, along with its first overtone. [Use a cosine function and assume $a = 1$ and $(h, k) = (0, 2)$.]

41. If a water wave has an amplitude of 3 ft and you count 10 waves/min, what are the wavelength, its phase velocity (in mi/h), and equation of the wave?

42. If a water wave has an amplitude of 35 ft with a period of 5 waves/min, what are its wavelength, its phase velocity (in mi/h), and its equation?

43. If a water wave has the equation

$$y = 12.0 \sin(x - 37.7t)$$

where x and y are measured in feet and t is measured in seconds, how high is the wave from trough to crest? What are the wavelength and phase velocity (in mi/h) of the wave?

44. If a water wave has the equation

$$y = 18.0 \sin 0.62(x - 75.4t)$$

where x and y are measured in feet and t is measured in seconds, how high is the wave from trough to crest? What are the wavelength and phase velocity (in mi/h)?

45. If two engines or motors are running side by side at almost the same speed, you can hear a series of beats. These beats are the result of adding the tonal ordinates of the separate engines. For example, consider $y_1 = \sin 7x$ and $y_2 = \sin 8x$.

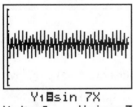

```
        Y₁⊟sin 7X
Xmin=0      Ymin=-3
Xmax=30     Ymax=3
Xscl=1      Yscl=.5
```

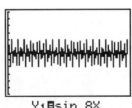

```
        Y₁⊟sin 8X
Xmin=0      Ymin=-3
Xmax=30     Ymax=3
Xscl=1      Yscl=.5
```

The result of $y = y_1 + y_2$ is a series of beats, as shown in Figure 3.57.

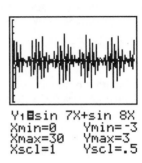

```
Y₁⊟sin 7X+sin 8X
Xmin=0      Ymin=-3
Xmax=30     Ymax=3
Xscl=1      Yscl=.5
```

FIGURE 3.57 Graph of $y = \sin 7x + \sin 8x$ shows "beats"

Graph $y = \sin 7x + \sin 9x$ and show the effect of beats.

46. The current I from two power sources is given by

$$I = 3 \sin 240\pi t + 6 \cos 120\pi t$$

where t is time in seconds. Sketch the graph of the current for $0 \le t \le 1/30$.

47. The current I for a circuit with two power sources is

$$I = 2 \sin 100\pi t + 4 \cos 120\pi t$$

Graph this current for $-0.2 \le t \le 0.2$.

Level 3 Drill

Calculator Window **?** **X**

***Problems* 48–53:** *On page 94, we graphed* $y = \dfrac{\sin x}{x}$, *and on page 340, we will graph curves such as* $y = x \sin x$. *Use your calculator to graph the given equations.*

48. a. $y = \dfrac{1}{\sin x}$ **b.** $y = \sin \dfrac{1}{x}$

49. a. $y = \dfrac{1}{\cos x}$ **b.** $y = \cos \dfrac{1}{x}$

50. $y = x \sin x$ **51.** $y = x \cos x$

52. $y = x^2 \sin x$ **53.** $y = x \sin \dfrac{1}{x}$

Problems 54–63: *Use a calculator to graph the curves defined by the given equations. You will need to pay attention to the domain to see the effect of the beats.*

54. $y = \cos 9x + \sin 11x$

55. $y = \cos 11x + \sin 9x$

56. $y = \cos 2x + \sin 3x$

57. $y = \cos 3x + \sin 2x$

58. $y = \sin 2x + \sin 3x$

59. $y = \cos 2x + \cos 3x$

60. a. $y = \cos x$ **b.** $y = \cos x + \dfrac{\cos 3x}{3}$

61. $y = \cos x + \dfrac{\cos 3x}{3} + \dfrac{\cos 5x}{5}$

62. a. $y = \sin x$ **b.** $y = \sin x + \dfrac{\sin 2x}{2}$

63. $y = \sin x + \dfrac{\sin 2x}{2} + \dfrac{\sin 3x}{3}$

65. IN YOUR OWN WORDS The system

$$\begin{cases} y = x \\ y = \cos x \end{cases}$$

has one solution (Example 5), and the system

$$\begin{cases} y = \frac{1}{5}x \\ y = \cos x \end{cases}$$

has three solutions (Example 6a). Find a solution of

$$\begin{cases} y = mx \\ y = \cos x \end{cases}$$

for m so that the system has the specified number of solutions.
a. Two solutions
b. Four solutions
c. Five solutions
d. Do you think you could answer this question for any number of solutions? Discuss.

Level 3 Theory

64. Find λ in terms of p by solving the system

$$\begin{cases} v = \sqrt{5.09\lambda} \\ v = \dfrac{\lambda}{p} \end{cases}$$

Level 3 Individual Projects

66. Write a paper on Joseph Fourier.

67. Give five real-world examples each of longitudinal and transverse waves.

SECTION 3.4   INVERSE TRIGONOMETRIC FUNCTIONS

PROBLEM OF THE DAY

Suppose you need to design parking spaces along a street. If the city council wants angle parking, at what angle should the parking spaces be laid out?

How would you answer this question?

When you check out the physical situation, you find that city codes require parking spaces to be rectangles 8 ft wide and 18 ft long. You also find that the street width allows only 5 ft for the smaller side of the triangle at the top of the rectangle, as shown in Figure 3.58.

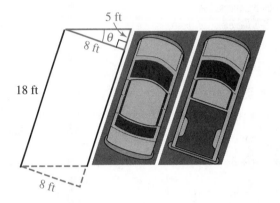

FIGURE 3.58   Parking lot Problem of the Day

We know (from the definition of tangent) that $\tan \theta = \dfrac{y}{x}$, and for this problem (see Figure 3.58), $x = 8$ and $y = 5$ so that

$$\tan \theta = \frac{5}{8}$$

We would like to find the angle θ whose tangent is $\frac{5}{8} = 0.625$. In Section 1.3, we defined this to be

$$\theta = \tan^{-1}\left(\frac{5}{8}\right) \approx 32°$$

The purpose of this section is to answer the general question of finding the value of an angle if we know the value of one of the trigonometric functions and we do not want to assume that the angle is an angle of a right triangle. We begin by considering the idea of an inverse.

The Idea of an Inverse

In mathematics, the ideas of "opposite operations" and "inverse properties" are very important. The basic notion of an opposite operation or an inverse property is to "undo" a previously performed operation. For example, pick a number, and call it x.

Pick number: x *Think: I pick 8.*

Add 5: $x + 5$ *Think: Now I have 8 + 5 = 13.*

The opposite operation returns you to x.

Subtract 5: $x + 5 - 5 = x$ *Think: 13 − 5 = 8, my original number.*

We now want to apply this idea to functions. Pick a number in the domain of a function f; call this number a.

$$a$$ *Think: I'll pick 4 this time.*

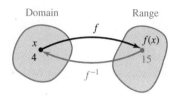

Domain   Range

Now evaluate f for the number you picked; suppose we let f be defined by $f(x) = 2x + 7$.

$$f(a) = 2a + 7 \quad \textit{Think:}\ \ f(4) = 2(4) + 7 = 15$$

The *inverse function* (call it f^{-1}), if it exists, is a function so that $f^{-1}(2a + 7)$ is a, the original number.

Think: I want to find a function such that $f^{-1}(15) = 4$, the original number.

Thus, if $f(a) = b$, then $f^{-1}(b) = a$, the original number. In symbols,

$$f^{-1}[f(a)] = a \qquad \text{for every number } a \text{ in the domain of } f$$

Of course, for f^{-1} to be an inverse function, it must "undo" the effect of f for *each and every member* of the domain. This may be impossible if f is a function such that two x-values give the same y-value. For example, if $g(x) = x^2$, then $g(2) = 4$ and $g(-2) = 4$, so we cannot find a function g^{-1} such that $g^{-1}(4)$ equals *both* 2 and −2 because that would violate the very definition of a function. Thus, it is necessary to limit the given function so that it is *one-to-one*: each y-value results from only *one* x-value. A function is one-to-one if it passes the so-called **horizontal line test.** This test supposes that a horizontal line sweeps the graph from top to bottom, and if it never crosses the function in more than one place, then we say the function *passes the horizontal line test.*

Inverse Functions

The symbol f^{-1} means the inverse of f and does NOT mean $\frac{1}{f}$.

For a given function f, we write $f(a) = b$ to indicate that f maps the number a in its domain into the corresponding number b in the range. If f has an inverse f^{-1}, it is the function that reverses the effect of f in the sense that

$$f^{-1}(b) = a$$

This means that

$$(f^{-1} \circ f)(a) = a$$

Furthermore, for every b in the domain of f^{-1},

$$(f \circ f^{-1})(b) = b$$

INVERSE FUNCTION

Let f be a function with domain D and range R. Then the function f^{-1} with domain R and range D is the **inverse of f** if

$$f^{-1}[f(a)] = a \qquad \text{for all } a \text{ in } D$$

and

$$f[f^{-1}(b)] = b \qquad \text{for all } b \text{ in } R$$

EXAMPLE 1 **Showing That Two Given Functions Are Inverses**

Show that f and g defined by $f(x) = 5x + 4$ and $g(x) = \dfrac{x - 4}{5}$ are inverse functions.

Solution We must show that f and g are inverse functions in two parts:

$$\begin{aligned}(g \circ f)(a) &= g(5a + 4) \\ &= \frac{(5a + 4) - 4}{5} \\ &= \frac{5a}{5} \\ &= a\end{aligned} \qquad \text{and} \qquad \begin{aligned}(f \circ g)(b) &= f\left(\frac{b - 4}{5}\right) \\ &= 5\left(\frac{b - 4}{5}\right) + 4 \\ &= (b - 4) + 4 \\ &= b\end{aligned}$$

Thus, $(g \circ f)(a) = a$ and $(f \circ g)(b) = b$, so f and g are inverse functions. ■

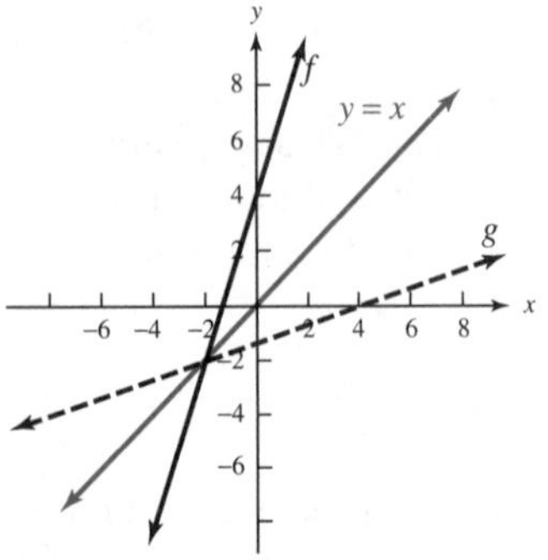

FIGURE 3.59
Graphs of f and g from Example 1

Graph of f^{-1}

The graphs of f and its inverse f^{-1} are closely related. Figure 3.59 shows the graphs of f and g, along with the line $y = x$. The relationship between a graph and its inverse is always such that the graphs are symmetric with respect to the line $y = x$.

To see why this is so, consider a point (a, b) on the graph of f. This point tells us that $f(a) = b$. If f has an inverse, f^{-1}, then $f^{-1}(b) = a$, which gives us

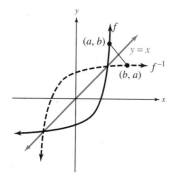

FIGURE 3.60
The graphs of f and f^{-1} are reflections in the line $y = x$.

the point (b, a) on the graph of f^{-1}. (See Figure 3.60.) Thus, for every ordered pair $P(a, b)$ on the graph of f, there is a corresponding ordered pair $Q(b, a)$ on the graph of f^{-1}. Now, two points P and Q are defined to be *symmetric with respect to a line* ℓ if line ℓ is the perpendicular bisector of the segment $\overline{PQ}$. It can be easily shown that the line $y = x$ is the perpendicular bisector of the segment $\overline{PQ}$ with endpoints $P(a, b)$ and $Q(b, a)$. Therefore, the points (a, b) and (b, a) are symmetric with respect to the line $y = x$, and it follows that the graphs of f and f^{-1} are symmetric with respect to the line $y = x$.

> **PROCEDURE FOR OBTAINING THE GRAPH OF f^{-1}**
>
> If f^{-1} exists, its graph may be obtained by reflecting the graph of f in the line $y = x$.

EXAMPLE 2

Evaluating a Function and Its Inverse Function by Looking at Graph

Consider the function f defined by the graph in Figure 3.61.

a. Find $f(5)$; what are the coordinates of point P?
b. Find $f^{-1}(6)$; what are the coordinates of point Q?
c. Draw the graph of f^{-1}. What points on f^{-1} correspond to the points P and Q?

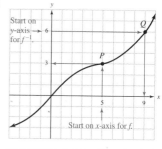

FIGURE 3.61   Graph of f

Solution   Use the graph shown in Figure 3.61.

This is a member of the domain of f; locate this on the x-axis.
↓
a. $f(5) = 3$
                        ↑
This is found by following the arrows in Figure 3.61.

The point P is $(3, 5)$.

This is a member of the domain of f^{-1}; locate on y-axis.
↓
b. $f^{-1}(6) = 9$
                        ↑
This is found by following the arrows in Figure 3.61.

The point Q is $(9, 6)$.

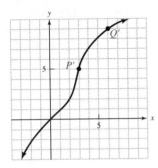

c. Draw the curve that is symmetric to the given curve with respect to the line $y = x$. That is, for each point of f, say $P(5, 3)$, there corresponds a point $P'(3, 5)$ on the graph of f^{-1}. For the point $Q(9, 6)$ on f, the corresponding point on f^{-1} is $Q'(6, 9)$. The graph is shown in the margin. ∎

Inverse Sine Function

Let us consider the sine function first. Figure 3.62 shows the graph of $y = \sin x$. Note the different x-values that lead to the same y-value; thus, the sine function is not a one-to-one function.

It is not necessary to draw the graph of the inverse to evaluate the inverse function, if you know the graph of the original function. Note that we did part b of Example 2 before we drew the graph.

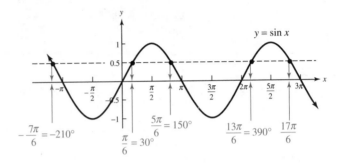

FIGURE 3.62   Graph of $y = \sin x$

We know, however, that the sine function is strictly increasing on the closed interval $\left[-\frac{\pi}{2}, \frac{\pi}{2}\right]$ and hence one-to-one. If we restrict $\sin x$ to this interval, it does have an inverse, as shown in Figure 3.63.

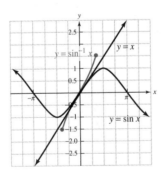

a. The graph of $\sin^{-1} x$ is obtained by reflecting the part of the sine on $\left[-\frac{\pi}{2}, \frac{\pi}{2}\right]$ about $y = x$.

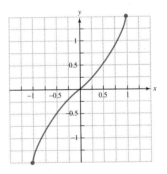

b. The graph of the inverse sine function $y = \sin^{-1}x$

FIGURE 3.63   Inverse sine function

INVERSE SINE FUNCTION

$$y = \sin^{-1} x \quad \text{if and only if} \quad x = \sin y \quad \text{and} \quad -\frac{\pi}{2} \le y \le \frac{\pi}{2}$$

STOP

The function $\sin^{-1} x$ *is NOT the reciprocal of* $\sin x$. *To denote the reciprocal, write* $(\sin x)^{-1}$.

Note that the domain for this inverse function is $-1 \le x \le 1$, the stated restriction on the range, namely, $-\pi/2 \le y \le \pi/2$ tells us the angle y must be in Quadrant I (measured from 0 to $\pi/2$) or in Quadrant IV (measured as a negative angle from $-\pi/2$ to 0).

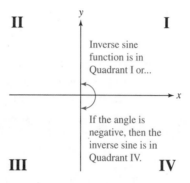

To *evaluate* $\sin^{-1} x$ means to find the angle θ so that $\sin \theta = x$.

EXAMPLE 3

Evaluating Inverse Sine Functions

Evaluate: **a.** $\sin^{-1}\left(\frac{1}{2}\sqrt{3}\right)$ **b.** $\sin^{-1}\left(-\frac{1}{2}\sqrt{3}\right)$

Solution

a. Let $\theta = \sin^{-1}\left(\frac{1}{2}\sqrt{3}\right)$. Remember, the inverse sine function represents the measurement of an angle. Therefore, we denote it by θ, and we find the angle or real number θ with sine equal to $\frac{1}{2}\sqrt{3}$ so that $-\pi/2 \le \theta \le \pi/2$. From the table of exact values (which you have memorized), we know $\sin(\pi/3) = \frac{1}{2}\sqrt{3}$, and since $\pi/3$ is between $-\pi/2$ and $\pi/2$, we have

$$\sin^{-1}\left(\tfrac{1}{2}\sqrt{3}\right) = \tfrac{\pi}{3}$$

b. You may want to work with reference angles when you find inverse trigonometric functions. That is because the table of exact values shows first quadrant angles only. Let

$$\theta' = \sin^{-1}\left(\tfrac{1}{2}\sqrt{3}\right) \quad \text{so } \theta' = \tfrac{\pi}{3} \text{ is the reference angle.}$$

↑
reference angle   ↑
absolute value of the given number

Now place θ in the appropriate quadrant. The sine is negative in both the third and fourth quadrants, but because $-\pi/2 \le \theta \le \pi/2$, we see that $\theta = -\pi/3$.

Inverse Trigonometric Functions

The following table summarizes the relationship between a given trigonometric function, its inverse function, and the notation we use for the inverse.

GIVEN FUNCTION	INVERSE	OTHER NOTATIONS
$y = \cos x$	$x = \cos y$	$y = \cos^{-1} x$; $\quad y = \arccos x$
$y = \sin x$	$x = \sin y$	$y = \sin^{-1} x$; $\quad y = \arcsin x$
$y = \tan x$	$x = \tan y$	$y = \tan^{-1} x$; $\quad y = \arctan x$
$y = \sec x$	$x = \sec y$	$y = \sec^{-1} x$; $\quad y = \text{arcsec } x$
$y = \csc x$	$x = \csc y$	$y = \csc^{-1} x$; $\quad y = \text{arccsc } x$
$y = \cot x$	$x = \cot y$	$y = \cot^{-1} x$; $\quad y = \text{arccot } x$

Given this notation for the inverse trigonometric functions, we now need to determine the domain and range for each inverse trigonometric function. For example, by restricting $\tan x$ to the open interval $\left(-\frac{\pi}{2}, \frac{\pi}{2}\right)$ where it is one-to-one, we can define the inverse tangent function as follows.

INVERSE TANGENT FUNCTION

$$y = \tan^{-1} x \quad \text{if and only if} \quad x = \tan y \quad \text{and} \quad -\frac{\pi}{2} < y < \frac{\pi}{2}$$

The graph of $y = \tan^{-1} x$ is shown in Figure 3.64.

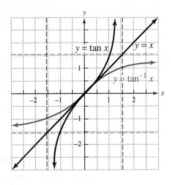

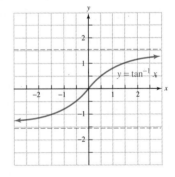

a. The graph of $\tan^{-1} x$ is obtained by reflecting the part of the tangent graph on $\left(-\frac{\pi}{2}, \frac{\pi}{2}\right)$ about the line $y = x$.

b. The graph of $\tan^{-1} x$

FIGURE 3.64 Graph of the inverse tangent

Definitions and graphs of the fundamental inverse trigonometric functions are given in Table 3.5 (pages 145–146).

TABLE 3.5 Definition of Inverse Trigonometric Functions

Inverse function	Domain	Range	Value of x can be			Graph
			Pos	Neg	Zero	
			Quadrant			
$y = \sin^{-1} x$	$-1 \leq x \leq 1$	$\underbrace{-\frac{\pi}{2} \leq y \leq \frac{\pi}{2}}_{\textbf{Quadrants I and IV}}$	I	IV	0	
$y = \cos^{-1} x$	$-1 \leq x \leq 1$	$\underbrace{0 \leq y \leq \pi}_{\textbf{Quadrants I and II}}$	I	II	$\frac{\pi}{2}$	
$y = \tan^{-1} x$	$-\infty < x < +\infty$	$\underbrace{-\frac{\pi}{2} < y < \frac{\pi}{2}}_{\textbf{Quadrants I and IV}}$	I	IV	0	

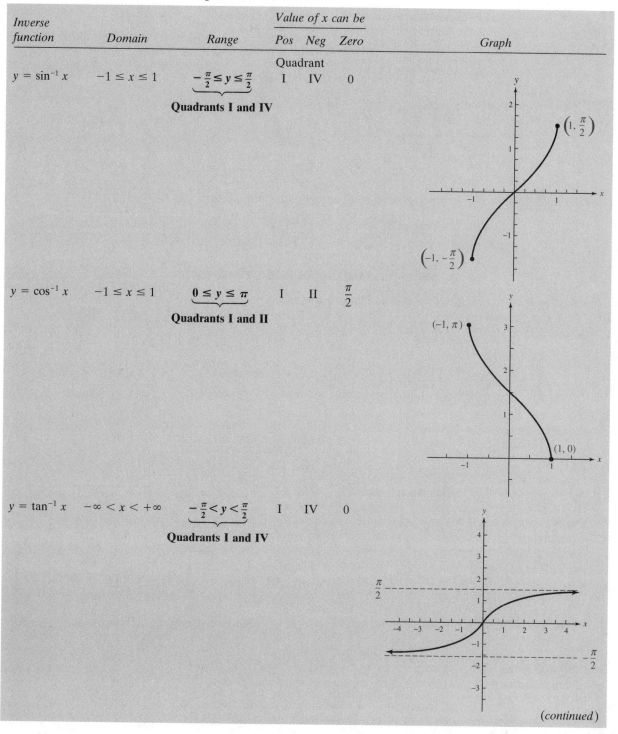

(*continued*)

TABLE 3.5 *Continued*

Inverse function	*Domain**	*Range*	*Value of x can be*			*Graph*
			Pos	Neg	Zero	
			Quadrant			
$y = \sec^{-1} x$	$x \geq 1$ or $x \leq -1$	$\underbrace{0 \leq y \leq \pi, y \neq \frac{\pi}{2}}_{\textbf{Quadrants I and II}}$	I	II	undefined	
$y = \csc^{-1} x$	$x \geq 1$ or $x \leq -1$	$\underbrace{-\frac{\pi}{2} \leq y \leq \frac{\pi}{2}, y \neq 0}_{\textbf{Quadrants I and IV}}$	I	IV	undefined	
$y = \cot^{-1} x$	$-\infty < x < +\infty$	$\underbrace{0 < y < \pi}_{\textbf{Quadrants I and II}}$	I	II	$\frac{\pi}{2}$	

* There are no standard domain restrictions for inverse secant and inverse cosecant. Some calculus textbooks use different restrictions from the ones we have chosen here. However, the ones we have chosen are the most convenient to use with standard calculators and follow the definition given in most trigonometry texts.

EXAMPLE 4   **Evaluating Inverse Trigonometric Functions**

Evaluate the given functions.

a. $\sin^{-1}\left(\frac{-\sqrt{2}}{2}\right)$ **b.** $\sin^{-1} 0.21$

c. $\cos^{-1} 0$ **d.** $\tan^{-1}\left(\frac{1}{\sqrt{3}}\right)$

Solution   **a.** $\sin^{-1}\left(\frac{-\sqrt{2}}{2}\right) = -\frac{\pi}{4}$;   *Think:* $x = \frac{-\sqrt{2}}{2}$ is negative, so y is in Quadrant IV; the reference angle is the angle whose sine is $\frac{\sqrt{2}}{2}$; it is $\frac{\pi}{4}$, so in Quadrant IV the angle is $-\frac{\pi}{4}$.

b. $\sin^{-1} 0.21 \approx 0.2115750$ *By calculator; be sure to use radian mode and inverse sine (not reciprocal).*

c. $\cos^{-1} 0 = \frac{\pi}{2}$ *Memorized exact value*

d. $\tan^{-1}\left(\frac{1}{\sqrt{3}}\right) = \frac{\pi}{6}$ *Think:* $x = \frac{1}{\sqrt{3}}$ is positive, so y is in Quadrant I; the reference angle is the same as the value of the inverse tangent in Quadrant I. ∎

Evaluating Inverses of the Reciprocal Functions

To evaluate arcsecant, arccosecant, or arccotangent functions when their exact values are not known, we need to use the reciprocal identities. First find the reciprocal, then find the inverse of the reciprocal. That is, find the inverse of the reciprocal function. This procedure is summarized by the *inverse identities* on page 148.*

---

* In Chapter 2, we presented identities numbered 1–8; we continue with that numbering in this section.

INVERSE IDENTITIES

$$
9. \ \cot^{-1} x = \begin{cases} \tan^{-1} \dfrac{1}{x} & \text{if } x \text{ is positive} \\[2mm] \tan^{-1} \dfrac{1}{x} + \pi & \text{if } x \text{ is negative} \\[2mm] \dfrac{\pi}{2} & \text{if } x = 0 \end{cases}
$$

$$
10. \ \sec^{-1} x = \cos^{-1} \frac{1}{x} \quad \text{if } x \geq 1 \text{ or } x \leq -1
$$

$$
11. \ \csc^{-1} x = \sin^{-1} \frac{1}{x} \quad \text{if } x \geq 1 \text{ or } x \leq -1
$$

We can derive the first part of identity 9. Let $\theta = \cot^{-1} x$ where x is positive. Then,

$$\cot \theta = x \qquad 0 < \theta < \tfrac{\pi}{2} \qquad \textit{Definition of } \cot^{-1} x$$

$$\frac{1}{\tan \theta} = x \qquad 0 < \theta < \tfrac{\pi}{2} \qquad \textit{Reciprocal identity}$$

$$\tan \theta = \frac{1}{x} \qquad 0 < \theta < \tfrac{\pi}{2} \qquad \textit{Solve for } \tan \theta.$$

$$\theta = \tan^{-1} \frac{1}{x} \qquad 0 < \theta < \tfrac{\pi}{2} \qquad \textit{Definition of } \tan^{-1} \theta$$

The derivation of the second part of identity 9 is left as an exercise. It is needed because the range of the inverse tangent function does not coincide with the range of the inverse cotangent function. The third part of the identity is obvious since $\cot \frac{\pi}{2} = 0$.

The derivations of identities 10 and 11 are also left as exercises.

EXAMPLE 5

Evaluating Inverse Reciprocal Functions

Evaluate the given inverse functions using the inverse identities and a calculator.

a. $\sec^{-1}(-3)$ **b.** $\csc^{-1} 7.5$ **c.** $\cot^{-1} 2.4747$ **d.** $\operatorname{arccot}(-4.852)$

Solution

a. Use identity 10: $\sec^{-1}(-3) \approx 1.910633236$
b. Use identity 11: $\csc^{-1} 7.5 \approx 0.1337315894$
c. Use identity 9 with x positive: $\cot^{-1} 2.4747 \approx 0.3840267299$
d. Use identity 9 with x negative: $\operatorname{arccot}(-4.852) \approx 2.938338095$ ∎

Functions of Inverse Functions

A word of caution is in order regarding the inverse trigonometric functions, especially when you use a calculator. Recall that for a function f with domain D and range R,

$$f^{-1}[f(a)] = a \quad \text{for all } a \text{ in } D$$
$$f[f^{-1}(b)] = b \quad \text{for all } b \text{ in } R$$

This means, for example, that for the function $\cos x$ with restricted domain $0 \le x \le \pi$ and range $-1 \le \cos x \le 1$, we have

$$\cos^{-1}(\cos x) = x \quad \text{for every } x \text{ in the interval } [0, \pi]$$
$$\cos(\cos^{-1} x) = x \quad \text{for every } x \text{ in the interval } [-1, 1]$$

Similarly,

$$\sin^{-1}(\sin x) = x \quad \text{for every } x \text{ in the interval } \left[-\tfrac{\pi}{2}, \tfrac{\pi}{2}\right]$$
$$\sin(\sin^{-1} x) = x \quad \text{for every } x \text{ in the interval } [-1, 1]$$

and

$$\tan^{-1}(\tan x) = x \quad \text{for every } x \text{ in the interval } \left(-\tfrac{\pi}{2}, \tfrac{\pi}{2}\right)$$
$$\tan(\tan^{-1} x) = x \quad \text{for all real } x$$

EXAMPLE 6

Evaluate Functions Involving Inverse Trigonometric Functions

Evaluate (correct to two decimal places):

a. $\cos^{-1}(\cos 2.2)$ **b.** $\cos^{-1}(\cos 4)$ **c.** $\cos(\cos^{-1} 2.2)$ **d.** $\cos(\cos^{-1} 0.22)$

e. $\sin^{-1}(\sin 2.2)$ **f.** $\sin^{-1}[\sin(-1)]$ **g.** $\sin(\sin^{-1} 0.46)$ **h.** $\sin(\sin^{-1} 2.46)$

i. $\tan(\tan^{-1} \sqrt{3})$ **j.** $\tan^{-1}(\tan \sqrt{3})$

Solution

a. $\cos^{-1}(\cos 2.2) = 2.2$ since 2.2 is on the interval $[0, \pi]$.

b. $\cos^{-1}(\cos 4) \approx 2.28$ *Note:* It is not 4 because 4 is not on the interval
  by calculator $[0, \pi]$.

c. $\cos(\cos^{-1} 2.2)$ is not defined.
           Note: 2.2 is not in the interval $[-1, 1]$.

d. $\cos(\cos^{-1} 0.22) = 0.22$

e. $\sin^{-1}(\sin 2.2) \approx 0.94$
  by calculator   *Note:* 2.2 is not on the interval $\left[-\tfrac{\pi}{2}, \tfrac{\pi}{2}\right]$.

f. $\sin^{-1}[\sin(-1)] = -1$ since -1 is on the interval $\left[-\tfrac{\pi}{2}, \tfrac{\pi}{2}\right]$.

g. $\sin(\sin^{-1} 0.46) = 0.46$ since 0.46 is on the interval $[-1, 1]$.

h. $\sin(\sin^{-1} 2.46)$ is not defined.
           Note: 2.46 is not on the interval $[-1, 1]$.

i. $\tan(\tan^{-1} \sqrt{3}) = \sqrt{3} \approx 1.73$, since $\tan(\tan^{-1} x) = x$ for all real numbers.

j. $\tan^{-1}(\tan \sqrt{3}) \approx -1.41$
   by calculator   *Note:* $\sqrt{3}$ is not on the interval $\left(-\frac{\pi}{2}, \frac{\pi}{2}\right)$. ■

EXAMPLE 7   **Using a Fundamental Identity to Evaluate a Function**

Evaluate $\cot[\csc^{-1}(-3)]$ without tables or a calculator.

Solution   Let $\theta = \csc^{-1}(-3)$.   *Negative value, so θ is in Quadrant* IV

Then $\csc \theta = -3$. We want to find $\cot[\csc^{-1}(-3)] = \cot \theta$.

$$\cot \theta = \pm\sqrt{\csc^2\theta - 1}$$ *Pythagorean identity*

$$= -\sqrt{(-3)^2 - 1}$$ *$\csc \theta = -3$; choose the negative value for the radical because θ is in Quadrant* IV *and cotangent is negative in Quadrant* IV.

$$= -\sqrt{8}$$

$$= -2\sqrt{2}$$ ■

EXAMPLE 8   **Deriving a Formula Involving an Inverse Function (in terms of x)**

Find $\sin(\cos^{-1} x)$.

Solution $$\sin(\cos^{-1} x) = \sin \theta$$ *Let $\theta = \cos^{-1} x$ so that $\cos \theta = x$.*

$$= \pm\sqrt{1 - \cos^2\theta}$$ *Pythagorean identity*

$$= \sqrt{1 - x^2}$$ *Substitute $\cos \theta = x$ and choose the positive value for the radical because θ is in Quadrant* I *or* II.   ■

PROBLEM SET 3.4

Essential Concepts

1. What is the range for the inverse cosine?

2. In which quadrants is the inverse cosine defined?

3. In which quadrants is the inverse sine defined?

4. What is the range for the inverse sine?

5. What is the range for the inverse tangent?

6. In which quadrants is the inverse tangent defined?

Problems 7–12: *State the keys you press on your calculator to evaluate each function.*

7. $y = \cos^{-1} x$ **8.** $y = \sin^{-1} x$ **9.** $y = \tan^{-1} x$

10. $y = \cot^{-1} x$ **11.** $y = \sec^{-1} x$ **12.** $y = \csc^{-1} x$

13. Let $c(x) = \cos x$ for $0 \leq x \leq \pi$. Using the graph shown here, find the functional values for c and c^{-1} as requested.

a. $c\left(\frac{\pi}{4}\right)$

b. $c\left(\frac{\pi}{2}\right)$

c. $c(\pi)$

d. $c\left(\frac{3\pi}{4}\right)$

e. $c^{-1}(1)$

f. $c^{-1}(0)$

g. $c^{-1}(0.5)$

h. $c^{-1}(-1)$

i. Using this graph, draw the graph of $c^{-1}(x)$ by plotting points.

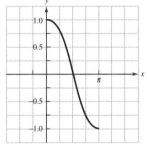

14. Let $s(x) = \sin x$ for $-\frac{\pi}{2} \le x \le \frac{\pi}{2}$. Using the graph shown here, find the functional values for s and s^{-1} as requested.

a. $s\left(\frac{\pi}{4}\right)$

b. $s\left(\frac{\pi}{2}\right)$

c. $s\left(-\frac{\pi}{4}\right)$

d. $s\left(-\frac{\pi}{2}\right)$

e. $s^{-1}(1)$

f. $s^{-1}(0)$

g. $s^{-1}(0.5)$

h. $s^{-1}(-1)$

i. Using this graph, draw the graph of $s^{-1}(x)$ by plotting points.

15. Let $t(x) = \tan x$ for $-\frac{\pi}{2} < x < \frac{\pi}{2}$. Using the graph shown here, find the functional values for t and t^{-1} as requested.

a. $t\left(\frac{\pi}{4}\right)$

b. $t\left(\frac{\pi}{8}\right)$

c. $t(0)$

d. $t\left(-\frac{\pi}{4}\right)$

e. $t^{-1}(1)$

f. $t^{-1}(2)$

g. $t^{-1}(0.5)$

h. $t^{-1}(-1)$

i. Using this graph, draw the graph of $t^{-1}(x)$ by plotting points.

Level 1 Drill

Problems **16–21:** *Obtain the given angle (in radians) from memory.*

16. a. $\tan^{-1}\left(\frac{\sqrt{3}}{3}\right)$ **b.** $\cos^{-1} 1$ **c.** $\sin^{-1}\frac{1}{2}$ **d.** $\sin^{-1} 1$

17. a. $\cos^{-1}\left(\frac{\sqrt{2}}{2}\right)$ **b.** $\cot^{-1} 1$ **c.** $\sin^{-1}(-1)$ **d.** $\sin^{-1} 0$

18. a. $\cot^{-1} \sqrt{3}$ **b.** $\cos^{-1}\left(\frac{\sqrt{3}}{2}\right)$

 c. $\tan^{-1} 1$ **d.** $\sin^{-1}\left(\frac{\sqrt{2}}{2}\right)$

19. a. $\tan^{-1} \sqrt{3}$ **b.** $\csc^{-1}(-1)$

 c. $\sin^{-1}\left(-\frac{\sqrt{3}}{2}\right)$ **d.** $\cot^{-1}(-1)$

20. a. $\arccos(-1)$ **b.** $\text{arccot}(-\sqrt{3})$

 c. $\arctan(-1)$ **d.** $\arcsin\left(-\frac{1}{2}\sqrt{3}\right)$

21. a. $\arctan\left(-\frac{1}{3}\sqrt{3}\right)$ **b.** $\arctan 0$

 c. $\arccos\frac{1}{2}$ **d.** $\arcsin\left(\frac{1}{2}\sqrt{3}\right)$

Problems **22–24:** *Use a calculator to find the values (in radians correct to the nearest hundredth).*

22. a. $\arcsin 0.20846$ **b.** $\cos^{-1} 0.83646$

 c. $\arctan 1.1156$ **d.** $\cot^{-1}(-0.08097)$

23. a. $\arctan(-3.7712)$ **b.** $\cos^{-1}(-0.94604)$

 c. $\arctan 2$ **d.** $\tan^{-1} 1.489$

24. a. $\text{arccot } 3.451$ **b.** $\sec^{-1} 4.315$

 c. $\text{arccsc } 2.461$ **d.** $\sec^{-1} 2.894$

Problems **25–27:** *Use a calculator to find the values (to the nearest degree).*

25. a. $\sin^{-1} 0.3584$ **b.** $\arccos 0.9455$

 c. $\cos^{-1} 0.3584$ **d.** $\arcsin(-0.4696)$

26. a. $\tan^{-1} 1.036$ **b.** $\text{arccot } 0.0875$

 c. $\csc^{-1} 2.816$ **d.** $\text{arccot}(-1)$

27. a. $\cot^{-1}(-2)$ **b.** $\text{arccsc } 3.945$

 c. $\sec^{-1}(-6)$ **d.** $\arctan(-3)$

WHAT IS WRONG, *if anything, with each of the statements in Problems 28–40? Explain your reasoning.*

28. In $y = \cos x$, the angle is y.

29. In $y = \cos^{-1} x$, the angle is y.

30. $\tan^{-1}(-2.5)$ is in Quadrant II.

31. $\cot^{-1}(-2.5)$ is in Quadrant IV.

32. The domain in $y = \sin x$ is $[-1, 1]$.

33. The domain in $y = \sin^{-1} x$ is $[-1, 1]$.

34. $\tan^{-1} x = \dfrac{1}{\tan x}$ **35.** $\sec^{-1} x = \cos^{-1}\left(\dfrac{1}{x}\right)$

36. $\sin^{-1} x = \dfrac{1}{\sin x}$ **37.** $\csc^{-1} x = \cos^{-1}\left(\dfrac{1}{x}\right)$

38. $\sin(\sin^{-1} x) = x$ because sine and inverse sine are inverse functions.

39. $\cos^{-1}(\sec x) = x$ because cosine and secant are reciprocal functions.

40. $\cos^{-1}(\cos x) = \cos(\cos^{-1} x)$ for all values of x in the domain.

Problems 41–46: *Evaluate the expressions without using a calculator or tables.*

41. a. $\cot(\text{arccot } 1)$ **b.** $\arccos\left(\cos\frac{\pi}{6}\right)$

42. a. $\sin\left(\sin^{-1}\frac{1}{3}\right)$ **b.** $\tan^{-1}\left(\tan\frac{\pi}{15}\right)$

43. a. $\cos\left(\cos^{-1}\frac{2}{3}\right)$ **b.** $\sin^{-1}\left(\sin\frac{2\pi}{15}\right)$

44. a. $\cot^{-1}(\cot 35°)$ **b.** $\tan(\tan^{-1} 0.4163)$

45. a. $\cos(\tan^{-1}\sqrt{3})$ **b.** $\sin\left(\cos^{-1}\frac{2}{3}\right)$

46. a. $\cos\left[\sin^{-1}\left(-\frac{1}{3}\right)\right]$ **b.** $\sin\left[\cos^{-1}\left(-\frac{1}{3}\right)\right]$

Level 2 Drill

Problems 47–52: *Find a formula for each expression.*

47. $\cos(\sin^{-1} x)$ **48.** $\sin(\csc^{-1} x)$

49. $\cos(\sec^{-1} x)$ **50.** $\sec(\tan^{-1} x)$

51. $\tan(\sec^{-1} x)$ **52.** $\csc(\cot^{-1} x)$

Level 2 Applications

53. In the Problem of the Day, we were asked for the parking angle. To answer the question, we found an approximate value using a table and a graph. Find the angle correct to the nearest degree. Find the exact value of the angle.

54. Find the parking angle in the Problem of the Day if the city ordinances require 10 ft for the smaller side of the triangle at the top of the rectangle shown in Figure 3.58.

Problems 55–56: *Suppose a belt connects two different sizes of pulleys, as shown in Figure 3.65.*

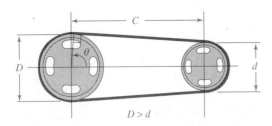

FIGURE 3.65 Two pulleys connected with a belt

The angle θ (measured in radians) is necessary to determine the length of the belt (see Problems 57 and 58). It can be shown that

$$\theta = \cos^{-1}\left(\frac{D - d}{2C}\right)$$

where D is the diameter of the larger pulley, d is the diameter of the smaller pulley, and C is the distance between the pulleys. Find the angle θ to the nearest degree.

55. $D = 10$ in., $d = 8$ in., and $C = 14$ in.

56. $D = 12$ in., $d = 4$ in., and $C = 2$ ft

57. In Problem 55, the length of the belt is found by the formula

$$L = 2C \sin \theta + \pi D + (D - d)\theta$$

Find the length of the belt in Problem 55.

58. In Problem 56, the length of the belt is found by the formula

$$L = 2C \sin \theta + \pi D + (D - d)\theta$$

Find the length of the belt in Problem 56.

Level 3 Drill

Problems 59–64: *Graph each curve.*

59. $y + 2 = \tan^{-1} x$ **60.** $y - 1 = \arcsin x$

61. $y = 2 \arccos x$ **62.** $y = 3 \sin^{-1} x$

63. $y = \arcsin(x - 2)$ **64.** $y = \sin^{-1}(x + 1)$

65. Prove identity 10:

$$\sec^{-1} x = \cos^{-1}\frac{1}{x} \quad \text{if } x \geq 1 \text{ or } x \leq -1$$

66. Prove identity 11:

$$\csc^{-1} x = \sin^{-1}\frac{1}{x} \quad \text{if } x \geq 1 \text{ or } x \leq -1$$

67. Prove identity 9 for $x < 0$:

$$\cot^{-1} x = \tan^{-1}\frac{1}{x} + \pi$$

3.5 CHAPTER THREE SUMMARY

Objectives

Section 3.1	Graphs of the Standard Trigonometric Functions
Objective 3.1	Graph the trigonometric functions, or variations, by plotting points.
Objective 3.2	Graph one period of each of the standard trigonometric functions. Know the period for each function.
Objective 3.3	Be able to draw a frame for the standard cosine, sine, and tangent functions.

Section 3.2	General Cosine, Sine, and Tangent Curves
Objective 3.4	Graph the general cosine and sine functions. Be able to state (h, k), amplitude (a), b, and the period for each function.
Objective 3.5	Graph the general tangent function. Be able to state (h, k), a, b, and the period.
Objective 3.6	Graph a trigonometric curve from given data points.

Section 3.3	Trigonometric Graphs
Objective 3.7	Graph trigonometric functions by adding ordinates.
Objective 3.8	Find an equation for water waves.
Objective 3.9	Solve a trigonometric system of equations by graphing.

Section 3.4	Inverse Trigonometric Functions
Objective 3.10	Evaluate inverse trigonometric functions.
Objective 3.11	Draw a quick sketch of each inverse trigonometric function.
Objective 3.12	Given the graph of a one-to-one function, estimate values for both the function and the inverse function by looking at the graph.
Objective 3.13	Know the domain and range for each inverse trigonometric function.
Objective 3.14	Evaluate functions involving the inverse trigonometric functions.

Terms

Adding ordinates [3.3]
Amplitude [3.1; 3.2]
Arccosecant [3.4]
Arccosine [3.4]
Arccotangent [3.4]
Arcsecant [3.4]
Arcsine [3.4]
Arctangent [3.4]

Asymptote [3.1]
Cosecant curve [3.1]
Cosine curve [3.1]
Cotangent curve [3.1]
Domain [3.1; 3.3]
Frame [3.1]
Frequency [3.1; 3.3]
Horizontal line test [3.4]

Inconsistent system [3.3]
Inverse of f [3.4]
Longitudinal wave [3.3]
Period [3.1]
Phase shift [3.2]
Range [3.1]
Secant curve [3.1]
Simultaneous solution [3.3]

Sine curve [3.1]
Sinusoidal [3.2]
System of equations [3.3]
Tangent curve [3.1]
Translation [3.2]
Transverse wave [3.3]
Vertical asymptote [3.1]
Wave equation [3.3]

Sample Test

All of the answers for this sample test are given in the back of the book.

1. Match each graph with a standard-form equation.

a.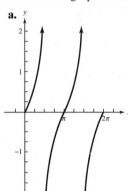

b.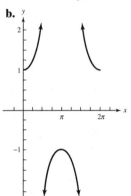

c.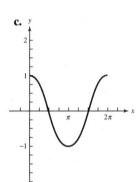

d.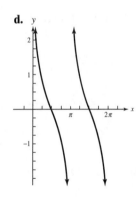

e.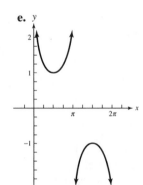

f.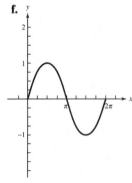

2. Plot the given point, and draw a frame using the given point as a starting point. Draw one period of the requested curve.

a. $\left(\frac{\pi}{6}, -\frac{3}{2}\right)$; sine curve   **b.** $\left(-\frac{3\pi}{2}, 1\right)$; cosine curve

c. $\left(-\frac{\pi}{3}, 2\right)$; tangent curve   **d.** $(-1, 1)$; sine curve

3. Evaluate the functions using exact values where possible.

a. $\arcsin\left(-\frac{1}{2}\sqrt{2}\right)$ **b.** $\arccos\left(-\frac{1}{2}\right)$

c. $\arccos\left(-\frac{1}{2}\sqrt{2}\right)$ **d.** $\mathrm{arccot}\left(-\frac{1}{3}\sqrt{3}\right)$

e. $\arcsin 0.75$ **f.** $\cos^{-1} 0.25$

g. $\tan^{-1} 2.050$ **h.** $\arctan(-3.732)$

4. Evaluate the given functions.

a. $\sin^{-1}(\sin 0.4)$ **b.** $\sin(\sin^{-1} 0.4)$

c. $\cos^{-1}(\cos 5.00)$ **d.** $\cos(\cos^{-1} 5.00)$

e. $\sin\left(\sin^{-1}\frac{3}{4}\right)$ **f.** $\cos\left(\cos^{-1}\frac{2}{7}\right)$

g. $\tan(\tan^{-1} 0.29)$ **h.** $\tan^{-1}(\tan 2.5)$

5. Fill in the blanks.

Standard forms: $y - k = a \cos b(x - h)$
$y - k = a \sin b(x - h)$
$y - k = a \tan b(x - h)$

a. The origin is translated to _____.

b. The height of the frame is _____.

c. The length of the sine and cosine frame is _____.

d. The length of the tangent frame is _____.

6. Graph $y = \sqrt{3} \cos \frac{1}{2}x$.

7. Graph $y + 2 = 2 \sin(3x + \pi)$.

8. Graph $y + 2 = \frac{1}{2} \tan\left(x - \frac{\pi}{4}\right)$.

9. Graph $y = \sin 2x + 3 \sin x$ by adding ordinates.

10. The sunrise times (A.M.) on the first day of each month in New Orleans, LA, are as follows:

Jan	Feb	Mar	Apr	May	Jun	Jul	Aug	Sep	Oct	Nov	Dec
6:56	6:52	6:26	5:51	5:18	5:00	5:02	5:17	5:35	5:51	6:12	6:38

Let x be the month (that is, Jan = 1, Feb = 2, . . .), and let y be the time of sunrise (in hours; for example, $6{:}56 = 6\frac{56}{60} \approx 6.93$ rounded to two places).

a. Plot the time of sunrise for this data set, and connect these points with a smooth curve.

b. Suppose we model the time of sunrise by the equation

$$y = a \cos b(x - h) + k$$

where

$$a = \frac{\text{LATEST TIME} - \text{EARLIEST TIME}}{2}$$

$$b = \frac{2\pi}{\text{PERIOD}} \text{ where the period is 12}$$

$$(h, k) = (0.5, 5.9)$$

Graph this equation.

c. Compare and contrast the graphs resulting from parts **a** and **b**.

Discussion Problems

1. **IN YOUR OWN WORDS** Explain the framing procedure for graphing.

2. Find as many examples of trigonometric curves as you can from magazines, newspapers, or books (other than mathematics books).

3. Write a paper on the impact of the calculator on trigonometry.

4. Write a 500-word paper on the relationship between music and sinusoidal curves.

Miscellaneous Problems—Chapters 1–3

1. **IN YOUR OWN WORDS** Draw the graphs of $y = \sin x$; $y = \frac{1}{2} \sin x$; and $y = 2 \sin x$ on the same coordinate axes. Discuss.

2. **IN YOUR OWN WORDS** Draw the graphs of $y = \sin x$; $y = \sin 2x$; and $y = \sin \frac{1}{2}x$ on the same coordinate axes. Discuss.

3. Fill in the blanks.
   a. As x increases from 0 to $\pi/2$, the sine function increases from _____ to _____.
   b. As x increases from $\pi/2$ to π, the sine function _____.
   c. As x increases from π to $3\pi/2$, the sine function _____.
   d. As x increases from $3\pi/2$ to 2π, the sine function _____.

4. Complete Problem 3 for the cosine function.

5. Evaluate using exact values:
   a. $\sin\left(\tan^{-1} \frac{2}{3}\right)$ b. $\cos(\sin^{-1} 0.4)$

6. Give the coordinates for points A–E on the curve $y = \cos x$.

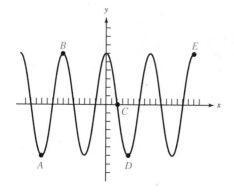

7. Give the coordinates for points A–E on the curve $y = \sin x$.

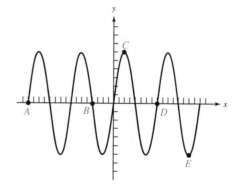

8. Give the coordinates for points A–E on the curve $y = \tan x$.

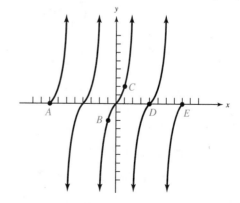

9. If β is an angle in standard position, what are the coordinates of the point of intersection of the terminal side of β and a circle with radius 5?

***Problems* 10–13:** *Determine the equation of a standard cosine or standard sine curve.*

10.

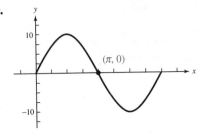

11.

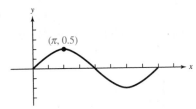

12.

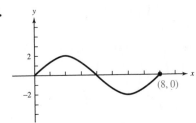

13.

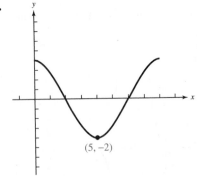

***Problems* 14–29:** *Graph one period of curves.*

14. $y = 3 \sin \frac{\pi x}{4}$

15. $y - 2 = \sin\left(x - \frac{\pi}{6}\right)$

16. $y = \cos\left(x + \frac{\pi}{4}\right)$

17. $y = \frac{1}{2} \tan \frac{1}{2} x$

18. $y = \tan\left(x - \frac{2\pi}{3}\right) - 2$

19. $y = 2 \cos\left(2x + \frac{\pi}{3}\right)$

20. $y = 2 \cos \frac{2}{3} x$

21. $y = \frac{3}{2} \tan \frac{3}{2} x$

22. $y = \sin 5x + 1$

23. $y = \sin(5x + 1)$

24. $y + 1 = \cos\left(x - \frac{\pi}{3}\right)$

25. $y + 3 = 3 \tan \frac{1}{2}\left(x - \frac{\pi}{3}\right)$

26. $y = \cot \frac{1}{2} x$

27. $y = \sec\left(x + \frac{\pi}{4}\right)$

28. $y = \csc 3x$

29. $y = \cot\left(x - \frac{\pi}{4}\right)$

***Problems* 30–51:** *Sketch each graph.*

30. a. $y = 5 \cos x - 3$ **b.** $y = 5 \cos(x - 3)$

31. a. $y = 5 \sin x - 3$ **b.** $y = 5 \sin(x - 3)$

32. a. $y = 5 \tan x - 3$ **b.** $y = 5 \tan(x - 3)$

33. $y = \tan(5x - 3)$

34. $y = \cos(5x - 3)$

35. $y = \sin(5x - 3)$

36. $y - 2 = \cos(3x - 1)$

37. $y + 1 = \sin(2x + 1)$

38. $y + \frac{1}{2} = \tan(4x - 1)$

39. $y = \cot x$

40. $y = \sec x$

41. $y = \sin^{-1} x$

42. $y = \cos^{-1} x$

43. $y = \csc x$

44. $y = 2x \sin x$

45. $y = \frac{3x}{4} + \cos x$

46. $y = 20 \sin(328\pi x)$

47. $y = \frac{\tan x}{5x}$

48. $y = \frac{x}{\cos x}$

49. $f(x) = \begin{cases} \sin x & \text{if } x \leq 0 \\ \tan x & \text{if } 0 < x \leq 1 \end{cases}$

50. $x = 4 \cos \theta,\ y = 4 \sin \theta,\ 0° \leq \theta < 360°$

51. $x = 2 - \sin t,\ y = 2 - \cos t,\ 0 \leq t < 2\pi$

***Problems* 52–55:** *Find the x-values $(0 \leq x < 2\pi)$ of the points of simultaneous solution for the given systems.*

52. $\begin{cases} y = 0.2 \\ y = \cos x \end{cases}$

53. $\begin{cases} y = 0.6 \\ y = \sin x \end{cases}$

54. $\begin{cases} y = x \\ y = 2 \cos x \end{cases}$

55. $\begin{cases} y = 2x \\ y = 0.5 \sin x \end{cases}$

Applications

56. Suppose a particle is moving in simple harmonic motion according to the equation

$$y = \sqrt{3} \sin\left(10\pi t - \frac{\pi}{2}\right)$$

What are the period and frequency, and what is the distance (use two decimal places) from the particle to the origin when $t = 0$?

57. Graphically, show that $y = \tan x$ and

$$y = \frac{\tan\left(x + \frac{\pi}{6}\right) - \tan\frac{\pi}{6}}{1 + \tan\left(x + \frac{\pi}{6}\right)\tan\frac{\pi}{6}}$$

are identical.

58. A 150-ft Ferris wheel (see Figure 3.66) completes one revolution every 4 minutes.

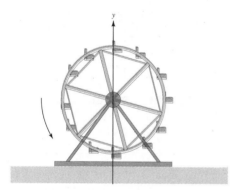

FIGURE 3.66   Ferris wheel

The table shows the height of a person riding on the Ferris wheel at a certain time.

Time, t	Height, h	Time, t	Height, h
0 min	6 ft	2.5 min	134 ft
0.5 min	28 ft	3.0 min	81 ft
1.0 min	81 ft	3.5 min	28 ft
1.5 min	134 ft	4.0 min	6 ft
2.0 min	156 ft		

Plot these points and write a possible equation to model this information.

59. One of the most famous types of waves is a *tsunami*, which is a seismic sea wave. These waves occur in shallow water. They have very long wave lengths (50–150 miles) and travel very fast (50–400 mi/h). (See Figure 3.67.) At sea, tsunami waves often go unnoticed because they have an amplitude of only 1–3 ft. As they approach shore and shallow water, however, they undergo a devastating change in amplitude (over 100 ft) that can do extensive damage to low-lying coastal areas. For example, the 1883 Krakatoa eruption caused a tsunami that killed 36,000 people. It is reported that this killer tsunami had a wavelength of 120 mi, traveled at 300 mi/h, and had a period of 30 min.

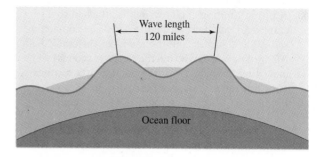

FIGURE 3.67   Tsunami wave

Use the wave equation $y = a\sin\frac{2\pi}{\lambda}(x - vt)$ to graph a tsunami with amplitude 2 ft at time $t = 0$.

60. The most westerly city in the continental United States is Ferndale, California, which is located at 124°W longitude, 41°N latitude. How far (to the nearest 100 miles) is Ferndale from the equator if you assume that the earth's radius is approximately 3,950 miles?

61. The angle between the magnetic and geographic north poles is measured by an angle that has its vertex at the center of the earth. Scientists estimate that the magnetic north pole is drifting in a westerly direction through an angle of approximately 0.0017 rad/century. How long will it take for the north pole to drift a total of 1°?

62. A lighthouse has a light that completes one revolution every 5 sec. The lighthouse is 50 ft from a point P on the shore, as shown in Figure 3.68. Let y be the distance of the light beam from point P at time x. Draw a graph (for the first 10 sec) that shows distance d at time x.

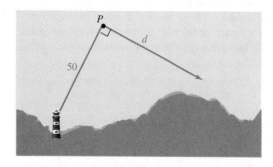

FIGURE 3.68   Light beam from a lighthouse

63. A piston rotating at 60 rev/min is attached to a crankshaft, as shown in Figure 3.69. The connecting rod AB is 6 inches, and the radius of the crankshaft is 2 in. Draw a graph (for the first 10 seconds) indicating the height y of the point B at x seconds after it reaches the uppermost position. [See Problem 63, Section 2.4 (page 82) for an equation.]

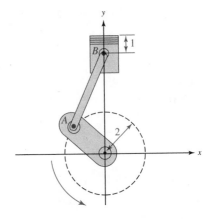

FIGURE 3.69   Crankshaft

Group Research Projects

1. Plot the sunrise and sunset for your town's latitude. On another graph, plot the length of daylight for your own town. Can you find equations for the plotted data?

2. In calculus, it is shown that if a particle is moving in simple harmonic motion according to the equation

$$y = a \cos(\omega \tau + h)$$

then the velocity of the particle at time t is given by

$$y = a\omega \sin(\omega \tau + h)$$

Graph this velocity curve, where the frequency is 0.25 cycle/unit of time, $h = \pi/6$, and $a = 120/\pi$. Find a maximum value and a minimum value for this velocity.

3. In the 1920s, Soviet economist Nikolai Kondratieff identified what he believed to be a long-term sinusoidal cycle governing economic activity. Economists today talk about cycles of bear markets and bull markets. Do some research on economic cycles.

4. A formula for the voltage V of a charged capacitor connected to a coil is a function of time, t. The voltage (in volts) gradually diminishes to 0 over time according to the formula

$$V(t) = e^{-1.5t} \cos \pi \tau$$

Graph V for $0 \le t \le 5$, and write a paper on the relationship between voltage and trigonometry.

5. A touch-tone phone's "dial" is arranged in four rows and three columns, with each row and each column set at a different frequency, as shown in Figure 3.70.

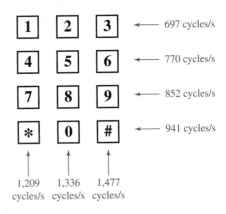

FIGURE 3.70   Touch-tone phone

When you press a particular button, the note sounded is the sum of two frequencies determined by the row and column. For example, if you press 1, then the sound emitted is

$$y = \sin 2\pi(697)x + \sin 2\pi(1,209)x$$

Graph this sound, and write a paper on touch-tone phone sounds.

CUMULATIVE REVIEW CHAPTERS 1-3

Essential Concepts

1. **IN YOUR OWN WORDS** Distinguish among equal angles, coterminal angle, and reference angles.

2. **IN YOUR OWN WORDS** Discuss the concept of arc length. What are the arc length formulas?

3. From memory, give the radian measure for each of the angles whose degree measure is stated. Also sketch the angle.
   a. $30°$ **b.** $45°$ **c.** $60°$ **d.** $90°$
   e. $180°$ **f.** $270°$ **g.** $360°$ **h.** $120°$
   i. $135°$ **j.** $150°$ **k.** $-45°$ **ℓ.** $-300°$
   m. $-240°$ **n.** $-225°$

4. From memory, give the degree measure for each of the angles whose radian measure is stated. Also sketch the angle.
   a. $\frac{\pi}{3}$ **b.** $\frac{\pi}{6}$ **c.** $\frac{\pi}{2}$ **d.** $\frac{\pi}{4}$
   e. π **f.** 2π **g.** $\frac{2\pi}{3}$ **h.** $\frac{3\pi}{4}$
   i. $\frac{11\pi}{6}$ **j.** $\frac{4\pi}{3}$ **k.** $-\frac{\pi}{4}$ **ℓ.** $-\frac{5\pi}{6}$
   m. $\frac{-11\pi}{6}$ **n.** $-\frac{5\pi}{4}$

5. **IN YOUR OWN WORDS** State the right-triangle definition of the trigonometric functions from memory.

6. **IN YOUR OWN WORDS** State the unit-circle definition of the trigonometric functions from memory.

7. **IN YOUR OWN WORDS** State the ratio definition of the trigonometric functions from memory.

8. **IN YOUR OWN WORDS** State the reduction principle and explain its meaning. Include some examples.

9. Name the reciprocal of each given trigonometric function.
   a. cosine   **b.** cotangent   **c.** sine
   d. secant   **e.** cosecant   **f.** tangent

10. State the two ratio identities.

11. State the three Pythagorean identities.

12. Write out the table of exact values.

Problems **13–21:** *Find the exact value for the given expressions.*

13. **a.** $\tan\frac{\pi}{4}$ **b.** $\cos 0$ **c.** $\cos 30°$
    d. $\sin\frac{\pi}{2}$ **e.** $\tan 0$ **f.** $\cos\frac{\pi}{4}$

14. **a.** $\tan 180°$ **b.** $\sin 45°$ **c.** $\sin \pi$
    d. $\sin 60°$ **e.** $\cos 270°$ **f.** $\tan\frac{\pi}{6}$

15. **a.** $\cos 60°$ **b.** $\sin\frac{\pi}{6}$ **c.** $\sin 0°$
    d. $\csc 90°$ **e.** $\sec\frac{3\pi}{2}$ **f.** $\sec 90°$

16. **a.** $\sec\frac{\pi}{6}$ **b.** $\csc 0$ **c.** $\sec\frac{\pi}{4}$
    d. $\sec 0°$ **e.** $\cot\frac{3\pi}{2}$ **f.** $\cot \pi$

17. **a.** $\sec\frac{\pi}{3}$ **b.** $\cot\frac{\pi}{6}$ **c.** $\cot 90°$
    d. $\cos \pi$ **e.** $\tan\frac{\pi}{3}$ **f.** $\cos\frac{\pi}{2}$

18. **a.** $\csc 30°$ **b.** $\csc \pi$ **c.** $\cot\frac{3\pi}{2}$
    d. $\cot\frac{\pi}{4}$ **e.** $\csc\frac{\pi}{4}$ **f.** $\sec 180°$

19. **a.** $\tan 90°$ **b.** $\sin\frac{3\pi}{2}$ **c.** $\tan\frac{3\pi}{2}$
    d. $\csc 60°$ **e.** $\cot\frac{\pi}{3}$ **f.** $\cot 0$

20. **a.** $\sin^{-1}\frac{1}{2}$ **b.** $\sin^{-1}\frac{\sqrt{2}}{2}$ **c.** $\cos^{-1}\frac{1}{2}$
    d. $\cos^{-1}\frac{\sqrt{2}}{2}$ **e.** $\tan^{-1}1$ **f.** $\tan^{-1}\sqrt{3}$

21. **a.** $\cot^{-1}\sqrt{3}$ **b.** $\csc^{-1}2$ **c.** $\sec^{-1}1$
    d. $\sec^{-1}\sqrt{2}$ **e.** $\cot^{-1}0$ **f.** $\cot^{-1}\frac{\sqrt{3}}{3}$

22. What are the amplitude and period of the standard cosine, sine, and tangent curves?

Problems **23–31:** *Draw a quick sketch of each curve from memory.*

23. $y = \sin x$ 24. $y = \cos x$ 25. $y = \tan x$

26. $y = \cot x$ 27. $y = \csc x$ 28. $y = \sec x$

29. $y = \cos^{-1}x$ 30. $y = \sin^{-1}x$ 31. $y = \tan^{-1}x$

32. What is a frame? What are the height, length, and starting point of the frames for the general form of the cosine, sine, and tangent curves?

33. IN YOUR OWN WORDS What is the general cosine form? Explain the meaning of each of the variables that you use.

34. IN YOUR OWN WORDS What is the general sine form? Explain the meaning of each of the variables that you use.

35. IN YOUR OWN WORDS What is the general tangent form? Explain the meaning of each of the variables that you use.

36. IN YOUR OWN WORDS Outline a procedure for graphing a general-form trigonometric equation.

37. In the table below, fill in the blanks indicated by the lowercase letters.

Function	Sign of x	Quadrant of θ	Approximate value of θ
$\theta = \cos^{-1}x$	positive	I	θ is between 0 and 1.57
$\theta = \cos^{-1}x$	negative	II	θ is between 1.57 and 3.14
$\theta = \sin^{-1}x$	positive	I	θ is between 0 and 1.57
$\theta = \sin^{-1}x$	negative	IV	θ is between −1.57 and 0
$\theta = \tan^{-1}x$	positive	**a.** _____	θ is between 0 and 1.57
$\theta = \tan^{-1}x$	negative	IV	**b.** _____
$\theta = \sec^{-1}x$	positive	**c.** _____	**d.** _____
$\theta = \sec^{-1}x$	negative	**e.** _____	θ is between 1.57 and 3.14
$\theta = \csc^{-1}x$	positive	I	**f.** _____
$\theta = \csc^{-1}x$	negative	**g.** _____	**h.** _____
$\theta = \cot^{-1}x$	positive	I	θ is between 0 and 1.57
$\theta = \cot^{-1}x$	negative	**i.** _____	**j.** _____

Level 1 Drill

Problems 38–56: *Choose the **best** answer from the choices given.*

38. If $\theta = -3$, then its reference angle is
A. 3
B. −3
C. $\pi - 3$
D. $3 - \pi$
E. none of these

39. The length of an arc with a central angle of 40° and radius of 1 is
A. less than 1
B. equal to 1
C. between 1 and 40
D. equal to 40
E. none of these

40. The size of θ in Figure 3.71 is about
A. $\dfrac{\pi}{3}$
B. −45°
C. −6
D. $-\dfrac{5\pi}{4}$
E. none of these

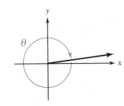

FIGURE 3.71 Angle θ

41. An angle with measure equal to $\dfrac{3\pi}{4}$ is
A. 45°
B. −45°
C. 135°
D. all are equal
E. none of these

42. The reciprocal of secant is
A. cosecant
B. sine
C. cosine
D. cotangent
E. none of these

43. $\cos \frac{5\pi}{6}$ is
A. $-\dfrac{\sqrt{3}}{2}$
B. −1
C. $-\sqrt{3}$
D. $\sqrt{3}$
E. none of these

44. csc 15.8° correct to four decimal places is
A. 3.6727
B. 0.9622
C. 1.0393
D. −10.8806
E. none of these

45. If θ is an angle in standard position and P_θ is a point on a unit circle on the terminal side of θ, then the coordinates of P_θ are
A. (0, 1)
B. (sin θ, cos θ)
C. (cos θ, sin θ)
D. impossible to find
E. none of these

46. 25 cot 38° ≈ ?
A. 19.53214066
B. 7.757741525
C. 80.565668471
D. 31.9985408
E. none of these

47. The length of the frame for $y = 3 \tan 2x$ is

A. $\frac{\pi}{2}$ B. π

C. $\frac{2\pi}{3}$ D. 2π

E. none of these

48. The graph of the cosecant curve passes through the point

A. $(0, 0)$ B. $\left(\frac{\pi}{2}, 1\right)$

C. $\left(\frac{3\pi}{2}, 1\right)$ D. all of the above

E. none of these

49. The center of the frame for $y - 1 = 2 \tan\left(x - \frac{\pi}{3}\right)$ is

A. $(1, 2)$ B. $\left(1, \frac{\pi}{3}\right)$

C. $\left(\frac{\pi}{3}, 1\right)$ D. $\left(-\frac{\pi}{3}, 1\right)$

E. none of these

50. When graphing the curve $y + 1 = 3 \sin\left(x + \frac{\pi}{3}\right)$, the point (h, k) is

A. $\left(-1, \frac{\pi}{3}\right)$ B. $\left(\frac{\pi}{3}, -1\right)$

C. $(0, 3)$ D. $\left(-\frac{\pi}{3}, 1\right)$

E. none of these

51. The top right-hand corner of the frame for $y = 2 \tan\left(x + \frac{3\pi}{4}\right)$ is

A. $\left(-\frac{\pi}{4}, 0\right)$ B. $\left(\frac{\pi}{4}, 0\right)$

C. $\left(\frac{\pi}{4}, 2\right)$ D. $\left(-\frac{\pi}{4}, 2\right)$

E. none of these

52. The leftmost point (yet still within the frame) of the graph of $y = -\cos \theta$ is

A. $\left(0, \frac{\pi}{2}\right)$ B. $\left(0, -\frac{\pi}{2}\right)$

C. $(0, 1)$ D. $(0, -1)$

E. none of these

53. The amplitude for the graph of $y = \sin \theta + \cos \theta$ is

A. 1   B. $\sqrt{2}$

C. $\frac{\pi}{4}$ D. $-\frac{\pi}{4}$

E. none of these

54. If $\sin \theta = -\sqrt{1 - \cos^2\theta}$, then θ could be in Quadrants

A. I and II   B. III and IV

C. I and IV   D. II and III

E. none of these

55. $\cot^2 173° - \csc^2 173°$ is

A. 16.35   B. 0

C. 1   D. 2

E. none of these

56. If $\csc \theta = -\sqrt{\frac{10}{3}}$ and $\cos \theta > 0$, then $\tan \theta$ is

A. $-\frac{1}{3}$ B. $\frac{3}{10}\sqrt{10}$

C. $-\frac{\sqrt{21}}{3}$ D. $-\frac{\sqrt{21}}{7}$

E. none of these

57. Using the definition of the trigonometric functions, find their values for an angle α whose terminal side passes through $(\sqrt{7}, 3)$.

58. Without using a calculator, decide which is larger: $\sin 1$ or $\sin 1°$?

Problems 59–62: Solve each triangle.

59. $a = 15.24$; $c = 28.0$; $\gamma = 90°$

60. $b = 15.26$; $c = 42.0$; $\gamma = 90.0°$

61. $c = 9.15$; $\alpha = 35.9°$; $\gamma = 90.0°$

62. $a = 30$; $b = 28$; $\gamma = 90°$

Problems 63–72: Sketch the graph of the given functions.

63. a. $y = 5 \cos x$ **b.** $y = \cos 5x$

64. a. $y = 5 \sin x$ **b.** $y = \sin 5x$

65. a. $y = 5 \tan x$ **b.** $y = \tan 5x$

66. a. $y = 2 \cos x - 1$ **b.** $y = 2 \cos(x - 1)$

67. $y = 2 \cos 3x$ **68.** $y = \frac{1}{2} \tan \frac{3}{2} x$

69. $y - 3 = \sin\left(x - \frac{\pi}{3}\right)$ **70.** $y + 2 = \cos(3x + 2\pi)$

71. $y = 2 \sin 3x + 3 \sin 2x$ **72.** $y = \sin \frac{x}{6} + 2 \cos \frac{x}{6}$

73. A UFO is sighted by people in two cities 12.6 miles apart. The UFO is between and in the same vertical plane as the two cities. The angle of elevation of the UFO from the first city is 11.9° and from the second is 35.8°. At what altitude (to the nearest 100 ft) is the UFO flying at the instant it is sighted?

74. A wheel moves with a speed of 20 rev/s. If the radius of the wheel is 3 in., find the angular velocity in radians/min and the linear velocity in ft/s.

75. If the large gear completes 20 revolutions, how many revolutions are made by the other gears in Figure 3.72? If the large gear is moving with an angular velocity of 20 rev/s, what is the angular velocity of the smallest gear?

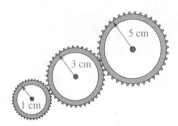

FIGURE 3.72   Gear mechanism

76. If $\cos \theta = \frac{4}{5}$ and $\sin \theta < 0$, what are the values of the other trigonometric functions?

77. Newton's laws of motion, along with some concepts from calculus, lead to the following parametric equations for the path of a projectile:

$$x = (v_0 \cos \theta)t, \quad y = (v_0 \sin \theta)t - 16t^2$$

where v_0 is the initial velocity of the projectile in the direction of θ with the horizontal. Suppose an arrow is released with an initial velocity of 160 ft/sec at an angle of 48°.
a. Sketch the path of the projectile.
b. What is the time of impact (to the nearest tenth of a second)?
c. What is the horizontal distance (to the nearest tenth of a foot)?

78. The wave displayed on an oscilloscope or synthesizer for the sound of a tuning fork vibrating f cycles/s with amplitude a is given by

$$y = a \sin 2\pi x$$

Sound is made up of fundamentals and **overtones,** which can be expressed as the sum of two trigonometric functions. For example,

Basic tone: $y = \sin \pi x$

First overtone: $y = \dfrac{\sin 3\pi x}{3}$

Second overtone: $y = \dfrac{\sin 5\pi x}{5}$

The complex sound wave, then, might be represented by the sum

$$y = \sin \pi x + \frac{\sin 3\pi x}{3} + \frac{\sin 5\pi x}{5} + \frac{\sin 7\pi x}{7} + \cdots$$

a. Draw the graph of the basic tone for $0 \le x \le 100$.
b. Draw the graph of the first overtone for $0 \le x \le 100$.
c. Draw the graph of the combination tone
$$y = \sin \pi x + \frac{\sin 3\pi x}{3} \text{ for } 0 \le x \le 100.$$
d. Draw the graph of the combination tone
$$y = \sin \pi x + \frac{\sin 3\pi x}{3} + \frac{\sin 5\pi x}{5} \text{ for } 0 \le x \le 100.$$

CHAPTER FOUR

TRIGONOMETRIC EQUATIONS AND IDENTITIES

PREVIEW

*To understand the content of this chapter, you need to know the difference between a conditional equation and an identity. The **solution** of a conditional trigonometric equation has the same meaning as the solution of any open equation—namely, the values of the variable that make a given equation true. Remember, **to solve an equation** means to find all replacements for the variable that make the equation true. If an open equation is true for all values in the domain of the variable, then it is called an **identity.***

In this chapter, we solve trigonometric equations. In addition, working through this chapter will give you confidence in proving and working with identities. We also develop the trigonometric identities you will need for your subsequent work in mathematics.

Thomas Jefferson regarded geometry and trigonometry as "most valuable to everyman."
   George Birkhoff
   "Fifty Years of American Mathematics," *Semicentennial Addresses of the American Mathematical Society, Vol. II* (New York: American Mathematical Society, 1938, p. 270).

SECTION 4.1

TRIGONOMETRIC EQUATIONS

PROBLEM OF THE DAY

What is the solution of the equation $\cos x = \frac{1}{2}$?

How would you answer this question?

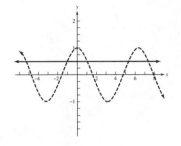

FIGURE 4.1
Graphs of $y = \frac{1}{2}$ and $y = \cos x$

A complete solution (from Section 3.3) is

$$x = \frac{\pi}{3} \pm 2\pi k,$$

$$x = \frac{5\pi}{3} \pm 2\pi k$$

If you think of the solution to this equation as the same as the solution of the system of equations

$$\begin{cases} y = \cos x \\ y = \frac{1}{2} \end{cases}$$

you might remember that we have already answered this question (see Example 6b, Section 3.3, page 131). The graph and solution are shown in Figure 4.1. In this section, we reconsider this problem in the context of solving equations. To understand the material here, you need to remember the notation and evaluation of inverse trigonometric functions from Section 1.3. In addition, some new terminology is necessary.

Suppose we wish to solve $\cos x = \frac{1}{2}$. This is an *open* equation because it contains a variable, whereas $\cos \frac{\pi}{3} = \frac{1}{2}$ is a *true* equation (no variable) and $\cos \frac{\pi}{4} = \frac{1}{2}$ is a *false* equation (no variable). When we say *solve* the open equation $\cos x = \frac{1}{2}$, we mean one of three things:

1. Find all values for which $x = \cos^{-1} \frac{1}{2}$ is true; we call this the **principal-value solution.** These are the values you obtain when you use a calculator. Using a calculator, we find

$$\cos^{-1} \frac{1}{2} \approx 1.05$$

2. Find all values that make the equation true within some given domain, for example, $0 \le x < 2\pi$; we call this a **restricted solution.** From the graph we see $\frac{\pi}{3}, \frac{5\pi}{3}$ as the restricted solution.
3. Find all values that make the original equation, $\cos x = \frac{1}{2}$, true; we call this the **complete solution.** This solution (from Example 6b, Section 3.3) is

$$x = \frac{\pi}{3} \pm 2\pi k, \quad x = \frac{5\pi}{3} \pm 2\pi k$$

 In this book we will assume radian measure of angles (not degree) unless otherwise specified.

We now consider procedures for solving trigonometric equations.

EXAMPLE 1

Distinguishing the Types of Solutions for a Trigonometric Equation

Solve $\sin x = \frac{1}{2}$ using

a. Principal-value solution (calculator solution to two decimal places).
b. Exact answer restricted to $0 \leq x < 2\pi$.
c. Complete solution (exact values).
d. Show these solutions graphically.

Solution

a. Principal-value solution (calculator):

$$x = \sin^{-1} \frac{1}{2} \approx 0.5235987756 \quad \textit{Calculator display}$$

Rounded to two decimal places, $x = 0.52$.

b. Here the answer is restricted to $0 \leq x < 2\pi$. From $\sin x = \frac{1}{2}$, we note that sine is positive, and we know that sine is positive in Quadrants I and II. So, we find the angles less than one revolution ($0 \leq x < 2\pi$) whose reference angle is $\pi/6$:

$$x = \frac{\pi}{6}, \frac{5\pi}{6}$$

c. Since sine is periodic with period 2π, we see that there are infinitely many values of x so that $\sin x = 1/2$, each found by adding multiples of 2π to the answers obtained in part **b**:

$$x = \frac{\pi}{6}, \frac{\pi}{6} + 2\pi, \frac{\pi}{6} + 4\pi, \ldots$$

and $\quad x = \frac{5\pi}{6}, \frac{5\pi}{6} + 2\pi, \ldots, \frac{5\pi}{6} + 2\pi, \frac{5\pi}{6} + 4\pi, \ldots$

We summarize by writing

$$x = \frac{\pi}{6} + 2k\pi; \frac{5\pi}{6} + 2k\pi \quad \text{where } k \text{ is any integer}$$

d. The graphs for the solutions are shown in Figure 4.2.

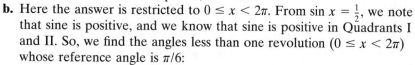

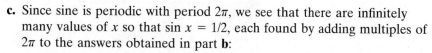

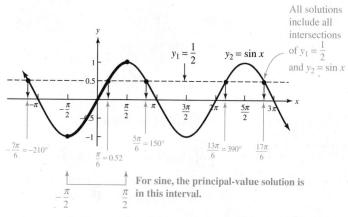

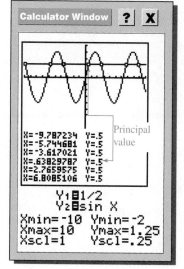

FIGURE 4.2   Graphical solution for $\sin x = \frac{1}{2}$

If you are using a calculator, you can let $y_1 = \frac{1}{2}$ and $y_2 = \sin x$ and then use the $\boxed{\text{TRACE}}$ function to find approximate values for the points of intersection. This method of solution is shown in the previous calculator window. Notice that the approximate value shown ($x \approx 0.63829787$) could be found with greater accuracy by using the $\boxed{\text{ZOOM}}$. ∎

When solving a trigonometric equation, you must distinguish the unknown from the angle and the angle from the function. Consider

The unknown is x.

$$\underbrace{\cos}\ \underbrace{(2x + 1)}$$

The function is cosine.

The angle, or argument, is $2x + 1$.

The steps in solving a trigonometric equation are now given.

You will need to remember this procedure, including the quadratic formula.

PROCEDURE FOR SOLVING TRIGONOMETRIC EQUATIONS

1. Solve for a single trigonometric function. You may use trigonometric identities, factoring, or the quadratic formula.
   a. Linear equation: $Ax + B = 0, A \neq 0$

   Procedure: Isolate the variable; $x = -\dfrac{B}{A}$

   b. Quadratic equation: $Ax^2 + Bx + C = 0, A \neq 0$
   Procedure: Factor, if possible. In this case, set each factor equal to zero and solve. If not factorable, then use the quadratic formula:

   $$x = \frac{-B \pm \sqrt{B^2 - 4AC}}{2A}$$

2. Solve for the angle. You will use the definition of the inverse trigonometric functions for this step.
3. Solve for the unknown.

Principal Values

EXAMPLE 2

Solving Linear Trigonometric Equations

Solve the equations for the principal values of θ. Compare the solutions of these similar problems.

a. $5 \sin(\theta + 1) = \frac{1}{2}$ **b.** $\sin 5(\theta + 1) = \frac{1}{2}$

Solution   **a.** $5 \sin(\theta + 1) = \frac{1}{2}$

$$\sin(\theta + 1) = \frac{1}{10} \qquad \textit{Solve for the function.}$$

$$\theta + 1 = \sin^{-1} 0.1 \qquad \textit{Solve for the angle.}$$

$$\theta = -1 + \sin^{-1} 0.1 \qquad \textit{Solve for the unknown.}$$

$$\approx -0.8998326 \qquad \textit{We use radians (real numbers)}$$
$$\textit{unless otherwise specified.}$$

b. $\sin 5(\theta + 1) = \frac{1}{2} \qquad \textit{Solve for the function.}$

$$5(\theta + 1) = \sin^{-1} \frac{1}{2} \qquad \textit{Solve for the angle.}$$

$$5\theta + 5 = \frac{\pi}{6} \qquad \textit{Principal value}$$

$$5\theta = -5 + \frac{\pi}{6} \qquad \textit{Solve for the unknown.}$$

$$\theta = -1 + \frac{\pi}{30} \qquad \textit{Divide both sides by 5.}$$

$$\approx -0.8952802 \qquad \textit{Calculator approximation}$$

Notice that these two answers are *not* the same.   ■

EXAMPLE 3   **Solving a Quadratic Trigonometric Equation by Factoring**

Find the principal values of θ for $15 \cos^2\theta - 2 \cos \theta - 8 = 0$.

Solution
$$15 \cos^2\theta - 2 \cos \theta - 8 = 0$$
$$(3 \cos \theta + 2)(5 \cos \theta - 4) = 0 \qquad \textit{Factor, if possible.}$$

This is solved by setting each factor equal to zero (zero-factor theorem):

$$3 \cos \theta + 2 = 0 \qquad\qquad 5 \cos \theta - 4 = 0$$
$$\cos \theta = -\frac{2}{3} \qquad\qquad \cos \theta = \frac{4}{5}$$
$$\theta = \cos^{-1}\left(-\frac{2}{3}\right) \qquad\qquad \theta = \cos^{-1} \frac{4}{5}$$
$$\approx 2.300524 \qquad\qquad \approx 0.6435011 \qquad ■$$

EXAMPLE 4   **Solving a Quadratic Trigonometric Equation by the Quadratic Formula**

Find the principal values of θ for $\tan^2\theta - 5 \tan \theta - 4 = 0$.

Solution   Because $\tan^2\theta - 5\tan\theta - 4$ cannot be factored, we use the quadratic formula.

$$\tan^2\theta - 5\tan\theta - 4 = 0$$

$$\tan\theta = \frac{5 \pm \sqrt{25 - 4(1)(-4)}}{2} \quad \textit{Quadratic formula if it cannot be factored}$$

$$= \frac{5 \pm \sqrt{41}}{2}$$

$$\theta = \tan^{-1}\left(\frac{5 \pm \sqrt{41}}{2}\right)$$

$$\approx \tan^{-1}(5.7015621), \tan^{-1}(-0.7015621)$$

$$\approx 1.3971718, -0.6117736 \qquad \blacksquare$$

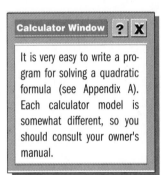

Calculator Window **?** **X**

It is very easy to write a program for solving a quadratic formula (see Appendix A). Each calculator model is somewhat different, so you should consult your owner's manual.

Complete Solution

To find the complete general solution of a trigonometric equation, we first find the reference angle. To do this, we use the table of exact values or a calculator. Then we find two values less than one revolution using the reference angle:

For $x = \arccos x$ or $y = \text{arcsec } x$:
If x is positive, then the angle y is in Quadrants I and IV.
If x is negative, then the angle y is in Quadrants II and III.

For $y = \arcsin x$ or $y = \text{arccsc } x$:
If x is positive, then the angle y is in Quadrants I and II.
If x is negative, then the angle y is in Quadrants III and IV.

For $y = \arctan x$ or $y = \text{arccot } x$:
If x is positive, then the angle y is in Quadrants I and III.
If x is negative, then the angle y is in Quadrants II and IV.

This information is summarized in Figure 4.3.

II **S**ine and Cosecant positive; others negative   **A**ll positive   **I**

Tangent and Cotangent positive; others negative   **C**osine and Secant positive; others negative

III   **IV**

FIGURE 4.3

Signs of the trigonometric functions

After you have found the two values less than one revolution, the entire solution is found by using the period of the function.

For cosine, sine, secant and cosecant: Add multiples of 2π.
For tangent and cotangent: Add multiples of π.

EXAMPLE 5   **Solving General Linear Trigonometric Equations**

a. Solve $\cos\theta = \frac{1}{2}$ **b.** Solve $\cos\theta = -\frac{1}{2}$

Solution **a.** Reference angle $\theta' = \arccos\left|\frac{1}{2}\right| = \frac{\pi}{3}$. Angles less than one revolution with a reference angle of $\frac{\pi}{3}$ are $\frac{\pi}{3}$ (Quadrant 1) and $\frac{5\pi}{3}$ (Quadrant IV). There are infinitely many values, so to find all solutions, add multiples of 2π to these values: $\theta = \frac{\pi}{3} + 2k\pi, \frac{5\pi}{3} + 2k\pi$ for any integral value of k.

b. $\theta = \arccos\left|-\frac{1}{2}\right| = \frac{\pi}{3}$. The angles are in Quadrants II and III:
$\theta = \frac{2\pi}{3} + 2k\pi, \frac{4\pi}{3} + 2k\pi.$ ∎

Restricted Solutions

With restricted solutions, give the values from the complete solution that fall within the restricted domain.

EXAMPLE 6   **Finding Approximate Values of a Linear Trigonometric Function**

Solve $\sin x = \frac{2}{\pi}$ for $0 \le x < 2\pi$.

Solution $x = \arcsin\left(\frac{2}{\pi}\right) \approx 0.6901071$. The solution is in Quadrants I and II $\left(\frac{2}{\pi}\right.$ is positive$\left.\right)$: $x \approx 0.6901071, 2.4514856.$ ∎

EXAMPLE 7   **Solving a Trigonometric Equation with a Double Angle**

Solve $\sin 2\theta = \frac{\sqrt{3}}{2}$ for $0 \le \theta < 2\pi$.

Solution   Since $0 \le \theta < 2\pi$, solve for 2θ such that $0 \le 2\theta < 4\pi$, and

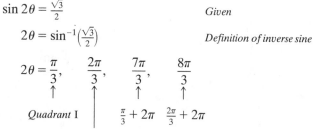

$$\sin 2\theta = \frac{\sqrt{3}}{2} \qquad \text{\textit{Given}}$$

$$2\theta = \sin^{-1}\left(\frac{\sqrt{3}}{2}\right) \qquad \text{\textit{Definition of inverse sine}}$$

$$2\theta = \frac{\pi}{3}, \qquad \frac{2\pi}{3}, \qquad \frac{7\pi}{3}, \qquad \frac{8\pi}{3}$$

$$\uparrow \qquad\qquad \uparrow \qquad\qquad \uparrow \qquad \uparrow$$

$$\textit{Quadrant } I \qquad\qquad \frac{\pi}{3} + 2\pi \quad \frac{2\pi}{3} + 2\pi$$

Quadrant II (*sine is positive in Quadrants* I *and* II)

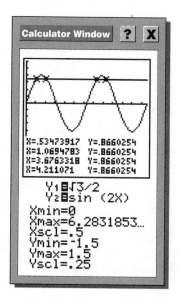

Mentally solve each equation for θ:

$$2\theta = \frac{\pi}{3} \qquad 2\theta = \frac{2\pi}{3} \qquad 2\theta = \frac{7\pi}{3} \qquad 2\theta = \frac{8\pi}{3}$$

As you solve these equations, notice that in each case, θ is between 0 and 2π (including 0 but not 2π). The solution is

$$\frac{\pi}{6}, \quad \frac{\pi}{3}, \quad \frac{7\pi}{6}, \quad \frac{4\pi}{3}$$

Find the approximate values 0.52, 1.05, 3.67, 4.19 (which, by the way, all satisfy $0 \leq \theta < 2\pi$), and check them graphically as shown in the margin. Note that the calculator solution is approximate, but if you use the $\boxed{\text{ZOOM}}$ feature, you can achieve a greater degree of calculator accuracy. ■

EXAMPLE 8

Trigonometric Equation with No Solution

Solve $\cos 3\theta = 1.2862$.

Solution

There is no solution because $-1 \leq \cos 3\theta \leq 1$. The graphical solution clearly shows that there is no solution:

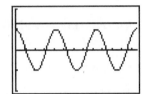 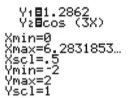

EXAMPLE 9

Solving a Trigonometric Equation with Multiple Angles

Solve $\cos 3\theta = -0.68222$ for $0 \leq \theta < 2\pi$ and approximate the answer to two decimal places. (*Note:* Since $0 \leq \theta < 2\pi$, $0 \leq 3\theta < 6\pi$.)

Solution

Begin with the reference angle θ' so that $\cos \theta' = 0.68222$. From the definition of the inverse cosine, we have

$$\theta' = \cos^{-1}|0.68222| \approx 0.8200016512$$

This is the reference angle for 3θ.

Since cosine is negative in Quadrants II and III, we have

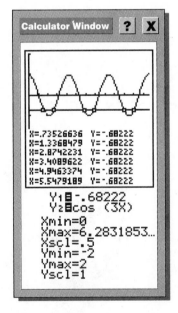

Quadrant II	Quadrant III
$\overbrace{\pi - 0.8200016512}$	$\overbrace{\pi + 0.8200016512}$
$3\theta = 2.321591002$	$3\theta = 3.961594305$
$3\theta = 8.60477631$ *Add 2π.*	$3\theta = 10.24477961$ *Add 2π to reference angle.*
$3\theta = 14.88796162$ *Add 4π.*	$3\theta = 16.52796492$ *Add 4π to reference angle.*

Solve each equation for θ (divide all six values by 3) to find the rounded solution:

0.77,   1.32,   2.87,   3.41,   4.96,   5.51

Notice that these values all satisfy $0 \leq \theta < 2\pi$. If you have a graphing calculator, you can find approximate x-values for the points of intersection of the curves $y_1 = -0.68222$ and $y_2 = \cos 3\theta$. This graphical solution is shown in the margin. ■

EXAMPLE 10

Solving a Trigonometric Equation
That Is Also a Radical Equation

Solve the equation $\sin x - 1 = \sqrt{1 - \sin^2 x}$ for $0 \le x < 2\pi$.

Solution   To avoid radicals, we square both sides of the given equation:

$$
\begin{aligned}
\sin x - 1 &= \sqrt{1 - \sin^2 x} && \textit{Given} \\
(\sin x - 1)^2 &= (\sqrt{1 - \sin^2 x})^2 && \textit{Square both sides, check required.} \\
\sin^2 x - 2\sin x + 1 &= 1 - \sin^2 x && \textit{Simplify.} \\
2\sin^2 x - 2\sin x &= 0 && \textit{Obtain a 0 on one side.} \\
\sin^2 x - \sin x &= 0 && \textit{Divide both sides by 2.} \\
\sin x(\sin x - 1) &= 0 && \textit{Common factor} \\
\sin x = 0 \qquad \text{or} \quad \sin x - 1 &= 0 && \textit{Zero-factor theorem} \\
x = \sin^{-1} 0 \qquad\qquad x &= \sin^{-1} 1 \\
= 0 \text{ or } \pi \qquad\qquad\quad &= \frac{\pi}{2}
\end{aligned}
$$

Check (because there might be extraneous roots).

Check $x = 0$: $\sin 0 - 1 = \sqrt{1 - \sin^2 0}$
$$0 - 1 = \sqrt{1 - 0^2}$$
$$-1 = 1 \qquad\qquad \text{False}$$

Check $x = \pi$: $\sin \pi - 1 = \sqrt{1 - \sin^2 \pi}$
$$-1 = 1 \qquad\qquad \text{False}$$

Check $x = \frac{\pi}{2}$: $\sin \frac{\pi}{2} - 1 = \sqrt{1 - \sin^2 \frac{\pi}{2}}$
$$1 - 1 = \sqrt{1 - 1^2}$$
$$0 = 0 \qquad\qquad \text{True}$$

The root is $\frac{\pi}{2}$.   ■

EXAMPLE 11

Solving an Equation That Does Not Have an Algebraic Solution

Solve $\cos 2x + x^2 - 2 = 0$ for $0 \le x < 2\pi$.

Solution   There is no easy algebraic solution for this equation, so we solve it numerically and graphically.

Many calculators have an INTERSECT *function. For this example, we find*

Intersection
$x = \pm1.719356, \quad y = 0$

GRAPHICAL SOLUTION:

We graph $y = \cos 2x + x^2 - 2$ and look to see where $y = 0$ (that is, where the graph crosses the *x*-axis).

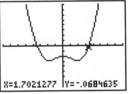

```
X=1.7021277  Y=-.0684635
```
Y₁⊟cos (2X)+X²-2
Xmin=-4   Ymin=-4
Xmax=4   Ymax=4
Xscl=.5   Yscl=.5

Using the TRACE, we see that the graph at the left crosses the *x*-axis at $x \approx 1.7$ and at $x \approx -1.7$. If we want a more accurate approximation, we can use the ZOOM feature to take a closer look in the neighborhood of $x = 1.7$:

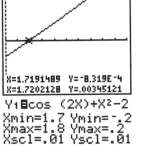

```
X=1.7191489  Y=-8.319E-4
X=1.7202128  Y=.00345121
```
Y₁⊟cos (2X)+X²-2
Xmin=1.7 Ymin=-.2
Xmax=1.8 Ymax=.2
Xscl=.01 Yscl=.01

From the graph at the right, we see that the root is between $x = 1.7191489$ (*y* is negative) and $x = 1.7202128$ (*y* is positive). We estimate the solution to be $x \approx 1.72$.

NUMERICAL SOLUTION:

We calculate $y = \cos 2x + x^2 - 2$ for values of *x* between 1.7 and 1.8:

X	Y₁
1.7	-.0768
1.71	-.0374
1.72	.00259
1.73	.04316
1.74	.08432
1.75	.12604
1.76	.16835

X=1.72

From the table of values at the left, the root is between $x = 1.71$ (*y* is negative) and $x = 1.72$ (*y* is positive). We now calculate values between $x = 1.716$ and $x = 1.722$ to find the following table:

X	Y₁
1.716	-.0135
1.717	-.0095
1.718	-.0055
1.719	-.0014
1.72	.00259
1.721	.00662
1.722	.01066

X=1.719

From the table of values at the right, we see that the root is between $x = 1.719$ (*y* is negative) and $x = 1.720$ (*y* is positive). We estimate the solution to be $x \approx 1.72$.

PROBLEM SET 4.1

Essential Concepts

1. a. What is an equation?
   b. What does it mean to solve an equation?

2. a. What is the procedure for solving a linear equation of the form $mx + b = 0, m \neq 0$?

   b. What is the procedure for solving a quadratic equation of the form $ax^2 + bx + c = 0, a \neq 0$?

   c. What is the procedure for solving a trigonometric equation?

3. What is an identity?

Level 1 Drill

Problems 4–13: *Solve each equation.*
a. *Give the principal calculator value(s), rounded to two decimal places.*
b. *Give the exact restricted solution.*
c. *Give the complete exact solution.*

4. $\cos x = \frac{1}{2}$
  b. $0 \le x \le \pi$

5. $\cos x = \frac{1}{2}$
  b. $0° \le x < 180°$

6. $\cos x = -\frac{1}{2}$
  b. $0 \le x \le \pi$

7. $\cos x = -\frac{1}{2}$
  b. $0 \le x \le 2\pi$

8. $\tan x = -1$
  b. $0 \le x < \pi$

9. $\tan x = 1$
  b. $0° \le x < 360°$

10. $\sec x = 2$
  b. $0 \le x < 2\pi$

11. $\sec x = -2$
  b. $0 \le x \le \pi$

12. $\sin x = -\frac{\sqrt{2}}{2}$
  b. $-\frac{\pi}{2} \le x < \frac{\pi}{2}$

13. $\sin x = \frac{\sqrt{2}}{2}$
  b. $0 \le x < 2\pi$

Problems 14–29: *Solve each equation.*
a. *Give the principal (calculator) values to two decimal places.*
b. *Give exact values such that $0 \le x < 2\pi$.*
c. *Give the complete exact solution.*

14. $\cos 2x = \frac{1}{2}$

15. $\sin 2x = -\frac{\sqrt{3}}{2}$

16. $\cos 3x = \frac{1}{2}$

17. $\sin 3x = \frac{\sqrt{2}}{2}$

18. $\cos 2x = -\frac{1}{2}$

19. $\sin 2x = -\frac{\sqrt{2}}{2}$

20. $\tan 3x = 1$

21. $\tan 3x = -1$

22. $\sec 2x = -\frac{2\sqrt{3}}{3}$

23. $\csc 4x = -1$

24. $(\sec x)(\tan x) = 0$

25. $(\sin x)(\tan x) = 0$

26. $(\cot x)(\cos x) = 0$

27. $(\sin x)(\cos x) = 0$

28. $(\csc x - 2)(2 \cos x - 1) = 0$

29. $(\sec x - 2)(2 \sin x - 1) = 0$

Level 2 Drill

Problems 30–35: *Solve the equations (exact answers) for $0 \le x < 2\pi$.*

30. $\tan^2 x = \tan x$

31. $\tan^2 x = \sqrt{3} \tan x$

32. $\sin^2 x = \frac{1}{2}$

33. $\cos^2 x = \frac{1}{2}$

34. $2 \sin x \cos x = \sin x$

35. $\sqrt{2} \cos x \sin x = \sin x$

Problems 36–41: *Find all solutions for each equation.*

36. $\tan x = -\sqrt{3}$

37. $\cos x = -\frac{\sqrt{3}}{2}$

38. $\sin x = -\frac{\sqrt{2}}{2}$

39. $\sin x = 0.3907$

40. $\cos x = 0.2924$

41. $\tan x = 1.376$

Problems 42–61: *Solve each equation for principal values to two decimal places.*

42. $\cos^2 x - 1 - \cos x = 0$

43. $\sin^2 x - \sin x - 2 = 0$

44. $\tan^2 x - 3 \tan x + 1 = 0$

45. $\csc^2 x - \csc x - 1 = 0$

46. $\sec^2 x - \sec x - 1 = 0$

47. $2 \cos 3x \sin 2x = \sin 2x$

48. $\cos 3x + 2 \sin 2x \cos 3x = 0$

49. $\sin 2x + 2 \cos x \sin 2x = 0$

50. $\cos(3x - 1) = \frac{1}{2}$

51. $\cos 3x + 1 = \sqrt{2}$

52. $\tan(2x + 1) = \sqrt{3}$

53. $\tan 2x + 1 = \sqrt{3}$

54. $\sin 2x + 1 = \sqrt{3}$

55. $1 - 2 \sin^2 x = \sin x$

56. $2 \cos^2 x - 1 = \cos x$

57. $2 \cos x \sin x + \cos x = 0$

58. $1 - \sin x = 1 - 2 \sin^2 x$

59. $\cos x = 2 \sin x \cos x$

60. $\sin^2 3x + \sin 3x + 1 = 1 - 4 \sin^2 3x$

61. $\sin^2 3x + \sin 3x = \cos^2 3x + 1$

Level 2 Applications

62. A tuning fork vibrating at 264 Hz (frequency $f = 264$) with an amplitude of 0.0050 cm produces C on the musical scale and can be described by an equation of the form

$$y = 0.0050 \sin 528\pi x$$

Find the smallest positive value of x (correct to three significant digits) for which $y = 0.0020$.

63. In a certain electrical circuit, the electromotive force V (in volts) and the time t (in seconds) are related by an equation of the form

$$V = \cos 2\pi t$$

Find the smallest positive value for t (correct to three significant digits) for which $V = 0.400$.

64. The orbit of a certain satellite alternates above and below the equator according to the equation

$$y = 4{,}000 \sin\left(\frac{\pi}{45}t + \frac{5\pi}{18}\right)$$

where t is the time (in minutes) and y is the distance (in kilometers) from the equator. Find the times at which the satellite crosses the equator during the first hour and a half (that is, for $0 \le t \le 90$).

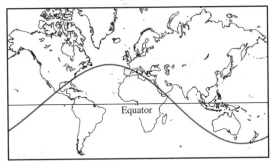

Equator

Level 3 Theory

Problems 65–72: *Solve the trigonometric equations (to the nearest tenth) for* $0 \le x < 6.3$.

65. $\cos 2x - 1 = \sin 2x$

66. $\sin 2x + x^2 = 0$

67. $x + \cos x = 0$

68. $\sin^{-1} x + 2\cos^{-1} x = \pi$

69. $\sin^{-1}\dfrac{x}{3} + \cos^{-1}\dfrac{x}{2} = \dfrac{\pi}{3}$

70. $\sin^{-1}\dfrac{x}{2} + \tan^{-1} x = \dfrac{\pi}{2}$

71. $\cos^3 x - \sin^3 x = \frac{1}{2}$

72. $\cos^4 x - \sin^4 x = \frac{1}{2}$

SECTION 4.2

TRIGONOMETRIC IDENTITIES

PROBLEM OF THE DAY

What is the graph of $y = \dfrac{1 - \cos^2 x}{\sin x}$?

How would you answer this question?

We begin by finding ordered pairs satisfying this equation, as shown in the margin. Next, we plot these points, and draw a smooth graph passing through those points, as shown in Figure 4.4.

Table of Values

x	$y = \dfrac{1 - \cos^2 x}{\sin x}$
0	undefined since $\sin 0 = 0$
1	$\dfrac{1 - \cos^2 1}{\sin 1} \approx 0.8415$
2	$\dfrac{1 - \cos^2 2}{\sin 2} \approx 0.9093$
3	$\dfrac{1 - \cos^2 3}{\sin 3} \approx 0.1411$
4	$\dfrac{1 - \cos^2 4}{\sin 4} \approx -0.7568$
5	$\dfrac{1 - \cos^2 5}{\sin 5} \approx -0.9589$
6	$\dfrac{1 - \cos^2 6}{\sin 6} \approx -0.2794$
7	$\dfrac{1 - \cos^2 7}{\sin 7} \approx 0.6570$

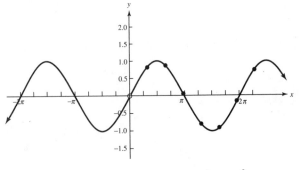

FIGURE 4.4 Graph of $y = \dfrac{1 - \cos^2 x}{\sin x}$

This graph looks familiar, doesn't it? Do you recognize it? It is the same as the graph of $y = \sin x$. That is, the graph of $y = \dfrac{1 - \cos^2 x}{\sin x}$ is the same as the graph of $y = \sin x$. In other words, the equation

$$\frac{1 - \cos^2 x}{\sin x} = \sin x$$

is true for all values of x that do not cause division by 0. In such a case, we call the equations an *identity*. When we work with expressions involving trigonometric functions, we can often rewrite the expressions in simpler form. The subject of this chapter is to practice writing trigonometric expressions in equivalent, but different algebraic forms.

In Section 2.3, we considered eight fundamental identities, which are used to simplify and change the form of a variety of trigonometric equations. These identities are repeated in Table 4.1 for easy reference. You may also wish to review some of the alternative forms of these identities; for example, $\sec \theta = 1/\cos \theta$ can also be written as $\cos \theta \sec \theta = 1$.

TABLE 4.1 Fundamental Identities

Identities	*Alternative forms*	
Reciprocal identities		
1. $\sec \theta = \dfrac{1}{\cos \theta}$	$\sec \theta \cos \theta = 1;$	$\cos \theta = \dfrac{1}{\sec \theta}$
2. $\csc \theta = \dfrac{1}{\sin \theta}$	$\csc \theta \sin \theta = 1;$	$\sin \theta = \dfrac{1}{\csc \theta}$
3. $\cot \theta = \dfrac{1}{\tan \theta}$	$\cot \theta \tan \theta = 1;$	$\tan \theta = \dfrac{1}{\cot \theta}$
Ratio identities		
4. $\tan \theta = \dfrac{\sin \theta}{\cos \theta}$	$\tan \theta \cos \theta = \sin \theta;$	$\cos \theta = \dfrac{\sin \theta}{\tan \theta}$
5. $\cot \theta = \dfrac{\cos \theta}{\sin \theta}$	$\cot \theta \sin \theta = \cos \theta;$	$\sin \theta = \dfrac{\cos \theta}{\cot \theta}$
Pythagorean identities		
6. $\cos^2 \theta + \sin^2 \theta = 1$	$\cos^2 \theta = 1 - \sin^2 \theta;$	$\sin^2 \theta = 1 - \cos^2 \theta$
7. $1 + \tan^2 \theta = \sec^2 \theta$	$\tan^2 \theta = \sec^2 \theta - 1;$	$\sec^2 \theta - \tan^2 \theta = 1$
8. $\cot^2 \theta + 1 = \csc^2 \theta$	$\cot^2 \theta = \csc^2 \theta - 1;$	$\csc^2 \theta - \cot^2 \theta = 1$

EXAMPLE 1   **Verifying a Fundamental Identity Graphically**

Show $\tan \theta = \dfrac{\sin \theta}{\cos \theta}$ graphically.

Solution   Graph $y = \tan \theta$, the left side of the equation; if you are using a graphing calculator, you will graph Y1 = tan X as shown in Figure 4.5a.

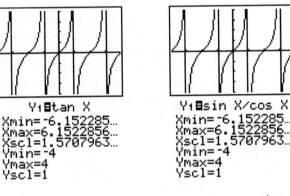

a. Graph of $y = \tan x$ b. Graph of $y = \dfrac{\sin x}{\cos x}$

FIGURE 4.5

Graphical verification that $\tan \theta = \dfrac{\sin \theta}{\cos \theta}$ is an identity

Now, graph Y2 = sin X/cos X, the right side of the equation, as shown in Figure 4.5b. Notice that even though the functions being graphed are different, the graphs are identical. This means that the equation is an identity.   ∎

In addition to these fundamental identities, you will need to remember (from arithmetic and algebra) the proper procedure for simplifying expressions, especially the concepts of common denominators and complex fractions. You can then combine algebraic simplification with a knowledge of these fundamental identities, as shown by the next example.

EXAMPLE 2   **Algebraically Simplifying an Expression with Trigonometric Functions**

Simplify $\dfrac{\sin\theta + \dfrac{\cos^2\theta}{\sin\theta}}{\dfrac{\cos\theta}{\sin\theta}}$ and leave your answer in terms of sine and cosine functions only.

Solution

$$\frac{\sin\theta + \dfrac{\cos^2\theta}{\sin\theta}}{\dfrac{\cos\theta}{\sin\theta}} = \frac{\dfrac{\sin^2\theta}{\sin\theta} + \dfrac{\cos^2\theta}{\sin\theta}}{\dfrac{\cos\theta}{\sin\theta}}$$

Common denominator

Algebraic form: $\dfrac{a + \dfrac{b^2}{a}}{\dfrac{b}{a}} = \dfrac{\dfrac{a^2}{a} + \dfrac{b^2}{a}}{\dfrac{b}{a}}$

$$= \frac{\sin^2\theta + \cos^2\theta}{\sin\theta} \cdot \frac{\sin\theta}{\cos\theta}$$

Dividing rational expressions

Algebraic form: $\dfrac{\dfrac{a^2}{a} + \dfrac{b^2}{a}}{\dfrac{b}{a}} = \dfrac{a^2 + b^2}{a} \cdot \dfrac{a}{b}$

$$= \frac{1}{\sin\theta} \cdot \frac{\sin\theta}{\cos\theta}$$

Fundamental identity

$$= \frac{1}{\cos\theta}$$

Reduce rational expressions. ∎

Procedure for Proving Identities

Suppose you are given a trigonometric equation such as

$$\tan\theta + \cot\theta = \sec\theta\,\csc\theta$$

and are asked to show that it is an identity, a process we call **proving an identity.** You must be careful not to treat this problem as though you were solving an algebraic equation. When asked to **prove** that a given equation is an identity, do *not* start with the given expression, because you cannot assume that it is true. *Begin* with what you know is true, and *end* with the given identity. There are three ways to proceed:

1. Reduce the left-hand side of the equation to the right-hand side by using algebra and the fundamental identities.
2. Reduce the right-hand side of the equation to the left-hand side.
3. Reduce both sides *independently* to the same expression.

EXAMPLE 3   **Proving an Identity**

Prove that $\tan \theta + \cot \theta = \sec \theta \csc \theta$.

Solution   Begin with either the left side or the right side (but begin with only *one* side). We begin with the left side:

$$\tan \theta + \cot \theta = \frac{\sin \theta}{\cos \theta} + \frac{\cos \theta}{\sin \theta}$$

Substitute one or more of the fundamental identities; in this case we are substituting $\tan \theta = \dfrac{\sin \theta}{\cos \theta}$ and $\cot \theta = \dfrac{\cos \theta}{\sin \theta}$.

$$= \frac{\sin^2 \theta + \cos^2 \theta}{\cos \theta \sin \theta}$$

Common denominator to add rational expressions.
Algebraic form: $\dfrac{a}{b} + \dfrac{b}{a} = \dfrac{a^2 + b^2}{ab}$

$$= \frac{1}{\cos \theta \sin \theta}$$

Fundamental identity; in this case, we substituted $\sin^2 \theta + \cos^2 \theta = 1$.

$$= \frac{1}{\cos \theta} \cdot \frac{1}{\sin \theta}$$

Break up the product.

$$= \sec \theta \csc \theta$$

Fundamental identities; here we use two reciprocal identities.

Since the left-hand side (eventually) equals the right-hand side, we say that we have proved that the given equation is an identity. ∎

Usually, it is easier to begin with what seems to be the more complicated side and try to reduce it to the simpler side. If both sides seem equally complex, you might change all the functions to cosines and sines and then simplify.

EXAMPLE 4   **Proving an Identity by Combining Fractions**

Prove that $\dfrac{1}{1 + \cos \lambda} + \dfrac{1}{1 - \cos \lambda} = 2 \csc^2 \lambda$ is an identity.

Solution   We begin with the more complicated side.

$$\frac{1}{1 + \cos \lambda} + \frac{1}{1 - \cos \lambda} = \frac{(1 - \cos \lambda) + (1 + \cos \lambda)}{(1 + \cos \lambda)(1 - \cos \lambda)}$$

Common denominators; algebraic form:
$$\frac{1}{1 + b} + \frac{1}{1 - b} = \frac{(1 - b) + (1 + b)}{(1 + b)(1 - b)}$$

$$= \frac{2}{1 - \cos^2\lambda}$$

Algebraic form:

$$\frac{(1 - b) + (1 + b)}{(1 + b)(1 - b)} = \frac{2}{1 - b^2}$$

Can you verify that this is an identity by graphing?

$$= \frac{2}{\sin^2\lambda}$$

$$= 2 \cdot \frac{1}{\sin^2\lambda}$$

$$= 2\csc^2\lambda \qquad \blacksquare$$

EXAMPLE 5

Proving an Identity by Separating a Fraction into a Sum

Prove that $\dfrac{\sec 2\beta + \cot 2\beta}{\sec 2\beta} = 1 + \csc 2\beta - \sin 2\beta$ is an identity.

Solution

Begin with the left-hand side. When you work with a fraction consisting of a single term as a denominator, it is often helpful to separate the fraction into the sum of several fractions.

$$\frac{\sec 2\beta + \cot 2\beta}{\sec 2\beta} = \frac{\sec 2\beta}{\sec 2\beta} + \frac{\cot 2\beta}{\sec 2\beta} \qquad \textit{Break up fractions.}$$

$$= 1 + \cot 2\beta \cdot \frac{1}{\sec 2\beta} \qquad \textit{Note } \frac{\sec 2\beta}{\sec 2\beta} = 1.$$

$$= 1 + \frac{\cos 2\beta}{\sin 2\beta} \cdot \cos 2\beta \qquad \textit{Fundamental identities}$$

$$= 1 + \frac{\cos^2 2\beta}{\sin 2\beta} \qquad \textit{Multiply fractions.}$$

$$= 1 + \frac{1 - \sin^2 2\beta}{\sin 2\beta} \qquad \textit{Fundamental identity}$$

$$= 1 + \frac{1}{\sin 2\beta} - \frac{\sin^2 2\beta}{\sin 2\beta} \qquad \textit{Break up fractions.}$$

$$= 1 + \csc 2\beta - \sin 2\beta \qquad \begin{array}{l}\textit{Fundmental identity and}\\ \textit{reduce the fraction.}\end{array} \qquad \blacksquare$$

EXAMPLE 6

Proving an Identity by Using a Conjugate

Prove that $\dfrac{\cos \alpha}{1 - \sin \alpha} = \dfrac{1 + \sin \alpha}{\cos \alpha}$ is an identity.

Solution   Sometimes, when there is a binomial in the numerator or denominator, the identity may be proved by multiplying the numerator and denominator by the conjugate of the binomial. That is, multiply both the numerator and the denominator of the left side by $1 + \sin\alpha$. (This does not change the value of the expression because you are multiplying by 1.)

$$\frac{\cos\alpha}{1 - \sin\alpha} = \frac{\cos\alpha}{1 - \sin\alpha} \cdot \frac{1 + \sin\alpha}{1 + \sin\alpha} \qquad \textit{Multiply by 1.}$$

$$= \frac{\cos\alpha(1 + \sin\alpha)}{1 - \sin^2\alpha} \qquad \textit{Multiply fractions.}$$

$$= \frac{\cos\alpha(1 + \sin\alpha)}{\cos^2\alpha} \qquad \textit{Fundamental identities}$$

$$= \frac{1 + \sin\alpha}{\cos\alpha} \qquad \textit{Reduce fraction.} \qquad ∎$$

EXAMPLE 7   **Proving an Identity by Factoring and Reducing**

Prove that $\dfrac{\sec^2 2\gamma - \tan^2 2\gamma}{\sec 2\gamma + \tan 2\gamma} = \dfrac{\cos 2\gamma}{1 + \sin 2\gamma}$ is an identity.

Solution   Sometimes the identity can be proved by factoring and reducing.

$$\frac{\sec^2 2\gamma - \tan^2 2\gamma}{\sec 2\gamma + \tan 2\gamma} = \frac{(\sec 2\gamma + \tan 2\gamma)(\sec 2\gamma - \tan 2\gamma)}{\sec 2\gamma + \tan 2\gamma}$$

$$= \sec 2\gamma - \tan 2\gamma \qquad \textit{If an expression seems}$$
$$= \frac{1}{\cos 2\gamma} - \frac{\sin 2\gamma}{\cos 2\gamma} \qquad \textit{simplified but is not in the form}$$
$$\textit{you want, consider changing}$$
$$= \frac{1 - \sin 2\gamma}{\cos 2\gamma} \qquad \textit{all functions to cosines and}$$
$$\textit{sines.}$$

$$= \frac{1 - \sin 2\gamma}{\cos 2\gamma} \cdot \frac{1 + \sin 2\gamma}{1 + \sin 2\gamma} \qquad \textit{Multiply by 1.}$$

$$= \frac{1 - \sin^2 2\gamma}{\cos 2\gamma(1 + \sin 2\gamma)} \qquad \textit{Multiply fractions.}$$

$$= \frac{\cos^2 2\gamma}{\cos 2\gamma(1 + \sin 2\gamma)} \qquad \textit{Fundamental identity:}$$
$$\cos^2 2\gamma = 1 - \sin^2 2\gamma$$

$$= \frac{\cos 2\gamma}{1 + \sin 2\gamma} \qquad \textit{Reduce fractions; common}$$
$$\textit{factor } \cos 2\gamma. \qquad ∎$$

EXAMPLE 8   **Proving an Identity by Multiplying by 1**

Prove that $\dfrac{-2\sin\theta\cos\theta}{1 - \sin\theta - \cos\theta} = 1 + \sin\theta + \cos\theta$ is an identity.

Solution   Sometimes, when there is a fraction on one side, the identity can be proved by multiplying the other side by 1, where 1 is written so that the desired denominator is obtained. Thus, for this example,

$1 + \sin\theta + \cos\theta$

$$= (1 + \sin\theta + \cos\theta) \cdot \frac{1 - \sin\theta - \cos\theta}{1 - \sin\theta - \cos\theta} \qquad \textit{Multiply by 1.}$$

$$= \frac{(1 + \sin\theta + \cos\theta)(1 - \sin\theta - \cos\theta)}{1 - \sin\theta - \cos\theta} \qquad \textit{Multiply fractions.}$$

$$= \frac{1 - \sin\theta - \cos\theta + \sin\theta - \sin^2\theta - \sin\theta\cos\theta + \cos\theta - \cos\theta\sin\theta - \cos^2\theta}{1 - \sin\theta - \cos\theta}$$

$$= \frac{1 - (\sin^2\theta + \cos^2\theta) - 2\sin\theta\cos\theta}{1 - \sin\theta - \cos\theta} \qquad \textit{Combine terms.}$$

$$= \frac{-2\sin\theta\cos\theta}{1 - \sin\theta - \cos\theta} \qquad \begin{array}{l}\textit{Fundamental identity:}\\ \sin^2\theta + \cos^2\theta = 1 \\ \textit{and } 1 - 1 = 0\end{array} \qquad ■$$

In summary, no single method is best for proving identities. However, the following hints should help.

PROCEDURES FOR PROVING IDENTITIES

1. If one side contains only one function, write all the trigonometric functions on the other side in terms of that function.
2. If the denominator of a fraction consists of only one function, break up the fraction.
3. Simplify by combining fractions.
4. Factoring is sometimes helpful.
5. Change all trigonometric functions to sines and cosines and simplify.
6. Multiply by the conjugate of either the numerator or the denominator.
7. If there are squares of functions, look for alternative forms of the Pythagorean identities.
8. Avoid the introduction of radicals.
9. Keep your destination in sight. Watch where you are going, and know when you are finished.

Disproving an Identity

If you suspect that a given equation is not an identity and want to prove that hypothesis, you may find a **counterexample.** That is, if you can find *one* *replacement* of the variable for which the functions are defined that will make the equation false, then you have proved that the equation is not an identity. In such a case, we say that you have **disproved an identity.**

EXAMPLE 9   **Disproving an Identity**

A common mistake for trigonometry students is to write $\sin 2\theta$ as $2 \sin \theta$. Show that

$$\sin 2\theta = 2 \sin \theta$$

is not an identity.

Solution   A graphing calculator can be used to show that the equation is not an identity.

Calculator Window ? X

A calculator makes it easy to show that a given equation, such as the one in Example 9, is not an identity. Graph $Y1 = \sin 2X$ and $Y2 = 2 \sin X$. The graphs are shown here.

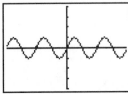

```
Y₁⊟sin (2X)
Xmin=-6.152285…
Xmax=6.1522856…
Xscl=1.5707963…
Ymin=-4
Ymax=4
Yscl=1
```

```
Y₂⊟2sin X
Xmin=-6.152285…
Xmax=6.1522856…
Xscl=1.5707963…
Ymin=-4
Ymax=4
Yscl=1
```

You can see these graphs are not the same, so the equation is not an identity. You can also graph both curves at the same time, as shown at the right.

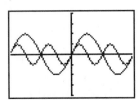

```
Y₁⊟sin (2X)
Y₂⊟2sin X
Xmin=-6.152285…
Xmax=6.1522856…
Xscl=1.5707963…
Ymin=-4
Ymax=4
Yscl=1
```

The identity can also be disproved by finding a counterexample. Let $\theta = \frac{\pi}{6}$. Then, by substitution,

$$\sin\left(2 \cdot \frac{\pi}{6}\right) = 2 \sin \frac{\pi}{6}$$

$$\frac{\sqrt{3}}{2} = 2\left(\frac{1}{2}\right) \qquad \textit{False equation}$$

Since this is a false equation, we have found a counterexample, namely, $x = \frac{\pi}{6}$. ∎

EXAMPLE 10

Deciding Whether an Equation Is an Identity

Is the equation $\cos^2 x - \sin^2 x = \sin x$ an identity?

Solution   If you suspect that an equation might not be an identity, you can compare the graphs of the left and right sides, or you can try to prove that it is not an identity by counterexample. Suppose we let $x = \frac{\pi}{6}$. Then, by substitution,

$$\cos^2 \frac{\pi}{6} - \sin^2 \frac{\pi}{6} = \sin \frac{\pi}{6} \qquad \textit{Substitute selected value.}$$

$$\left(\frac{\sqrt{3}}{2}\right)^2 - \left(\frac{1}{2}\right)^2 = \frac{1}{2} \qquad \textit{Evaluate the trigonometric functions.}$$

$$\frac{3}{4} - \frac{1}{4} = \frac{1}{2} \qquad \textit{True equation}$$

The fact that the value $x = \frac{\pi}{6}$ satisfies the equation does not answer the question. We have found a root for the equation, but we have not shown that the equation is an identity. Remember, for an equation to be an identity, it must be true for *all* values of the variable in the domain, not just one or two. *One* counterexample disproves the identity, but *one* true value does **not** prove an identity. In fact, if we solve this equation (we leave these details to you), we find $x = \frac{\pi}{6} + 2k\pi, \frac{5\pi}{6} + 2k\pi, \frac{3\pi}{2} + 2k\pi, k$ any integer. If you choose any of these values, the equation will be satisfied. If you choose any other value—for example, $x = \frac{\pi}{3}$—the equation is false.

$$\cos^2 \frac{\pi}{3} - \sin^2 \frac{\pi}{3} = \sin \frac{\pi}{3} \qquad \textit{Substitute selected value.}$$

$$\left(\frac{1}{2}\right)^2 - \left(\frac{\sqrt{3}}{2}\right)^2 = \frac{\sqrt{3}}{2} \qquad \textit{Evaluate the trigonometric functions.}$$

$$\frac{1}{4} - \frac{3}{4} = \frac{\sqrt{3}}{2} \qquad \textit{False equation}$$

We have now found a counterexample, and the identity is disproved. ∎

PROBLEM SET 4.2

Essential Concepts

1. **IN YOUR OWN WORDS** What is an equation? What is an identity?

2. **IN YOUR OWN WORDS** Distinguish the procedures for solving an equation and for proving that an equation is an identity.

3. **IN YOUR OWN WORDS** State the eight fundamental identities.

4. **IN YOUR OWN WORDS** List some of the procedures you can use to prove identities.

5. **IN YOUR OWN WORDS** If you have a graphing calculator, explain how it can help you prove or disprove that an equation is an identity.

Level 1 Drill

Problems 6–9: Use the fundamental identities to write the remaining five trigonometric values as functions of u. Assume u > 0, but pay close attention to the given quadrant for the angle θ.

6. $\cos \theta = \dfrac{u}{3}, \quad 0 < \theta < \dfrac{\pi}{2}$

7. $\sin \theta = -3u, \quad \dfrac{3\pi}{2} < \theta < 2\pi$

8. $\sin \theta = \dfrac{u}{\sqrt{3}}, \quad 0 < \theta < \dfrac{\pi}{2}$

9. $\cos \theta = -\dfrac{2u}{5}, \quad \dfrac{\pi}{2} < \theta < \pi$

Problems 10–33: Prove that the given equations are identities.

10. $\sec \theta = \sec \theta \sin^2\theta + \cos \theta$

11. $\tan \theta = \cot \theta \tan^2\theta$

12. $\dfrac{\sin \theta \cos \theta + \sin^2\theta}{\sin \theta} = \cos \theta + \sin \theta$

13. $\tan^2\theta - \sin^2\theta = \tan^2\theta \sin^2\theta$

14. $\cot^2\theta \cos^2\theta = \cot^2\theta - \cos^2\theta$

15. $\tan \theta + \cot \theta = \sec \theta \csc \theta$

16. $\dfrac{\cos \alpha + \tan \alpha \sin \alpha}{\sec \alpha} = 1$

17. $\dfrac{1 - \sec^2\beta}{\sec^2\beta} = -\sin^2\beta$

18. $(\sec \theta - \cos \theta)^2 = \tan^2\theta - \sin^2\theta$

19. $1 - \sin 2\theta = \dfrac{1 - \sin^2 2\theta}{1 + \sin 2\theta}$

20. $\sin \lambda = \dfrac{\sin^2\lambda + \sin \lambda \cos \lambda + \sin \lambda}{\sin \lambda + \cos \lambda + 1}$

21. $\dfrac{1 + \cos 2\lambda \sec 2\lambda}{\tan 2\lambda + \sec 2\lambda} = \dfrac{2 \cos 2\lambda}{\sin 2\lambda + 1}$

22. $\sin 3\beta \cos 3\beta(\tan 3\beta + \cot 3\beta) = 1$

23. $(\sin \alpha + \cos \alpha)^2 + (\sin \alpha - \cos \alpha)^2 = 2$

24. $\csc \beta - \cos \beta \cot \beta = \sin \beta$

25. $\dfrac{1 + \cot^2\gamma}{1 + \tan^2\gamma} = \cot^2\gamma$

26. $\dfrac{\sin^2\omega - \cos^2\omega}{\sin \omega + \cos \omega} = \sin \omega - \cos \omega$

27. $\tan^2\phi + \sin^2\phi + \cos^2\phi = \sec^2\phi$

28. $\dfrac{\tan \varphi + \cot \varphi}{\sec \varphi \csc \varphi} = 1$

29. $1 + \sin^2\theta = 2 - \cos^2\theta$

30. $\dfrac{\sin \theta}{\tan \theta} + \dfrac{\cos \theta}{\cot \theta} = \cos \theta + \sin \theta$

31. $\dfrac{1}{1 + \cos 2\theta} + \dfrac{1}{1 - \cos 2\theta} = 2 \csc^2 2\theta$

32. $2 \sin^2 2\alpha - 1 = 1 - 2 \cos^2 2\alpha$

33. $\dfrac{1}{\tan 2\beta} + \dfrac{\cos 2\beta}{\cot 2\beta} = \cot 2\beta + \sin 2\beta$

Problems 34–39: Find a counterexample to show that each equation is not an identity. Be sure you do not choose a value that makes any of the functions undefined.

34. $\sec^2 x - 1 = \sqrt{3} \tan x$

35. $2 \cos 2\theta \sin 2\theta = \sin 2\theta$

36. $\sin \theta = \cos \theta$

37. $\cos^2\theta - 3 \sin \theta + 3 = 0$

38. $\sin^2\theta + \cos \theta = 0$

39. $2 \sin^2\theta - 2 \cos^2\theta = 1$

Problems 40–43: *Use graphs to decide whether the equations are identities. For each problem, show the graphs and the input functions.*

40. $\cos^4 x = \sin^3 x + \cos^2 x - \sin x$

41. $\dfrac{\tan^2 x - 2 \tan x}{2 \tan x - 4} = \dfrac{1}{2} \tan x$

42. $\dfrac{\cos^2 x - \cos x \csc x}{\cos^2 x \csc x - \sin^2 x \csc^2 x} = \sin x$

43. $\dfrac{\sin x}{\csc x} + \dfrac{\cos x}{\sec x} = 1$

WHAT IS WRONG, *if anything, with each of the statements in Problems 44–46? Explain your reasoning.*

44. Prove

$$\sin x + \cos x = \frac{\sec x + \csc x}{\csc x \sec x}$$

is an identity.
  Multiply both sides by $\csc x \sec x$:

$$(\csc x \sec x)(\sin x + \cos x)$$

$$= \frac{\sec x + \csc x}{\csc x \sec x}(\csc x \sec x)$$

$$\csc x \sec x \sin x + \csc x \sec x \cos x$$

$$= \sec x + \csc x$$

$$\sec x + \csc x = \sec x + \csc x$$

Since this is true, the identity is proved.

45. Prove or disprove that

$$\frac{\sin 2\theta}{2} = \sin \theta$$

It is an identity, by dividing: $\dfrac{\sin 2\theta}{2} = \sin \theta$

46. Prove or disprove that

$$\cos(\alpha - \beta) = \cos \alpha - \cos \beta$$

It is an identity because of the distributive property.

Problems 47–64: *Prove that the equations are identities.*

47. $\dfrac{1 + \tan \alpha}{1 - \tan \alpha} = \dfrac{\sec^2 \alpha + 2 \tan \alpha}{2 - \sec^2 \alpha}$

48. $(\cot x + \csc x)^2 = \dfrac{\sec x + 1}{\sec x - 1}$

49. $\dfrac{\sin^3 x - \cos^3 x}{\sin x - \cos x} = 1 + \sin x \cos x$

50. $\dfrac{1 - \cos \theta}{1 + \cos \theta} = \left(\dfrac{1 - \cos \theta}{\sin \theta}\right)^2$

51. $\dfrac{(\sec^2 \gamma + \tan^2 \gamma)^2}{\sec^4 \gamma - \tan^4 \gamma} = 1 + 2 \tan^2 \gamma$

52. $(\sec \gamma + \csc \gamma)^2 = \dfrac{1 + 2 \sin \gamma \cos \gamma}{\cos^2 \gamma \sin^2 \gamma}$

53. $\dfrac{1}{\sec \theta + \tan \theta} = \sec \theta - \tan \theta$

54. $\dfrac{1 + \tan^3 \theta}{1 + \tan \theta} = \sec^2 \theta - \tan \theta$

55. $\dfrac{1 - \sec^3 \theta}{1 - \sec \theta} = \tan^2 \theta + \sec \theta + 2$

56. $\dfrac{\tan \theta}{\cot \theta} - \dfrac{\cot \theta}{\tan \theta} = \sec^2 \theta - \csc^2 \theta$

57. $\sqrt{(3 \cos \theta - 4 \sin \theta)^2 + (3 \sin \theta + 4 \cos \theta)^2} = 5$

58. $\dfrac{\csc \theta + 1}{\cot^2 \theta + \csc \theta + 1} = \dfrac{\sin^2 \theta + \sin \theta \cos \theta}{\sin \theta + \cos \theta}$

59. $\dfrac{\cos^4 \theta - \sin^4 \theta}{(\cos^2 \theta - \sin^2 \theta)^2} = \dfrac{\cos \theta}{\cos \theta + \sin \theta} + \dfrac{\sin \theta}{\cos \theta - \sin \theta}$

60. $(\cos \alpha - \cos \beta)^2 + (\sin \alpha - \sin \beta)^2$
$\qquad = 2 - 2(\cos \alpha \cos \beta + \sin \alpha \sin \beta)$

61. $(\sec \alpha + \sec \beta)^2 - (\tan \alpha - \tan \beta)^2$
$\qquad = 2 + 2(\sec \alpha \sec \beta + \tan \alpha \tan \beta)$

62. $\sin \theta + \cos \theta + 1 = \dfrac{2 \sin \theta \cos \theta}{\sin \theta + \cos \theta - 1}$

63. $\dfrac{2 \tan^2 \theta + 2 \tan \theta \sec \theta}{\tan \theta + \sec \theta - 1} = \tan \theta + \sec \theta + 1$

64. $(\cos \alpha \cos \beta \tan \alpha + \sin \alpha \sin \beta \cot \beta) \csc \alpha \sec \beta = 2$

Level 2 Applications

PROBLEMS FROM CALCULUS *Trigonometric identities are often used in calculus. One example involves radicals of the form*

$$\sqrt{a^2 - u^2}, \quad \sqrt{u^2 - a^2}, \quad \text{and} \quad \sqrt{a^2 + u^2}$$

Problems 65–67 deal with these radicals.

65. In the expression $\sqrt{4 - u^2}$, let $u = 2 \sin x$ for $-\pi/2 \le x \le \pi/2$. Make this trigonometric substitution so you can write this expression as a function of x in a final form that is free from radicals.

66. In the expression $\sqrt{u^2 - 9}$, let $u = 3 \sec x$ for $0 \le x < \pi/2$. Make this trigonometric substitution so you can write this expression as a function of x in a final form that is free from radicals.

67. In the expression $\sqrt{1 + u^2}$, let $u = \tan x$ for $0 \le x < \pi/2$. Make this trigonometric substitution so you can write this expression as a function of x in a final form that is free from radicals.

Level 3

68. JOURNAL PROBLEM [*The Mathematics Student Journal* of the National Council of Teachers of Mathematics, April 1973, 20(4).] For $0 \le \theta < \frac{\pi}{2}$, solve

$$\left(\frac{16}{81}\right)^{\sin^2\theta} + \left(\frac{16}{81}\right)^{1-\sin^2\theta} = \frac{26}{27}$$

69. Determine the number of solutions for the equation

$$\frac{x}{100} = \sin x$$

ADDITION LAWS

PROBLEM OF THE DAY

Simplify $\cos(\alpha - \beta)$.
Since $\cos(\alpha - \beta) \ne \cos \alpha - \cos \beta$, what does $\cos(\alpha - \beta)$ equal?

How would you answer this question?

When proving identities, it is sometimes necessary to simplify the functional values of the sum or difference of two angles. If α and β represent any two angles,

$$\cos(\alpha - \beta) \ne \cos \alpha - \cos \beta$$

For example, if $\alpha = 60°$ and $\beta = 30°$, then

$$\cos(60° - 30°) = \cos 60° - \cos 30° \qquad \textit{Substitute values.}$$
$$\cos 30° = \cos 60° - \cos 30° \qquad \textit{Parentheses first}$$
$$\frac{\sqrt{3}}{2} = \frac{1}{2} - \frac{\sqrt{3}}{2} \qquad \textit{This is a false equation.}$$

Thus, by counterexample, $\cos(60° - 30°) \ne \cos 60° - \cos 30°$. To solve the Problem of the Day, we must find a formula (in terms of functions of α and β) for $\cos(\alpha - \beta)$.

Historical Note

Early references to trigonometry were concerned with the solutions of triangles. The first systematic European book on plane and spherical trigonometry was written by Johann Müller (1436–1476), who was known as Regiomontanus. It was called De triangulis omnimodis, and a large part of the first two volumes was devoted to solving triangles. According to historian Howard Eves, Regiomontanus had many interests, including algebra, astronomy, and printing. He constructed a mechanical eagle that flapped its wings and was considered one of the marvels of the age. He died suddenly at the age of 40; rumor had it that he was poisoned by an enemy.

Distance Formula

To find a formula for $\cos(\alpha - \beta)$, we need the **distance formula** from algebra.

YIELD

DISTANCE FORMULA

If $P(x_1, y_1)$ and $P_2(x_2, y_2)$ are any two points, then the **distance** d between P_1 and P_2 is

$$d = \sqrt{(x_2 - x_1)^2 + (y_2 - y_1)^2}$$

The proof of this formula follows directly from the Pythagorean theorem and is found in most algebra books.

EXAMPLE 1   **Using the Distance Formula**

Find the distance between $(1, -2)$ and $(-3, 4)$.

Solution

$d = \sqrt{(-3 - 1)^2 + (4 + 2)^2}$ *Substitute into the distance formula.*

$\quad = \sqrt{16 + 36} = \sqrt{52} = 2\sqrt{13}$ *Simplify the radical expression.*   ∎

EXAMPLE 2   **Using the Distance Formula with Trigonometric Functions**

Find the distance between $(\cos \theta, \sin \theta)$ and $(1, 0)$.

Solution

$d = \sqrt{(1 - \cos \theta)^2 + (0 - \sin \theta)^2}$ *Substitute into the distance formula.*

$\quad = \sqrt{1 - 2 \cos \theta + \cos^2\theta + \sin^2\theta}$ *Expand squared expressions.*

$\quad = \sqrt{1 - 2 \cos \theta + 1}$ *Fundamental identity,* $\cos^2\theta + \sin^2\theta = 1$

$\quad = \sqrt{2 - 2 \cos \theta}$ ∎

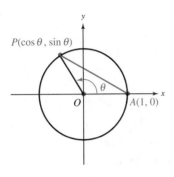

FIGURE 4.6

Length of a chord determined by an angle θ

Do you remember the Problem of the Day? Don't forget what it asks us to find. We are seeking a formula for $\cos(\alpha - \beta)$.

Chord Length

The solution in Example 2 can be used to find the length of any chord (in a unit circle) whose corresponding arc subtends or is intercepted by a central angle θ, where θ is in standard position. Let A be the point $(1, 0)$ and P the point on the intersection of the terminal side of angle θ and the unit circle (see Figure 4.6). This means (from the definition of cosine and sine) that the coordinates of P are $(\cos \theta, \sin \theta)$. For chord AP, the **chord length**, denoted by $|AP|$, can be found using the formula from Example 2.

CHORD LENGTH

The length of chord AP is

$$|AP| = \sqrt{2 - 2\cos\theta}$$

where θ is the central angle between OA and OP.

EXAMPLE 3

Finding the Length of a Chord in a Unit Circle

Find $|P_\alpha P_\beta|$, where P_α and P_β are points on a unit circle determined by the angles α and β, respectively.

Solution

Because $P_\alpha P_\beta$ forms a chord in a unit circle and since the central angle θ is $\alpha - \beta$ (see Figure 4.7),

$$|P_\alpha P_\beta| = \sqrt{2 - 2\cos(\alpha - \beta)} \qquad \blacksquare$$

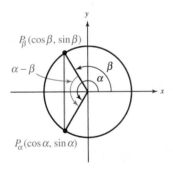

FIGURE 4.7

Distance between P_α and P_β

Let's rework Example 3 using the distance formula instead of the chord length formula:

$$|P_\alpha P_\beta| = \sqrt{(\cos\beta - \cos\alpha)^2 + (\sin\beta - \sin\alpha)^2}$$
$$= \sqrt{(\cos^2\beta - 2\cos\alpha\cos\beta + \cos^2\alpha + \sin^2\beta - 2\sin\alpha\sin\beta + \sin^2\alpha}$$
$$= \sqrt{(\cos^2\beta + \sin^2\beta) + (\cos^2\alpha + \sin^2\alpha) - 2(\cos\alpha\cos\beta + \sin\alpha\sin\beta)}$$
$$= \sqrt{2 - 2(\cos\alpha\cos\beta + \sin\alpha\sin\beta)}$$

This result and the result from Example 2 look very different, yet they both represent the same distance. Have you kept in mind the Problem of the Day? It can be solved by equating these expressions (since they are equivalent):

$$\sqrt{2 - 2\cos(\alpha - \beta)} = \sqrt{2 - 2(\cos\alpha\cos\beta + \sin\alpha\sin\beta)}$$
$$2 - 2\cos(\alpha - \beta) = 2 - 2(\cos\alpha\cos\beta + \sin\alpha\sin\beta)$$
$$-2\cos(\alpha - \beta) = -2(\cos\alpha\cos\beta + \sin\alpha\sin\beta)$$
$$\cos(\alpha - \beta) = \cos\alpha\cos\beta + \sin\alpha\sin\beta$$

YIELD

This identity is the formula we are seeking, and we summarize it in the following box. Later we include it in our list of identities, but because it provides the cornerstone for building a great many additional identities, we discuss it here first.

DIFFERENCE OF ANGLES IDENTITY

$$\cos(\alpha - \beta) = \cos \alpha \cos \beta + \sin \alpha \sin \beta$$

We can use this identity to find the exact values of functions of angles that are multiples of 15°.

EXAMPLE 4

Finding the Exact Value for a Cosine That Is a Multiple of 15°

Find the exact value of cos 345°.

Solution

$$\cos 345° = \cos 15°$$ *Reduction principle; cosine is positive in Quadrant IV.*

$$= \cos(45° - 30°)$$ *Since 45° − 30° = 15°*

$$= \cos 45° \cos 30° + \sin 45° \sin 30°$$ *Difference of angles identity*

$$= \left(\tfrac{1}{2}\sqrt{2}\right)\left(\tfrac{1}{2}\sqrt{3}\right) + \left(\tfrac{1}{2}\sqrt{2}\right)\left(\tfrac{1}{2}\right)$$ *Exact values*

$$= \frac{\sqrt{6} + \sqrt{2}}{4}$$ *Simplify.*   ■

It is also possible to derive the exact value of cos 15° by a geometrical argument using right triangles; this is left as a problem in the problem set.

Cofunction Identities

Even though the difference of angles identity is helpful in making evaluations (as in Example 4), its real value lies in the fact that it is true for *any* choice of α and β. By making some particular choices for α and β, we find several useful special cases for this identity.

EXAMPLE 5

Deriving a Cofunction Identity

Prove $\cos\left(\tfrac{\pi}{2} - \theta\right) = \sin \theta$.

Solution

This proof is based on the identity:

$$\cos(\alpha - \beta) = \cos \alpha \cos \beta + \sin \alpha \sin \beta$$ *Difference of angles identity*

$$\cos\left(\tfrac{\pi}{2} - \theta\right) = \cos \tfrac{\pi}{2} \cos \theta + \sin \tfrac{\pi}{2} \sin \theta$$ *Let $\alpha = \tfrac{\pi}{2}$ and $\beta = \theta$.*

$$\cos\left(\tfrac{\pi}{2} - \theta\right) = 0 \cdot \cos \theta + 1 \cdot \sin \theta$$ *Exact values: $\cos \tfrac{\pi}{2} = 0$, $\sin \tfrac{\pi}{2} = 1$*

$$\cos\left(\tfrac{\pi}{2} - \theta\right) = \sin \theta$$ *Simplify.*   ■

Example 5 is one of three identities known as the **cofunction identities.**

These identities tell us that we can change a trigonometric function to the cofunction of its complement.

COFUNCTION IDENTITIES
For any real number (or angle) θ,
12. $\cos\left(\frac{\pi}{2} - \theta\right) = \sin\theta$
13. $\sin\left(\frac{\pi}{2} - \theta\right) = \cos\theta$
14. $\tan\left(\frac{\pi}{2} - \theta\right) = \cot\theta$

Identity 12 was proved in Example 5, and the proof of identity 13 depends on identity 12, as shown here:

$$\cos\theta = \cos\left[\frac{\pi}{2} - \left(\frac{\pi}{2} - \theta\right)\right] = \sin\left(\frac{\pi}{2} - \theta\right)$$

Identities involving the tangent are usually proved after proving similar identities for cosine and sine. The fundamental identity $\tan\theta = \sin\theta/\cos\theta$ is applied first; this allows us to use the appropriate identities for cosine and sine. We illustrate this process with the following derivation:

$$\tan\left(\frac{\pi}{2} - \theta\right) = \frac{\sin\left(\frac{\pi}{2} - \theta\right)}{\cos\left(\frac{\pi}{2} - \theta\right)} = \frac{\cos\theta}{\sin\theta} = \cot\theta$$

The cofunction identities allow us to change a trigonometric function to the cofunction of its complement.

EXAMPLE 6

Writing a Trigonometric Function in Terms of Its Cofunction

Write each function in terms of its cofunction.

a. $\sin 38°$ **b.** $\cos 53°$ **c.** $\cot 4°$ **d.** $\tan\frac{\pi}{6}$ **e.** $\sec\alpha$ **f.** $\cos\left(\frac{\pi}{2} - \beta\right)$

Solution

a. $\sin 38° = \cos(90° - 38°) = \cos 52°$

b. $\cos 53° = \sin(90° - 53°) = \sin 37°$

c. $\cot 4° = \tan(90° - 4°) = \tan 86°$

d. $\tan\frac{\pi}{6} = \cot\left(\frac{\pi}{2} - \frac{\pi}{6}\right) = \cot\left(\frac{3\pi}{6} - \frac{\pi}{6}\right) = \cot\frac{2\pi}{6} = \cot\frac{\pi}{3}$

e. $\sec\alpha = \csc\left(\frac{\pi}{2} - \alpha\right)$

f. $\cos\left(\frac{\pi}{2} - \beta\right) = \sin\beta$ ∎

Opposite-Angle Identities

Suppose the given angle in Example 6 is larger than 90°; for example,

$$\cos 125° = \sin(90° - 125°) = \sin(-35°)$$

This result can be further simplified using the following **opposite-angle identities.**

> **OPPOSITE-ANGLE IDENTITIES**
>
> For any real number (or angle) θ,
>
> **15.** $\cos(-\theta) = \cos\theta$
> **16.** $\sin(-\theta) = -\sin\theta$
> **17.** $\tan(-\theta) = -\tan\theta$

Using identity 16, we see $\sin(-35°) = -\sin 35°$. The opposite-angle identities remind us of the definitions for even and odd functions.

> **EVEN FUNCTION; ODD FUNCTION**
>
> A function f is an **even function** if
>
> $$f(-x) = f(x)$$
>
> and an **odd function** if
>
> $$f(-x) = -f(x)$$

If $c(x) = \cos x$, then $c(-x) = \cos(-x) = \cos x = c(x)$, so cosine is an even function. Recall that this means that the graph of $y = \cos x$ is symmetric with respect to the y-axis. If $s(x) = \sin x$, then $s(-x) = \sin(-x) = -\sin x = -s(x)$, so sine is an odd function. This means that the graph of $y = \sin x$ is symmetric with respect to the origin. Is tangent an even or an odd function?

The proof for identity 15 depends on the difference of angles identity with the following choice; let $\alpha = 0$ and $\beta = \theta$:

$$\cos(0 - \theta) = \cos 0 \cos\theta + \sin 0 \sin\theta = 1 \cdot \cos\theta + 0 \cdot \sin\theta = \cos\theta$$

Thus, $\cos(-\theta) = \cos\theta$. The proofs of identities 16 and 17 are similar and are left as exercises.

EXAMPLE 7

Writing a Trigonometric Function As a Function of a Positive Angle

Write each function as a function of a positive angle or number:

a. $\cos(-19°)$ **b.** $\sin(-19°)$ **c.** $\tan(-2)$

Write each function in terms of its cofunction:

d. $\cos 125°$ **e.** $\sin 102°$ **f.** $\cot 2.5$

Solution   **a.** $\cos(-19°) = \cos 19°$

b. $\sin(-19°) = -\sin 19°$

c. $\tan(-2) = -\tan 2$

d. $\cos 125° = \sin(90° - 125°) = \sin(-35°) = -\sin 35°$

e. $\sin 102° = \cos(90° - 102°) = \cos(-12°) = \cos 12°$

f. $\cot 2.5 = \tan\left(\frac{\pi}{2} - 2.5\right) \approx \tan(-0.9292) = -\tan 0.9292$ ∎

Sometimes we use opposite-angle identities with other identities. For example, you know that

$$a - b \quad \text{and} \quad b - a$$

are opposites. This means that $(a - b) = -(b - a)$. In trigonometry, we often see angles such as $\frac{\pi}{2} - \theta$ and want to write $\theta - \frac{\pi}{2}$. This means that $\frac{\pi}{2} - \theta = -\left(\theta - \frac{\pi}{2}\right)$. In particular,

$$\cos\left(\frac{\pi}{2} - \theta\right) = \cos\left[-\left(\theta - \frac{\pi}{2}\right)\right] = \cos\left(\theta - \frac{\pi}{2}\right)$$

EXAMPLE 8

Using an Opposite-Angle Identity

Write the given functions using the opposite-angle identities:

a. $\sin\left(\frac{\pi}{2} - \theta\right)$ **b.** $\tan\left(\frac{\pi}{2} - \theta\right)$ **c.** $\cos(\pi - \theta)$

Solution   **a.** $\sin\left(\frac{\pi}{2} - \theta\right) = \sin\left[-\left(\theta - \frac{\pi}{2}\right)\right] = -\sin\left(\theta - \frac{\pi}{2}\right)$

b. $\tan\left(\frac{\pi}{2} - \theta\right) = \tan\left[-\left(\theta - \frac{\pi}{2}\right)\right] = -\tan\left(\theta - \frac{\pi}{2}\right)$

c. $\cos\left(\frac{\pi}{2} - \theta\right) = \cos\left[-\left(\theta - \frac{\pi}{2}\right)\right] = \cos\left(\theta - \frac{\pi}{2}\right)$ ∎

Graphing with Opposites

The procedure for graphing $y = -\cos x$ is identical to the procedure for graphing $y = \cos x$, except that, after the frame is drawn, the endpoints are at the bottom of the frame instead of at the top, and the midpoint is at the top, as shown in Figure 4.8.

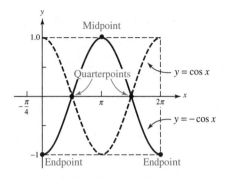

FIGURE 4.8   One period of $y = -\cos x$

EXAMPLE 9   **Using an Opposite-Angle Identity to Graph a Sine**

Graph $y = \sin(-x)$.

Solution   If necessary, use an opposite-angle identity first: $\sin(-x) = -\sin x$. To graph $y = -\sin x$, build the frame as before. The endpoints and midpoints are the same, but the quarterpoints are reversed, as shown in Figure 4.9.

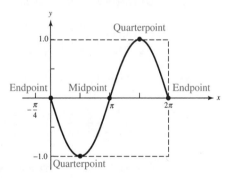

FIGURE 4.9   One period of $y = -\sin x$ ■

Addition Laws

The difference of angles identity proved at the beginning of this section is one of six identities known as the **addition laws.** Since subtraction can easily be written as a sum, the designation *addition laws* refers to both addition and subtraction.

ADDITION LAWS

18. $\cos(\alpha + \beta) = \cos\alpha\cos\beta - \sin\alpha\sin\beta$

19. $\cos(\alpha - \beta) = \cos\alpha\cos\beta + \sin\alpha\sin\beta$

20. $\sin(\alpha + \beta) = \sin\alpha\cos\beta + \cos\alpha\sin\beta$

21. $\sin(\alpha - \beta) = \sin\alpha\cos\beta - \cos\alpha\sin\beta$

22. $\tan(\alpha + \beta) = \dfrac{\tan\alpha + \tan\beta}{1 - \tan\alpha\tan\beta}$

23. $\tan(\alpha - \beta) = \dfrac{\tan\alpha - \tan\beta}{1 + \tan\alpha\tan\beta}$

EXAMPLE 10   **Using an Addition Law**

Write $\cos\!\left(\frac{2\pi}{3} + \theta\right)$ as a function of θ only.

Solution

$$
\begin{aligned}
\cos\!\left(\tfrac{2\pi}{3} + \theta\right) &= \cos\tfrac{2\pi}{3}\cos\theta - \sin\tfrac{2\pi}{3}\sin\theta \qquad \textit{Addition law}\\[2mm]
&= \left(-\tfrac{1}{2}\right)\cos\theta - \left(\tfrac{\sqrt{3}}{2}\right)\sin\theta \qquad \textit{Exact values}\\[2mm]
&= -\tfrac{1}{2}\left(\cos\theta + \sqrt{3}\sin\theta\right) \qquad \textit{Common factor}
\end{aligned}
$$
■

EXAMPLE 11   **Using an Addition Law to Evaluate a Trigonometric Expression**

Evaluate $\dfrac{\tan 18° - \tan 40°}{1 + \tan 18° \tan 40°}$ using a calculator (two-place accuracy).

Solution   You can do a lot of calculator arithmetic, or you can use an addition law:

$$
\frac{\tan 18° - \tan 40°}{1 + \tan 18° \tan 40°} = \tan(18° - 40°) = \tan(-22°) \approx -0.4040262258
$$

The result (to two-place accuracy) is -0.40. ■

EXAMPLE 12   **Proving an Addition Law**

Prove $\cos(\alpha + \beta) = \cos\alpha\cos\beta - \sin\alpha\sin\beta$.

Solution

$$
\begin{aligned}
\cos(\alpha + \beta) &= \cos[\alpha - (-\beta)] \qquad\qquad \textit{Definition of subtraction}\\[2mm]
&= \cos\alpha\cos(-\beta) + \sin\alpha\sin(-\beta) \qquad \textit{Difference of angles identity}\\[2mm]
&= \cos\alpha\cos\beta - \sin\alpha\sin\beta \qquad\quad \textit{Opposite-angle identities}
\end{aligned}
$$
■

EXAMPLE 13   **Using an Addition Law to Evaluate a Function of Inverse Functions**

Evaluate $\cos\left[\sin^{-1}\left(\frac{4}{5}\right) + \cos^{-1}\left(\frac{3}{5}\right)\right]$ without tables or calculator.

Solution   Let α and β be Quadrant I angles (both inverse sine and inverse cosine are positive in the first quadrant) so that

$$\alpha = \sin^{-1}\left(\frac{4}{5}\right) \text{ or } \sin \alpha = \frac{4}{5} \quad \text{and} \quad \beta = \cos^{-1}\left(\frac{3}{5}\right) \text{ or } \cos \beta = \frac{3}{5}$$

We now need to evaluate

$$\cos\left[\sin^{-1}\left(\tfrac{4}{5}\right) + \cos^{-1}\left(\tfrac{3}{5}\right)\right] = \cos(\alpha + \beta)$$

$$= \cos \alpha \cos \beta - \sin \alpha \sin \beta$$

$$= \left(\sqrt{1 - \sin^2\alpha}\right)\cos \beta - \sin \alpha\left(\sqrt{1 - \cos^2\beta}\right)$$

$$= \left(\sqrt{1 - \left(\tfrac{4}{5}\right)^2}\right)\left(\tfrac{3}{5}\right) - \left(\tfrac{4}{5}\right)\left(\sqrt{1 - \left(\tfrac{3}{5}\right)^2}\right)$$

$$= \tfrac{3}{5}\sqrt{1 - \tfrac{16}{25}} - \tfrac{4}{5}\sqrt{1 - \tfrac{9}{25}}$$

$$= \tfrac{3}{5}\left(\tfrac{3}{5}\right) - \tfrac{4}{5}\left(\tfrac{4}{5}\right)$$

$$= -\tfrac{7}{25}$$ ∎

PROBLEM SET 4.3

Essential Concepts

1. What is the distance formula?
2. List the three cofunction identities.
3. List the three opposite-angle identities.
4. List the six addition laws.

Level 1

WHAT IS WRONG, *if anything, with each of the statements in Problems 5–8? Explain your reasoning.*

5. $\sin(30° + 45°) = \sin 30° + \sin 45°$

6. $\cos(2 \cdot 45°) = \cos 45° + \cos 45°$

7. $\sin\left(\frac{\pi}{2} - \theta\right) = \sin\frac{\pi}{2} - \sin \theta$

8. $\sin\left(\frac{\pi}{2} - \theta\right) = \sin\left(\theta - \frac{\pi}{2}\right)$

Level 1 Drill

Problems **9–16:** *Change each expression to a function of θ only.*

9. $\sin(\theta + 45°)$
10. $\cos\left(\frac{\pi}{6} + \theta\right)$
11. $\cos(\theta - 45°)$
12. $\cos\left(\frac{\pi}{3} - \theta\right)$
13. $\tan(45° + \theta)$
14. $\tan(\theta + \theta)$
15. $\sin(\theta + \theta)$
16. $\cos(\theta + \theta)$

Problems **17–22:** *Write each function in terms of its cofunction by using the cofunction identities.*

17. $\sin 15°$
18. $\sin 38°$
19. $\tan 62°$
20. $\cos \frac{\pi}{6}$
21. $\cos \frac{5\pi}{6}$
22. $\cot \frac{2\pi}{3}$

Problems **23–28:** *Write each function as a function of a positive angle.*

23. $\sin(-23°)$
24. $\tan(-54°)$
25. $\cos(-57°)$
26. $\cos(-19°)$
27. $\tan(-29°)$
28. $\sin(-83°)$

Problems 29–34: Evaluate each expression to four decimal places.

29. $\sin 158° \cos 92° - \cos 158° \sin 92°$

30. $\sin 18° \cos 23° + \cos 18° \sin 23°$

31. $\cos 30° \cos 48° + \sin 30° \sin 48°$

32. $\cos 114° \cos 85° + \sin 114° \sin 85°$

33. $\dfrac{\tan 32° + \tan 18°}{1 - \tan 32° \tan 18°}$

34. $\dfrac{\tan 59° - \tan 25°}{1 + \tan 59° \tan 25°}$

Problems 35–40: Find the exact values for the sine, cosine, and tangent of each angle.

35. $15°$ **36.** $-15°$ **37.** $75°$

38. $105°$ **39.** $165°$ **40.** $195°$

41. Write each as a function of $\theta - \frac{\pi}{3}$.

  a. $\cos\left(\frac{\pi}{3} - \theta\right)$ **b.** $\sin\left(\frac{\pi}{3} - \theta\right)$ **c.** $\tan\left(\frac{\pi}{3} - \theta\right)$

42. Write each as a function of $\alpha - \beta$.

  a. $\cos(\beta - \alpha)$ **b.** $\sin(\beta - \alpha)$ **c.** $\tan(\beta - \alpha)$

Problems 43–48: Graph each function.

43. $y = -2 \cos x$ **44.** $y = -3 \sin x$

45. $y = \tan(-x)$ **46.** $y = \cos(-2x)$

47. $y - 1 = \cos(2\pi - x)$ **48.** $y - 2 = -\cos(\pi - x)$

49. Decide whether each function is even, odd, or neither.

  a. $f(x) = \sec x$ **b.** $g(x) = \dfrac{\sin x}{x}$

  c. $h(x) = \dfrac{x}{\cos x}$ **d.** $F(x) = \sin x \tan x$

Level 2 Theory

Problems 50–55: Prove the identities.

50. $\sin(\alpha - \beta) = \sin \alpha \cos \beta - \cos \alpha \sin \beta$

51. $\tan(\alpha - \beta) = \dfrac{\tan \alpha - \tan \beta}{1 + \tan \alpha \tan \beta}$

52. $\dfrac{\cos 5\theta}{\sin \theta} - \dfrac{\sin 5\theta}{\cos \theta} = \dfrac{\cos 6\theta}{\sin \theta \cos \theta}$

53. $\dfrac{\sin 6\theta}{\sin 3\theta} - \dfrac{\cos 6\theta}{\cos 3\theta} = \sec 3\theta$

54. $\sin(\alpha + \beta)\cos \beta - \cos(\alpha + \beta)\sin \beta = \sin \alpha$

55. $\cos(\alpha - \beta)\cos \beta - \sin(\alpha - \beta)\sin \beta = \cos \alpha$

Problems 56–59: Evaluate each expression without using tables or a calculator.

56. $\sin\left[\sin^{-1}\left(\frac{4}{5}\right) - \cos^{-1}\left(\frac{3}{5}\right)\right]$

57. $\cos\left[\sin^{-1}\left(\frac{3}{5}\right) + \cos^{-1}\left(\frac{4}{5}\right)\right]$

58. $\cos\left[\cos^{-1}\left(\frac{1}{4}\right) - \sin^{-1}\left(\frac{1}{3}\right)\right]$

59. $\sin\left[\cos^{-1}\left(\frac{2}{3}\right) + \sin^{-1}\left(\frac{1}{2}\right)\right]$

60. Show (without a calculator) that

$$\tan^{-1}\left(\tfrac{1}{2}\right) + \tan^{-1}\left(\tfrac{1}{3}\right) = \tfrac{\pi}{4}$$

Level 2 Applications

Problems 61–64: Let $\triangle ABC$ be a triangle labeled with angles α, β, and γ. Prove or disprove the following identities concerning interior angles in a triangle.

61. $\tan(\alpha + \beta) = -\tan \gamma$

62. $\sin \alpha = \cos\left(\frac{1}{2}\alpha + \frac{3}{2}\beta + \frac{1}{2}\gamma\right)$

63. $\sin\left(\dfrac{\alpha + \beta}{2}\right) = \cos\left(\dfrac{\gamma}{2}\right)$

64. $\tan \alpha + \tan \beta + \tan \gamma = \tan \alpha \tan \beta \tan \gamma$

65. In Example 4, we found the exact value of $\cos 15°$ by using the difference of angles identity. In this problem, we find the exact value of $\cos 15°$ geometrically. Draw isoceles $\triangle ABC$ with $|\overline{AC}| = |\overline{BC}| = 2$, and $\angle ACB = 150°$, as shown in Figure 4.10. Next, construct right $\triangle ABD$. Use this figure and the definition of the trigonometric functions to find the exact value of $\cos 15°$.

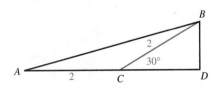

FIGURE 4.10
Triangles for finding $\cos 15°$

PROBLEMS FROM CALCULUS *An important expression, called a difference quotient, is defined to be*

$$\frac{f(x+h) - f(x)}{h}$$

Problems 66–67: Prove each identity.

66. Let $f(x) = \sin x$; prove that the difference quotient

$$\frac{\sin(x+h) - \sin x}{h}$$

is equivalent to

$$\cos x \left(\frac{\sin h}{h}\right) - \sin x \left(\frac{1 + \cos h}{h}\right)$$

67. Let $f(x) = \cos x$; prove that the difference quotient

$$\frac{\cos(x+h) - \cos x}{h}$$

is equivalent to

$$-\sin x \left(\frac{\sin h}{h}\right) - \cos x \left(\frac{1 - \cos h}{h}\right)$$

68. JOURNAL PROBLEM (From *The Mathematics Teacher,* May 1989, p. 384.) Prove the addition law

$$\tan(\alpha + \beta) = \frac{\tan \alpha + \tan \beta}{1 - \tan \alpha \tan \beta}$$

using a geometrical argument for the illustration given in Figure 4.11.

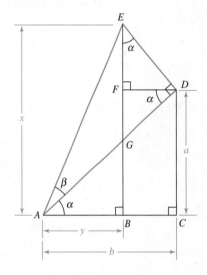

FIGURE 4.11
Geometric proof of an addition law

DOUBLE-ANGLE AND HALF-ANGLE IDENTITIES

PROBLEM OF THE DAY

Suppose a rectangle is inscribed in a semicircle of radius 1. What are the dimensions of the rectangle with the largest area?

How would you answer this question? Can you draw such a figure?

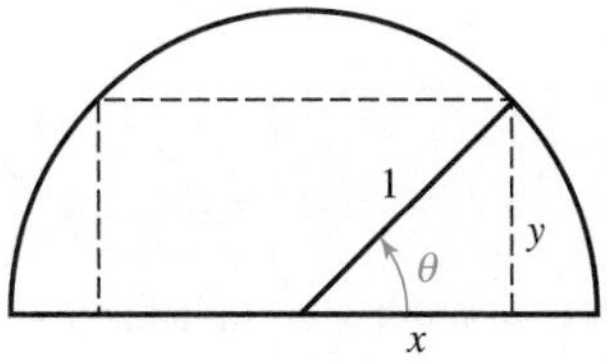

We can use trigonometry to answer this question. We draw a semicircle of radius 1 and then inscribe a rectangle, as shown in Figure 4.12. The area of a rectangle is found with the formula $A = \ell w$, so for this rectangle, the length is $2x$ and the width is y. Therefore

$$A = 2xy$$

FIGURE 4.12
What is the area of the inscribed rectangle?

As a function of the angle θ, we know (from the definition of the trigonometric functions) that $\cos \theta = x/1 = x$ and $\sin \theta = y/1 = y$, so the area as a function of θ is

$$A = 2 \cos \theta \sin \theta$$

To complete the answer, we need an identity known as a *double-angle identity*.

Double-Angle Identities

We now consider a special case of the addition laws.

$$
\begin{aligned}
\cos 2\theta &= \cos(\theta + \theta) \\
&= \cos \theta \cos \theta - \sin \theta \sin \theta \quad \textit{Addition law for } \cos(\alpha + \beta) \\
&= \cos^2\theta - \sin^2\theta \\
\sin 2\theta &= \sin(\theta + \theta) \\
&= \sin \theta \cos \theta + \cos \theta \sin \theta \quad \textit{Addition law for } \sin(\alpha + \beta) \\
&= 2 \sin \theta \cos \theta \\
\tan 2\theta &= \tan(\theta + \theta) \\
&= \frac{\tan \theta + \tan \theta}{1 - \tan \theta \tan \theta} \quad \textit{Addition law for } \tan(\alpha + \beta) \\
&= \frac{2 \tan \theta}{1 - \tan^2\theta}
\end{aligned}
$$

These identities are known as the **double-angle identities.**

DOUBLE-ANGLE IDENTITIES

24. $\cos 2\theta = \cos^2\theta - \sin^2\theta$

25. $\sin 2\theta = 2 \sin \theta \cos \theta$

26. $\tan 2\theta = \dfrac{2 \tan \theta}{1 - \tan^2\theta}$

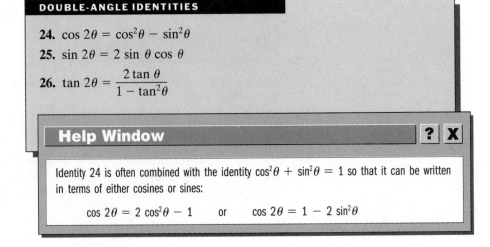

Help Window ? X

Identity 24 is often combined with the identity $\cos^2\theta + \sin^2\theta = 1$ so that it can be written in terms of either cosines or sines:

$$\cos 2\theta = 2 \cos^2\theta - 1 \qquad \text{or} \qquad \cos 2\theta = 1 - 2 \sin^2\theta$$

EXAMPLE 1   **Using a Double-Angle Identity for a Multiple Greater Than 2**

Write $\cos 3\theta$ in terms of $\cos \theta$.

Solution

$$\begin{aligned}
\cos 3\theta &= \cos(2\theta + \theta) \\
&= \cos 2\theta \cos \theta - \sin 2\theta \sin \theta \\
&= (\cos^2\theta - \sin^2\theta)\cos \theta - (2 \sin \theta \cos \theta)\sin \theta \\
&= \cos^3\theta - \sin^2\theta \cos \theta - 2 \sin^2\theta \cos \theta \\
&= \cos^3\theta - 3 \sin^2\theta \cos \theta \\
&= \cos^3\theta - 3(1 - \cos^2\theta)\cos \theta \\
&= \cos^3\theta - 3 \cos \theta + 3 \cos^3\theta \\
&= 4 \cos^3\theta - 3 \cos \theta
\end{aligned}$$
■

EXAMPLE 2   **Using a Double-Angle Identity to Evaluate an Expression**

Evaluate $\dfrac{2 \tan \frac{\pi}{16}}{1 - \tan^2 \frac{\pi}{16}}$ correct to four decimal places.

Solution   You could do this with a great deal of calculator work, or you can notice that this is the right-hand side of identity 26, so

$$\frac{2 \tan \frac{\pi}{16}}{1 - \tan^2 \frac{\pi}{16}} = \tan\left(2 \cdot \tfrac{\pi}{16}\right) = \tan \tfrac{\pi}{8} \approx 0.4142135624$$

The requested value is 0.4142.
■

We now return to the Problem of the Day. Remember that we want to find the maximum area of a rectangle inscribed in a semicircle with radius 1. We know that the area as a function of θ (see Figure 4.12) is

$$A = 2 \cos \theta \sin \theta$$

We now apply a double-angle identity so we can rewrite this formula as

$$A = \sin 2\theta$$

We know that the maximum value of a sine function is 1 and that this value occurs at 90°. We have

$$2\theta = 90°$$
$$\theta = 45°$$

Thus, the maximum area of 1 occurs for a rectangle with the dimensions

$$x = \cos 45° = \frac{\sqrt{2}}{2} \quad \text{and} \quad y = \sin 45° = \frac{\sqrt{2}}{2}$$

EXAMPLE 3   **Finding the Maximum Range for a Projectile**

Suppose an object is propelled upward at an angle θ to the horizontal with an initial velocity of 224 ft/s. Find the angle θ for which the range is a maximum, and then find that maximum downrange point of impact. (Ignore air resistance.)

Solution   Recall parametric equations for the path of a projectile (Example 5, Section 2.3, page 72):

$$x = (v_0 \cos \theta)t \qquad y = (v_0 \sin \theta)t - 16t^2$$

We know that the maximum range will be measured when the projectile hits the ground, that is, when $y = 0$:

$(v_0 \sin \theta)t - 16t^2 = y$	*Given*
$t[v_0 \sin \theta - 16t] = 0$	*Set $y = 0$ and factor the left side.*
$v_0 \sin \theta - 16t = 0$	*Set each factor equal to 0; $t = 0$ and $v_0 \sin \theta - 16t = 0$.*
$v_0 \sin \theta = 16t$	*Add 16t to both sides.*
$t = \dfrac{v_0 \sin \theta}{16}$	*Divide both sides by 16.*

Thus, the range (which is measured by the variable x) is

$x = (v_0 \cos \theta)t$	*Given*
$= (v_0 \cos \theta)\left(\dfrac{v_0 \sin \theta}{16}\right)$	*Write t as a function of θ.*
$= \frac{1}{16} v_0^2 \cos \theta \sin \theta$	*Simplify.*
$= \frac{1}{32} v_0^2 \sin 2\theta$	*Double-angle identity $\sin 2\theta = 2 \sin \theta \cos \theta$*

Now we can solve the problem. The maximum range (measured by x) will occur when $\sin 2\theta$ is the greatest, and we know that this occurs when $\sin 2\theta = 1$, or when $2\theta = 90°$ and $\theta = 45°$. We also find

$x = \frac{1}{32} v_0^2 \sin 2\theta$	*Given*
$= \frac{1}{32}(224)^2 \sin 90°$	*Substitute known values.*
$\approx 1{,}570$	*Answer to three significant figures.*

The answer is that the point of impact of the projectile will be 1,570 feet downrange if the angle of elevation is set at 45°, and furthermore, this is the maximum downrange distance. We check this path using a calculator, as shown in the margin.   ■

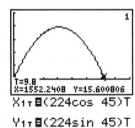

T=9.8
X=1552.2408 Y=15.600806
X₁ᴛ⊟(224cos 45)T

Y₁ᴛ⊟(224sin 45)T
-16T²
    Tmin=0
    Tmax=10
    Tstep=.1
Xmin=0   Ymin=-100
Xmax=2000 Ymax=500
Xscl=250 Yscl=100

EXAMPLE 4   **Using a Double-Angle Identity to Draw a Graph**

Graph $y = \sin^2 x$.

Solution   Since $\sqrt{x^2} = |x|$, some students *incorrectly* graph $y = \sin^2 x$ by drawing the graph of $y = |\sin x|$, as shown in Figure 4.13.

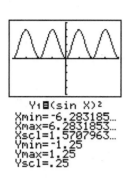

Y₁⧉(sin X)²
Xmin=-6.283185...
Xmax=6.2831853...
Xscl=1.5707963...
Ymin=-1.25
Ymax=1.25
Yscl=.25

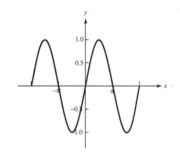

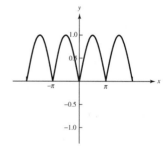

a. Calculator graph   **b.** Graph of $y = \sin x$ **c.** Graph of $y = |\sin x|$

FIGURE 4.13   Comparison of the graphs of $y = \sin^2 x$ (by calculator), $y = \sin x$, and $y = |\sin x|$

To graph $y = \sin^2 x$ correctly (other than by plotting points or by using a graphing calculator as shown in Figure 4.13a), consider using a double-angle identity:

$$\cos 2x = 1 - 2\sin^2 x \quad \textit{Double-angle identity}$$
$$2\sin^2 x = 1 - \cos 2x \quad \textit{Solve for } \sin^2 x.$$
$$\sin^2 x = \tfrac{1}{2} - \tfrac{1}{2}\cos 2x$$

We see that the curve $y = \sin^2 x$ is a cosine with $a = -\tfrac{1}{2}$, $b = 2$, and $(h, k) = \left(0, \tfrac{1}{2}\right)$, as shown in Figure 4.14. Note that the period p is found by $p = (2\pi)/b = (2\pi)/\pi = \pi$. The graph of $y = |\sin x|$ is shown as a dashed curve. Note that $y = \sin^2 x$ and $y = |\sin x|$ are not the same.

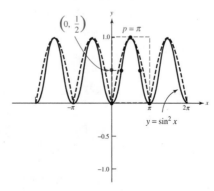

FIGURE 4.14   Graph of $y = \sin^2 x$ ∎

EXAMPLE 5

Finding Trigonometric Functions of Double Angles, Given a Single Angle

If $\cos\theta = \frac{3}{5}$ and θ is in Quadrant IV ($270° < \theta < 360°$), find $\cos 2\theta$, $\sin 2\theta$, and $\tan 2\theta$.

Solution

Since $\cos 2\theta = 2\cos^2\theta - 1$,

$$\cos 2\theta = 2\left(\tfrac{3}{5}\right)^2 - 1 = -\tfrac{7}{25}$$

For the other functions of 2θ, we also need to know $\sin\theta$. To find this value, we use

$$\sin\theta = -\sqrt{1 - \cos^2\theta} \qquad \textit{Negative since sine is negative in Quadrant IV}$$

$$= -\sqrt{1 - \left(\tfrac{3}{5}\right)^2} = -\tfrac{4}{5}$$

Now, we finish with the other double-angle formulas:

$$\sin 2\theta = 2\sin\theta\cos\theta = 2\left(-\tfrac{4}{5}\right)\left(\tfrac{3}{5}\right) = -\tfrac{24}{25}$$

$$\tan 2\theta = \frac{\sin 2\theta}{\cos 2\theta} = \frac{-\tfrac{24}{25}}{-\tfrac{7}{25}} = \tfrac{24}{7}$$

■

Half-Angle Identities

Sometimes, as in Example 5, we want to write $\cos 2\phi$ in terms of cosines, and at other times, we want to write it in terms of sines:

$$\cos 2\phi = 2\cos^2\phi - 1 \qquad \text{and} \qquad \cos 2\phi = 1 - 2\sin^2\phi$$

These forms of the double-angle identity lead us to what are called the **half-angle identities.** We do this by letting $2\phi = \theta$ so that $\phi = \frac{1}{2}\theta$.* We will show the derivation of the first half-angle identity and leave the second one as an exercise.

$$\cos 2\phi = 2\cos^2\phi - 1 \qquad \textit{Start with double-angle identity.}$$

$$\cos\theta = 2\cos^2\tfrac{1}{2}\theta - 1 \qquad \textit{Substitute } 2\phi = \theta.$$

$$2\cos^2\tfrac{1}{2}\theta = 1 + \cos\theta \qquad \textit{Solve for } \cos\tfrac{1}{2}\theta.$$

$$\cos^2\tfrac{1}{2}\theta = \frac{1 + \cos\theta}{2}$$

$$\cos\tfrac{1}{2}\theta = \pm\sqrt{\frac{1 + \cos\theta}{2}}$$

---

* Remember from algebra that $\frac{1}{2}\theta$ and $\frac{\theta}{2}$ mean the same thing, so $\cos\frac{1}{2}\theta$ and $\cos\frac{\theta}{2}$ are the same. Do not confuse this with $\dfrac{\cos\theta}{2}$, which is different.

The + or − sign is chosen according to the quadrant in which $\frac{1}{2}\theta$ is located. The formula requires either + or −, but not both. This use of $\pm$ is different from its use in algebra. For example, when we use $\pm$ in the quadratic formula, we are indicating *two* possible correct roots. In this trigonometric identity, we obtain *one* correct sign, + or − (but not both), depending on the quadrant of $\frac{1}{2}\theta$.

HALF-ANGLE IDENTITIES

27. $\cos\frac{1}{2}\theta = \pm\sqrt{\dfrac{1 + \cos\theta}{2}}$

28. $\sin\frac{1}{2}\theta = \pm\sqrt{\dfrac{1 - \cos\theta}{2}}$

29. $\tan\frac{1}{2}\theta = \dfrac{\sin\theta}{1 + \cos\theta}$

To help you remember the correct sign for identities 27 and 28, remember "sinus is minus."

To prove the half-angle identity, you can use

$$\tan\frac{1}{2}\theta = \frac{\sin\frac{1}{2}\theta}{\cos\frac{1}{2}\theta}$$

and then use identities 27 and 28 before you simplify. You can also use a geometric derivation, which is left for the problem set.

EXAMPLE 6

Using a Half-Angle Identity

Find the exact value of $\cos\frac{9\pi}{8}$.

Solution

$$\cos\frac{9\pi}{8} = \cos\left(\frac{1}{2}\cdot\frac{9\pi}{4}\right) = -\sqrt{\frac{1 + \cos\frac{9\pi}{4}}{2}}$$

Choose a negative sign because $9\pi/8$ is in Quadrant III, and the cosine is negative in Quadrant III.

$$= -\sqrt{\frac{1 + \cos\frac{\pi}{4}}{2}}$$

Reduction principle

$$= -\sqrt{\frac{1 + \frac{\sqrt{2}}{2}}{2}}$$

Exact value

$$= -\sqrt{\frac{2 + \sqrt{2}}{4}} = -\frac{1}{2}\sqrt{2 + \sqrt{2}}$$

■

EXAMPLE 7   **Finding Trigonometric Functions of Single Angles, Given a Double Angle**

If $\cot 2\theta = \frac{3}{4}$, find $\cos \theta$, $\sin \theta$, and $\tan \theta$, where 2θ is in Quadrant III.

Solution   We need to find $\cos 2\theta$ so that we can use it in the half-angle identities. To do this, we find $\tan 2\theta = 4/3$. Next, find $\sec 2\theta$:

$$\sec 2\theta = \pm \sqrt{1 + \tan^2 2\theta} \quad \textit{Pythagorean identity}$$

$$= -\sqrt{1 + \left(\frac{4}{3}\right)^2} \quad \begin{array}{l}\textit{Negative because } 2\theta \textit{ is in} \\ \textit{Quadrant III; Substitute } \tan 2\theta = 4/3.\end{array}$$

$$= -\frac{5}{3}$$

Finally, using a reciprocal relationship, $\cos 2\theta = -\frac{3}{5}$. We are now ready to use the half-angle identities (shown side by side):

Since $\pi < 2\theta < \frac{3\pi}{2}$,

$\quad \frac{\pi}{2} < \theta < \frac{3\pi}{4}$

so θ is in Quadrant II.

$$\cos \theta = \pm \sqrt{\frac{1 + \cos 2\theta}{2}} \qquad \sin \theta = \pm \sqrt{\frac{1 - \cos 2\theta}{2}}$$

$$= -\sqrt{\frac{1 + \left(-\frac{3}{5}\right)}{2}} \qquad\qquad = +\sqrt{\frac{1 - \left(-\frac{3}{5}\right)}{2}}$$

$$\underset{\textit{Quadrant II}}{\uparrow} \qquad\qquad\qquad \underset{\textit{Quadrant II}}{\uparrow}$$

$$= -\frac{1}{\sqrt{5}} \qquad\qquad\qquad = \frac{2}{\sqrt{5}}$$

$$\tan \theta = \frac{\sin \theta}{\cos \theta} = \frac{2/\sqrt{5}}{-1/\sqrt{5}} = -2 \qquad\qquad\qquad\blacksquare$$

EXAMPLE 8   **Using a Double-Angle Identity to Prove an Identity**

Prove that $\sin \theta = \dfrac{2 \tan \frac{1}{2}\theta}{1 + \tan^2 \frac{1}{2}\theta}$ is an identity.

Solution   When you prove identities involving functions of different angles, you should write all the trigonometric functions in the problem as functions of a single angle.

$$\frac{2 \tan \frac{1}{2}\theta}{1 + \tan^2 \frac{1}{2}\theta} = \frac{2 \dfrac{\sin \frac{1}{2}\theta}{\cos \frac{1}{2}\theta}}{\sec^2 \frac{1}{2}\theta} = 2 \frac{\sin \frac{1}{2}\theta}{\cos \frac{1}{2}\theta} \cdot \cos^2 \frac{1}{2}\theta = 2 \sin \frac{1}{2}\theta \cos \frac{1}{2}\theta = \sin 2\left(\frac{1}{2}\theta\right) = \sin \theta$$

$$\blacksquare$$

PROBLEM SET 4.4

Essential Concepts

WHAT IS WRONG, *if anything, with each of the statements in Problems 1–11? Explain your reasoning.*

1. $\cos 2\theta = \cos^2\theta + \sin^2\theta = 1$

2. $\dfrac{\cos 2\theta}{2} = \cos \theta$

3. $\cos \frac{1}{2}\theta = \dfrac{\cos \theta}{2}$

4. $\cos(\theta + \phi) = \cos \theta + \cos \phi$

5. $\tan(45° + \theta) = 1 + \tan \theta$

6. $\sin 2\theta = 2 \sin \theta(2 \cos \theta)$

7. $\tan 2\theta = \tan(\theta + \theta) = \tan \theta + \tan \theta$

8. Since $\sin \frac{1}{2}\theta = \pm \sqrt{\dfrac{1 - \cos \theta}{2}}$,
there are two values for θ (because of $\pm$).

9. $\tan \frac{1}{2}\theta = \dfrac{1 - \cos \theta}{\sin \theta}$

10. If $\cos \dfrac{\theta}{2} = \pm \sqrt{\dfrac{1 + \cos \theta}{2}}$, then choose $+$ if θ is in Quadrant I or IV and $-$ if θ is in Quadrant II or III.

11. $\dfrac{2 \tan \frac{1}{2}\theta}{1 - \tan^2\frac{1}{2}\theta} = \dfrac{2}{1 + \tan \theta}$

Problems 12–18: Assume that θ is positive and less than one revolution. Give the possible quadrant(s) for the indicated angles.

12. a. $\frac{1}{2}\theta$ if θ is in Quadrant I
   b. 2θ if θ is in Quadrant I

13. a. 2θ if θ is in Quadrant II
   b. $\frac{1}{2}\theta$ if θ is in Quadrant II

14. a. $\frac{1}{2}\theta$ if θ is in Quadrant III
   b. $\frac{1}{2}\theta$ if θ is in Quadrant IV

15. a. $\frac{1}{2}\theta$ in $\cos \frac{1}{2}\theta = \sqrt{\dfrac{1 + \cos \theta}{2}}$

   b. $\frac{1}{2}\theta$ in $\sin \frac{1}{2}\theta = \sqrt{\dfrac{1 - \cos \theta}{2}}$

16. a. 2θ if θ is in Quadrant III
   b. 2θ if θ is in Quadrant III and $\cos 2\theta = -\frac{5}{9}$

17. θ if $\sec 2\theta = \sqrt{1 + \tan^2 2\theta}$ and $\sin 2\theta < 0$

18. θ if $\cos \theta = -\frac{3}{5}$, and 2θ is in Quadrant III

Level 1 Drill

Problems 19–22: Use double-angle or half-angle identities to evaluate using exact values.

19. a. $2 \cos^2 22.5° - 1$ **b.** $\tan 22.5°$

20. a. $\sin 22.5°$ **b.** $\cos \dfrac{\pi}{8}$

21. a. $\dfrac{2 \tan \frac{\pi}{8}}{1 - \tan^2\frac{\pi}{8}}$ **b.** $\sqrt{\dfrac{1 - \cos 60°}{2}}$

22. a. $1 - 2 \sin^2 90°$ **b.** $-\sqrt{\dfrac{1 - \cos 420°}{2}}$

Problems 23–28: Find $\cos \theta$, $\sin \theta$, and $\tan \theta$ when θ is in Quadrant I and $\cot 2\theta$ is given, $0 \le 2\theta < \pi$.

23. $\cot 2\theta = -\frac{3}{4}$ **24.** $\cot 2\theta = \frac{1}{\sqrt{3}}$

25. $\cot 2\theta = -\frac{4}{3}$ **26.** $\cot 2\theta = 0$

27. $\cot 2\theta = -\frac{1}{\sqrt{3}}$ **28.** $\cot 2\theta = \frac{4}{3}$

Problems 31–34: Find the exact values of the cosine, sine, and tangent of 2θ, $0 \le 2\theta < 2\pi$.

29. $\sin \theta = \frac{3}{5}$; θ in Quadrant I

30. $\sin \theta = \frac{5}{13}$; θ in Quadrant II

31. $\tan \theta = -\frac{5}{12}$; θ in Quadrant IV

32. $\tan \theta = -\frac{3}{4}$; θ in Quadrant II

33. $\cos \theta = \frac{5}{9}$; θ in Quadrant I

34. $\cos \theta = -\frac{5}{13}$; θ in Quadrant III

Level 2 Drill

Problems 35–40: *Find the exact values of the cosine, sine, and tangent of $\frac{1}{2}\theta$ where $0 \le \theta \le \pi$.*

35. $\sin \theta = \frac{3}{5}$; θ in Quadrant I

36. $\sin \theta = \frac{5}{13}$; θ in Quadrant II

37. $\tan \theta = -\frac{5}{12}$; θ in Quadrant II

38. $\tan \theta = -\frac{3}{4}$; θ in Quadrant II

39. $\cos \theta = \frac{5}{9}$; θ in Quadrant I

40. $\cos \theta = -\frac{5}{13}$; θ in Quadrant II

Problems 41–46: *Sketch each graph.*

41. $y = \cos^2 x$

42. $y = |\cos x|$

43. $y = \dfrac{\sin x}{1 + \cos x}$

44. $y = \dfrac{1 - \cos x}{\sin x}$

45. $y = \pm\sqrt{\dfrac{1 + \cos x}{2}}$

46. $y = \pm\sqrt{\dfrac{1 - \cos x}{2}}$

Level 2 Applications

47. An airplane flying faster than the speed of sound is said to have a speed greater than Mach 1. In such a case, a sonic boom, created by sound waves that form a cone with vertex angle θ (see Figure 4.15), is heard. The Mach number is the ratio of the plane's speed to the speed of sound and is denoted by M. It can be shown that, if $M > 1$, then $\sin \frac{1}{2}\theta = M^{-1}$.
  a. If $\theta = \frac{\pi}{6}$, find the Mach number to the nearest tenth.
  b. Find the exact Mach number for part **a**.
  c. Solve for θ; that is, suppose you know the Mach number, how can you determine the angle?

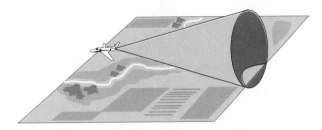

FIGURE 4.15 Pattern of sound waves

48. If a boat is moving at a constant rate that is faster than the water waves it produces, then the boat sends out waves in the shape of a cone with a vertex angle θ, as shown in Figure 4.16. If r represents the ratio of the boat's speed to the wave's speed and if $r > 1$, then

$$\sin \frac{1}{2}\theta = r^{-1}$$

  a. If $\theta = \frac{\pi}{4}$, find r to the nearest tenth.
  b. Find the exact value of r for part **a**.
  c. Solve for θ. That is, suppose you know r; how can you determine the angle?

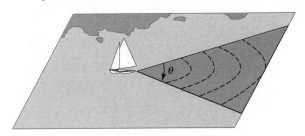

FIGURE 4.16 Pattern of boat waves

49. Suppose an object is propelled upward at an angle θ to the horizontal with an initial velocity of 256 ft/s. Find the angle θ for which the range is a maximum, and then find that maximum downrange point of impact. (Ignore air resistance.)

50. Suppose an object is propelled upward at an angle θ to the horizontal with an initial velocity of 384 ft/s. Find the angle θ for which the range is a maximum, and then find that maximum downrange point of impact. (Ignore air resistance.)

Level 3 Theory

Problems 51–58: *Prove each identity.*

51. $\sin \alpha = 2 \sin \dfrac{\alpha}{2} \cos \dfrac{\alpha}{2}$

52. $\cos 4\theta = \cos^2 2\theta - \sin^2 2\theta$

53. $\sin 2\theta = \dfrac{2 \tan \theta}{1 + \tan^2 \theta}$

54. $\tan \dfrac{3}{2}\beta = \dfrac{2 \tan \frac{3}{4}\beta}{1 - \tan^2 \frac{3}{4}\beta}$

55. $\tan \dfrac{1}{2}\theta = \dfrac{1 - \cos \theta}{\sin \theta}$

56. $\tan \dfrac{1}{2}\theta = \dfrac{\sin \theta}{1 + \cos \theta}$

57. $\sin 2\theta \sec \theta = 2 \sin \theta$

58. $\sin 2\theta \tan \theta = 2 \sin^2 \theta$

Level 3
Individual Projects

59. JOURNAL PROBLEM (From "Classroom Capsules" by Roger B. Nelsen, Lewis and Clark College, Portland, Oregon, *College Mathematics Journal,* January 1989, p. 51.) Use Figure 4.17 to prove

$$\sin 2\theta = 2 \sin \theta \cos \theta$$

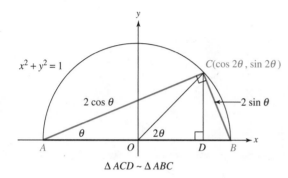

$\triangle ACD \sim \triangle ABC$

FIGURE 4.17 Geometric proof of $\sin 2\theta = 2 \sin \theta \cos \theta$

60. Use Figure 4.17 to prove

$$\cos 2\theta = 2 \cos^2\theta - 1$$

61. Draw a unit circle with center at O and with radii $\overline{AO}$ and $\overline{BO}$. Let θ be an angle of right $\triangle BCO$, as shown in Figure 4.18.
  a. Show $\alpha = \angle BAO$ is equal to $\frac{1}{2}\theta$.
  b. Use Figure 4.18 to prove

$$\tan\frac{1}{2}\theta = \frac{\sin\theta}{1 + \cos\theta}$$

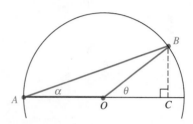

FIGURE 4.18 Geometric proof of $\tan\frac{1}{2}\theta = \dfrac{\sin\theta}{1 + \cos\theta}$

PRODUCT AND SUM IDENTITIES

PROBLEM OF THE DAY

Graph $y = \sin x + \cos x$. Is this graph sinusoidal? If so, what are the period and amplitude of this graph?

How would you answer this question?

We begin by using a graphing calculator. The graph is shown in the margin. It looks sinusoidal, so we use the methods of Section 3.2 to estimate the period and the amplitude. First, we select some point on the curve as a starting point, and draw a frame:

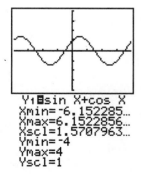

```
Y₁⊟sin X+cos X
Xmin=-6.152285…
Xmax=6.1522856…
Xscl=1.5707963…
Ymin=-4
Ymax=4
Yscl=1
```

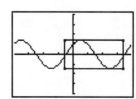

The period appears to be about 6.25, and the amplitude is about 1.4. Suppose we assume that the Problem of the Day is asking for the exact values for the period and the amplitude. Is there a formula for these properties of the graph? In this section, we find a formula that gives us exact values for the period and amplitude of $y = \sin x + \cos x$.

Product-to-Sum Identities

It is sometimes convenient, and even necessary, to write a trigonometric sum as a product or a product as a sum. These are known as the **product-to-sum identities.**

> **PRODUCT-TO-SUM IDENTITIES**
>
> **30.** $2 \cos \alpha \cos \beta = \cos(\alpha - \beta) + \cos(\alpha + \beta)$
> **31.** $2 \sin \alpha \sin \beta = \cos(\alpha - \beta) - \cos(\alpha + \beta)$
> **32.** $2 \sin \alpha \cos \beta = \sin(\alpha + \beta) + \sin(\alpha - \beta)$
> **33.** $2 \cos \alpha \sin \beta = \sin(\alpha + \beta) - \sin(\alpha - \beta)$

The proofs of the product-to-sum identities involve systems of equations and the addition laws for cosine:

$$\cos(\alpha - \beta) = \cos \alpha \cos \beta + \sin \alpha \sin \beta$$
$$\cos(\alpha + \beta) = \cos \alpha \cos \beta - \sin \alpha \sin \beta$$

Add these equations:

$$\cos(\alpha - \beta) + \cos(\alpha + \beta) = 2 \cos \alpha \cos \beta \quad \textit{This is identity 30.}$$

Subtract these equations:

$$\cos(\alpha - \beta) - \cos(\alpha + \beta) = 2 \sin \alpha \sin \beta \quad \textit{This is identity 31.}$$

Starting over with the addition laws for sine:

$$\sin(\alpha + \beta) = \sin \alpha \cos \beta + \cos \alpha \sin \beta$$
$$\sin(\alpha - \beta) = \sin \alpha \cos \beta - \cos \alpha \sin \beta$$

Add these equations:

$$\sin(\alpha + \beta) + \sin(\alpha - \beta) = 2 \sin \alpha \cos \beta \quad \textit{This is identity 32.}$$

Subtract these equations:

$$\sin(\alpha + \beta) - \sin(\alpha - \beta) = 2 \cos \alpha \sin \beta \quad \textit{This is identity 33.}$$

EXAMPLE 1

Using Product-to-Sum Identities

Write each product as the sum of two functions.

a. 2 sin 3 sin 1   **b.** sin 40° cos 12°

Solution   **a.** Let $\alpha = 3$ and $\beta = 1$ in identity 31:

$$2 \sin 3 \sin 1 = \cos(3 - 1) - \cos(3 + 1) = \cos 2 - \cos 4$$

b. Let $\alpha = 40°$ and $\beta = 12°$ in identity 32:

$$2 \sin 40° \cos 12° = \sin(40° + 12°) + \sin(40° - 12°) = \sin 52° + \sin 28°$$

But what about the coefficient of 2 in this problem? Since you know that the preceding is an *equation* that is true, you can divide both sides by 2 to obtain

$$\sin 40° \cos 12° = \tfrac{1}{2}(\sin 52° + \sin 28°)$$ ∎

Sum-to-Product Identities

YIELD

Identities 30–33 change a product to a sum. They can also be used to change a sum to a product. To do this, we recast them as follows. Let $\boldsymbol{x = \alpha + \beta}$ and $\boldsymbol{y = \alpha - \beta}$. To rewrite identity 30 in sum form, substitute x for $\alpha + \beta$ and y for $\alpha - \beta$:

$$2 \cos \alpha \cos \beta = \cos(\boldsymbol{\alpha - \beta}) + \cos(\boldsymbol{\alpha + \beta})$$
$$= \cos y + \cos x$$
$$= \cos x + \cos y$$

Now, we need to eliminate α and β:

$$\begin{cases} x = \alpha + \beta \\ y = \alpha - \beta \end{cases} \qquad\qquad \begin{cases} x = \alpha + \beta \\ y = \alpha - \beta \end{cases}$$

Add: $x + y = 2\alpha$ Subtract: $x - y = 2\beta$

$$\frac{x + y}{2} = \alpha \qquad\qquad\qquad \frac{x - y}{2} = \beta$$

Thus, by substitution:

$$2 \cos \alpha \cos \beta = \cos x + \cos y$$

$$2 \cos\left(\frac{x + y}{2}\right) \cos\left(\frac{x - y}{2}\right) = \cos x + \cos y$$

This is called a **sum-to-product identity.** If we make the same substitutions into identities 31, 32, and 33, we will obtain the other sum-to-product identities.

SUM-TO-PRODUCT IDENTITIES

34. $\cos x + \cos y = 2 \cos\left(\dfrac{x+y}{2}\right)\cos\left(\dfrac{x-y}{2}\right)$

35. $\cos x - \cos y = -2 \sin\left(\dfrac{x+y}{2}\right)\sin\left(\dfrac{x-y}{2}\right)$

36. $\sin x + \sin y = 2 \sin\left(\dfrac{x+y}{2}\right)\cos\left(\dfrac{x-y}{2}\right)$

37. $\sin x - \sin y = 2 \sin\left(\dfrac{x-y}{2}\right)\cos\left(\dfrac{x+y}{2}\right)$

EXAMPLE 2 **Using a Sum-to-Product Identity**

Write $\sin 35° + \sin 27°$ as a product.

Solution $\quad x = 35°, y = 27°, \dfrac{x+y}{2} = \dfrac{35° + 27°}{2} = 31°,$ and $\dfrac{x-y}{2} = \dfrac{35° - 27°}{2} = 4°$

Therefore, $\quad \sin 35° + \sin 27° = 2 \sin 31° \cos 4° \quad$ *Identity 36* ∎

EXAMPLE 3 **Using a Sum-to-Product Identity to Prove an Identity**

Prove that $\dfrac{\sin 7\theta + \sin 5\theta}{\cos 7\theta - \cos 5\theta} = -\cot \theta$ is an identity.

Solution

$$\frac{\sin 7\theta + \sin 5\theta}{\cos 7\theta - \cos 5\theta} = \frac{2 \sin\left(\dfrac{7\theta + 5\theta}{2}\right)\cos\left(\dfrac{7\theta - 5\theta}{2}\right)}{-2 \sin\left(\dfrac{7\theta + 5\theta}{2}\right)\sin\left(\dfrac{7\theta - 5\theta}{2}\right)}$$

$$= \frac{2 \sin 6\theta \cos \theta}{-2 \sin 6\theta \sin \theta}$$

$$= -\frac{\cos \theta}{\sin \theta}$$

$$= -\cot \theta$$

∎

We now return to the Problem of the Day. Remember the question? What are the exact period and amplitude for this graph?

$$y = \sin x + \cos x$$

We begin by using a cofunction identity followed by a sum-to-product identity.

$$y = \sin x + \cos x \qquad \qquad \textit{Given}$$

$$= \sin x + \sin\left(\tfrac{\pi}{2} - x\right) \qquad \textit{Cofunction identity (identity 13)}$$

$$= 2 \sin\left(\frac{x + \frac{\pi}{2} - x}{2}\right)\cos\left(\frac{x - \frac{\pi}{2} + x}{2}\right) \qquad \textit{Sum-to-product identity (identity 36)}$$

$$= 2 \sin\frac{\pi}{4}\cos\left(\frac{2x - \frac{\pi}{2}}{2}\right) \qquad \textit{Simplify.}$$

$$= 2 \sin\frac{\pi}{4}\cos\left(x - \frac{\pi}{4}\right)$$

$$= 2\left(\frac{\sqrt{2}}{2}\right)\cos\left(x - \frac{\pi}{4}\right) \qquad \textit{Exact value of } \sin\frac{\pi}{4} \textit{ is } \frac{\sqrt{2}}{2}.$$

$$= \sqrt{2}\cos\left(x - \tfrac{\pi}{4}\right)$$

From this form, we see the amplitude is $\sqrt{2}$ and the period is 2π.

EXAMPLE 4   **Using a Trigonometric Identity to Solve a Trigonometric Equation**

Solve $\cos 5x - \cos 3x = 0$ for $0 \le x < 2\pi$.

Solution

$$\cos 5x - \cos 3x = 0 \qquad \textit{Given}$$

$$-2 \sin\left(\frac{5x + 3x}{2}\right)\sin\left(\frac{5x - 3x}{2}\right) = 0 \qquad \textit{Use sum-to-product identity.}$$

$$\sin 4x \sin x = 0 \qquad \textit{Divide both sides by } -2, \textit{ and simplify.}$$

$\sin 4x = 0 \qquad \qquad \qquad \sin x = 0 \qquad \textit{Use the zero-factor theorem.}$

$4x = \sin^{-1} 0 \qquad \qquad \quad x = \sin^{-1} 0$

$= 0, \pi, 2\pi, 3\pi, 4\pi, 5\pi, 6\pi, 7\pi \qquad = 0, \pi$

$x = 0, \frac{\pi}{4}, \frac{\pi}{2}, \frac{3\pi}{4}, \pi, \frac{5\pi}{4}, \frac{3\pi}{2}, \frac{7\pi}{4}$

∎

A Reduction Identity

This Problem of the Day is a special case of an identity that is occasionally used in graphing, calculus, and in certain engineering applications. This identity, called the **reduction identity,** is stated in the box on page 212.

> **REDUCTION IDENTITY**
>
> **38.** $a \sin \theta + b \cos \theta = \sqrt{a^2 + b^2} \sin(\theta + \alpha)$
>
> where α is chosen so that
>
> $$\cos \alpha = \frac{a}{\sqrt{a^2 + b^2}} \quad \text{and} \quad \sin \alpha = \frac{b}{\sqrt{a^2 + b^2}}$$

The proof of this identity is straightforward after you multiply by 1, written in the form

$$\frac{\sqrt{a^2 + b^2}}{\sqrt{a^2 + b^2}}$$

The proof is left as a problem.

EXAMPLE 5 | **Using the Reduction Identity**

Change $\dfrac{\sqrt{3}}{2} \sin \theta + \dfrac{1}{2} \cos \theta$ using the reduction identity.

Solution | By inspection, $a = \frac{\sqrt{3}}{2}$ and $b = \frac{1}{2}$, so $\sqrt{a^2 + b^2} = \sqrt{\frac{3}{4} + \frac{1}{4}} = 1$. Find α so that $\cos \alpha = \frac{\sqrt{3}}{2}$ and $\sin \alpha = \frac{1}{2}$. The reference angle α' is $\frac{\pi}{6}$, and since cosine and sine are both positive, α is in Quadrant I: $\alpha = \frac{\pi}{6}$. Therefore

$$\frac{\sqrt{3}}{2} \sin \theta + \frac{1}{2} \cos \theta = \sin\left(\theta + \frac{\pi}{6}\right)$$

∎

EXAMPLE 6 | **Using the Reduction Identity to Graph an Equation**

Graph $y = \sqrt{3} \cos \theta - \sin \theta$. What are the period, amplitude, and starting point for this curve?

Solution | A calculator graph is shown in the margin. To answer the questions, compare $y = \sqrt{3} \cos \theta - \sin \theta$ to $y = a \sin \theta + b \cos \theta$ to see that $a = -1$ and $b = \sqrt{3}$. Thus, $\sqrt{a^2 + b^2} = \sqrt{1 + 3} = 2$. Find α so that

$$\cos \alpha = \frac{-1}{2} \quad \text{and} \quad \sin \alpha = \frac{\sqrt{3}}{2}$$

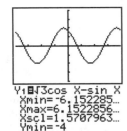

The reference angle for α is $\pi/3$. Sine is positive while cosine is negative, so α must be in Quadrant II: $\alpha = 2\pi/3$. Thus

$$y = \sqrt{3} \cos \theta - \sin \theta = 2 \sin\left(\theta + \frac{2\pi}{3}\right)$$

The graph has starting point $\left(-\frac{2\pi}{3}, 0\right)$, $a = 2$, and period 2π, as shown in Figure 4.19.

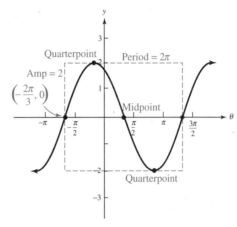

FIGURE 4.19
Graph of $y = \sqrt{3} \cos\theta - \sin\theta$

PROBLEM SET 4.5

Level 1 Drill

Problems 1–8: *Write each product as a sum (or difference).*

1. $2 \sin 35° \sin 24°$

2. $2 \cos 46° \cos 18°$

3. $2 \cos 70° \sin 24°$

4. $2 \sin 41° \cos 19°$

5. $\sin 2\theta \sin 4\theta$

6. $\cos 3\theta \sin 2\theta$

7. $\cos\theta \cos 3\theta$

8. $\sin 2\theta \cos 5\theta$

Problems 9–16: *Write each expression as a product of two functions.*

9. $\sin 43° + \sin 64°$

10. $\cos 79° - \cos 77°$

11. $\sin 15° + \sin 30°$

12. $\cos 25° - \cos 100°$

13. $\cos 6x + \cos 2x$

14. $\sin 3x - \sin 2x$

15. $\cos 5y + \cos 9y$

16. $\sin 6z - \sin 9z$

Problems 17–22: *Change each expression by using the reduction identity to write it as a sine function.*

17. $\sin\theta + \cos\theta$

18. $\sin\theta - \cos\theta$

19. $-\sin\frac{\theta}{2} + \cos\frac{\theta}{2}$

20. $\frac{1}{2}\sin\theta + \frac{\sqrt{3}}{2}\cos\theta$

21. $\sqrt{3}\cos\pi\theta - \sin\pi\theta$

22. $\cos 2\theta - \sqrt{3}\sin 2\theta$

Level 2 Drill

Problems 23–34: *Prove each identity.*

23. $\dfrac{\sin\theta + \sin 3\theta}{2\sin 2\theta} = \cos\theta$

24. $\dfrac{\cos\theta + \cos 3\theta}{2\cos 2\theta} = \cos\theta$

25. $\dfrac{\sin\theta + \sin 3\theta}{4\cos^2\theta} = \sin\theta$

26. $\dfrac{\cos 3\theta - \cos\theta}{4\sin^2\theta} = -\cos\theta$

27. $\dfrac{\sin 5\theta + \sin 3\theta}{\cos 5\theta + \cos 3\theta} = \tan 4\theta$

28. $\dfrac{\cos 3\theta - \cos\theta}{\sin\theta - \sin 3\theta} = \tan 2\theta$

29. $\dfrac{\cos 5\omega + \cos\omega}{\cos\omega - \cos 5\omega} = \dfrac{\cot 2\omega}{\tan 3\omega}$

30. $\dfrac{\cos 3\phi + \cos\phi}{\cos 3\phi - \cos\phi} = -\dfrac{\cot 2\phi}{\tan\phi}$

31. $\dfrac{\sin x + \sin y}{\cos x + \cos y} = \tan\left(\dfrac{x+y}{2}\right)$

32. $\dfrac{\sin x + \sin y}{\sin x - \sin y} = \tan\left(\dfrac{x+y}{2}\right)\cot\left(\dfrac{x-y}{2}\right)$

33. $\cos^2\frac{\theta}{2} - \sin^2\frac{\theta}{2} = \cos\theta$

34. $\left(\sin\frac{\theta}{2} + \cos\frac{\theta}{2}\right)^2 = 1 + \sin\theta$

Problems 35–40: Use the reduction identity to graph each equation. State the amplitude, period, and starting point.

35. $y = \cos\theta - \sin\theta$ **36.** $y = -\sin\theta - \cos\theta$

37. $y = 2\sin\theta - 3\cos\theta$ **38.** $y = 3\sin\theta + 2\cos\theta$

39. $y = 8\sin\theta + 15\cos\theta$ **40.** $y = -8\sin\theta - 15\cos\theta$

Problems 41–46: Solve each equation for $0 \le x < 2\pi$. You might consider first using a trigonometric identity.

41. $\sin x - \sin 3x = 0$ **42.** $\cos 5y + \cos 3y = 0$

43. $\cos 5\alpha - \cos 3\alpha = 0$ **44.** $\sin 4\theta + \sin 4\theta = 0$

45. $\cos^2 x - 3\sin x + 3 = 0$ **46.** $\cos 2x + \cos x = 0$

Level 2 Theory

47. Show that identities 32 and 33 are really the same identity.

48. Prove identity 35.   **49.** Prove identity 36.

50. Prove identity 37.

Level 2 Applications

51. Graph $y = 3\sin x + 3\cos x$. Is this graph sinusoidal? If so, what are the period and amplitude of this graph?

52. Graph $y = \sin x - \cos x$. Is this graph sinusoidal? If so, what are the period and amplitude of this graph?

53. Graph $y = \sqrt{2}\left(\sin\frac{x}{2} + \cos\frac{x}{2}\right)$. Is this graph sinusoidal? If so, what are the period and amplitude of this graph?

54. Graph $y = \sin x + \cos x + \sin x$. Is this graph sinusoidal? If so, what are the period and amplitude of this graph?

55. A touch-tone phone is arranged into four rows and three columns, with each row and each column set at a different frequency, as shown in Figure 4.20. When you press a particular button, the sound emitted is the sum of two frequencies determined by the row and column. For example, if you press 1, then the sound emitted is

$$y = \sin 2\pi(697)x + \sin 2\pi(1,209)x$$

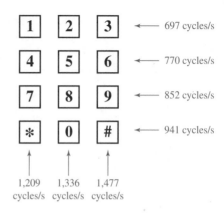

FIGURE 4.20
Touch-tone phone

In the Chapter 3 Group Research Projects, you were asked to graph this sound. Write this sound as a product of sines and/or cosines.

56. Write the sound emitted when you press the numeral 5 on a touch-tone phone as a product of sines and/or cosines. (See Problem 55.)

57. If two instruments are not in tune and each plays a note at the same time, a beat is heard. The respective equations for sound waves with frequencies 112 Hz and 122 Hz are

$$y = \cos 224\pi x \quad\text{and}\quad y = -\cos 244\pi x$$

For these problems, let $0 \le x \le 0.1$ and $-2 \le y \le 2$.
a. Graph $y = \cos 224\pi x$.
b. Graph $y = -\cos 244\pi x$.
c. Graph $y = \cos 224\pi x - \cos 244\pi x$.
d. Convert the sound wave to a product and graph the resulting sound.

Level 3 Theory

58. Graph $f(x) = \dfrac{\sin x + \sin 2x + \sin 3x}{\cos x + \cos 2x + \cos 3x}$.

From the graph, find another, simpler curve that has the same graph.

59. Graph $g(x) = \dfrac{\cos x + \cos 4x + \cos 7x}{\sin x + \sin 4x + \sin 7x}$.

From the graph, find another, simpler curve that has the same graph.

60. Find the exact value of

$$\cos \frac{\pi}{7} \cos \frac{2\pi}{7} \cos \frac{3\pi}{7}$$

    a. First evaluate this using your calculator and write the answer as a fraction.

    b. Use trigonometric identities to derive the answer.

61. PROBLEM FROM CALCULUS In calculus, it is shown that if a particle is moving in simple harmonic motion according to the equation

$$y = a \cos(\omega t + h)$$

then the velocity of the particle at any time t is given by

$$v = a\omega \sin(\omega t + h)$$

    a. Graph this velocity curve, where the frequency is $\omega^{-1} = \frac{1}{4}$ cycle per unit of time, $h = \frac{\pi}{6}$, and $a = \frac{120}{\pi}$.

    b. Find a maximum value and a minimum value for the velocity.

4.6 CHAPTER FOUR SUMMARY

Objectives

Section 4.1	Solving Equations
Objective 4.1	Solve trigonometric equations in linear form, by factoring, or by using the quadratic formula.
Objective 4.2	Solve trigonometric equations with multiple angles.

Section 4.2	Proving Identities
Objective 4.3	State and prove the eight fundamental identities. (This is repeated from Chapter 2.)
Objective 4.4	Simplify trigonometric expressions.
Objective 4.5	Prove identities using algebraic simplification and the eight fundamental identities.
Objective 4.6	Prove a given identity by using various "tricks of the trade": • combining fractions • writing all functions as sines and cosines • breaking up a fraction into a sum of fractions • using the conjugate • factoring
Objective 4.7	Prove that an equation is not an identity by finding a counterexample, or by graphing.

Objective 4.8	Prove or disprove that a given equation is an identity.

Section 4.3	Addition Laws
Objective 4.9	Know and use the distance formula.
Objective 4.10	Find the chord length for a given angle.
Objective 4.11	Know the difference of angles identity.
Objective 4.12	Use the cofunction identities.
Objective 4.13	Use the opposite-angle identities.
Objective 4.14	Use the opposite-angle identities as an aid to graph certain trigonometric functions.
Objective 4.15	Use the addition laws.
Objective 4.16	Find exact values using the addition laws.
Objective 4.17	Prove identities using the addition laws.

Section 4.4	Double-Angle and Half-Angle Identities
Objective 4.18	Use the double-angle identities.
Objective 4.19	Use a double-angle identity to graph certain trigonometric equations.

Objective 4.20 Given $\cos\theta$, $\sin\theta$, or $\tan\theta$ and the quadrant, find $\cos 2\theta$, $\sin 2\theta$, and $\tan 2\theta$.

Objective 4.21 Use the half-angle identities.

Objective 4.22 Given $\cot 2\theta$, find $\cos\theta$, $\sin\theta$, and $\tan\theta$.

Objective 4.23 Prove identities using the half-angle and double-angle identities.

Section 4.5   Product and Sum Identities

Objective 4.24 Use the product-to-sum identities to write a product as a sum.

Objective 4.25 Use the sum-to-product identities to write a sum as a product.

Objective 4.26 Use the sum-to-product identities to find the amplitude and period of certain trigonometric sums.

Objective 4.27 Prove identities using the product and sum identities.

Objective 4.28 Use the reduction identity to simplify certain trigonometric equations.

Objective 4.29 Use the reduction identity to graph certain trigonometric equations.

Terms

Addition laws [4.3]	Double-angle identities
Chord length [4.3]	[4.4]
Cofunction identities [4.3]	Equation [4.1]
Conjugate [4.2]	Even function [4.3]
Counterexample [4.2]	Half-angle identities [4.4]
Disprove an identity [4.2]	Identity [Preview; 4.2]
Distance formula [4.3]	

Odd function [4.3]	Reduction identity [4.5]
Opposite-angle identities	Sum-to-product identities
[4.3]	[4.5]
Product-to-sum identities	Solution [4.1]
[4.5]	Solve an equation
Proving an identity [4.2]	[Preview]

Sample Test

All of the answers for this sample test are given in the back of the book.

1. Simplify $\dfrac{\sec^2\theta + \tan^2\theta + 1}{\sec\theta}$.

2. Find the exact value of $\cos 105°$.

Problems 3–4: *Evaluate using exact values.*

3. $\dfrac{2\tan\frac{\pi}{6}}{1 - \tan^2\frac{\pi}{6}}$

4. $-\sqrt{\dfrac{1 + \cos 240°}{2}}$

Problems 5–9: *Prove each identity.*

5. $\dfrac{\csc^2\alpha}{1 + \cot^2\alpha} = 1$

6. $\dfrac{\sin^2\beta - \cos^2\beta}{\sin\beta + \cos\beta} = \sin\beta - \cos\beta$

7. $\dfrac{1}{\sin\gamma + \cos\gamma} + \dfrac{1}{\sin\gamma - \cos\gamma} = \dfrac{2\sin\gamma}{\sin^4\gamma - \cos^4\gamma}$

8. $\dfrac{1 - \tan^2 3\theta}{1 - \tan 3\theta} = 1 + \tan 3\theta$

9. $\dfrac{\sin 5\theta + \sin 3\theta}{\cos 5\theta - \cos 3\theta} = -\cot\theta$

Problems 10–12: *Prove or disprove that the given equations are identities.*

10. $\tan\theta + \csc\theta = \cot\theta$

11. $(\sin t + \cos t)^2 = 1$

12. $\cos x \tan x \csc x \sec x = 1$

Problems 13–16: *Solve each equation for $0 \le \theta < 2\pi$. Find exact values, if possible; otherwise give the answer correct to two decimal places.*

13. $4\cos^2\theta = 1$

14. $\tan^2\theta - 2\tan\theta - 3 = 0$

15. $\cos^2\theta = 3\sin\theta$

16. $2\sin^2 2\theta - 2\cos^2 2\theta = 1$

Problems 17–20: *Graph each equation.*

17. $y - 1 = 2\sin\left(\frac{\pi}{6} - x\right)$

18. $y + 1 = 2 \cos(2 - x)$ **19.** $y = \cos^2 x$

20. $y = 12 \sin x - 5 \cos x$

21. If $\cot 2\theta = -\frac{4}{3}$, find the exact value of $\tan \theta$ (θ is in Quadrant I).

22. Evaluate $\tan(\sec^{-1} 3)$ without using tables or a calculator.

23. Suppose the circle given in Figure 4.21 is a unit circle. Find the distance between P_α and P_β.

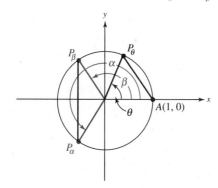

FIGURE 4.21   Points on a unit circle

24. Suppose the circle given in Figure 4.21 is a unit circle. Find the length of chord AP_θ. This is the length of any chord in a unit circle subtended by a central angle θ.

25. Prove $\cos(\alpha - \beta) = \cos \alpha \cos \beta + \sin \alpha \sin \beta$. [*Hint:*   You may use the results of Problems 23 and 24.]

26. Use the trigonometric substitution $u = 10 \sin \theta$ to rewrite the algebraic expression in terms of θ and then simplify:

$$\frac{\sqrt{100 - u^2}}{u}$$

27. If $\cos \theta = -\frac{4}{5}$ and 2θ is in Quadrant IV, find the exact value of $\sin 2\theta$.

28. Write $\sin(x + h) - \sin x$ as a product.

29. Write $\sin 3\theta \cos \theta$ as a sum.

30. Use identity 38 to write $12 \sin \theta - 5 \cos \theta$ as a function of sine; approximate α to the nearest tenth.

Discussion Problems

1. What are the fundamental identities, and why do you think they are called "fundamental"?

2. Outline your "method of attack" for proving a trigonometric identity.

3. Derive (prove) identities 1–20. (They are listed on the quick reference card at the back of this book.)

4. Derive (prove) identities 21–34. (They are listed on the quick reference card at the back of this book.)

5. Write a 500-word paper on cartography (mapmaking) and its relationship to trigonometry.

Miscellaneous Problems—Chapters 1–4

1. From memory, draw a standard sine curve.

2. From memory, draw a standard cosine curve.

3. From memory, draw a standard tangent curve.

4. From memory, draw a standard secant curve.

5. From memory, draw a standard cosecant curve.

6. From memory, draw a standard cotangent curve.

Problems **7–10:** *The graph at the top of page* 218 *was obtained from the February* 1991 *issue of* Scientific American. *Use it to work the problems. Assume that the amplitude of the cycles is* 1 *and that the distance between each vertical line is* 10 *days.*

7. Write a possible equation for the excitatory neurons.

8. Write a possible equation for the inhibitory neurons.

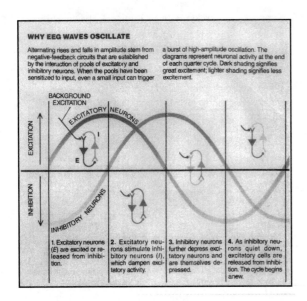

WHY EEG WAVES OSCILLATE

Alternating rises and falls in amplitude stem from negative-feedback circuits that are established by the interaction of pools of excitatory and inhibitory neurons. When the pools have been sensitized to input, even a small input can trigger a burst of high-amplitude oscillation. The diagrams represent neuronal activity at the end of each quarter cycle. Dark shading signifies great excitement; lighter shading signifies less excitement.

1. Excitatory neurons (E) are excited or released from inhibition.

2. Excitatory neurons stimulate inhibitory neurons (I), which dampen excitatory activity.

3. Inhibitory neurons further depress excitatory neurons and are themselves depressed.

4. As inhibitory neurons quiet down, excitatory cells are released from inhibition. The cycle begins anew.

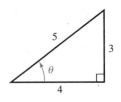

FIGURE 4.22
A right triangle

9. Add these two equations and draw the graph of the sum of the excitatory and inhibitory neurons.

10. Graph $y = 6 \cos^2 \dfrac{x}{2}$ for $0 \le x \le 4\pi$.

11. If $\sin \theta = \frac{4}{5}$ and θ is in Quadrant II, find the exact value of $\cos 2\theta$.

12. If $\cot 2\theta = \frac{4}{3}$, find the exact value of $\cos \theta$, $90° < \theta < 180°$.

Problems 13–15: *Evaluate the given functions using exact values.*

$$c(\theta) = \cos \theta \quad s(\theta) = \sin \theta \quad t(\theta) = \tan \theta$$

13. **a.** $c(30°)$ **b.** $s(45°)$ **c.** $t(-60°)$

14. Let $\theta = 60°$:
   a. $s(2\theta)$ **b.** $c\left(\frac{\theta}{2}\right)$ **c.** $t(2\theta)$

15. Let $\theta = 15°$:
   a. $t(-2\theta)$ **b.** $s(10\theta)$ **c.** $[c(4\theta)]^2$

Problems 16–19: *Use Figure 4.22 to evaluate each function.*

16. **a.** $\cos \theta$ **b.** $\sin \theta$ **c.** $\tan \theta$

17. **a.** $\cos 2\theta$ **b.** $\sin 2\theta$ **c.** $\tan 2\theta$

18. **a.** $\cos \frac{\theta}{2}$ **b.** $\sin \frac{\theta}{2}$ **c.** $\tan \frac{\theta}{2}$

19. **a.** $\cot 2\theta$ **b.** $\sec 2\theta$ **c.** $\csc 2\theta$

Problems 20–25: *Solve the equations for principal values. Find exact values, if possible; otherwise approximate the solution correct to two decimal places.*

20. $3 \tan 2\theta - \sqrt{3} = 0$

21. $\sin^2\theta + 2 \cos \theta = 1 + 3 \cos \theta + \sin^2\theta$

22. $3 \cos^2\theta = 1 + \cos \theta$

23. $\tan^2 2\theta = 3 \tan 2\theta$

24. $\cos 2x = \sin x$

25. $3 \sin 4x - \sin 3x + \sqrt{3} = 2 \sin 2x - \sin 3x + 3 \sin 4x$

Problems 26–44: *Prove each equation is an identity.*

26. $\dfrac{\sec^2 u}{\tan^2 u} = \csc^2 u$

27. $\dfrac{\sec u}{\tan^2 u} = \cot u \csc u$

28. $\csc(-\theta) = -\csc \theta$

29. $\sec(-\theta) = \sec \theta$

30. $\dfrac{\sin \alpha}{\tan \alpha} + \dfrac{\cos \alpha}{\cot \alpha} = \cos \alpha + \sin \alpha$

31. $\dfrac{\sin \beta}{\cos \beta} + \cot \beta = \sec \beta \csc \beta$

32. $(\sin \beta - \cos \beta)^2 + (\sin \beta + \cos \beta)^2 = 2$

33. $\sin \alpha = 2 \sin \frac{1}{2}\alpha \cos \frac{1}{2}\alpha$

34. $\cos 4\theta = \cos^2 2\theta - \sin^2 2\theta$

35. $\sin 2\theta \sec \theta = 2 \sin \theta$

36. $\sin 2\theta \tan \theta = 2 \sin^2\theta$

37. $\dfrac{\csc \gamma + 1}{\csc \gamma \cos \gamma} = \sec \gamma + \tan \gamma$

38. $\sec 3\theta - \cos 3\theta = \tan 3\theta \sin 3\theta$

39. $\dfrac{\cot 3\theta - \sin 3\theta}{\sin 3\theta} = \dfrac{\cos^2 3\theta + \cos 3\theta - 1}{\sin^2 3\theta}$

40. $\cot \beta - \sin \beta = (\cos^2\beta + \cos \beta - 1)\csc \beta$

41. $(\sin \alpha \cos \alpha \cos \beta + \sin \beta \cos \beta \cos \alpha)\sec \alpha \sec \beta$
   $= \sin \alpha + \sin \beta$

42. $(\cos \alpha \cos \beta \tan \alpha + \sin \alpha \sin \beta \cot \beta)\csc \alpha \sec \beta = 2$

43. $\dfrac{\tan(\alpha + \beta) - \tan \beta}{1 + \tan(\alpha + \beta)\tan \beta} = \tan \alpha$

44. $\dfrac{\tan \alpha - \tan(\alpha - \beta)}{1 + \tan \alpha \tan(\alpha - \beta)} = \tan \beta$

*Problems **45–48:** Find the exact value of each expression.*

45. $\cos 345°$

46. $2 \cos^2 15° - 1$

47. $\sqrt{\dfrac{1 - \cos 270°}{2}}$

48. $-\sqrt{\dfrac{1 + \cos 270°}{2}}$

*Problems **49–50:** Use the trigonometric substitution $u = a \sin \theta$ for $-\frac{\pi}{2} \le \theta \le \frac{\pi}{2}$ to rewrite the algebraic expression in terms of θ and simplify.*

49. $\dfrac{1}{\sqrt{36 - u^2}}; \quad a = 6$

50. $\dfrac{u}{\sqrt{9 - u^2}}; \quad a = 3$

*Problems **51–52:** Use the trigonometric substitution $u = a \tan \theta$ for $-\frac{\pi}{2} < \theta < \frac{\pi}{2}$ to rewrite the algebraic expression in terms of θ and simplify.*

51. $\sqrt{u^2 + 100}; \quad a = 10$

52. $\dfrac{1}{\sqrt{u^2 + 1}}; \quad a = 1$

*Problems **53–54:** Use the trigonometric substitution $u = a \sec \theta$ for $0 < \theta < \frac{\pi}{2}, a > 0$, to rewrite the algebraic expression in terms of θ and simplify.*

53. $\sqrt{u^2 - 64}; \quad a = 8$

54. $\dfrac{1}{(u^2 - 81)^{1/2}}; \quad a = 9$

Applications

55. If $30°$ is the measure of an angle α in standard position, what are the coordinates of the point of intersection of the terminal side of α and a unit circle?

56. Use a graphing calculator to find the graph of

$$f(x) = \sin(1 + \sin x)$$

Using the trace, find the maximum and minimum values of f for $0 \le x \le 2\pi$.

57. Use a graphing calculator to find the graph of

$$g(x) = \cos(\cos(\cos x))$$

Using the trace, find the maximum and minimum values of g for $0 \le x \le 2\pi$.

58. A regular pentagon is inscribed in a unit circle. Find the area of the pentagon. Give both the approximate answer correct to four decimal places and the exact answer.

59. The slope of a line can be defined using a trigonometric function, as we did in Section 2.4, p. 82. The *angle of inclination* θ of a line is the smallest positive angle that line makes with the x-axis (see Figure 4.23a). Let ϕ be the smallest positive angle between two nonvertical lines L_1 and L_2. Show that

$$\tan \phi = \frac{m_2 - m_1}{1 + m_1 m_2}$$

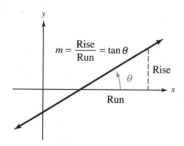

a. m is the slope of a line

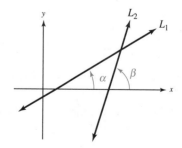

b. Lines L_1 and L_2

FIGURE 4.23
Finding the angle between two lines

60. In a remote area, the population of the endangered marbled murrelet is found to change according to the formula

$$y = 45 + 6 \sin\left(\tfrac{\pi}{4}t\right)$$

where t is the time (in years) after the beginning of the study in 1994.

a. Graph the equation for a 12-year period starting in 1994.

b. What is the maximum expected population over the 12-year period?

c. What is the minimum expected population over the 12-year period?

Group Research Projects

Working in small groups is typical of most work environments, and this book seeks to develop skills with group activities. At the end of each chapter, we present a list of suggested projects, and even though they could be done as individual projects, we suggest that these projects be done in groups of three or four students.

1. Two dimes touch the side of a rectangle at the same point, but one is on the inside and the other is on the outside. The coins are rolled in the plane along the perimeter of the rectangle until they come back to their initial positions. The height of the rectangle is twice the circumference of the dimes, and its width is twice its height. How many revolutions will each coin make?

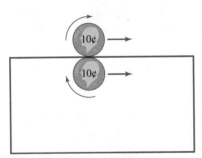

2. What are the precise times when the minute hand of an ordinary clock crosses the hour hand?

3. JOURNAL PROBLEM [This problem is adapted from "Geometric Proofs of Multiple Angle Formulas," by Wayne Dancer in *American Mathematical Monthly*, Vol. 44 (1937), pp. 366–367.] In this chapter, we considered several identities from both algebraic and geometric standpoints. Use Figure 4.24 to prove each identity:

$$\cos 2\theta = 2\cos^2\theta - 1$$
$$\cos 3\theta = 4\cos^3\theta - 3\cos\theta$$
$$\sin 2\theta = 2\sin\theta\cos\theta$$
$$\sin 3\theta = 3\sin\theta - 4\sin^3\theta$$

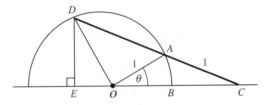

FIGURE 4.24
Geometric proofs of multiple-angle identities

4. JOURNAL PROBLEM (This problem is from the September 1995 issue of *The Mathematics Teacher*.) Robert H. Becker, in a letter to the editor, challenges us with sixteen surprising identities. See how many of these your group can prove. Suppose that α, β, and γ are angles of a triangle, and assume that in (11) and (16) no angle may have measure 90°, and in (3) no angle may have measure 45° or 135°.

(1) $\sin 2\alpha + \sin 2\beta + \sin 2\gamma = 4\sin\alpha\sin\beta\sin\gamma$

(2) $\cos 2\alpha + \cos 2\beta + \cos 2\gamma$
$= -1 - 4\cos\alpha\cos\beta\cos\gamma$

(3) $\sin\alpha + \sin\beta + \sin\gamma = 4\cos\tfrac{1}{2}\alpha\cos\tfrac{1}{2}\beta\cos\tfrac{1}{2}\gamma$

(4) $\cos\alpha + \cos\beta + \cos\gamma = 1 + 4\sin\tfrac{1}{2}\alpha\sin\tfrac{1}{2}\beta\sin\tfrac{1}{2}\gamma$

(5) $\sin^2\alpha + \sin^2\beta + \sin^2\gamma = 2(1 + \cos\alpha\cos\beta\cos\gamma)$

(6) $\cos^2\alpha + \cos^2\beta + \sin^2\gamma = 1 - 2\cos\alpha\cos\beta\cos\gamma$

(7) $\sin^2\frac{1}{2}\alpha + \sin^2\frac{1}{2}\beta + \sin^2\frac{1}{2}\gamma$

$= 1 - 2\sin\frac{1}{2}\alpha \sin\frac{1}{2}\beta \sin\frac{1}{2}\gamma$

(8) $\cos^2\frac{1}{2}\alpha + \cos^2\frac{1}{2}\beta + \cos^2\frac{1}{2}\gamma$

$= 2\left(1 + \sin\frac{1}{2}\alpha \sin\frac{1}{2}\beta \sin\frac{1}{2}\gamma\right)$

(9) $\sin^2 2\alpha + \sin^2 2\beta + \sin^2 2\gamma$

$= 2(1 - \cos 2\alpha \cos 2\beta \cos 2\gamma)$

(10) $\cos^2 2\alpha + \cos^2 2\beta + \cos^2 2\gamma$

$= 1 + 2\cos 2\alpha \cos 2\beta \cos 2\gamma$

(11) $\tan\alpha + \tan\beta + \tan\gamma = \tan\alpha \tan\beta \tan\gamma$

(12) $\cot\frac{1}{2}\alpha + \cot\frac{1}{2}\beta + \cot\frac{1}{2}\gamma = \cot\frac{1}{2}\alpha \cot\frac{1}{2}\beta \cot\frac{1}{2}\gamma$

(13) $\tan 2\alpha + \tan 2\beta + \tan 2\gamma = \tan 2\alpha \tan 2\beta \tan 2\gamma$

(14) $\cot\alpha \cos\beta + \cot\beta \cot\gamma + \cot\alpha \cot\gamma = 1$

(15) $\tan\frac{1}{2}\alpha \tan\frac{1}{2}\beta + \tan\frac{1}{2}\beta \tan\frac{1}{2}\gamma + \tan\frac{1}{2}\alpha \tan\frac{1}{2}\gamma = 1$

(16) $\cot 2\alpha + \cot 2\beta + \cot 2\beta \cos 2\gamma + \cot 2\alpha \cos 2\gamma = 1$

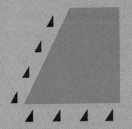

CHAPTER FIVE

OBLIQUE TRIANGLES AND VECTORS

5.1 Law of Cosines
5.2 Law of Sines
5.3 Areas and Volumes
5.4 Vector Triangles
5.5 Vector Operations
5.6 Chapter 5 Summary

PREVIEW

*In the first chapter, we solved right triangles. In this chapter, we turn to the solution of **oblique triangles** (triangles that are not right triangles). We will derive two trigonometric laws, the* law of cosines *and the* law of sines, *that allow us to solve certain triangles using a calculator.*

We conclude this chapter by considering vector triangles and vector operations, a topic that is important to physics, calculus, and more advanced work.

The science of figures is most glorious and beautiful.
  N. Frischlinus
  Dialog 1

SECTION 5.1

LAW OF COSINES

PROBLEM OF THE DAY

The Riddler has just committed the crime of the century. Thirty minutes later, he is traveling 460 mi/h in his Learjet with a heading of 120° (measured clockwise from North). At this instant, Batman is located 150 miles from this location on a bearing of 140°. How far apart are the two planes (assuming a constant speed)?

How would you answer this question?

North

120°

140°

S

230

150

20°

D

B

x

FIGURE 5.1
Finding an unknown distance

We might begin by drawing a diagram, as shown in Figure 5.1. Fill in the known distances. If the Learjet travels for 30 min $= \frac{1}{2}$ h at 460 mi/h, the distance is found by the formula $d = rt$, so that

$$d = \tfrac{1}{2}(460) = 230 \text{ mi}$$

We have determined that the angle between the planes is 20°. The unknown distance we seek is labeled x, and we see that the triangle formed is *not* a right triangle. In this section we find unknown parts of a triangle that is not a right triangle.

Oblique Triangles

In the previous section, we solved right triangles. We now extend that study to triangles with no right angles. Such triangles are called **oblique triangles.** Consider a triangle labeled as in Figure 5.2, and notice that now γ is not restricted.

In general, we will be given three parts of a triangle and be asked to find the remaining three parts. But can you do so given *any* three parts? Consider the possibilities:

SSS   Three sides are given.

SAS   Two sides and an included angle are given.

SSA   Two sides and the angle opposite one of them are given.

ASA   Two angles and an included side are given. AAS or SAA are considered to be the same as ASA, because if we know two angles, then we know all three angles (since the sum of the angles of a triangle is 180°).

AAA   Three angles are given.

We consider these possibilities, one at a time, in this section and Section 5.2. At the end of Section 5.2, we will again summarize all these possibilities.

SSS

To solve a triangle given SSS, it is necessary for the sum of the lengths of the two smaller sides to be greater than the length of the largest side. We use a generalization of the Pythagorean theorem called the **law of cosines.**

LAW OF COSINES

In any $\triangle ABC$

$$c^2 = a^2 + b^2 - 2ab \cos \gamma$$

It will be helpful to remember this law.

The proof of the law of cosines involves the distance formula. Let γ be an angle in standard position with A on the positive x-axis, as shown in Figure 5.3. The coordinates of the vertices are as follows:

$C(0, 0)$ *Since C is in standard position*

$A(b, 0)$ *Since A is on the positive x-axis and a distance of b units from the origin*

$B(a \cos \gamma, a \sin \gamma)$ *Let B = (x, y); then by definition of the trigonometric functions, cos γ = x/a and sin γ = y/a. Thus, x = a cos γ and y = a sin γ.*

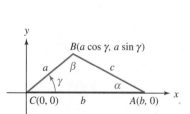

FIGURE 5.3 $\triangle ABC$ with angle γ in standard position

FIGURE 5.2
Correctly labeled triangle

Historical Note

Nicholas Copernicus (1473–1543) is probably best known as the astronomer who revolutionized the world with his heliocentric theory of the universe. However, in his book *De revolutionibus orbium coelestium,* he also developed a substantial amount of trigonometry. This book was published in the year of his death; as a matter of fact, the first copy off the press was rushed to him as he lay on his deathbed. It was on Copernicus' work that his student Rheticus based his ideas, which soon brought trigonometry into full use. In a two-volume work, *Opus palatinum de triangulis,* Rheticus used and calculated elaborate tables for all six trigonometric functions.

Use the distance formula to find the distance between the point $A(b, 0)$ and the point $B(a \cos \gamma, a \sin \gamma)$:

$$
\begin{aligned}
c^2 &= (a \cos \gamma - b)^2 + (a \sin \gamma - 0)^2 \\
&= a^2 \cos^2 \gamma - 2ab \cos \gamma + b^2 + a^2 \sin^2 \gamma \\
&= a^2(\cos^2 \gamma + \sin^2 \gamma) + b^2 - 2ab \cos \gamma \\
&= a^2 + b^2 - 2ab \cos \gamma \quad \textit{Since } \cos^2 \gamma + \sin^2 \gamma = 1
\end{aligned}
$$

We can show that this law of cosines is a generalization of the Pythagorean theorem. This means that if $\gamma = 90°$:

$$
c^2 = a^2 + b^2 \quad \textit{Because } \cos 90° = 0, \, 2ab \cos 90° = 0.
$$

By letting A and B, respectively, be in standard position, it can also be shown that

$$
a^2 = b^2 + c^2 - 2bc \cos \alpha \quad \text{and} \quad b^2 = a^2 + c^2 - 2ac \cos \beta
$$

To find the angle when you are given three sides, solve for α, β, or γ.

LAW OF COSINES

$$
a^2 = b^2 + c^2 - 2bc \cos \alpha \quad \text{or} \quad \cos \alpha = \frac{b^2 + c^2 - a^2}{2bc}
$$

$$
b^2 = a^2 + c^2 - 2ac \cos \beta \quad \text{or} \quad \cos \beta = \frac{a^2 + c^2 - b^2}{2ac}
$$

$$
c^2 = a^2 + b^2 - 2ab \cos \gamma \quad \text{or} \quad \cos \gamma = \frac{a^2 + b^2 - c^2}{2ab}
$$

EXAMPLE 1

Finding a Part of a Triangle Given Three Sides (SSS)

What is the smallest angle of a triangular patio whose sides measure 25, 18, and 21 ft?

Solution   If γ represents the smallest angle, then c (the side opposite γ) must be the smallest side, so $c = 18$. Then

$$
\begin{aligned}
\cos \gamma &= \frac{a^2 + b^2 - c^2}{2ab} \quad \textit{Law of cosines} \\
&= \frac{25^2 + 21^2 - 18^2}{2(25)(21)} \\
\gamma &= \cos^{-1}\left[\frac{25^2 + 21^2 - 18^2}{2(25)(21)}\right] \approx 45.03565072°
\end{aligned}
$$

To two significant digits, the answer is $45°$.   ■

SAS

The next possibility for solving oblique triangles is that of being given two sides and an included angle. It is necessary that the given angle be less than 180°. Again, use the law of cosines for this possibility.

EXAMPLE 2   **Finding One Part of a Triangle Given Two Sides and an Included Angle (SAS)**

Find c where $a = 52.0$, $b = 28.3$, and $\gamma = 28.5°$.

Solution $c^2 = a^2 + b^2 - 2ab \cos \gamma$ *Law of cosines*

$\qquad = (52.0)^2 + (28.3)^2 - 2(52.0)(28.3)\cos 28.5° \approx 918.355474$

Thus, $c \approx 30.30438044$; to three significant digits, $c = 30.3$.   ■

EXAMPLE 3   **Problem of the Day**

The Riddler has just committed the crime of the century and left 30 minutes ago traveling 460 mi/h in his Learjet with a heading of 120° (measured clockwise from North). At this instant, Batman is located 150 mi from this location on a bearing of 140°. How far apart are the two planes?

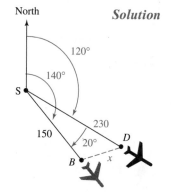

Solution   We seek the distance x, and from the law of cosines,

$$x^2 = |SB|^2 + |SD|^2 - 2|SB||SD|\cos(\angle DSB)$$

$$= 150^2 + 230^2 - 2(150)(230) \cos 20°$$

$$\approx 10{,}561.20917 \quad \text{*By calculator*}$$

Thus, $x \approx 102.7677438$, so the planes are 100 miles apart (to two significant digits).   ■

SSA

Suppose we know two sides and an angle that is not included. If we use the law of cosines, the resulting equation is second degree, which requires the quadratic formula. We stated the quadratic formula in Section 4.1, and repeat it here for your convenience.

Quadratic Formula   **If $Ax^2 + Bx + C = 0$, $A \neq 0$, then $x = \dfrac{-B \pm \sqrt{B^2 - 4AC}}{2A}$.**

EXAMPLE 4   **Solving a Triangle Given Two Sides and an Angle That Is Not Included (SSA)**

Solve the triangle where $a = 3.0$, $b = 2.0$, and $\alpha = 110°$.

$$a^2 = b^2 + c^2 - 2bc \cos \alpha \quad \text{*Law of cosines*}$$

$$9 = 4 + c^2 - 4c \cos 110° \quad \text{*Substitute given values.*}$$

$$c^2 - 4 \cos 110° c - 5 = 0 \qquad \text{*Quadratic equation needs a 0 on one side; subtract 9 from both sides.*}$$

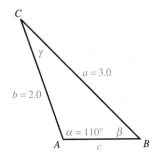

The variable is c, and we use the quadratic formula where $A = 1$, $B = -4 \cos 110°$, and $C = -5$:

$$c = \frac{-(-4 \cos 110°) \pm \sqrt{(-4 \cos 110°)^2 - 4(1)(-5)}}{2(1)}$$

$$\approx 1.654316212, \; -3.022396785 \qquad \textit{Use a calculator.}$$

Even though the quadratic formula gives two solutions, we reject the negative value (because c represents a length). We can now use the positive value to find the angles (using a calculator):

$$\cos \beta = \frac{a^2 + c^2 - b^2}{2ac} \approx \frac{9 + 1.654316212^2 - 4}{2(3)(1.654316212)} \approx 0.7794521661$$

Therefore, $\beta \approx 38.78955642°$. We state the solution (to the required degree of accuracy):

$$a = 3.0, \quad b = 2.0, \quad c = 1.7; \qquad \alpha = 110°, \quad \beta = 39°, \quad \gamma = 31° \qquad ■$$

It may have occurred to you when working Example 4 that, since we are using the quadratic formula, it is certainly possible to obtain two solutions. We consider this possibility with the next example.

EXAMPLE 5

Solving an SSA Triangle for Which There Are Two Solutions

Solve the triangle where $a = 1.50$, $b = 2.00$, and $\alpha = 40.0°$.

Solution

$$a^2 = b^2 + c^2 - 2bc \cos \alpha \qquad \textit{Law of cosines}$$

$$(1.50)^2 = (2.00)^2 + c^2 - 2(2)c \cos 40°$$

$$c^2 - 4 \cos 40° c + 1.75 = 0 \qquad \begin{array}{l}\textit{Quadratic equation in c; use}\\ \textit{the quadratic formula where}\\ A = 1, B = -4 \cos 40°,\\ \textit{and } C = 1.75.\end{array}$$

$$c = \frac{4 \cos 40° \pm \sqrt{(-4 \cos 40°)^2 - 4(1)(1.75)}}{2}$$

$$\approx 2.30493839, \; 0.7592393826$$

How can this be? How can we have two solutions? Look at Figure 5.4. There are two possible values of c and consequently two triangles that satisfy the conditions of this problem.

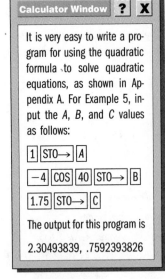

It is very easy to write a program for using the quadratic formula to solve quadratic equations, as shown in Appendix A. For Example 5, input the A, B, and C values as follows:

$\boxed{1}\;\boxed{\text{STO} \rightarrow}\;\boxed{A}$

$\boxed{-}\;\boxed{4}\;\boxed{\text{COS}}\;\boxed{40}\;\boxed{\text{STO} \rightarrow}\;\boxed{B}$

$\boxed{1.75}\;\boxed{\text{STO} \rightarrow}\;\boxed{C}$

The output for this program is

2.30493839, .7592393826

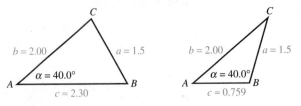

FIGURE 5.4   Two possible triangles with $a = 1.50$, $b = 2.00$, and $\alpha = 40.0°$

Since there are two triangles, we must now double our work (answers are shown to the correct number of significant digits):

SOLUTION I	SOLUTION II

$a = 1.50$ *Given* $a = 1.50$ *Given*

$b = 2.00$ *Given* $b = 2.00$ *Given*

$c_1 = 2.30$ *1st solution* $c_2 = 0.759$ *2nd solution*

 Do not work with rounded values when finding the other parts. This means that you should use the calculator output without rounding, not the rounded values shown above.

$\alpha = 40.0°$ *Given* $\alpha = 40.0°$ *Given*

$\beta = \cos^{-1}\left(\dfrac{a^2 + c^2 - b^2}{2ac}\right)$ $\beta = \cos^{-1}\left(\dfrac{a^2 + c^2 - b^2}{2ac}\right)$

$\approx 58.98696953°$ $\approx 121.0130305°$

$\beta_1 = 59.0°$ *3 sig. digits* $\beta_2 = 121.0°$ *3 sig. digits*

$\gamma_1 = 81.0°$ $\gamma_2 = 19.0°$ ∎

Since there is more than one solution in Example 5, this possibility (namely two sides with an angle that is not included) is sometimes referred to as the **ambiguous case,** because under certain circumstances two triangles result. However, as long as you solve this problem by the law of cosines, you can simply interpret the results of the quadratic formula to determine whether there are one or two solutions.

EXAMPLE 6   **Solving an SSA Triangle with No Solution**

Solve the triangle where $a = 3.0$, $b = 1.5$, and $\beta = 40°$.

Solution   We are given two sides and an angle that is not included. There *may* be two solutions, so we use the law of cosines.

$$b^2 = a^2 + c^2 - 2ac \cos \beta$$
$$(1.5)^2 = (3.0)^2 + c^2 - 2(3.0)c \cos 40°$$
$$c^2 - 6 \cos 40° c + 6.75 = 0$$
$$c = \frac{6 \cos 40° \pm \sqrt{(-6 \cos 40°)^2 - 4(1)(6.75)}}{2(1)}$$

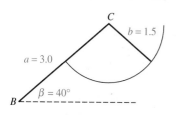

FIGURE 5.5
No solution

Since the discriminant (the number under the radical) is negative, we see there is no solution. Look at Figure 5.5 to see why there is no solution. The side opposite the 40° angle is not as long as the height of the triangle with a 40° angle and a side of length 3. ∎

EXAMPLE 7   **Solving an SSA Triangle with One Solution**

Solve the triangle given by $a = 2.0$, $b = 3.0$, and $\beta = 40°$.

Solution   We are given two sides and an angle that is not included. There *may* be two solutions, so we use the law of cosines:

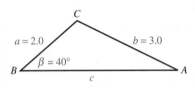

$$b^2 = a^2 + c^2 - 2ac\cos\beta$$
$$(3.0)^2 = (2.0)^2 + c^2 - 2(2.0)c\cos 40°$$
$$c^2 - 4\cos 40°c - 5 = 0$$
$$c = \frac{4\cos 40° \pm \sqrt{(-4\cos 40°)^2 - 4(1)(-5)}}{2(1)}$$
$$\approx 4.242678556, -1.178500783$$

We disregard the negative value of c (c is a length). Finally, we need to find another angle of the triangle:

$$\cos\alpha = \frac{b^2 + c^2 - a^2}{2bc} = \frac{(3.0)^2 + c^2 - 2.0^2}{2(3.0)c}$$

so that

$$\alpha = \cos^{-1}\left[\frac{(3.0)^2 + c^2 - (2.0)^2}{2(3.0)c}\right] \approx 25.37399394°$$

We now state the answer to the correct number of significant digits:

$\alpha = 25°$	$a = 2.0$
$\beta = 40°$	$b = 3.0$
$\gamma = 115°$ $(\gamma = 180° - \alpha - 40°)$	$c = 4.2$

■

PROBLEM SET 5.1

Essential Concepts

1. Draw a correctly labeled $\triangle ABC$ so that β is an angle in standard position with C on the positive x-axis. Specify the coordinates of A, B, and C.

2. Draw a correctly labeled $\triangle ABC$ so that α is an angle in standard position with B on the positive x-axis. Specify the coordinates of A, B, and C.

3. IN YOUR OWN WORDS How do you know if there are two solutions when solving triangles?

4. IN YOUR OWN WORDS In the text we said that to form a triangle, the sum of the lengths of any two sides of the triangle must be greater than the remaining side. Explain this concept. How does this relate to the familiar saying, "The shortest distance between two points is a straight line"?

Level 1 Drill

Problems 5–16: Solve $\triangle ABC$.

5. $a = 7.0$; $b = 8.0$; $c = 2.0$. Find α.

6. $a = 7.0$; $b = 5.0$; $c = 4.0$. Find β.

7. $a = 10$; $b = 4.0$; $c = 8.0$. Find γ.

8. $a = 11$; $b = 9.0$; $c = 8.0$. Find the largest angle.

9. $a = 12$; $b = 6.0$; $c = 15$. Find the smallest angle.

10. $a = 4.0$; $b = 5.0$; $c = 6.0$. Find the middle-sized angle.

11. $a = 18$; $b = 25$; $\gamma = 30°$. Find c.

12. $a = 18$; $c = 11$; $\beta = 63°$. Find b.

13. $a = 15$; $b = 8.0$; $\gamma = 38°$. Find c.

14. $b = 21$; $c = 35$; $\alpha = 125°$. Find a.

15. $b = 14$; $c = 12$; $\alpha = 82°$. Find a.

16. $a = 31$; $b = 24$; $\gamma = 120°$. Find c.

Problems 17–34: *Solve △ABC; if two triangles are possible, give both solutions. If the triangle cannot be solved, explain why.*

17. $a = 14.2$; $b = 16.3$; $\beta = 115.0°$

18. $a = 14.2$; $c = 28.2$; $\gamma = 135.0°$

19. $a = 5.0$; $b = 4.0$; $\alpha = 125°$

20. $a = 5.0$; $b = 4.0$; $\alpha = 80°$

21. $b = 82.5$; $c = 52.2$; $\gamma = 32.1°$

22. $a = 151$; $b = 234$; $c = 416$

23. $a = 10.2$; $b = 11.8$; $\alpha = 47.0°$

24. $a = 82.5$; $b = 16.9$; $\gamma = 80.6°$

25. $a = 123$; $b = 225$; $c = 351$

26. $a = 27.2$; $c = 35.7$; $\alpha = 43.7°$

27. $b = 5.2$; $c = 3.4$; $\alpha = 54.6°$

28. $a = 81$; $c = 53$; $\beta = 85.2°$

29. $a = 214$; $b = 320$; $\gamma = 14.8°$

30. $a = 18$; $b = 12$; $c = 23.3$

31. $a = 140$; $b = 85.0$; $c = 105$

32. $b = 428$; $c = 395$; $\gamma = 28.4°$

33. $a = 36.9$; $b = 20.45$; $\gamma = 90.0°$

34. $a = 50.2$; $c = 29.0$; $\gamma = 30.0°$

Level 2 Applications

35. New York City is approximately 210 miles N9°E of Washington, D.C., and Buffalo, New York, is N49°W of Washington, D.C. How far is Buffalo from New York City if the distance from Buffalo to Washington, D.C., is approximately 285 miles?

36. New Orleans, Louisiana, is approximately 1,080 miles S56°E of Denver, Colorado, and Chicago, Illinois, is N76°E of Denver. How far is Chicago from New Orleans if it is approximately 950 miles from Denver to Chicago?

37. An artillery-gun observer must determine the distance to a target at a point T. The observer knows that the target is 5.20 miles from point I on a nearby island. The observer knows that the gun location at H is 4.30 miles from point I. If $\angle HIT$ is 68.4°, how far is the gun from the target? (See Figure 5.6.)

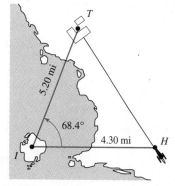

FIGURE 5.6   Distance to a target

38. A buyer is interested in purchasing a triangular lot with vertices LOT, but unfortunately, the marker at point L has been lost. The deed indicates that TO is 453 ft and LO is 112 ft, and that the angle at L is 82.6°. What is the distance from L to T?

39. A TV antenna is attached to the gable of a roof with pitch and dimensions as shown in Figure 5.7. How long is the cable from point P to point Q?

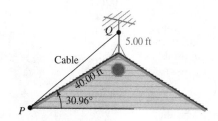

FIGURE 5.7   Guy wire on a TV antenna

40. A tower is built on a hill that has an inclination of 5.25° with the horizontal. From point P on the hill, the angle of elevation of a 150.0-ft tower is 29.45°. How much cable is necessary to reach the tower at point Q, as shown in Figure 5.8 (page 230)?

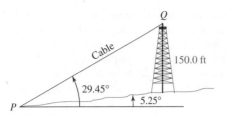

FIGURE 5.8 Cable of a power tower

41. A weight is supported by ropes that are attached to both ends of a 12-ft balance beam (as shown in Figure 5.9). What are the angles between the balance beam and the ropes?

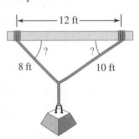

FIGURE 5.9
Supported weight

42. Find the distance $|AB|$ between the ends of a construction crane, as shown in Figure 5.10.

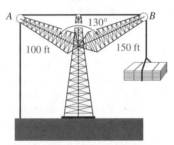

FIGURE 5.10
A construction crane

43. A dime, a penny, and a quarter are placed on a table so that they just touch each other, as shown in Figure 5.11. Let D, P, and Q be the respective centers. Solve $\triangle DPQ$. [*Hint:* If you measure the coins, the diameters are 1.75, 2.00, and 2.50 cm.]

FIGURE 5.11
Three coins determine a triangle

44. What is the perimeter of an equilateral triangle inscribed in a circle of radius 5.0 in.?

45. What is the perimeter of a square inscribed in a circle of radius 5.0 in.?

46. What is the perimeter of a regular pentagon inscribed in a circle of radius 5.0 in.?

47. The Pentagon in Arlington, Virginia, is a regular pentagon constructed inside a circle with radius 783.5 ft. What is the perimeter of the Pentagon?

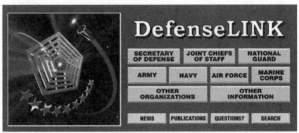

© The Image Works Archives

***Problems* 48–51:** *A boat going from Newport Beach to Catalina has been traveling at a constant speed of 15 mi/h for 60 min before the crew discovers that they need to correct course by 15°. Assume Avalon is located 26 miles S45°W of Newport Beach.*

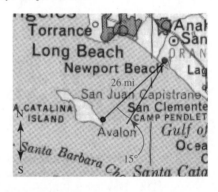

48. a. How far is the boat from Newport Beach?

   b. How far is the boat from Avalon?

49. How long will it take for the boat to reach Avalon?

50. How much time was added to the trip because of this navigational error? (Assume a constant speed of 15 mi/h for the entire trip.)

51. What was the original heading, and what was the correct original heading? After traveling for an hour along the incorrect heading, what is the necessary new heading to reach Avalon?

52. A plane flying from Anchorage to Fairbanks (a distance of about 250 miles) has been traveling for 30 min before the pilot discovers that she is off course by 10.0°. If her plane maintains an average speed of 180 mi/h, through what angle should the pilot turn to reach Fairbanks?

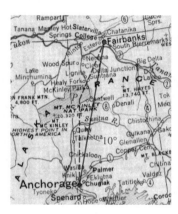

53. In Problem 52, if the pilot wishes to reach Fairbanks at the scheduled time, she must increase the

airspeed. Find the airspeed that enables the plane to reach Fairbanks on time.

Level 3 Theory

54. Prove that $a^2 = b^2 + c^2 - 2bc \cos \alpha$.

55. Prove that $b^2 = a^2 + c^2 - 2ac \cos \beta$.

56. Let a and b be the equal sides of an isosceles triangle. Prove

$$c^2 = 2a^2(1 - \cos \gamma)$$

57. In $\triangle ABC$, use the law of cosines to prove

$$a = b \cos \gamma + c \cos \beta$$

58. In $\triangle ABC$, use the law of cosines to prove

$$b = a \cos \gamma + c \cos \alpha$$

59. In $\triangle ABC$, use the law of cosines to prove

$$c = a \cos \beta + b \cos \alpha$$

60. From the equation given in Problem 54, show that

$$\cos \alpha + 1 = \frac{(b + c - a)(a + b + c)}{2bc}$$

61. From the equation given in Problem 55, show that

$$\cos \beta + 1 = \frac{(a + c - b)(a + b + c)}{2ac}$$

SECTION 5.2

LAW OF SINES

PROBLEM OF THE DAY

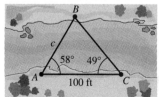

Two points A and B on opposite sites of a river are the endpoints for a proposed footbridge. In order to find the length of this proposed bridge, a point C is located 100 ft from point A. It is then determined that $\alpha = 58°$, $\gamma = 49°$. What is the length of the footbridge?

How would you answer this question?

_____

You begin as usual, by drawing a diagram. Find the third angle of the triangle (since the sum of the angles of a triangle is 180°):

$$\beta = 180° - 58° - 49° = 73°$$

Since this is not a right triangle, you might try the law of cosines, but in every form there will be more than one unknown. In other words, the law of cosines will not solve this problem. To answer the question posed with the Problem of the Day, we need a new result, called the *law of sines,* which is based on a right-triangle solution of a triangle.

ASA

Suppose that two angles and a side are given. For such a triangle to be formed, the angles must be positive, and the sum of the two given angles must be less than 180°. Also, the length of the given side must be greater than zero. In this case, we state and prove a result called the **law of sines.**

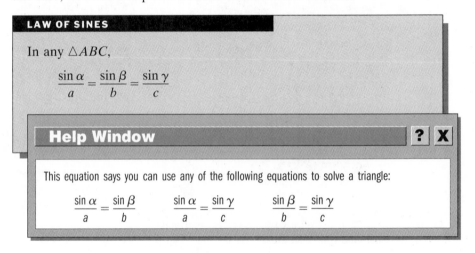

LAW OF SINES

In any $\triangle ABC$,

$$\frac{\sin \alpha}{a} = \frac{\sin \beta}{b} = \frac{\sin \gamma}{c}$$

Help Window   ? X

This equation says you can use any of the following equations to solve a triangle:

$$\frac{\sin \alpha}{a} = \frac{\sin \beta}{b} \qquad \frac{\sin \alpha}{a} = \frac{\sin \gamma}{c} \qquad \frac{\sin \beta}{b} = \frac{\sin \gamma}{c}$$

To prove this law, we consider any oblique triangle, as shown in Figure 5.12.

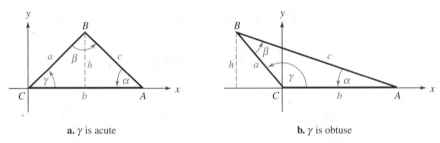

a. γ is acute   **b.** γ is obtuse

FIGURE 5.12   Oblique triangles for law of sines

Let h be the height of the triangle with base CA. Then

$$\sin \alpha = \frac{h}{c} \qquad \text{and} \qquad \sin \gamma = \frac{h}{a}$$

Solving for h,

$$h = c \sin \alpha \quad \text{and} \quad h = a \sin \gamma$$

Thus

$$c \sin \alpha = a \sin \gamma$$

Dividing both sides by ac,

$$\frac{\sin \alpha}{a} = \frac{\sin \gamma}{c}$$

Repeat these steps for the height of the same triangle with base AB to find

$$\frac{\sin \alpha}{a} = \frac{\sin \beta}{b}$$

If we put both of these equations together, we have

$$\frac{\sin \alpha}{a} = \frac{\sin \beta}{b} = \frac{\sin \gamma}{c}$$

EXAMPLE 1

Solving a Triangle Given Two Angles (ASA)

Solve the triangle in which $a = 20$, $\alpha = 38°$, and $\beta = 121°$.

Solution

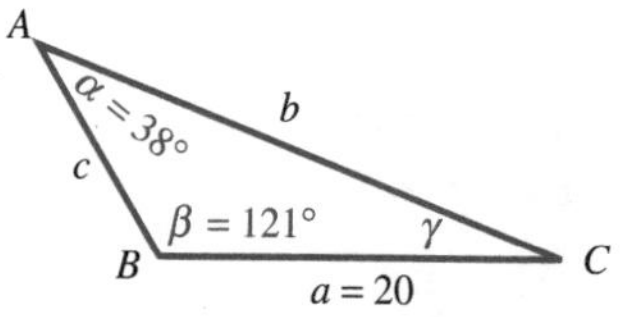

We are given AAS, but since two given angles ensures we know all three angles, we call this ASA.

$\alpha = 38°$	*Given*
$\beta = 121°$	*Given*
$\gamma = 21°$	*Since $\alpha + \beta + \gamma = 180°$, $\gamma = 180° - 38° - 121° = 21°$*
$a = 20$	*Given*

b: Use the law of sines:

$$\frac{\sin 38°}{20} = \frac{\sin 121°}{b} \qquad \text{\textit{Side b associated with opposite angle}}$$

$$b = \frac{20 \sin 121°}{\sin 38°} \qquad \text{\textit{Solve for b.}}$$

$$\approx 27.8454097 \qquad \text{\textit{By calculator}}$$

$b = 28$ *State answer to two significant digits.*

c: Use the law of sines:

$$\frac{\sin 38°}{20} = \frac{\sin 21°}{c} \qquad \text{\textit{Side c associated with opposite angle}}$$

$$c = \frac{20 \sin 21°}{\sin 38°} \qquad \text{\textit{Solve for c.}}$$

$$\approx 11.64172078 \qquad \text{\textit{By calculator}}$$

$c = 12$ *State answers to two significant digits.* ∎

EXAMPLE 2

Problem of the Day: Finding the Length of a Footbridge

Points A and B on opposite sides of a river are the endpoints for a proposed footbridge. To find the length of this proposed bridge, a point C is located 100 ft from point A. It is then determined that $\alpha = 58°$, $\gamma = 49°$. Find the length of the footbridge.

Solution

We know $\alpha = 58°$ and $\gamma = 49°$, so $\beta = 73°$. We also know $b = 100$, so using the law of sines,

$$\frac{\sin \beta}{b} = \frac{\sin \gamma}{c} \qquad \text{Law of sines}$$

$$\frac{\sin 73°}{100} = \frac{\sin 49°}{c} \qquad \text{Substitute known values.}$$

$$c = \frac{100 \sin 49°}{\sin 73°} \qquad \begin{array}{l}\text{Multiply both sides by 100c, and}\\ \text{divide both sides by } \sin 73°.\end{array}$$

$$\approx 78.91935866 \qquad \text{By calculator}$$

To two significant digits, the length of the footbridge is 79 feet. ■

EXAMPLE 3

Using Trigonometry to Find a Rate

A boat traveling at a constant rate due west passes a buoy that is 1.0 km from a lighthouse. The lighthouse is N30°W of the buoy. After the boat has traveled for one-half hour, its bearing to the lighthouse is N74°E. How fast is the boat traveling?

Solution

The angle at the lighthouse (see Figure 5.13) is $180° - 60° - 16° = 104°$.

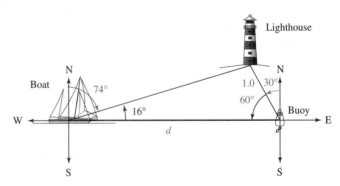

FIGURE 5.13   Find the boat's rate of travel

Use the law of sines:

$$\frac{\sin 104°}{d} = \frac{\sin 16°}{1.0}$$

$$d = \frac{\sin 104°}{\sin 16°} \approx 3.520189502$$

After one hour, the distance is $2d$, so the rate of the boat is about 7.0 kilometers per hour. ■

AAA

Another possible case supposes that three angles are given. However, from what we know of similar triangles (see Figure 5.14), they have the same shape but are not necessarily the same size. This means their corresponding angles have equal measure, so we conclude that the triangle cannot be solved without knowing the length of at least one side.

FIGURE 5.14
AAA implies similar triangles (no solution)

Summary

An important skill to be learned from this section is how to select the proper trigonometric law when given a particular problem. Refer to Table 5.1 for a summary of the procedure.

TABLE 5.1 Procedure for Solving Triangles

Given	Conditions on the given information	Method of solution
SSS	**More than one side given:** **a.** The sum of the lengths of the two smaller sides is less than or equal to the length of the larger side. **b.** The sum of the lengths of the two smaller sides is greater than the length of the larger side.	**Law of cosines** No solution Law of cosines
SAS	**a.** The given angle is greater than or equal to 180°. **b.** The given angle is less than 180°.	No solution Law of cosines
SSA	**a.** The given angle is greater than or equal to 180°. **b.** There are no solutions, one solution, or two solutions, as determined by the quadratic formula.	No solution Law of cosines
ASA or AAS	**More than one angle given:** **a.** The sum of the given angles is greater than or equal to 180°. **b.** The sum of the given angles is less than 180°.	**Law of sines** No solution **Law of sines**
AAA	The given triangles are similar; however, this does not provide enough information to solve the triangles.	No solution

---

EXAMPLE 4   **Determining the Time to Reach a Destination**

An airplane is 100 miles N40°E of a Loran station, traveling due west at 240 mi/h. How long will it take (to the nearest minute) before the plane is 90 miles from the Loran station?

Solution   We begin by drawing a picture and labeling the parts as shown in Figure 5.15 (page 236).

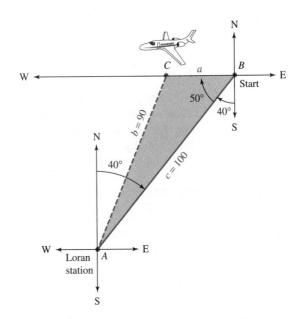

FIGURE 5.15 Time to reach a specified location

We are given more than one side, so we can use the law of cosines. (We do not want you to think that every problem in this section is an application of the law of sines.) Furthermore, we have SSA, so we expect to use the quadratic formula. We find $\beta = 50°$, so we have

$$b^2 = a^2 + c^2 - 2ac \cos \beta$$
$$90^2 = a^2 + 100^2 - 2a(100)\cos 50°$$
$$a^2 - 200 \cos 50° \, a + 1{,}900 = 0$$
$$a = \frac{200 \cos 50° \pm \sqrt{(-200 \cos 50°)^2 - 4(1)(1{,}900)}}{2}$$
$$\approx 111.5202587, \ 17.0372632$$

We see that there are two solutions. The one shown in Figure 5.15 is $a \approx 17.04$. To find the times, use the distance formula $d = rt$, for the two distances found in the quadratic formula and a given rate of 240 mi/h:

SOLUTION I

$$t = \frac{d}{r}$$
$$= \frac{17.0372632}{240}$$
$$\approx 0.0709885967$$

SOLUTION II

$$t = \frac{d}{r}$$
$$= \frac{111.5202587}{240}$$
$$\approx 0.4646677447$$

To convert these times (which are in hours) to minutes, multiply each by 60, and round to the nearest minute:

$$t_1 = 4 \text{ min} \qquad t_2 = 28 \text{ min}$$

The plane will be 90 miles from the Loran station in 4 minutes (location shown in Figure 5.15), and again in 28 minutes. Can you draw the location of the plane at this time? ■

Computer Window ? X

There are many computer programs that will help you solve triangles. One such program is *CONVERGE* from JEMware and is typical. You begin with the choice on the pull-down menu called *Alg/Trig—Solving Triangles* and then input the given values. Next, YOU must choose the correct method of solution:

1. SSS   2. SAS   3. SSA   4. AAS or ASA   5. AAA

Then, if you make the correct choice, the program forces you to select one of the following:

1. No solution   2. Infinitely many solutions   3. One solution using the law of cosines
4. One solution using the law of sines   5. Two solutions using the law of sines

Finally, after making the correct choice here (or being tutored if you do not), you can request not only the solution, but also a diagram of the given triangle.

Some problems can be worked by either the law of cosines or the law of sines. If you read the preceding Computer Window, you might have noticed that this software uses the law of sines for the ambiguous case. Consider the following example where we use both solutions. Our goal is to show that, even though both methods may work, one method is preferable.

EXAMPLE 5   **Comparing Solutions by Laws of Cosine and Sine**

Suppose we know the sides of a triangle are $a = 6.5$, $b = 8.3$, and $c = 12.2$. Also, $\alpha = 29.94°$. Find γ to the nearest tenth of a degree.

Solution   The triangle is shown in Figure 5.16.

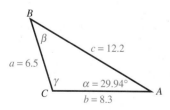

FIGURE 5.16
Given triangle

METHOD I: Use the law of cosines (as suggested in this text).

$$\cos \gamma = \frac{a^2 + b^2 - c^2}{2ab} \qquad \textit{Law of cosines}$$

$$= \frac{6.5^2 + 8.3^2 - 12.2^2}{2(6.5)(8.3)} \qquad \textit{Substitute known values.}$$

$$\approx -0.3493975904 \qquad \textit{By calculator}$$

$$\gamma = \cos^{-1}(-0.3493975904) \approx 110.4504735°$$

To the nearest tenth of a degree, the solution is $\gamma = 110.5°$.

METHOD II: Use the law of sines.

$$\frac{\sin \alpha}{a} = \frac{\sin \gamma}{c} \qquad \textit{Law of sines}$$

$$\frac{\sin 29.94°}{6.5} = \frac{\sin \gamma}{12.2} \qquad \textit{Substitute known values.}$$

$$\sin \gamma = \frac{12.2 \sin 29.94°}{6.5} \qquad \textit{Multiply both sides by 12.2.}$$

$$\approx 0.9367588433 \qquad \textit{By calculator}$$

$$\gamma = \sin^{-1}(0.9367588433) \approx 69.51418175°$$

To the nearest tenth of a degree, the solution is 69.5°. The solutions do not agree. Why? Remember, when we solved equations, we made a point of distinguishing the solution of an equation from the principal value. In method II, $\gamma = 69.5°$. Recall that the sine function is positive in the second quadrant as well as the first quadrant. Therefore, γ may be an obtuse angle as well as an acute angle. In the second quadrant, the angle γ may be found by

$$\gamma = 180° - 69.51418175° = 110.4858183°$$

From the problem stated, we clearly want the obtuse solution, so from the law of sines, $\gamma = 110.5°$ (to the nearest tenth of a degree). ∎

Situations arise when a rough sketch is not sufficient to decide which value from the law of sines is appropriate. For this reason, we suggest you use the law of cosines whenever possible. If you follow the recommendations given in Table 5.1, you will be led to the most appropriate method of solution.

PROBLEM SET 5.2

Essential Concepts

1. **IN YOUR OWN WORDS** State the law of cosines, and explain when you would use the law of cosines.

2. **IN YOUR OWN WORDS** State the law of sines, and explain when you would use the law of sines.

3. **IN YOUR OWN WORDS** Describe a procedure for deciding whether you would use a right-triangle solution, law of cosines, or law of sines to solve a triangle.

Level 1 Drill

Problems 4–15: Draw $\triangle ABC$ with the given parts. From the figure, do you think that there will be one, two, or no solutions? State the number of significant digits in the solution if there is at least one solution.

4. $a = 15.8$; $b = 14.8$, $\alpha = 90°$

5. $a = 5.8$; $b = 14.8$, $\alpha = 90°$

6. $b = 5.50$; $c = 9.30$; $\beta = 54.5°$

7. $b = 8.5$; $c = 9.3$; $\beta = 55.2°$

8. $b = 45.3$; $c = 56.4$; $\gamma = 75.4°$

9. $b = 45.3$; $c = 110.3$; $\gamma = 75.4°$

10. $b = 45$; $c = 40.3$; $\gamma = 75.4°$

11. $b = 45.3$; $c = 36$; $\beta = 75.4°$

12. $a = 56.7$; $c = 85.4$; $\alpha = 110.5°$

13. $a = 6.5$; $b = 8.5$; $\alpha = 35°$

14. $a = 7.43$; $b = 8.52$; $\alpha = 35.00°$

15. $a = 4.5$; $b = 8.5$; $\alpha = 35.5°$

***Problems* 16–38:** *Solve △ABC with the given parts. If two triangles are possible, give both solutions. If the triangle cannot be solved, explain why.*

16. $a = 10$; $\alpha = 48°$; $\beta = 62°$

17. $a = 18$; $\alpha = 65°$; $\gamma = 115°$

18. $a = 30$; $\beta = 50°$; $\gamma = 100°$

19. $b = 23$; $\alpha = 25°$; $\beta = 110°$

20. $b = 40$; $\alpha = 50°$; $\gamma = 60°$

21. $b = 90$; $\beta = 85°$; $\gamma = 25°$

22. $c = 53$; $\alpha = 82°$; $\beta = 19°$

23. $c = 43$; $\alpha = 120°$; $\gamma = 7°$

24. $c = 115$; $\beta = 81.0°$; $\gamma = 64.0°$

25. $\alpha = 18.3°$; $\beta = 54.0°$; $a = 107$

26. $\beta = 85°$; $\gamma = 24°$; $b = 223$

27. $a = 10.8$; $b = 8.80$; $\beta = 21.9°$

28. $a = 59.4$; $b = 71.7$; $\alpha = 27.0°$

29. $b = 55.0$; $c = 92.0$; $\alpha = 98.0°$

30. $b = 58.3$; $\alpha = 120°$; $\gamma = 68.0°$

31. $c = 123$; $\alpha = 85.2°$; $\beta = 38.7°$

32. $a = 26$; $b = 71$; $c = 88$

33. $\alpha = 48°$; $\beta = 105°$; $\gamma = 27°$

34. $a = 25.0$; $\beta = 81.0°$; $\gamma = 25.0°$

35. $a = 25.0$; $b = 45.0$; $c = 89.0$

36. $a = 27.6$; $c = 35.0$; $\alpha = 20.0°$

37. $a = 1.1$; $b = 2.1$; $\alpha = 25°$

38. $a = 481$; $\beta = 28.6°$; $\gamma = 103.0°$

Level 2 Applications

39. In San Francisco, a certain hill makes an angle of 20° with the horizontal and has a tall building at the top. At a point 100 ft down the hill from the base of the building, the angle of elevation to the top of the building is 72°. What is the height of the building?

40. A vertical tower is located on a hill whose inclination is 6°. From a point *P* located 100 ft down the hill from the base of the tower, the angle of elevation to the top of the tower is 28°. What is the height of the tower?

41. In the classic movie *Star Wars*, the hero, Luke, must hit a small target on the Death Star by flying a horizontal distance to reach the target. When the target is sighted, the onboard computer calculates the angle of depression to be 28.0°. If after 150 km, the target (which has not yet been passed) has an angle of depression of 42.0°, how far is the target from Luke's spacecraft at that instant?

42. The Galactic Empire's computers on the Death Star are monitoring the positions of the invading forces. At a particular instant, two observation points 2,500 m apart make a fix on Luke's incoming spacecraft, which is between the observation points and in the same vertical plane. If the angle of elevation from the first observation point is 3.00° and the angle of elevation from the second is 1.90°, find the distance from Luke's spacecraft to each observation point.

43. Solve Problem 42 where both observation points are on the same side of the spacecraft and all the other information is unchanged.

44. At 500 ft in the direction the Tower of Pisa is leaning, the angle of elevation is 20.24°. If the tower leans at an angle of 5.45° from the vertical, what is the length of the tower?

45. What is the angle of elevation of the leaning Tower of Pisa (described in Problem 44) if you measure from a point 500 ft in the opposite direction from which it is leaning?

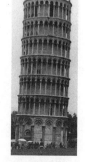

© Stock Boston, Inc.
1992/Peter Menzel

46. From a blimp, the angle of depression is 23.2° to the top of the Eiffel Tower and 64.6° to the bottom. After flying over the tower at the same height but at a distance of 1,000 ft from the first location, you determine that the angle of depression to the top of the tower is now 31.4°. What is the height of the Eiffel Tower, given that these measurements are in the same vertical plane, as shown in Figure 5.17?

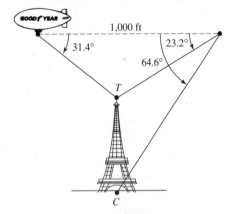

FIGURE 5.17 Determining the height of the Eiffel Tower

47. A ski lift is planned for the south slope of Mt. Frissell in Connecticut. Point B is located directly below the top (point T), as shown in Figure 5.18. A surveyor determines that the angle of elevation from the start of the lift, S, to the end of the lift, T, is 34.06°. Next, 1,000 ft is measured on level ground to determine point P. If the angle of elevation to T at P is 27.77°, what is the length of the ski lift (from point S to point T)?

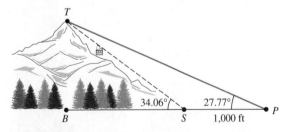

FIGURE 5.18 Mt. Frissell in Connecticut

48. What is the approximate height of Mt. Frissell? (See Problem 47.)

49. *Sputnik I,* the first artificial satellite, was launched on October 4, 1957. When *Sputnik* was directly above point B, the angle at S measured 70.9° to point A, and the distance (on the surface of the earth) between A and B was 1,048 mi, as shown in Figure 5.19. Find the height of *Sputnik* above the point B. [The actual height of *Sputnik* varied from its low point (perigee) of 156 mi to its high point (apogee) of 560 mi.]

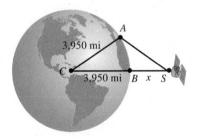

FIGURE 5.19 Orbiting satellite

50. How does a sundial work? Consider a sundial labeled as shown in Figure 5.20.

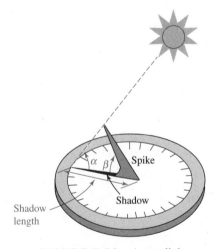

FIGURE 5.20 A sundial

To work properly, the angle (β in Figure 5.20) of the spike of a sundial must be the same as the latitude where the sundial is used. Suppose on a certain day in Columbia, South Carolina (latitude 34.0°), the angle of elevation of the sun (α in Figure 5.20) is 68.0° at 1:00 P.M. If the length of the spike is 10.0 in., what is the length of the shadow?

Level 3 Theory

51. Prove that $\dfrac{\sin \alpha}{a} = \dfrac{\sin \beta}{b}$.

52. Prove that $\dfrac{\sin \beta}{b} = \dfrac{\sin \gamma}{c}$.

53. Prove that $\dfrac{\sin \alpha}{\sin \beta} = \dfrac{a}{b}$.

54. Using Problem 51, show that

$$\frac{\sin \alpha - \sin \beta}{\sin \alpha + \sin \beta} = \frac{a - b}{a + b}$$

55. Using Problem 54, show that

$$\frac{2 \cos \frac{1}{2}(\alpha + \beta) \sin \frac{1}{2}(\alpha - \beta)}{2 \sin \frac{1}{2}(\alpha + \beta) \cos \frac{1}{2}(\alpha - \beta)} = \frac{a - b}{a + b}$$

56. Using Problem 55, show that

$$\frac{\tan \frac{1}{2}(\alpha - \beta)}{\tan \frac{1}{2}(\alpha + \beta)} = \frac{a - b}{a + b}$$

This result is known as the *law of tangents*.

57. Another form of the law of tangents (see Problem 56) is

$$\frac{\tan \frac{1}{2}(\beta - \gamma)}{\tan \frac{1}{2}(\beta + \gamma)} = \frac{b - c}{b + c}$$

Derive this formula.

58. A third form of the law of tangents (see Problem 56) is

$$\frac{\tan \frac{1}{2}(\alpha - \gamma)}{\tan \frac{1}{2}(\alpha + \gamma)} = \frac{a - c}{a + c}$$

59. *Newton's formula* involves all six parts of a triangle. It is not useful in solving a triangle, but it is helpful in checking your results. Show that

$$\frac{a + b}{c} = \frac{\cos \frac{1}{2}(\alpha - \beta)}{\sin \frac{1}{2}\gamma}$$

60. *Mollweide's formula* involves all six parts of a triangle. It is not useful in solving a triangle, but in helping to check your results. Show that

$$\frac{a - b}{c} = \frac{\sin \frac{1}{2}(\alpha - \beta)}{\cos \frac{1}{2}\gamma}$$

Level 3 Applications

61. A harpsichord maker, reproducing a particular historical instrument, needs to join two pieces of wood of different thicknesses to give a neat miter at a certain angle. The pieces of wood have thickness w and W, and the necessary angle at the joint is θ.

Find the angles γ and ϕ, as shown in Figure 5.21, at which the two pieces must be cut, in terms of w, W, and θ. [*Hint:* $\theta = \gamma + \phi$]

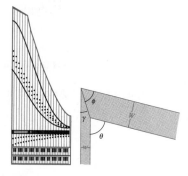

FIGURE 5.21 Construction of a harpsichord

62. INDIVIDUAL RESEARCH PROJECT An article, "Calculating the Distance to the Sun by Observing the Trail of a Meteor" by Jearl Walker, is discussed in the March 1987 issue of *Scientific American* (p. 122). In it, Walker solved several triangles. Using this article as a basis, give one triangle from the article that must be solved, and then show the solution of that triangle.

63. HISTORICAL QUESTION The transition from Renaissance mathematics to modern mathematics was aided, in a large part, by the Frenchman François Viète (1540–1603). He was a lawyer for whom mathematics was a hobby. In his book *Canon mathematicus*, he solved oblique triangles by breaking them into right triangles. He was probably the first person to develop and use the law of tangents. He was also the first to prove (in 1579) the law of cosines. When he proved this law, he did not use the form we used in this text. Instead, he used

$$\frac{2ab}{a^2 + b^2 - c^2} = \frac{1}{\sin(90° - C)}$$

Show that Viète's formula is equivalent to the law of cosines.

SECTION 5.3

AREAS AND VOLUMES

PROBLEM OF THE DAY

What is the area of a triangle with sides 43, 89, and 120 ft?

How would you answer this question?

You might begin by reviewing what you know about areas of triangles. In elementary school, you learned that the area, K, of a triangle is

Notice this formula:

$$K = \tfrac{1}{2}bh^*$$

Area of a Triangle

EXAMPLE 1 **Area of a Triangle with Known Base and Height**

Find the area of each triangle. Note that the height of a triangle is drawn from a vertex perpendicular to a base. The height of a triangle may be interior or exterior to the shaded triangle.

a.

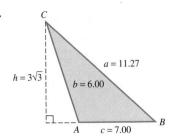

b.

Solution Note that for both triangles, $h = 3\sqrt{3}$ and the base is $c = 7.00$, so both triangles have the same area—namely

$$K = \tfrac{1}{2}(7.00)(3\sqrt{3}) = \tfrac{21}{2}\sqrt{3} \approx 18.19$$

NOTE *Also note that area is given in square units. When units of measurement are given for the sides, then use units of measurement for the area. For example, if the sides in this problem are given in centimeters, then the answer is in square centimeters: 18.19 cm². On the other hand, it is not uncommon to have side measurements given without particular units, as in this example. In such a case, it is implied that the area is 18.19 square units.* ∎

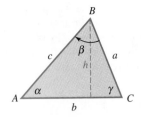

Sometimes we are given side measurements to find the area of a triangle, as in the Problem of the Day. Other times, we are given one side and two angles or two sides and one angle, but not the height. In such cases, we need trigonometry. Consider a $\triangle ABC$, as shown in Figure 5.22.

---

* In elementary school, you no doubt used A for area. In trigonometry, we use K because we use A to represent a vertex of a triangle. Also note that the base b in this formula might be labeled a, b, or c for a particular triangle.

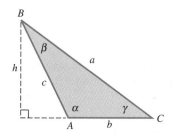

FIGURE 5.22
Arbitrary triangles

$$K = \tfrac{1}{2}bh \qquad \text{and} \qquad \sin \alpha = \frac{h}{c} \quad \text{or} \quad h = c \sin \alpha$$

Thus

$$K = \tfrac{1}{2}bc \sin \alpha$$

Any other pair of sides could be used to derive the following area formulas.

AREA OF A TRIANGLE (SAS)

Two sides and an included angle are known.

$$K = \tfrac{1}{2}bc \sin \alpha \qquad K = \tfrac{1}{2}ac \sin \beta \qquad K = \tfrac{1}{2}ab \sin \gamma$$

Help Window   ? X

In other words, the area of a triangle is equal to half the product of the lengths of two sides times the sine of the included acute angle.

EXAMPLE 2   **Finding the Area of a Triangle**

Find the area of $\triangle ABC$ where $\alpha = 18.4°$, $b = 154$ ft, and $c = 211$ ft.

Solution $K = \tfrac{1}{2}(154)(211)\sin 18.4° \approx 5{,}128.349903$

To three significant digits, the area is $5{,}130 \text{ ft}^2$.   ■

EXAMPLE 3   **Area Enclosed by the Pentagon**

The largest ground area covered by any administrative building in the world is that of the Pentagon in Arlington, Virginia. If the radius of the circumscribed circle is 783.5 ft, find the ground area enclosed by the exterior walls of the building.

© Stock Boston, Inc. 1995/Spencer Grant III.

Solution   The Pentagon forms what is called a *regular pentagon,* which is a five-sided polygon with sides of equal length. Label the vertices as shown in Figure 5.23. The area we seek is 5 times the area of $\triangle AOB$. We know $|AO| = |BO| = 783.5$, and we also know that $\angle AOB = 72°$, since $(1/5)(360°) = 72°$. The area of $\triangle AOB$ is

$$K = \tfrac{1}{2}(783.5)(783.5)\sin 72° \qquad K = \tfrac{1}{2}|AO||BO|\sin \angle AOB$$

$$\approx 291{,}913.6018$$

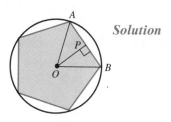

FIGURE 5.23
Schematic for the Pentagon

Next, find the total ground area:

$$5K \approx 5(291{,}913.6018) = 1{,}459{,}568.009$$

The ground area enclosed is about 1.460 million square feet. By the way, since the Pentagon is 5 stories high, it actually has a floor area of about 7.3 million square feet. ■

In the previous section, right triangles were used to find the area of a triangle when two sides and an included angle are known. Now suppose we know the angles, but only one side. If we know only side a, then we use the formula $K = \frac{1}{2}ab \sin \gamma$ and the law of sines:

$$\frac{\sin \alpha}{a} = \frac{\sin \beta}{b} \qquad \text{so} \qquad b = \frac{a \sin \beta}{\sin \alpha}$$

Therefore

$$K = \tfrac{1}{2}ab \sin \gamma = \tfrac{1}{2}a\,\frac{a \sin \beta}{\sin \alpha}\sin \gamma = \frac{a^2 \sin \beta \sin \gamma}{2 \sin \alpha}$$

The other area formulas can be similarly derived.

AREA OF A TRIANGLE (ASA)

Two angles and an included side are known.

$$K = \frac{a^2 \sin \beta \sin \gamma}{2 \sin \alpha} \qquad K = \frac{b^2 \sin \alpha \sin \gamma}{2 \sin \beta} \qquad K = \frac{c^2 \sin \alpha \sin \beta}{2 \sin \gamma}$$

If three sides (but none of the angles) are known, you will need yet another formula to find the area of a triangle. The formula is derived from the law of cosines; this derivation is left as an exercise but is summarized here. The result is known as **Heron's** (or **Hero's**) formula.

AREA OF A TRIANGLE (SSS)

Three sides are known.

$$K = \sqrt{s(s-a)(s-b)(s-c)} \qquad \text{where } s = \tfrac{1}{2}(a+b+c)$$

We now return to the Problem of the Day.

EXAMPLE 4

Finding the Area of an SSS Triangle

Find the area of a triangle having sides 43, 89, and 120 ft.

Solution   Let $a = 43$, $b = 89$, and $c = 120$. Then $s = \frac{1}{2}(43 + 89 + 120) = 126$. Thus

$$K = \sqrt{126(126 - 43)(126 - 89)(126 - 120)} \approx 1{,}523.704696$$

To two significant digits, the area is $1{,}500$ ft^2.   ∎

Area of a Sector

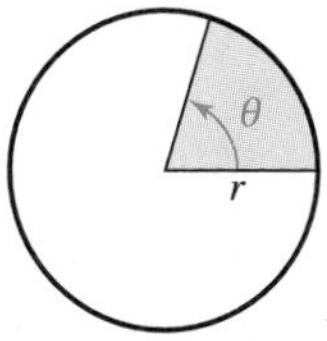

A **sector** of a circle is the portion of the interior of a circle subtended by a central angle θ, as shown in Figure 5.24. Since the area of a circle is $A = \pi r^2$, the area of a sector is the fraction $\theta/(2\pi)$ of the entire circle.

AREA OF A SECTOR = (FRACTIONAL PART OF THE CIRCLE)(AREA OF CIRCLE)

We repeat this result in the following box.

FIGURE 5.24
Sector of a circle (shaded portion)

AREA OF A SECTOR

Area of a sector with radius r and θ measured in radians is

$$K = \frac{1}{2}\theta r^2$$

EXAMPLE 5   **Find the Area of a Sector**

a. What is the area of a sector of a circle of radius 12 in. whose central angle is $2°$?
b. What is the area of a sector of a circle of radius 420 cm whose central angle is 2?

Solution   Note that the central angle may be given in degrees or in radians. If it is given in degrees, you must first convert to radians.

a. $2° = 2\left(\dfrac{\pi}{180}\right) = \dfrac{\pi}{90}$ rad;

$$K = \frac{1}{2}\left(\frac{\pi}{90}\right)(12 \text{ in.})^2 = \frac{4}{5}\pi \text{ in.}^2 \approx 2.5 \text{ in.}^2$$

b. $K = \frac{1}{2}(2)(420 \text{ cm})^2 = 176{,}400 \text{ cm}^2$ ∎

Volume of a Cone

A third application in this section is finding the volume of a cone. Recall from geometry,

$$V = \frac{\pi r^2 h}{3}$$

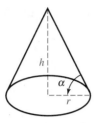

FIGURE 5.25
Volume of a cone

Suppose we know the radius of the base, but not the height. If we know the angle of elevation α (see Figure 5.25), then

$$\tan \alpha = \frac{h}{r} \quad \text{or} \quad h = r \tan \alpha$$

Thus, the volume is

$$V = \frac{\pi r^2 h}{3} = \frac{\pi r^2 (r \tan \alpha)}{3} = \tfrac{1}{3}\pi r^3 \tan \alpha$$

VOLUME OF A CONE

The volume of a circular cone with base radius r and height h is

$$V = \tfrac{1}{3} r^2 h$$

If the angle of elevation from the base to the vertex is α, then

$$V = \tfrac{1}{3}\pi r^3 \tan \alpha$$

EXAMPLE 6   **Estimate the Volume of a Sandpile**

If sand is dropped from the end of a conveyor belt, the sand will fall in a conical heap such that the angle of elevation is about 33°, as shown in Figure 5.26. Find the volume of sand when the radius is 10 ft.

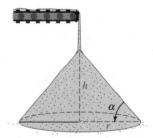

FIGURE 5.26
Estimate the volume of this sandpile

Solution $V = \tfrac{1}{3}\pi r^3 \tan \alpha = \tfrac{1}{3}\pi (10)^3 \tan 33° \approx 680.0580413$

The volume of the pile is about 680 ft³. Such volumes are often converted to cubic yards. To convert cubic feet to cubic yards, divide by 27: The pile is approximately 25 yd³.   ∎

PROBLEM SET 5.3

Essential Concepts

Problems 1–5: *State the area formula or formulas you would use to find the area of a triangle with the given information.*

1. Given a base and the height

2. Three sides given

3. Two sides and an included angle given

4. Two angles and included side given

5. Two sides and a given angle that is not the included angle

6. **IN YOUR OWN WORDS** Outline a procedure for finding the area of a triangle.

7. What is the formula for the area of a sector?

8. What is the formula for the volume of a cone?

Level 1 Drill

Problems 9–40: *Find the area of each triangle. If no triangle is formed, so state.*

9. $b = 15.6$; $h = 2.51$

10. $a = 145$; $h = 95.0$

11. $c = 6.81$; $h = 4.00$

12. $b = 23$; $h = \frac{1}{2}\sqrt{3}$ (two significant figures)

13. $c = 12\sqrt{2}$; $h = 3\sqrt{3}$ (two significant figures)

14. $a = \sqrt{2}$; $h = 4\sqrt{2}$

15. $a = 15$; $b = 8.0$; $\gamma = 38°$

16. $a = 18$; $c = 11$; $\beta = 63°$

17. $b = 14$; $c = 12$; $\alpha = 82°$

18. $b = 21$; $c = 35$; $\alpha = 125°$

19. $a = 30$; $\beta = 50°$; $\gamma = 100°$

20. $b = 23$; $\alpha = 25°$; $\beta = 110°$

21. $b = 40$; $\alpha = 50°$; $\gamma = 60°$

22. $b = 90$; $\beta = 85°$; $\gamma = 25°$

23. $a = b = c = 5$ (exact answer)

24. $a = b = c = 1$ (exact answer)

25. $a = 7.0$; $b = 8.0$; $c = 2.0$

26. $a = 10$; $b = 4.0$; $c = 8.0$

27. $a = 11$; $b = 9.0$; $c = 8.0$

28. $a = 12$; $b = 6.0$; $c = 15$

29. $a = 12.0$; $b = 9.00$; $\alpha = 52.0°$

30. $a = 7.0$; $b = 9.0$; $\alpha = 52°$

31. $a = 10.2$; $b = 11.8$; $\alpha = 47.0°$

32. $a = 8.629973679$; $b = 11.8$; $\alpha = 47.0°$

33. $b = 82.5$; $c = 52.2$; $\gamma = 32.1°$

34. $a = 352$; $b = 230$; $c = 418$

35. $\beta = 15.0°$; $\gamma = 18.0°$; $b = 23.5$

36. $b = 45.7$; $\alpha = 82.3°$; $\beta = 61.5°$

37. $a = 68.2$; $\alpha = 145°$; $\beta = 52.4°$

38. $a = 151$; $b = 325$; $c = 351$

39. $a = 124$; $b = 325$; $c = 351$

40. $a = 27.2$; $b = 83.4$; $c = 55.2$

Problems 41–46: *Find the area of each sector.*

41. $r = 12$ in., $\theta = 1.0$ 42. $r = 24$ cm, $\theta = \frac{\pi}{6}$

43. $r = 3.5$ m, $\theta = 2.3$ 44. $r = 150$ ft, $\theta = 120°$

45. $r = 10$ in., $\theta = 30°$ 46. $r = 23$ in., $\theta = 210°$

Level 2 Drill

Problems 47–50: *Find the area of the interior of the regular polygons inscribed in a circle of radius 1. Give both the exact answer and an answer correct to two decimal places.*

47.   48.

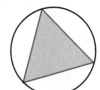 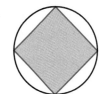

49.   **50.**

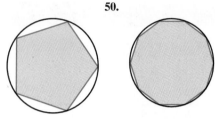

51. Refer to Figure 5.27, which shows a unit circle with center at O and arc AC defining a central angle θ.

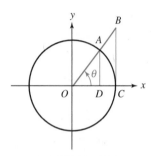

FIGURE 5.27
Area problem

a. Find the area of $\triangle AOD$.
b. Find the area of $\triangle BOC$.
c. Find the area of sector AOC.

Level 2 Applications

Problems 52–54: *A line segment drawn from the center of a circle perpendicular to one side of an inscribed polygon is called an* apothem *of the polygon, as shown in Figure 5.23.*

52. Find the radius of the circle inscribed in the Pentagon in Arlington, Virginia. (See Example 3.) This is the apothem of the Pentagon.
53. What is the apothem (correct to two decimal places) of the pentagon shown in Problem 49?
54. Find the apothem (correct to two decimal places) of the decagon shown in Problem 50.
55. If the central angle subtended by the arc of a segment of a circle is 1.78 and the area is 54.4 cm², what is the radius of the circle?
56. If the area of a sector of a circle is 162.5 cm² and the angle of the sector is 0.52, what is the radius of the circle?

57. A level lot has the dimensions shown in Figure 5.28. What is the total cost of treating the area for poison oak if the fee is \$45/acre? (1 acre = 43,560 ft²)

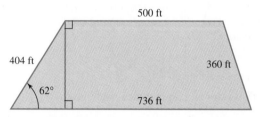

FIGURE 5.28   Area of a level lot

Calculator Window **?** **X**

58. A trough is to be constructed from 1-ft sheets of aluminum by bending up the ends at an angle θ, as shown in Figure 5.29.

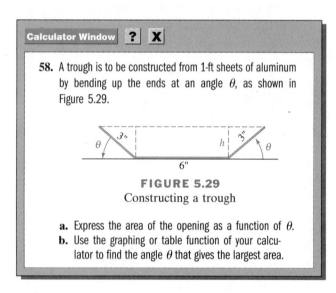

FIGURE 5.29
Constructing a trough

a. Express the area of the opening as a function of θ.
b. Use the graphing or table function of your calculator to find the angle θ that gives the largest area.

59. If vulcanite is dropped from the end of a conveyor belt, it will fall into a conical heap such that the angle of elevation α is about 36°. Find the volume (in yd³) of vulcanite when the radius is 30 ft.

60. If a pup tent is 7 ft long with the height at the center 4 ft, how much material is needed to make the tent? (See Figure 5.30.) Assume that one end is a door that is an equilateral triangle and is made of double material. The other end is a single-material equilateral triangle.

FIGURE 5.30   Two-person pup tent

61. The volume of a cylinder is found by the formula

$$V = (\text{HEIGHT})(\text{AREA OF THE BASE})$$
$$= h(\pi r^2) = \pi r^2 h$$

The volume of a slice cut from a cylinder (see Figure 5.31) is found by the formula

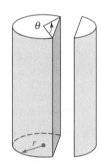

$$V = hr^2\left(\frac{\theta}{2} - \sin\frac{\theta}{2}\cos\frac{\theta}{2}\right)$$

for θ measured in radians such that $0 \le \theta \le \pi$.

a. Derive this formula.
b. Find the volume of a slice cut from a log with a 6.0 in. radius that is 3.0 ft long when $\theta = 60°$.

FIGURE 5.31

62. A 50-ft culvert carries water under a road. If there is 2 ft of water in a culvert with a 3-ft radius, as shown in Figure 5.32, use Problem 61 to find the volume of water in the culvert. Give your answer to the nearest gallon. (Use 1 ft³ ≈ 7.48 gal.)

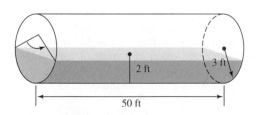

FIGURE 5.32   Circular culvert

Level 3 Theory

63. Express the area of the shaded region in Figure 5.33 as a function of θ.

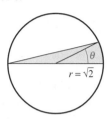

FIGURE 5.33   Area of part of a circle

64. Historical Question This problem derives the formula of Brahmagupta.* Given a quadrilateral with sides measuring a, b, c, and d. Let θ and ϕ be the angles, as shown in Figure 5.34. Let $\mu = \theta + \phi$.

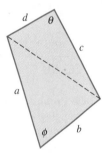

FIGURE 5.34   Brahmagupta's formula

a. Show that the area of the quadrilateral is

$$K = \frac{ab}{2}\sin\phi + \frac{cd}{2}\sin\theta$$

b. Show that

$$a^2 + b^2 - 2ab\cos\phi = c^2 + d^2 - 2cd\cos\theta$$

c. Using parts **a** and **b**, derive Brahmagupta's formula:

$$K^2 = (p-a)(p-b)(p-c)(p-d) - abcd\cos^2\left(\frac{\mu}{2}\right)$$

where $p = \dfrac{a+b+c+d}{2}$.

* This elegant problem was given to me over lunch with my friend Les Lang. I extend many thanks for this and many other mathematical ideas he has given me over the years.

65. Use Brahmagupta's formula to derive Heron's formula for the area of a triangle, to be used when three sides are known:

$$K = \sqrt{s(s-a)(s-b)(s-c)}$$

where $s = \dfrac{a+b+c}{2}$.

66. Inscribed Circle Problem Consider $\triangle ABC$ where $s = \frac{1}{2}(a+b+c)$. The lines that bisect each angle of $\triangle ABC$ meet at a point that is the center of a circle called the *inscribed circle*. (See Figure 5.35.)

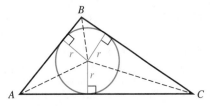

FIGURE 5.35 $\triangle ABC$ with inscribed circle

Show that the radius r of the inscribed circle is

$$r = \sqrt{\dfrac{(s-a)(s-b)(s-c)}{s}}$$

67. Circumscribed Circle Problem Consider $\triangle ABC$ and the circle passing through points A, B, and C. Such a circle is called the *circumscribed circle*. (See Figure 5.36.)

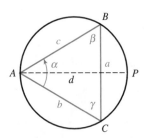

FIGURE 5.36 $\triangle ABC$ with circumscribed circle

Show that the diameter d of the circumscribed circle satisfies

$$d = \dfrac{a}{\sin \alpha}$$

[*Hint:* Let O be the center of the circle and consider $\angle BOD$ where D is on BC so that

$OD \perp BC$. Then use the fact from geometry that $\angle BOC = 2\alpha$.]

68. Show that the diameter d of the circumscribed circle satisfies

$$d = \dfrac{b}{\sin \beta}$$

(See Problem 67.)

69. Suppose that $\triangle ABC$ is an equilateral triangle and each side has length 3 cm. Divide each side into two segments 1 cm and 2 cm, as shown in Figure 5.37.

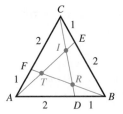

FIGURE 5.37   Area of $\triangle TRI$ and $\triangle ABC$

a. Calculate the area of $\triangle ABC$ using exact values.
b. Calculate the area of $\triangle TRI$ using exact values.

70. IN YOUR OWN WORDS Show that the area of a regular n-gon inscribed in a circle of radius 1 is given by

$$\dfrac{n}{2} \sin \dfrac{360°}{n}$$

Explain why the successive values approach π as n approaches ∞.

71. Show that in every right triangle

$$\dfrac{a+b}{\sqrt{2}} < c < a+b$$

72. Suppose the central angle of a sector is 2θ (instead of θ, as shown in Figure 5.24). Suppose the perimeter of the sector is 24 in. (That is, the perimeter of a sector of a circle of radius r and arc length s is $P = 2r + s$.)
a. Express the radius r of the circle as a function of θ.
b. Express the area K of the sector as a function of θ.

VECTOR TRIANGLES

PROBLEM OF THE DAY

How would you devise a graphical representation for each of these ideas?

a. An airplane traveling with an airspeed of 790 mi/h in the direction of 230°
b. A car traveling at 42 mi/h due west
c. A wind blowing a sailboat in the direction of S63°W with a gale force of 46.1 mi/h
d. An object with mass 850 kg on an inclined plane
e. A person lifting a 130-lb weight by using a pulley and a rope being pulled through the pulley at a 42° angle from the vertical

What do all these representations have in common?

How would you answer this question?

Many applications of mathematics, engineering, and physics involve quantities that have *both* **magnitude** (length) and **direction,** such as forces, velocities, accelerations, and displacements. We represent such quantities on a plane using a directed line segment, called a *vector.*

NOTE

In aviation, direction is measured clockwise from the north.

a. An airplane travels with an airspeed of 790 mi/h and a direction of 230°. The direction is called the **heading.** The *true course* is the direction of the path over the ground (which is measured clockwise from the north).

Magnitude: 790
Direction: 230°

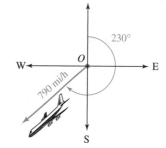

b. A car travels due west at 42 mi/h.

Magnitude: 42
Direction: west

c. The wind blows at 46.1 mi/h in the direction of S63°W.

Magnitude: 46.1
Direction: S63°W

Measure angle to the west of the south direction.

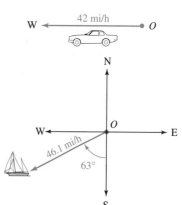

d. An object of an inclined plane has mass 850 kg.

> Magnitude: 850
> Direction: down (gravity)

850 kg

e. A person is lifting a 130-lb weight by using a pulley and a rope being pulled through the pulley at a 42° angle from the vertical.

> Magnitude: 130
> Direction: 42° from the vertical

What do these situations all have in common? They all express an idea that includes both magnitude and direction.

Terminal point

Q

$\mathbf{v} = \vec{PQ}$

P

Initial point

FIGURE 5.38
A vector

Vector

If P and Q are any two points in the plane, then the **directed line segment** from P to Q, denoted by $\vec{PQ}$, is the line segment joining P to Q with an arrowhead placed at Q to indicate the direction is from P to Q. Point P is called the **initial point,** and Q is called the **terminal point.** (See Figure 5.38.)

> **VECTOR**
>
> A **vector** is a directed line segment specifying both a magnitude and a direction. The **magnitude** of the vector represents the length of the quantity being represented, and two vectors have the same **direction** if they are parallel and point in the same direction. Two vectors are said to be **equal** if they have the same magnitude and direction.

Suppose we choose a point O in the plane and call it the origin. A vector may be represented by a directed line segment from O to a point $P(x, y)$ in the plane. This vector is denoted by $\vec{OP}$ or $\mathbf{v}$. In the text we use $\mathbf{v}$; in your work you will write $\vec{v}$. The magnitude of $\vec{OP}$ is denoted by $|\mathbf{OP}|$ or $|\mathbf{v}|$. The vector from O to O is called the zero vector, $\mathbf{0}$.

If $\mathbf{v}$ and $\mathbf{w}$ represent two vectors that have different (but not opposite) directions, they can be drawn so that they have the same initial point. They thus determine a parallelogram, as shown in Figure 5.39.

Segments $\overline{OB}$ and $\overline{AC}$ are called the *diagonals* of the parallelogram, and angles having a common side are called *adjacent angles*. It is easy to show that the sum of two adjacent angles in the parallelogram is 180°.

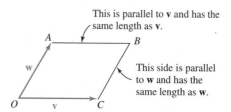

FIGURE 5.39   A parallelogram formed by vectors **v** and **w**

Sum of Vectors (resultant vector)

The **sum** or **resultant** of two vectors **v** and **w** is the vector that is formed when the initial point of **w** is placed to coincide with the terminal point of **v**. (See Figure 5.40a.) We will call this the *end-to-end method.* Another (equivalent) method is to place the initial points of both **v** and **w** so that they coincide, and then draw the parallelogram having **v** and **w** as the adjacent sides; the diagonal is the sum (as shown in Figure 5.40b). We will call this the *parallelogram method.* The vectors **v** and **w** are called **components.**

There are basically two types of vector problems dealing with addition of vectors. The first problem requires us to find the resultant vector. To do this, we use a right triangle, law of sines, or law of cosines. The second problem requires us to *resolve* a given vector into two component vectors. If the two vectors form a right angle, then they are called **rectangular components.** We usually resolve a vector into rectangular components.

One of the most common applications of vector addition is with force vectors. A **force vector** is a vector that represents the direction and magnitude of an applied force. If two forces act on an object, then if the resultant force replaces the acting forces on an object, the behavior of the object will not change. Consider the following example.

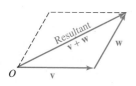

a. End-to-end method

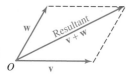

b. Parallelogram method

FIGURE 5.40
Resultant of vectors **v** and **w**

EXAMPLE 1

Finding the Resultant Force

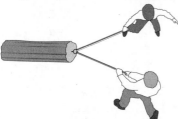

Al and his son are moving a heavy log by attaching two ropes to one end. Al is pulling with a force of 250 lb and his son is pulling with a force of 100 lb. What is the resultant force when the forces are applied so that the angle between the forces is 45°? Determine the direction the log will move by finding the angle of the log's path with the two tow ropes.

Solution   We draw a vector diagram, with vector **a** representing the force applied by Al. This means that the magnitude of this vector is 250 lb; we write this as $|\mathbf{a}| = 250$. The vector **b** represents the force applied by the son; $|\mathbf{b}| = 100$, as shown in Figure 5.41 (page 254).

Let **v** be the vector representing the resultant force. The magnitude of **v** is written $|\mathbf{v}|$ and is found using the law of cosines (remember that adjacent angles in a parallelogram are supplementary).

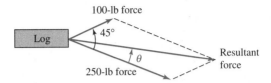

FIGURE 5.41 Force diagram for a resultant vector

$$|\mathbf{v}|^2 = 100^2 + 250^2 - 2(100)(250)\cos 135°$$
$$= 72{,}500 + 25{,}000\sqrt{2} \approx 107{,}855.3391$$

Thus, $|\mathbf{v}| \approx 328.4133661$, or to two significant figures, 330 lb.

The direction of the log can be found by finding θ using the law of sines:

$$\frac{\sin \theta}{100} = \frac{\sin 135°}{|\mathbf{v}|}$$

The sum of adjacent angles in a parallelogram is 180°, and 180° − 45° = 135°.

$$\theta = \sin^{-1}\left(\frac{100 \sin 135°}{|\mathbf{v}|}\right)$$

Remember, $|\mathbf{v}|$ is known from the first calculation.

$$\approx 12.43371428°$$

The log is moving so that it makes an angle of 12° with the 250-lb force, and $(45° - \theta) \approx 33°$ with the 100-lb force. ∎

The general form for finding a resultant vector using the parallelogram method is summarized in the following box.

FINDING A RESULTANT VECTOR

If **a** and **b** are two given vectors with an angle of ϕ between them, then the resultant vector **v** has magnitude $|\mathbf{v}|$, which is found by

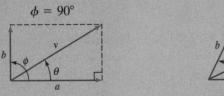

$\phi = 90°$ $0 < \phi < 180°$

$$|\mathbf{v}| = \sqrt{|\mathbf{a}|^2 + |\mathbf{b}|^2} \qquad |\mathbf{v}| = \sqrt{|\mathbf{a}|^2 + |\mathbf{b}|^2 - 2|\mathbf{a}||\mathbf{b}|\cos(180° - \phi)}$$

The angle θ that the resultant vector makes with the **a** vector is found by

$$\tan \theta = \frac{|\mathbf{b}|}{|\mathbf{a}|} \qquad \frac{\sin \theta}{|\mathbf{b}|} = \frac{\sin(180° - \phi)}{|\mathbf{v}|}$$

The angle the resultant vector makes with the **b** vector is found by $\phi - \theta$.

EXAMPLE 2

Finding a Resultant Vector

Consider two forces, one with magnitude 3.0 in a N20°W direction and the other with magnitude 7.0 in a S50°W direction. Find the resultant vector.

Solution   Sketch the given vectors and draw the parallelogram formed by these vectors. The diagonal is the resultant vector, as shown in Figure 5.42.

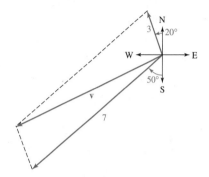

FIGURE 5.42   Find a resultant vector

We can easily find $\phi = 180° - 50° - 20° = 110°$. Then,

$$|\mathbf{v}| = \sqrt{3^2 + 7^2 - 2(3)(7)\cos(180° - 110°)}$$
$$= \sqrt{9 + 49 - 42\cos 70°}$$
$$\approx 6.60569103$$

The direction of $\mathbf{v}$ can be found using the law of sines to find θ, the angle $\mathbf{v}$ makes with the south–west vector:

$$\frac{\sin\theta}{3} = \frac{\sin 70°}{|\mathbf{v}|}$$

$$\sin\theta = \frac{3}{|\mathbf{v}|}\sin 70° \approx 0.4267650197$$

$$\theta = \sin^{-1}(0.4267650197) \approx 25.26243465°$$

Since $\theta + 50° \approx 75°$, we see that the direction of $\mathbf{v}$ should be measured from the south. Thus, the magnitude of $\mathbf{v}$ is 6.6 and the direction is S75°W.

∎

For some applications, it is more convenient to use the end-to-end method. As an example of such an application, we consider the effect the wind (or current) direction has on the direction of an airplane (or boat). The direction of the plane (or boat) is called the *heading*. The magnitude of the vector is the air speed (or boat speed). The effect of the wind or current changes both the direction and speed. The actual direction is the same as the direction of the diagonal of the parallelogram, and the magnitude of the diagonal is the **ground speed.** These relationships are shown in Figure 5.43 on page 256.

Let **p** be the vector representing the plane's speed and direction and **w** be the vector representing the wind's speed and direction. If ϕ is the angle between **p** and **w** when they are placed end-to-end, then

$$|\mathbf{v}| = \sqrt{|\mathbf{p}|^2 + |\mathbf{w}|^2 - 2|\mathbf{p}||\mathbf{w}|\cos\phi}$$

and the drift angle can be found by

$$\frac{\sin\theta}{|\mathbf{w}|} = \frac{\sin\phi}{|\mathbf{v}|}$$

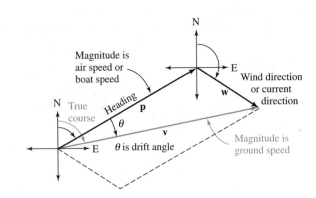

FIGURE 5.43   Navigation chart

EXAMPLE 3

Finding Direction and Ground Speed in a Navigational Example

An airplane flying at 260 mi/h on a heading of 55° is being blown by a headwind of 50 mi/h in a direction of 202°. Find the ground speed and actual direction of a flight.

Solution   We first draw a vector diagram, as shown in Figure 5.44. Remember that heading is measured clockwise from north.

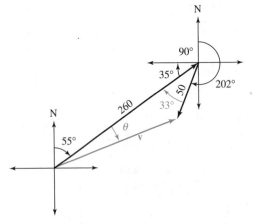

Then we find the angle ϕ:

$$\phi = 360° - 202° - 35° - 90°$$
$$= 33°$$

FIGURE 5.44   Vector navigational diagram

We use the end-to-end method to find the ground speed, which is the magnitude of vector **v**.

$$|\mathbf{v}| = \sqrt{50^2 + 260^2 - 2(50)(260)\cos 33°} \quad \text{\textit{Law of cosines}}$$
$$\approx 219.7602449 \quad \text{\textit{By calculator}}$$

Thus, the ground speed is approximately 220 mi/h.

Next, find the drift angle θ.

$$\frac{\sin \theta}{50} = \frac{\sin \phi}{|\mathbf{v}|} \qquad \textit{Law of sines}$$

$$\sin \theta = \frac{50 \sin 33°}{|\mathbf{v}|} \qquad \textit{Multiply both sides by 50.}$$

$$\theta = \sin^{-1}\left(\frac{50 \sin 33°}{|\mathbf{v}|}\right) \qquad \textit{Solve for } \theta.$$

$$\approx 7.118197578°$$

Since the direction is measured from the north, we see that the actual direction (to two significant digits) is $55° + 7° = 62°$.

Thus, the ground speed is 220 mi/h and the heading is 62°.   ■

Resolve a Vector

This next problem is one in which a vector is given yet needs to be written as the sum of two perpendicular vectors.

EXAMPLE 4

Resolving a Vector

Suppose a vector $\mathbf{v}$ has magnitude 5.00 and a direction given by $\theta = 30.0°$, where θ is the angle the vector makes with the positive x-axis. Resolve this vector into horizontal and vertical components.

Solution Let $\mathbf{v}_x$ be the horizontal component and $\mathbf{v}_y$ the vertical component, as shown in Figure 5.45. Then

$$\cos \theta = \frac{|\mathbf{v}_x|}{|\mathbf{v}|} \qquad\qquad \sin \theta = \frac{|\mathbf{v}_y|}{|\mathbf{v}|}$$

$$|\mathbf{v}_x| = |\mathbf{v}|\cos \theta \qquad\qquad |\mathbf{v}_y| = |\mathbf{v}|\sin \theta$$

$$= 5 \cos 30° \qquad\qquad\quad = 5 \sin 30°$$

$$= \tfrac{5}{2}\sqrt{3} \approx 4.33 \qquad\qquad = \tfrac{5}{2} = 2.50$$

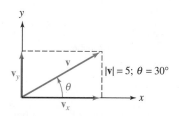

FIGURE 5.45
Resolving a vector

Thus, $\mathbf{v}_x$ is a horizontal vector with magnitude 4.33 and $\mathbf{v}_y$ is a vertical vector with magnitude 2.50.   ■

In the case of resolving a vector, we almost always resolve it into its horizontal and vertical components as shown in Example 4. We summarize this procedure in the following box.

RESOLVING A VECTOR

Given a vector **v** with magnitude $|\mathbf{v}|$, which makes an angle of θ with the x-axis: to **resolve** the vector **v** means to find a horizontal vector **a** and a vertical vector **b**, so that $\mathbf{v} = \mathbf{a} + \mathbf{b}$. It can be found as follows:

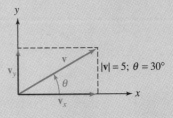

$|\mathbf{v}| = 5; \ \theta = 30°$

$$|\mathbf{a}| = |\mathbf{v}| \cos \theta \qquad |\mathbf{b}| = |\mathbf{v}| \sin \theta$$

Inclined Plane

One application involves forces acting on an object that is resting on an inclined plane. If the object is resting on a horizontal plane, the weight of the object has the same magnitude as the force pressing against the plane, as shown in Figure 5.46a.

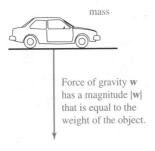

Force of gravity **w** has a magnitude $|\mathbf{w}|$ that is equal to the weight of the object.

a. Force of gravity

Force down the plane, **d**

Force against the plane, **p**

Weight of the object, $|\mathbf{w}|$

b. The weight of the vehicle is the resultant of the force down the plane and the one against the plane.

FIGURE 5.46 Inclined-plane application

However, if the object is on an inclined plane, the weight, $|\mathbf{w}|$, of the object is the resultant of two forces—the downward pull, **d**, along the inclined plane and the force pressing against the plane, **p**, as shown in Figure 5.46b. If the angle of inclination, ϕ, is the angle the roadway makes with the horizontal, and α is the angle the roadway makes with the vertical, then the vector **w**, which is vertical, forms an angle of α with the roadway because of alternate interior angles. Thus, since the vector **p** is perpendicular to the roadway, it follows that the angle between **p** and **w** is also ϕ, as shown in Figure 5.46. You can use this information as shown in Example 5.

EXAMPLE 5   **Inclined-Plane Application**

What is the smallest force necessary to keep a 3,420-lb car from sliding down a hill that makes an angle of 5.25° with the horizontal? Also, what is the force against the hill? Assume that friction is ignored.

Solution   The weight of the car, $|\mathbf{OC}|$, acts vertically downward. This vector can be resolved into two vectors: **OH**, the force against the hill, and **OD**, the force pushing down the hill, as shown in Figure 5.47.

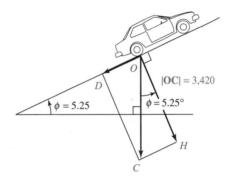

FIGURE 5.47   Inclined-plane solution

$$\sin 5.25° = \frac{|\mathbf{CH}|}{|\mathbf{OC}|} \qquad\qquad \cos 5.25° = \frac{|\mathbf{OH}|}{|\mathbf{OC}|}$$

$$
\begin{aligned}
|\mathbf{CH}| &= |\mathbf{OC}|\sin 5.25° & |\mathbf{OH}| &= |\mathbf{OC}|\cos 5.25° \\
&= 3{,}420 \sin 5.25° & &= 3{,}420 \cos 5.25° \\
&\approx 312.9 & &\approx 3{,}406
\end{aligned}
$$

Since $|\mathbf{OD}| = |\mathbf{CH}|$, we see that a force of 312.9 lb is necessary to keep the car from sliding down the hill, and the force against the hill is 3,406 lb.   ■

PROBLEM SET 5.4

Essential Concepts

1. **IN YOUR OWN WORDS** What is a vector?

2. **IN YOUR OWN WORDS** What is the resultant vector, and how do you find it?

3. **IN YOUR OWN WORDS** What does it mean to resolve a vector?

Problems 4–15: *Draw the listed vectors.*

4. 50 km/h due west

5. 65 mi/h due south

6. 35 mi/h due east

7. 40 km/h due north

8. 475 mi/h in a heading of 350°

9. 185 mi/h in a heading of 200°

10. 200 ft/s in a heading of 45°

11. 850 ft/s in a heading of 100°

12. 8.5 m/s in a heading of N55°E

13. 10.5 cm/s in a heading of N25°W

14. 2 in./min in a heading of S10°W

15. 5 in./h in a heading of S80°E

Level 1 Drill

Problems 16–21: *Find the resultant vector.*

16. **v** with direction west and magnitude 3.0 and **w** with direction north and magnitude 4.0

17. **v** with direction south and magnitude 4.0 and **w** with direction east and magnitude 3.0

18. **v** with direction east and magnitude 14 and **w** with direction north and magnitude 18

19. **v** with heading 45° and magnitude 100 and **w** with heading 135° with magnitude 200

20. **v** with heading 10° and magnitude 50 and **w** with heading 300° and magnitude 100

21. **v** with heading 110° and magnitude 75 and **w** with heading 240° and magnitude 200

Problems 22–27: *Resolve the vectors.*

22. $|\mathbf{v}| = 10$, $\theta = 30°$ 23. $|\mathbf{v}| = 20$, $\theta = 60°$

24. $|\mathbf{v}| = 23$, $\theta = 18°$ 25. $|\mathbf{v}| = 125$, $\theta = 27°$

26. $|\mathbf{v}| = 120$, $\theta = 250°$ 27. $|\mathbf{v}| = 525$, $\theta = 130.0°$

Level 1 Applications

28. A woman sets out in a rowboat on a due-west heading and rows at 4.8 mi/h. The current is carrying the boat due south at 12 mi/h. What is the true course of the rowboat, and how fast is the boat traveling relative to the ground?

29. An airplane is headed due west at 240 mi/h. The wind is blowing due south at 43 mi/h. What is the true course of the plane, and how fast is it traveling relative to the ground?

30. A cannon (see Figure 5.48) is fired at an angle of 10.0° above the horizontal, with an initial speed of

the cannonball at 2,120 ft/s. Find the magnitude of the vertical and horizontal components of the initial velocity.

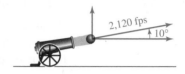

FIGURE 5.48   Projectile trajectory

31. Reconsider Problem 30 for an angle of 17.8°.

32. An airplane is heading 215° with a velocity of 723 mi/h. How far south has it traveled in one hour?

33. An airplane is heading 43.0° with a velocity of 248 mi/h. How far north has it traveled in two hours?

Level 2 Applications

34. Two forces of 220 lb and 180 lb are acting on the same point in directions that differ by 52°. What is the magnitude of the resultant, and what is the angle that it makes with each of the given forces?

35. If two sides of a parallelogram are 50 cm and 70 cm, and the diagonal connecting the ends of those sides is 40 cm, find the angles of the parallelogram.

36. If the resultant of two forces of 30 lb and 50 lb is a force of 60 lb, find the angle the resultant makes with each component force.

37. A boat leaves a dock at noon and travels N53°W at 10 km/h. When will the boat be 12 km from a lighthouse located 15 km due west of the dock?

38. At noon a plane leaves an airport and flies with a course of 61° at 200 mi/h. When will the plane be 35 miles from a point located 50 miles due east of the airport?

39. Two people are moving a tree by attaching cables to one end. (See Figure 5.49.) One person is pulling with a force of 200 lb and the other is pulling with a force of 100 lb. What is the resultant force when the forces are applied so that the angle between the forces is 35°? Determine the direction the tree will move by finding the angle of the path of the tree with the two tow cables.

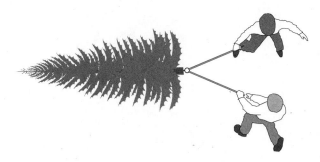

FIGURE 5.49 Two people dragging a tree

40. Two children are pulling a sled, with directions and forces as shown in Figure 5.50. What is the combined weight of the sled and the child being pulled?

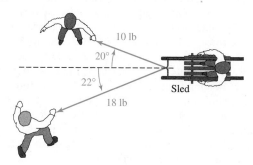

FIGURE 5.50   Children pulling a sled

41. A boat moving at 25.0 knots heads in the direction of N68.0°W across a river whose current is traveling at 12.0 knots in the direction of N16.0°W. Give the true course of the boat. How fast is it traveling as a result of these forces?

42. Two boats leave a dock at the same time; one travels S15°W at 8 mi/h, and the other travels at N28°W at 12 mi/h. How far apart are they in three hours?

43. Two airplanes leave an airport at the same time; one travels at 180 mi/h with a course of 280°, while the other travels at 260 mi/h with a course of 35°. How far apart are the planes in one hour?

44. A boat traveling at a constant rate in the direction N25°W passes a 3-mile marker from a lighthouse whose direction from the marker is due west. Twenty minutes after the boat passes the marker, it

is N58°E from the lighthouse. How fast is the boat traveling?

45. Two boats leave port at the same time. The first travels 24.1 mi/h in the direction of S58.5°W, while the other travels 9.80 mi/h in the direction of N42.1°E. How far apart are the boats after two hours?

46. A sailboat (see Figure 5.51) is in a 5.30 mi/h current in the direction of S43.2°W, and where the wind is blowing at 3.20 mi/h in a direction of S25.3°W. Find the course and speed of the sailboat.

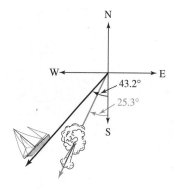

FIGURE 5.51   Course and speed of a sailboat

47. A tightrope walker who weighs 140 lb is walking across from point *A* to point *B* as shown in Figure 5.52. If the sag on either side of the tightrope walker is as shown, find the magnitudes of the tension in the rope toward each of the ends.

FIGURE 5.52   Tightrope walker

48. The world's steepest standard-gauge railroad grade measures 5.2° and is located between the Samala River Bridge and Zunil in Guatemala. What is the minimum force necessary to keep a boxcar weighing 52.0 tons from sliding down this incline? What is the force against the hill? Assume that friction is ignored.

49. What is the minimum force necessary to keep a 250-lb barrel from sliding down an inclined plane (see Figure 5.53) making an angle of 12° with the horizontal? What is the force against the plane? Assume that friction is ignored.

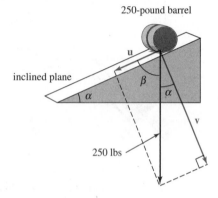

250-pound barrel

inclined plane

250 lbs

FIGURE 5.53   Inclined plane

50. A force of 253 lb is necessary to keep a weight of exactly 400 lb from sliding down an inclined plane. What is the angle of inclination of the plane? Assume that friction is ignored.

51. A force of 486 lb is necessary to keep a weight of exactly 800 lb from sliding down an inclined plane. What is the angle of inclination of the plane? Assume that friction is ignored.

52. A cable that can withstand 5,250 lb is used to pull cargo up an inclined ramp for storage. If the inclination of the ramp is 25.5°, find the heaviest piece of cargo that can be pulled up the ramp. Assume that friction is ignored.

53. Answer the question posed in Problem 52 for a ramp whose angle of inclination is 18.2°.

54. Some children are playing on a slide as shown in Figure 5.54. The slide is constructed at an angle of 45.0°, and the child on the slide weighs 53.5 lb. What is the force necessary to keep the child from sliding down the slide? Assume that friction is ignored.

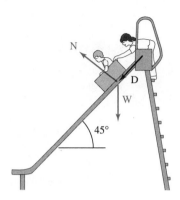

FIGURE 5.54   Children playing on a slide

55. Two children are playing on a swing, as shown in Figure 5.55. Suppose that the child in the swing weighs 53.5 lb and that a second older child is holding him back at an angle of 40.0°. There are three forces acting on the child in the swing: **W**, which is the force of gravity pulling straight down toward the center of the earth; **T**, the tension in the cable holding the swing's chair; and **H**, the force with which the second child is pulling back horizontally. Find the tension on the cable (the magnitude of **T**) and the force with which the second child is pulling back (the magnitude of **H**). Assume that friction is ignored.

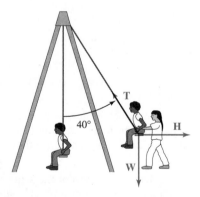

FIGURE 5.55   Children playing on a swing

56. The weight of astronauts on the moon is about 1/6 of their weight on earth. This has a marked effect on such simple acts as walking, running, and jumping. To study these effects and to train astronauts for working under reduced gravity conditions, scientists at NASA's Langley Research Center have designed an inclined-plane apparatus to simulate reduced gravity (see Figure 5.56). The apparatus consists of a sling that holds the astronaut in a position perpendicular to an inclined plane. The sling is attached to one end of a long cable that runs parallel to the inclined plane. The other end of the cable is attached to a trolley that runs along an overhead track. This device allows the astronaut to move freely in a plane perpendicular to the inclined plane. Let W be the weight of the astronaut and θ the angle between the inclined plane and the ground. Make a vector diagram showing the tension in the cable and the force exerted by the inclined plane against the feet of the astronaut.

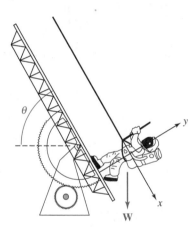

FIGURE 5.56   Apparatus to simulate reduced gravity

57. From the point of view of the astronaut in Problem 56, the inclined plane is the ground and the astronaut's simulated mass (that is, the downward force against the inclined plane) is $W \cos \theta$. What value of θ is required to simulate lunar gravity?

Level 3 Applications

58. A race car going around a racetrack is subject to a sideways force called the *centrifugal force*. If m is the mass of the race car traveling at a velocity v along a circular racetrack of radius r, then it can be shown that the centrifugal force has magnitude mv^2/r. Also, the force of the car pressing down (due to gravity) is mg (the weight of the car). A properly constructed track has a bank angle θ so that the vectors **V** and **W** shown in Figure 5.57 have equal magnitude. Find a formula for θ in terms of v, g, and r.

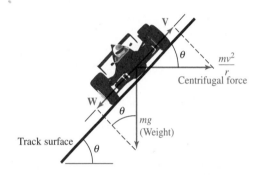

FIGURE 5.57   Bank of a racetrack

59. At what angle should a track be banked (see Problem 58) to eliminate sideways forces parallel to the track, if we assume that the average speed of the race cars is 200 mi/h ($\approx$ 293 ft/s) and that the racetrack is circular with a radius of 100 yd ($= 300$ ft). The force of gravity is 32 ft/s^2, and the weight of the race car is mg.

SECTION 5.5

VECTOR OPERATIONS

PROBLEM OF THE DAY

Is it possible to represent the magnitude and direction of a vector in an algebraic manner so that vector operations can be done algebraically as well as geometrically?

How would you answer this question?

In the last section, we first defined a vector geometrically and then defined addition of vectors geometrically. The Problem of the Day raises the possibility that we might be able to define a vector algebraically so that we can perform certain vector operations in a symbolic manner. The key to this representation is the definition of two very special vectors—namely, the **i** and **j** vectors:

i is the vector of unit length in the direction of the positive x-axis.

j is the vector of unit length in the direction of the positive y-axis.

To complete the necessary steps to answer the question raised in the Problem of the Day, we need an idea called *scalar multiplication*. When used in the context of vectors, the word **scalar** refers to a real number. If we write $c\mathbf{v}$, where c is a real number and **v** is a vector, we refer to c as a *scalar*; the expression $c\mathbf{v}$ is a vector in the same direction as **v** with magnitude c times the magnitude of **v**. We call the operation of multiplying a real number and a vector **scalar multiplication,** and it is defined geometrically, as shown in Figure 5.58. If c is a negative real number, then the scalar multiplication results in a vector in exactly the *opposite* direction as **v**—again with length c times the magnitude of **v**, as shown in Figure 5.58. If $c = 0$, then $c\mathbf{v}$ is the **0** vector.

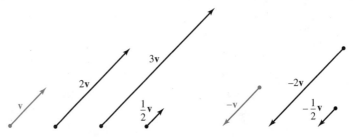

FIGURE 5.58 Examples of scalar multiplication

Scalar multiplication allows us to define **subtraction** for vectors:

$$\mathbf{v} - \mathbf{w} = \mathbf{v} + (-\mathbf{w})$$

Vector subtraction is shown geometrically in Figure 5.59.

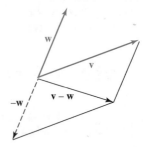

FIGURE 5.59 Vector subtraction

We now answer the question asked in the Problem of the Day with the following definition.

Algebraic Representation of a Vector

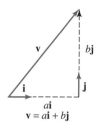

FIGURE 5.60 Algebraic representation of a vector

> **ALGEBRAIC REPRESENTATION OF A VECTOR**
>
> If **i** is the vector of unit length in the direction of the positive x-axis, and **j** is the vector of unit length in the direction of the positive y-axis, then any vector **v** can be written as
>
> $$\mathbf{v} = a\mathbf{i} + b\mathbf{j}$$
>
> where a and b are the magnitudes of the horizontal and vertical components, respectively. This is called the **algebraic representation of a vector.** (See Figure 5.60.)

If you know the magnitude and direction, then you can find the algebraic representation by using Figure 5.61.

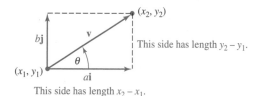

FIGURE 5.61 Writing a vector algebraically

So, we know that $a = |a\mathbf{i}|$ and $b = |b\mathbf{j}|$, and from the definition of the trigonometric functions, we know

$$\cos\theta = \frac{a}{|\mathbf{v}|} \qquad \sin\theta = \frac{b}{|\mathbf{v}|}$$

$$a = |\mathbf{v}|\cos\theta \qquad b = |\mathbf{v}|\sin\theta$$

Thus, since $\mathbf{v} = a\mathbf{i} + b\mathbf{j}$, we know

$$\mathbf{v} = (|\mathbf{v}|\cos\theta)\mathbf{i} + (|\mathbf{v}|\sin\theta)\mathbf{j}$$

If a vector has initial point (x_1, y_1) and terminal point (x_2, y_2) then

$$a = x_2 - x_1 \qquad \text{and} \qquad b = y_2 - y_1$$

so that

$$\mathbf{v} = (x_2 - x_1)\mathbf{i} + (y_2 - y_1)\mathbf{j}$$

EXAMPLE 1

Representing a Vector Algebraically Given Magnitude and Angle

Find the algebraic representation for a vector **v** with magnitude 10 making an angle of 60° with the positive x-axis.

Solution   From Figure 5.61,

$$a = |\mathbf{v}|\cos\theta \qquad b = |\mathbf{v}|\sin\theta$$
$$= 10\cos 60° \qquad = 10\sin 60°$$
$$= 5 \qquad\qquad = 5\sqrt{3}$$

Therefore, $\mathbf{v} = 5\mathbf{i} + 5\sqrt{3}\mathbf{j}$.   ■

EXAMPLE 2   **Representing a Vector Algebraically Given Two Points**

Find the algebraic representation for a vector $\mathbf{v}$ with initial point $(4, -3)$ and terminal point $(-2, 4)$.

Solution   From Figure 5.61,

$$a = x_2 - x_1 \qquad b = y_2 - y_1$$
$$= -2 - 4 \qquad = 4 - (-3)$$
$$= -6 \qquad\qquad = 7$$

Therefore, $\mathbf{v} = -6\mathbf{i} + 7\mathbf{j}$.   ■

Algebraic Characterizations of Vector Operations

The magnitude, or length, of a vector $\mathbf{v}$ is found by using the Pythagorean theorem.

> **MAGNITUDE OF A VECTOR**
>
> The **magnitude** of a vector $\mathbf{v} = a\mathbf{i} + b\mathbf{j}$ is given by
>
> $$|\mathbf{v}| = \sqrt{a^2 + b^2}$$

EXAMPLE 3   **Finding Magnitude**

Find the magnitude of the vectors $\mathbf{v}_1 = 5\mathbf{i} + 5\sqrt{3}\mathbf{j}$ and $\mathbf{v}_2 = -6\mathbf{i} + 7\mathbf{j}$.

Solution
$$|\mathbf{v}_1| = \sqrt{5^2 + (5\sqrt{3})^2} \qquad |\mathbf{v}_2| = \sqrt{(-6)^2 + 7^2}$$
$$= \sqrt{25 + 25(3)} \qquad\qquad = \sqrt{36 + 49}$$
$$= 10 \qquad\qquad\qquad\quad = \sqrt{85}$$

■

We now define the algebraic characteristics for vector operations.

> ### ALGEBRAIC DEFINITION OF VECTOR OPERATIONS
>
> Let $\mathbf{v} = a\mathbf{i} + b\mathbf{j}$ and $\mathbf{w} = c\mathbf{i} + d\mathbf{j}$; then
>
> | | |
> |---|---|
> | Addition | $\mathbf{v} + \mathbf{w} = (a + c)\mathbf{i} + (b + d)\mathbf{j}$ |
> | Subtraction | $\mathbf{v} - \mathbf{w} = (a - c)\mathbf{i} + (b - d)\mathbf{j}$ |
> | Scalar multiplication | $c\mathbf{v} = ca\mathbf{i} + cb\mathbf{j}$ |

EXAMPLE 4

Performing Vector Operations

Let $\mathbf{v} = 6\mathbf{i} + 4\mathbf{j}$ and $\mathbf{w} = -2\mathbf{i} + 3\mathbf{j}$.
Find **a.** $|\mathbf{v}|$ **b.** $|\mathbf{w}|$ **c.** $\mathbf{v} + \mathbf{w}$ **d.** $\mathbf{v} - \mathbf{w}$ **e.** $-\mathbf{v}$ **f.** $-2\mathbf{w}$

Solution

a.
$$|\mathbf{v}| = \sqrt{6^2 + 4^2}$$
$$= \sqrt{36 + 16}$$
$$= 2\sqrt{13}$$

b.
$$|\mathbf{w}| = \sqrt{(-2)^2 + 3^2}$$
$$= \sqrt{4 + 9}$$
$$= \sqrt{13}$$

c. $\mathbf{v} + \mathbf{w} = (6 - 2)\mathbf{i} + (4 + 3)\mathbf{j}$
$\qquad\quad = 4\mathbf{i} + 7\mathbf{j}$

d. $\mathbf{v} - \mathbf{w} = (6 + 2)\mathbf{i} + (4 - 3)\mathbf{j}$
$\qquad\quad = 8\mathbf{i} + \mathbf{j}$

e. $-\mathbf{v} = (-1)6\mathbf{i} + (-1)4\mathbf{j}$
$\qquad = -6\mathbf{i} - 4\mathbf{j}$

f. $-2\mathbf{w} = (-2)(-2)\mathbf{i} + (-2)(3)\mathbf{j}$
$\qquad\quad = 4\mathbf{i} - 6\mathbf{j}$ ■

As seen in Example 4, the algebraic representation of a vector makes it easy to handle vectors and their operations. Another advantage of the algebraic representation is that it specifies a direction and a magnitude, but not a particular location. When you draw a particular vector, begin by placing the initial point at any desired location; then from that initial point, draw a vector with a particular direction and magnitude.

Scalar Product

In arithmetic, the words *multiplication* and *product* mean the same thing. When working with vectors, these words are used to denote different ideas. Recall that scalar multiplication does not tell us how to multiply vectors; instead, it tells us how to multiply a scalar and a vector. Now, we define an operation called **scalar product** in order to multiply two vectors to obtain a number. It is called *scalar* product because we obtain a number, or scalar, as an answer. Sometimes scalar product is called *dot product* or *inner product*. In subsequent courses, you will define another vector multiplication, called *vector product,* in which you multiply vectors and obtain a vector as an answer.

> **SCALAR PRODUCT**
>
> Let $\mathbf{v} = a\mathbf{i} + b\mathbf{j}$ and $\mathbf{w} = c\mathbf{i} + d\mathbf{j}$. Then the **scalar product,** written $\mathbf{v} \cdot \mathbf{w}$, is defined by
>
> $$\mathbf{v} \cdot \mathbf{w} = ac + bd$$

EXAMPLE 5

Finding the Scalar Product

Find the scalar product of

a. $\mathbf{v} = 2\mathbf{i} + 5\mathbf{j}$ and $\mathbf{w} = 6\mathbf{i} - 3\mathbf{j}$

b. $\mathbf{v} = \cos 30°\mathbf{i} + \sin 30°\mathbf{j}$ and $\mathbf{w} = \cos 60°\mathbf{i} - \sin 60°\mathbf{j}$

c. $\mathbf{v} = -\sqrt{3}\mathbf{i} + \sqrt{2}\mathbf{j}$ and $\mathbf{w} = 3\sqrt{3}\mathbf{i} + 5\sqrt{2}\mathbf{j}$

d. $\mathbf{v} = 2\mathbf{i} - 3\mathbf{j}$ and $\mathbf{w} = 4\mathbf{i} + a\mathbf{j}$

Solution

a. $\mathbf{v} \cdot \mathbf{w} = (2\mathbf{i} + 5\mathbf{j}) \cdot (6\mathbf{i} - 3\mathbf{j}) = 2(6) + 5(-3) = -3$

b. $\mathbf{v} \cdot \mathbf{w} = (\cos 30°\mathbf{i} + \sin 30°\mathbf{j}) \cdot (\cos 60°\mathbf{i} - \sin 60°\mathbf{j})$
$$= \cos 30° \cos 60° - \sin 30° \sin 60°$$
$$= \cos(30° + 60°)$$
$$= 0$$

c. $\mathbf{v} \cdot \mathbf{w} = (-\sqrt{3}\mathbf{i} + \sqrt{2}\mathbf{j}) \cdot (3\sqrt{3}\mathbf{i} + 5\sqrt{2}\mathbf{j})$
$$= (-\sqrt{3})(3\sqrt{3}) + (\sqrt{2})(5\sqrt{2})$$
$$= -9 + 10$$
$$= 1$$

d. $\mathbf{v} \cdot \mathbf{w} = (2\mathbf{i} - 3\mathbf{j}) \cdot (4\mathbf{i} + a\mathbf{j})$
$$= 2(4) + (-3)a$$
$$= 8 - 3a$$

Angle Between Vectors

There is a very useful geometric property for scalar product that is apparent if we find an expression for the angle between two vectors.

> **ANGLE BETWEEN VECTORS**
>
> The angle θ between nonzero vectors $\mathbf{v}$ and $\mathbf{w}$ is found by
>
> $$\cos \theta = \frac{\mathbf{v} \cdot \mathbf{w}}{|\mathbf{v}||\mathbf{w}|}$$

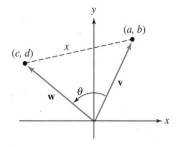

FIGURE 5.62 Finding the angle between two vectors

To derive this formula, let $\mathbf{v} = a\mathbf{i} + b\mathbf{j}$ and $\mathbf{w} = c\mathbf{i} + d\mathbf{j}$ be drawn with their bases at the origin, as shown in Figure 5.62. Let x be the distance between the endpoints of the vectors. Then, by the law of cosines,

$$
\begin{aligned}
\cos \theta &= \frac{|\mathbf{v}|^2 + |\mathbf{w}|^2 - x^2}{2|\mathbf{v}||\mathbf{w}|} \\
&= \frac{(\sqrt{a^2 + b^2})^2 + (\sqrt{c^2 + d^2})^2 - (\sqrt{(a-c)^2 + (b-d)^2})^2}{2|\mathbf{v}||\mathbf{w}|} \\
&= \frac{a^2 + b^2 + c^2 + d^2 - (a^2 - 2ac + c^2 + b^2 - 2bd + d^2)}{2|\mathbf{v}||\mathbf{w}|} \\
&= \frac{2ac + 2bd}{2|\mathbf{v}||\mathbf{w}|} \\
&= \frac{ac + bd}{|\mathbf{v}||\mathbf{w}|} \\
&= \frac{\mathbf{v} \cdot \mathbf{w}}{|\mathbf{v}||\mathbf{w}|}
\end{aligned}
$$

EXAMPLE 6

Finding the Angle Between Two Vectors

Find the angle (to the nearest degree) between the vectors $\mathbf{v} = 2\mathbf{i} + 3\mathbf{j}$ and $\mathbf{w} = 5\mathbf{i} - 4\mathbf{j}$.

Solution

If θ is the angle between $\mathbf{v}$ and $\mathbf{w}$, then

$$
\begin{aligned}
\cos \theta &= \frac{\mathbf{v} \cdot \mathbf{w}}{|\mathbf{v}||\mathbf{w}|} \\
&= \frac{2(5) + 3(-4)}{\sqrt{2^2 + 3^2}\,\sqrt{5^2 + (-4)^2}} \\
&= \frac{-2}{\sqrt{13}\sqrt{41}} \\
\theta &= \cos^{-1}\left(\frac{-2}{\sqrt{13}\sqrt{41}}\right) \approx 94.96974073°
\end{aligned}
$$

The angle between the vectors is about $95°$. ∎

Orthogonal Vectors

There is a useful geometric property of vectors whose directions differ by $90°$. If you are dealing with lines, they are called **perpendicular lines,** but if they are vectors forming a $90°$ angle, they are called **orthogonal.** Notice, if $\theta = 90°$, then $\cos 90° = 0$; therefore, from the angle between vectors formula,

$$
0 = \frac{\mathbf{v} \cdot \mathbf{w}}{|\mathbf{v}||\mathbf{w}|}
$$

If you multiply both sides by $|\mathbf{v}||\mathbf{w}|$, the result is the following important property.

> **PROPERTY OF ORTHOGONAL VECTORS**
>
> Nonzero vectors **v** and **w** are **orthogonal** if and only if $\mathbf{v} \cdot \mathbf{w} = 0$.

EXAMPLE 7

Showing That Given Vectors Are Orthogonal

Show that $\mathbf{v} = 3\mathbf{i} - 2\mathbf{j}$ and $\mathbf{w} = 6\mathbf{i} + 9\mathbf{j}$ are orthogonal.

Solution $\mathbf{v} \cdot \mathbf{w} = (3\mathbf{i} - 2\mathbf{j}) \cdot (6\mathbf{i} + 9\mathbf{j}) = 3(6) + (-2)(9) = 18 - 18 = 0$

Since the scalar product is zero, the vectors are orthogonal. ■

EXAMPLE 8

Finding a Constant so That Orthogonal Vectors Result

Find a so that $\mathbf{v} = 3\mathbf{i} + a\mathbf{j}$ and $\mathbf{w} = \mathbf{i} - 2\mathbf{j}$ are orthogonal.

Solution   We know $\mathbf{v} \cdot \mathbf{w} = 3 - 2a$, so if the vectors are orthogonal, $3 - 2a = 0$ or $a = 1.5$. ■

Work

We have been looking at various *forces* that push, pull, compress, distort, or in some other way change the state of rest or the state of motion of a body. When a constant force of magnitude F is applied to an object through a distance d, the *work* performed is defined to be the product of the magnitude of the force and the distance.

> **WORK DONE BY A CONSTANT FORCE**
>
> If a body moved a distance d in the direction of an applied constant force **F**, the **work** W done by the force is
>
> $$W = |\mathbf{F}|d$$
>
> In terms of vectors, the work done by a constant force **F** as its point of application moves along a vector **D**, where $|\mathbf{D}| = d$, is
>
> $$W = \mathbf{F} \cdot \mathbf{D}$$

Common Units of Work and Force

Mass	Dist.		WORK
kg	m	newton (N)	joule
g	cm	dyne (dyn)	erg
slug	ft	pound	ft-lb

In the U.S. system of measurements, work is typically expressed in *foot-pounds, inch-pounds,* or *foot-tons.* In the International System (SI), work is expressed in newton-meters (called *joules*), and in the centimeter-gram-second (CGS) system, the basic unit of work is the dyne-centimeter (called an *erg*).

For example, the work done to lift a 90-lb bag of concrete 3 ft is

$$W = |\mathbf{F}|d = (90 \text{ lb})(3 \text{ ft}) = 270 \text{ ft-lb}$$

Notice that this definition does not conform to everyday use of the word *work.* You may labor at lifting concrete all day, but if you are not able to move the sack of concrete, then no work has been done.

It is not necessary for **F** and *d* to be constants, but if they are not, then calculus is used to find the work done. In this book, we will consider examples where both **F** and *d* are constants.

EXAMPLE 9   **Finding the Work Done Mowing a Lawn**

Consider the vector diagram in Figure 5.63.

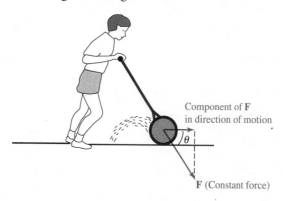

Component of **F** in direction of motion

θ

F (Constant force)

FIGURE 5.63   Vector diagram for a lawn mower

The *displacement d*, the distance the lawn mower is moved, is 2,450 ft and the angle in which the boy is pressing down on the handle is 40.0°. What is the amount of work done by the boy if the magnitude of the vector **F** is 480 lb?

Solution   The vector **F**, the constant force, has magnitude 48.0 lb. The work, however, is done in the direction of motion, so we can see from Figure 5.63 that **F** has magnitude 48 cos 40°. We can now calculate the amount of work:

$W = ($COMPONENT OF FORCE IN THE DIRECTION OF MOTION$)($DISPLACEMENT$)$

$= (48 \cos 40.0°)(2,450)$

$= 90,086.82651$

The amount of work done mowing the lawn is (to three significant digits) 90,100 ft-lb.   ■

EXAMPLE 10   **Finding Work Done in a Plane**

How much work is done by a force $\mathbf{F} = 6\mathbf{i} - 2\mathbf{j}$ that moves an object from the origin to point (4, 5)? (Force is in pounds and displacement is in feet.)

Solution $W = \mathbf{F} \cdot \mathbf{D}$

$= (6\mathbf{i} - 2\mathbf{j}) \cdot (4\mathbf{i} + 5\mathbf{j}) = 24 - 10 = 14$

The amount of work done is 14 ft-lb.   ■

PROBLEM SET 5.5

Essential Concepts

1. What is the algebraic representation of a vector?

2. What is the formula for the magnitude of a vector?

3. **IN YOUR OWN WORDS** How do you find the scalar product?

4. **IN YOUR OWN WORDS** What is the dot product?

5. What is the formula for the angle between vectors?

6. What does it mean for vectors to be orthogonal?

7. What is the property of orthogonal vectors?

Level 1 Drill

Problems 8–19: Find the exact algebraic representation for each vector. Use exact values when possible, and otherwise approximate answers to four decimal places. The number $|\mathbf{v}|$ is the magnitude of the vector $\mathbf{v}$, and θ is the angle the vector makes with the positive x-axis. The points A and B are the endpoints of the vector $\mathbf{v}$, and A is the base point. Draw each vector.

8. $|\mathbf{v}| = 12, \theta = 60°$ 9. $|\mathbf{v}| = 8, \theta = 30°$

10. $|\mathbf{v}| = 5\sqrt{2}, \theta = 45°$ 11. $|\mathbf{v}| = 5\sqrt{2}, \theta = 315°$

12. $|\mathbf{v}| = 4.2561, \theta = 112°$

13. $|\mathbf{v}| = 10.4813, \theta = 214.5°$

14. $A(4, 1), B(2, 3)$ 15. $A(-1, -2), B(4, 5)$

16. $A(-2, 4), B(0, 0)$ 17. $A(5, -3), B(-8, -3)$

18. $A(0, 0), B(750, 500)$ 19. $A(0, 0), B(4, 1250)$

Problems 20–27: Find the magnitude of each vector.

20. $\mathbf{v} = 3\mathbf{i} + 4\mathbf{j}$ 21. $\mathbf{v} = 5\mathbf{i} - 12\mathbf{j}$

22. $\mathbf{v} = 6\mathbf{i} - 7\mathbf{j}$ 23. $\mathbf{v} = -3\mathbf{i} + 5\mathbf{j}$

24. $\mathbf{v} = -2\mathbf{i} + 2\mathbf{j}$ 25. $\mathbf{v} = 5\mathbf{i} - 8\mathbf{j}$

26. $\mathbf{v} = 2\sqrt{3}\mathbf{i} + 4\mathbf{j}$ 27. $\mathbf{v} = 6\sqrt{2}\mathbf{i} - 5\sqrt{2}\mathbf{j}$

Problems 28–33: State whether the given pairs of vectors are orthogonal.

28. $\mathbf{v} = 3\mathbf{i} - 2\mathbf{j}; \mathbf{w} = 6\mathbf{i} + 9\mathbf{j}$

29. $\mathbf{v} = 2\mathbf{i} + 3\mathbf{j}; \mathbf{w} = 6\mathbf{i} - 9\mathbf{j}$

30. $\mathbf{v} = 4\mathbf{i} - 5\mathbf{j}; \mathbf{w} = 8\mathbf{i} + 10\mathbf{j}$

31. $\mathbf{v} = 5\mathbf{i} + 4\mathbf{j}; \mathbf{w} = 8\mathbf{i} - 10\mathbf{j}$

32. $\mathbf{v} = \cos 30°\mathbf{i} + \sin 30°\mathbf{j}; \mathbf{w} = \cos 60°\mathbf{i} + \sin 60°\mathbf{j}$

33. $\mathbf{v} = \cos 20°\mathbf{i} + \sin 20°\mathbf{j}; \mathbf{w} = \cos 70°\mathbf{i} + \sin 70°\mathbf{j}$

Level 2 Drill

Problems 34–43: Find $\mathbf{v} \cdot \mathbf{w}$, $|\mathbf{v}|$, $|\mathbf{w}|$, and $\cos \theta$, where θ is the angle between $\mathbf{v}$ and $\mathbf{w}$.

34. $\mathbf{v} = 3\mathbf{i} + 4\mathbf{j}; \mathbf{w} = 5\mathbf{i} + 12\mathbf{j}$

35. $\mathbf{v} = 8\mathbf{i} - 6\mathbf{j}; \mathbf{w} = -5\mathbf{i} + 12\mathbf{j}$

36. $\mathbf{v} = 2\mathbf{i} + \sqrt{5}\mathbf{j}; \mathbf{w} = 3\sqrt{5}\mathbf{i} - 3\mathbf{j}$

37. $\mathbf{v} = 7\mathbf{i} - \sqrt{15}\mathbf{j}; \mathbf{w} = 2\sqrt{15}\mathbf{i} + 14\mathbf{j}$

38. $\mathbf{v} = 2\mathbf{i} + 3\mathbf{j}; \mathbf{w} = 6\mathbf{i} + 5\mathbf{j}$

39. $\mathbf{v} = 3\mathbf{i} + 9\mathbf{j}; \mathbf{w} = 2\mathbf{i} - 5\mathbf{j}$

40. $\mathbf{v} = \mathbf{i}; \mathbf{w} = \mathbf{i}$ 41. $\mathbf{v} = \mathbf{j}; \mathbf{w} = \mathbf{j}$

42. $\mathbf{v} = \mathbf{i}; \mathbf{w} = -\mathbf{j}$ 43. $\mathbf{v} = -\mathbf{j}; \mathbf{w} = -\mathbf{i}$

Problems 44–49: Find the angle θ to the nearest degree, $0° \leq \theta \leq 180°$, between the vectors $\mathbf{v}$ and $\mathbf{w}$.

44. $\mathbf{v} = \frac{1}{2}\mathbf{i} + \frac{\sqrt{3}}{2}\mathbf{j}; \mathbf{w} = \frac{1}{2}\mathbf{i} + \frac{1}{2}\mathbf{j}$

45. $\mathbf{v} = \sqrt{2}\mathbf{i} - \sqrt{2}\mathbf{j}; \mathbf{w} = \frac{\sqrt{3}}{2}\mathbf{i} + \frac{1}{2}\mathbf{j}$

46. $\mathbf{v} = \mathbf{j}; \mathbf{w} = \frac{1}{2}\mathbf{i} - \frac{\sqrt{3}}{2}\mathbf{j}$

47. $\mathbf{v} = -\mathbf{i}; \mathbf{w} = -2\sqrt{2}\mathbf{i} + 2\sqrt{2}\mathbf{j}$

48. $\mathbf{v} = 2\mathbf{i} + 3\mathbf{j}; \mathbf{w} = -\mathbf{i} + 4\mathbf{j}$

49. $\mathbf{v} = -3\mathbf{i} + 2\mathbf{j}; \mathbf{w} = 6\mathbf{i} + 9\mathbf{j}$

Problems 50–53: Find a number a so that the given vectors are orthogonal.

50. $\mathbf{v} = 2\mathbf{i} + 3\mathbf{j}; \mathbf{w} = 5\mathbf{i} + a\mathbf{j}$

51. $\mathbf{v} = 4\mathbf{i} - a\mathbf{j}; \mathbf{w} = -2\mathbf{i} + 5\mathbf{j}$

52. $\mathbf{v} = a\mathbf{i} + 5\mathbf{j}; \mathbf{w} = a\mathbf{i} - 15\mathbf{j}$

53. $\mathbf{v} = 3\mathbf{i} + a\mathbf{j}; \mathbf{w} = -6\mathbf{i} + a\mathbf{j}$

Problems 54–57: What is the amount of work done by a force $\mathbf{F}$ in moving an object from the origin to the given point? (Force is in pounds and displacement is in feet.)

54. $\mathbf{F} = 2\mathbf{i} + 3\mathbf{j}; (5, 8)$ 55. $\mathbf{F} = 4\mathbf{i} + 5\mathbf{j}; (4, 3)$

56. $\mathbf{F} = 8\mathbf{i} - 3\mathbf{j}; (-3, 4)$ 57. $\mathbf{F} = \sqrt{2}\mathbf{i} + 4\mathbf{j}; (-6, 8)$

58. A 40-lb weight is hanging on two cables, as shown in Figure 5.64. Write the vectors $\mathbf{F}_1$ and $\mathbf{F}_2$ in algebraic form. Write a vector $\mathbf{W}$ representing the weight in algebraic form.

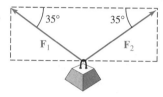

40-lb weight

FIGURE 5.64 Hanging weight problem

59. A 140-lb weight is hanging on two cables, as shown in Figure 5.65. Write the vectors $\mathbf{F}_1$ and $\mathbf{F}_2$ in algebraic form. Write a vector $\mathbf{W}$ representing the weight in algebraic form.

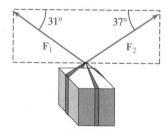

140-lb weight

FIGURE 5.65 Hanging weight problem

60. What is the work done by lifting a 90-lb bag of concrete 10 ft?

61. What is the work done by moving a 65-lb chest of drawers horizontally 5 ft?

62. What is the work done by pushing a 30-lb barrel 10 ft?

63. What is the work done by pushing a car 30 ft along a level road while exerting a constant forward force of 80 lb?

64. Suppose a person pushes a lawn mower a total of 950 ft back and forth across a level lawn with a constant force of 51.0 lb directed down the handle, which makes a 45.0° angle relative to the horizontal. Calculate the amount of work done in mowing this lawn.

65. A dad pulls his child in a wagon (see Figure 5.66) for 150 ft by exerting a constant force of 25 lb along the handle. How much work is done?

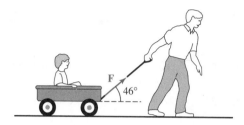

FIGURE 5.66 Work done by pulling a wagon

66. Repeat Problem 65, except assume that the dad is pulling the wagon weighing 52 lb (for both the wagon and child) up a hill that makes a constant angle of 12° with the horizontal.

67. Suppose the wind is blowing with a force $\mathbf{F}$ of magnitude 500 lb in the direction of N30°E over a boat's sail (see Figure 5.67). How much work does the wind perform in moving the boat in a northerly direction a distance of 100 ft? Give your answer in ft-lb.

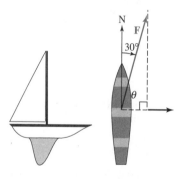

Side view   Top view

FIGURE 5.67 Work done by the wind

Level 3 Theory

68. Suppose the vectors $\mathbf{v}$ and $\mathbf{w}$ are sides of an equilateral triangle with area $25\sqrt{3}$. Find $\mathbf{v} \cdot \mathbf{w}$.

69. Show that
$$(\mathbf{v} + \mathbf{w}) \cdot (\mathbf{v} + \mathbf{w}) = |\mathbf{v}|^2 + |\mathbf{w}|^2 + 2(\mathbf{v} \cdot \mathbf{w})$$

70. Use Problem 69 to prove the *triangle inequality:*
$$|\mathbf{v} + \mathbf{w}| \leq |\mathbf{v}| + |\mathbf{w}|$$

5.6 CHAPTER FIVE SUMMARY

Objectives

Section 5.1   Law of Cosines

Objective 5.1   State the law of cosines and tell when it is used.

Objective 5.2   Solve SSS triangles using the law of cosines.

Objective 5.3   Solve SAS triangles using the law of cosines.

Objective 5.4   Solve SSA triangles using the law of cosines.

Objective 5.5   Solve applied problems using the law of cosines.

Section 5.2   Law of Sines

Objective 5.6   State the law of sines and tell when it is used.

Objective 5.7   Solve ASA triangles using the law of sines.

Objective 5.8   Solve AAS triangles using the law of sines.

Objective 5.9   Solve applied problems using the law of sines.

Objective 5.10   Solve triangles by using the law of sines.

Section 5.3   Areas and Volumes

Objective 5.11   Find the area of a triangle given SAS, ASA, and SSS.

Objective 5.12   Find the area of a sector of a circle.

Objective 5.13   Find the volume of a cone when the height is not known.

Section 5.4   Vector Triangles

Objective 5.14   Given two nonzero vectors, find the resultant vector.

Objective 5.15   Solve applied problems by finding the resultant vector.

Objective 5.16   Resolve a vector into horizontal and vertical components.

Objective 5.17   Solve applied problems involving inclined planes.

Section 5.5   Vector Operations

Objective 5.18   Write the algebraic representation of a vector.

Objective 5.19   Find the sum, difference, and magnitude of vectors.

Objective 5.20   Find the scalar product of two vectors.

Objective 5.21   Find the angle between vectors.

Objective 5.22   Determine whether two vectors are orthogonal.

Objective 5.23   Solve applied work problems when work is applied as a constant force.

Terms

Sample Test

All of the answers for this sample test are given in the back of the book.

1. Consider Figure 5.68.

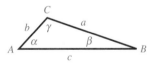

FIGURE 5.68 Standard oblique triangle

a. Using the law of cosines, find a^2.
b. Using the law of cosines, find $\cos \beta$.
c. State the law of sines.

Problems 2–4: Solve each triangle.

2. $a = 14$, $b = 27$, and $c = 19$

3. $b = 7.2$, $c = 15$, and $\alpha = 113°$

4. $a = 92.6$, $\alpha = 18.3°$, and $\beta = 112.4°$

Problems 5–7: Find the area of each triangle.

5. $b = 16$, $c = 43$, and $\alpha = 113°$

6. $\alpha = 40.0°$, $\beta = 51.8°$, and $c = 14.3$

7. $a = 121$, $b = 46$, and $c = 92$

8. a. Consider two forces, one with magnitude 5.0 in a S30°E direction and another with magnitude 12.0 in a N40°W direction. Find a resultant vector.
  b. Resolve the vector with magnitude 4.5 and $\theta = 51°$ into horizontal and vertical components.

9. Find the algebraic representation of each vector.
  a. $|\mathbf{v}| = \sqrt{2}$, $\theta = 45°$
  b. $|\mathbf{v}| = 9.3$, $\theta = 118°$
  c. $A(1, -2)$, $B(-5, -7)$; write $\overrightarrow{AB}$

10. Let $\mathbf{v} = 3\mathbf{i} - 2\mathbf{j}$ and $\mathbf{w} = \mathbf{i} - \mathbf{j}$. Find:
  a. $|\mathbf{v}|$ and $|\mathbf{w}|$ **b.** $\mathbf{v} \cdot \mathbf{w}$
  c. What is the angle between $\mathbf{v}$ and $\mathbf{w}$?
  d. If $\mathbf{t} = \mathbf{i} - a\mathbf{j}$, find a so that $\mathbf{v}$ and $\mathbf{t}$ are orthogonal.

11. A triangularly shaped garden has two angles measuring 46.5° and 105.8°, with the side opposite the 46.5° angle measuring 38.0 ft. How much fencing is needed to enclose the garden?

12. A farmer plows a circular field with radius 1,000 ft into separate equal sections. If one section has a central angle of 15°, what is the area of one section?

13. Ferndale is 7.0 miles N50°W of Fortuna. If I leave Fortuna at noon and travel due west at 2.0 mi/h, when will I be exactly 6.0 miles from Ferndale?

14. What is the smallest force necessary to keep a 3-ton boxcar from sliding down a hill that makes an angle of 4° with the horizontal? Assume that friction is ignored.

15. A mine shaft is dug into the side of a sloping hill. The shaft is dug horizontally for 485 ft. Next, a turn is made so that the angle of elevation of the second shaft is 58.0°, thus forming a 58° angle between the shafts. The shaft is then continued for 382 ft before exiting, as shown in Figure 5.69. How far is it along a straight line from the entrance to the exit, assuming that all tunnels are in a single plane?

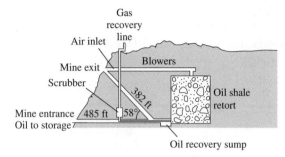

FIGURE 5.69   Mine shaft schematic

Discussion Problems

1. Describe what is meant by "the ambiguous case."

2. Outline your "method of attack" for solving triangles.

3. Describe a process for finding the area of a triangle.

4. What is meant by a "trigonometric substitution"?

5. Write a computer program for solving triangles.

6. Write a 500-word paper on the importance and use of vectors.

7. Discuss the mathematical concept of "work."

Miscellaneous Problems—Chapters 1–5

Problems 1–4: Draw a unit circle and label the points as shown in the figure. Find:

a. The coordinates of A and P
b. $|OA|$ **c.** $|PA|$
d. $|OB|$ **e.** Area of $\triangle AOB$

1.   **2.**

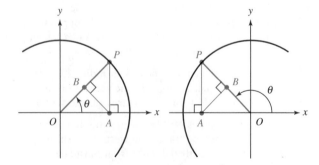

3.

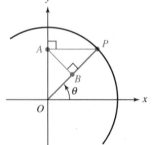

4.

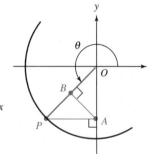

5. State the following information about the standard sine function.
   a. What is the domain?
   b. What is the range?
   c. What are the amplitude and period?

d. What is its maximum value?

e. Where does the graph cross the x-axis?

6. State the following information about the standard cosecant function.

a. What is the domain?

b. What is the range?

c. What is the period?

d. Where are its vertical asymptotes?

e. Where does the graph cross the x-axis?

Problems 7–15: *Decide whether the equations are identities for $0 \leq \theta < 2\pi$. If not, show a counterexample and then solve the equation, correct to two decimal places or by using exact values whenever possible.*

7. $\sin^2\theta - \sin\theta - 1 = 0$

8. $\cos^2\theta - \cos\theta - 1 = 0$

9. $\tan\theta + \cot\theta = \sec\theta\csc\theta$

10. $\dfrac{\sin\theta}{\tan\theta} + \dfrac{\cos\theta}{\cot\theta} = \cos\theta + \sin\theta$

11. $\sin\theta + 1 = \sqrt{3}$

12. $\cot\theta + \sqrt{3} = \csc\theta$

13. $\sin^2\theta + \cos\theta = 0$

14. $2\cos^2\theta + \sin^2\theta = 1$

15. $\dfrac{\tan(\alpha + \beta) - \tan\beta}{1 + \tan(\alpha + \beta)\tan\beta} = \tan\alpha$

16. Given: $a\sin\theta + b\cos\theta = \sqrt{a^2 + b^2}\,\sin(\theta + \alpha)$ where α is chosen so that

$$\cos\alpha = \frac{a}{\sqrt{a^2 + b^2}} \quad \text{and} \quad \sin\alpha = \frac{b}{\sqrt{a^2 + b^2}}$$

Use this identity to graph

$$y = \sqrt{3}\cos\theta - \sin\theta$$

Problems 17–22: *Graph each function.*

17. $y = \sin x$

18. $y = \cos x$

19. $y = \tan x$

20. $y = 3\sin\left(3x - \frac{\pi}{2}\right) + 1$

21. $y = -2\cos(2x + 1) - 1$

22. $y = \tan(1 - x) + 2$

Problems 23–52: *Solve each triangle.*

23. $a = 24$; $c = 61$; $\beta = 58°$

24. $b = 34$; $c = 21$; $\gamma = 16°$

25. $a = 30$; $\beta = 50°$; $\gamma = 100°$

26. $b = 23$; $\alpha = 25°$; $\beta = 110°$

27. $b = 40$; $\alpha = 50°$; $\gamma = 60°$

28. $b = 90$; $\beta = 85°$; $\gamma = 25°$

29. $a = 15$; $b = 8.0$; $\gamma = 38°$

30. $a = 18$; $c = 11$; $\beta = 63°$

31. $b = 14$; $c = 12$; $\alpha = 82°$

32. $b = 21$; $c = 35$; $\alpha = 125°$

33. $a = 7.0$; $b = 8.0$; $c = 2.0$

34. $a = 10$; $b = 4.0$; $c = 8.0$

35. $a = 6.80$; $b = 12.2$; $c = 21.5$

36. $b = 4.6$; $\alpha = 108°$; $\gamma = 38°$

37. $a = 14.5$; $b = 17.2$; $\alpha = 35.5°$

38. $\alpha = 48.0°$; $b = 25.5$; $c = 48.5$

39. $a = 121$; $b = 315$; $\gamma = 50.0°$

40. $a = 6.50$; $b = 8.30$; $c = 12.6$

41. $a = 11$; $b = 9.0$; $c = 8.0$

42. $a = 12$; $b = 6.0$; $c = 15$

43. $a = 12.4$; $b = 32.5$; $c = 35.1$

44. $a = 2.72$; $c = 3.57$; $\alpha = 43.7°$

45. $b = 68.2$; $\alpha = 145°$; $\beta = 52.4°$

46. $a = 151$; $b = 235$; $c = 412$

47. $\beta = 15.0°$; $\gamma = 18.0°$; $b = 23.5$

48. $b = 45.7$; $\alpha = 82.3°$; $\beta = 61.5°$

49. $b = 82.5$; $c = 52.2$; $\gamma = 23.1°$

50. $a = 352$; $b = 230$; $c = 418$

51. $a = 10.2$; $b = 11.8$; $\alpha = 47.0°$

52. $a = 8.62$; $b = 11.8$; $\alpha = 37.0°$

Applications

53. A surveyor needs to estimate the distance of a tunnel through a hill. The distances to the ends are 1,000 ft and 1,236 ft with an angle of 50.6° between these sides, as shown in Figure 5.70 (page 278). What is the length of the tunnel?

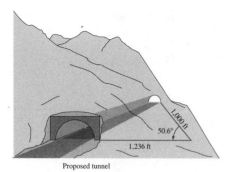

Proposed tunnel

FIGURE 5.70   Determining tunnel length

54. A conical pile of sawdust has a radius of 135 ft and an angle of elevation of 35.0°. What is the volume of the pile? How many cubic yards?

55. If a light ray passes from one medium to another of greater or lesser density, the ray will bend. A measure of this tendency is called the *index of refraction, R,* and is defined by

$$R = \frac{\sin \alpha}{\sin \beta}$$

For example, if the angle of incidence, α, of a light ray entering a diamond is 30.0° and the angle of refraction, β, is 11.9°, then the diamond's index of refraction is

$$R = \frac{\sin 30.0°}{\sin 11.9°} \approx 2.424781044$$

Since the index of refraction for any particular substance is a constant, we can compare our calculated number to the known index number to determine whether a particular stone is an imitation. The index of refraction of an imitation diamond is rarely greater than 2. If a stone is a diamond and the angle of incidence is 45.0°, what should be its refracted angle?

$$R = \frac{\sin \alpha}{\sin \beta} \quad \text{where } R \approx 2.42, \alpha = 45°$$

Solve for β and substitute known values to find the refracted angle β:

$$\sin \beta = \frac{\sin \alpha}{R}$$

$$\beta = \sin^{-1}\left(\frac{\sin \alpha}{R}\right) \approx 16.95°$$

To the nearest tenth, the refracted angle is 17.0°.
a. If a stone is a topaz, with an index of refraction of 1.63 and an angle of incidence of 20.0°, what should be its refracted angle (correct to the nearest tenth of a degree)?
b. If a stone is a garnet with an index of refraction of 1.63 and an angle of incidence of 35.0°, what should be its refracted angle (correct to the nearest tenth of a degree)?

56. The world's largest pyramid is located 63 mi S45°E of Mexico City. If I leave Mexico City in a jeep and travel due east, how far from Mexico City will I be when I am 50 mi from the pyramid?

57. A pilot leaves Santa Rosa and flies 250 miles with a heading of 128° to the Nut Tree. The pilot then leaves the Nut Tree at noon and flies north at 160 mi/h. When the pilot is exactly 200 mi from Santa Rosa after leaving the Nut Tree, a log entry showing the time must be made. What time (to the nearest minute) should the log show?

58. San Francisco and Los Angeles are approximately 450 miles apart. A pilot flying from San Francisco finds that she is 15.0° off course after 150 miles. How far is she from Los Angeles at this time?

59. A fire is simultaneously spotted from two observation towers that are located 10,520 ft apart, as shown in Figure 5.71. The first observation post reports the fire at N54.8°E, and the second post reports the fire at N48.2°W. How far is the fire from each observation location? Assume that all measurements are in the same plane, and that the towers are on an east–west line.

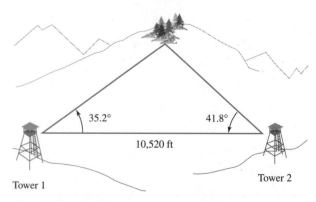

FIGURE 5.71   Triangulation for the location of a fire

Group Research Projects

Working in small groups is typical of most work environments, and this book seeks to develop skills with group activities. At the end of each chapter, we present a list of suggested projects, and even though they could be done as individual projects, we suggest that these projects be done in groups of three or four students.

THIRTY-SEVEN-YEAR-OLD PUZZLE

For more than 37 years I have tried, on and off, to solve a problem that appeared in *Popular Science*.

All right, I give up. What's the answer? In the September 1939 issue (page 14), the following letter appeared:

"Have enjoyed monkeying with the problems you print, for the last couple of years.

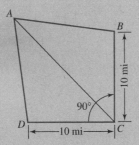

A man always drives at the same speed (his car probably has a governor on it). He makes it from *A* direct to *C* in 30 minutes; from *A* through *B* to *C* in 35 minutes; and from *A* through *D* to *C* in 40 minutes. How fast does he drive?

D.R.C., Sacramento, Calif."

It certainly looked easy, and I started to work on it. By 3 A.M., I had filled a lot of sheets of paper, both sides, with notations. In subsequent issues of *Popular Science*, all I could find on the subject was the following, in the November 1939 issue:

"In regard to the problem submitted by D.R.C. of Sacramento, Calif., in the September issue. According to my calculations, the speed of the car must be 38.843 miles an hour.

W.L.B., Chicago, Ill."

But the reader gave no hint of how he had arrived at that figure. It may or may not be correct. Over the years, I probably have shown the problem to more than 500 people. The usual reaction was to nod the head knowingly and start trying to find out what the hypotenuse of the right triangle is. But those few who remembered the formula then found there is little you can do with it when you know the hypotenuse distance.

Now I'd like to challenge *Popular Science*. If your readers cannot solve this little problem (and prove the answer), how about your finding the answer, if there is one, editors?

R.F. Davis, Sun City, Ariz.

Darrell Huff, master of the pocket calculator and frequent PS contributor [Feb. '76, June '76, Dec. '74] has come up with a couple of solutions. We'd like to see what you readers can do before we publish his answers.

1. **JOURNAL PROBLEM** The preceding letter to the editor appeared in the February 1977 issue of *Popular Science*. (Reprinted with permission from *Popular Science*. © 1977.) See whether you can solve the puzzle.

2. Table 5.1 outlines a procedure for solving triangles. To help you understand this procedure, write a flowchart or use pseudocode to write a computer program that will solve any triangle. Even if you do not know how to program computers, you can write a

set of directions stating exactly how to proceed with solving triangles.

3. Consider a unit circle with inscribed $\triangle ABC$, as shown in Figure 5.72. Express the area of $\triangle ABC$ as a function of x.

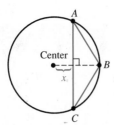

FIGURE 5.72   Unit circle

4. Pythagorean triples are well known and are defined to be integers that satisfy the Pythagorean theorem—namely, integers a, b, and c that satisfy

$$c^2 = a^2 + b^2$$

Similarly, a *primitive quadruple* is defined to be integers a, b, and c, along with an angle θ (measured in degrees that is also an integer) that satisfy the law of cosines. For example, if $a = 57$, $b = 55$, $c = 97$, and $\theta = 120°$, then

$$c^2 = a^2 + b^2 - 2ab \cos \theta$$

is satisfied because

$$97^2 = 57^2 + 55^2 - 2(57)(55)\cos 120°$$

Find another primitive quadruple.

5. **COMPUTER PROBLEM** Use a computer to print out the primitive quadruples for positive integers a, b, and c less than 100. (See Problem 4.)

6. Write a paper on the relationship between music and sine curves.

7. **JOURNAL PROBLEM** In *The Mathematics Teacher* (November 1958, pp. 545–546), Howard Eves published an article, "Pappus's Extension of the Pythagorean Theorem." In this article, he points out that if you combine Propositions 12 and 13 of Book II and Proposition 47 of Book I of Euclid's *Elements,* you can derive the law of cosines. Do some research and write a paper showing how this can be done.

CHAPTER SIX

COMPLEX NUMBERS AND POLAR-FORM GRAPHING

PREVIEW

*In mathematics, several applications give rise to equations such as $x^2 + 1 = 0$, which cannot be solved in the set of real numbers. A set of **complex numbers** has been defined that includes not only the real numbers, but also square roots of negative numbers. The number i, defined in Section 6.1, is called the **imaginary unit,** although it is not more "imaginary" than the number 5. When working with complex and imaginary numbers, you must keep in mind that all numbers are concepts, and that the number i is just as much a number as any of the "real" numbers. The trigonometric form of a complex number, introduced in Section 6.2, is a very useful form you will use not only in this course, but also in your future mathematical work. We conclude our work with complex numbers in Section 6.3 by looking at roots and powers of numbers.*

In the second part of this chapter, we introduce another method of representing a point in a plane: a polar coordinate system. *Polar coordinates open up avenues of investigation that would be very difficult if we were limited to rectangular coordinates. Polar coordinates were used to plot global locations on the floor of the Paris Observatory (see Problems 61 and 62 of Problem Set 6.3). We will be discussing curves called* cardioids *(shaped like hearts),* rose curves *(we count the number of petals), and* lemniscates *(figure-8 curves).*

If you tell students that the marks which they make for numbers are numerals and not numbers, a natural reply is, "What are numbers then?" The student writes "Empire State Building" and can point to the Empire State Building, whereas the student who writes "5" cannot point to 5. One of the immediate advantages of pointing out the distinction between numbers and numerals is that it raises the question: What are numbers?

UICSM Project Staff

("Words, 'words', "words"," *The Mathematics Teacher,* March 1957, p. 195.)

THE IMAGINARY UNIT AND COMPLEX NUMBERS

PROBLEM OF THE DAY

What is the solution to the equation $x^2 + 1 = 0$?

How would you answer this question?

To find the roots of certain equations, such as the equation in the Problem of the Day, we must sometimes consider the square root of a negative number. Since the set of real numbers does not allow for such a possibility, we need to define a number that is *not a real number*. This number is denoted by the symbol i.

IMAGINARY UNIT

The number i, called the **imaginary unit,** is defined as a number with the following properties:
$$i^2 = -1 \quad \text{and} \quad \sqrt{-a} = i\sqrt{a} \quad \text{if } a > 0$$

With this number available, the square root of any negative number can be written as the product of a real number and the number i:

$$\sqrt{-1} = i\sqrt{1} = i$$
$$\sqrt{-4} = i\sqrt{4} = 2i$$

 NOTE

You should remember this.

Check: $(2i)^2 = 4i^2 = 4(-1) = -4$

So, by definition $\sqrt{-4} = 2i$.
Also, note that

$$\sqrt{-13} = i\sqrt{13} \quad \text{\textit{If you write this as }} \sqrt{13}\,i, \text{\textit{ make sure}}$$
*that the **i** is not included under the radical.*

$$\sqrt{-b} \quad \text{is simplified if } b < 0$$
$$\sqrt{-b} = i\sqrt{b} \quad \text{if } b > 0$$

Now we consider numbers of the form $a + bi$.

If $b = 0$, then $a + bi = a + 0i = a$ is a *real* number.
If $a = 0$ and $b = 1$, then $a + bi = 0 + 1i = i$ is the *imaginary unit*.
If $a = 0$ and $b \neq 0$, then $a + bi = 0 + bi$ is called a *pure imaginary number*.

Numbers of the form $a + bi$ are called *complex numbers*.

> **COMPLEX NUMBERS**
>
> The set of numbers of the form
>
> $\quad a + bi$
>
> where a and b are real numbers and i is the imaginary unit, is called the set of **complex numbers.**

A complex number is **simplified** when it is written in the form $a + bi$, where the real numbers a and b are separately written in simplified form:

$5 + 3i$, $3 - i$, and $\sqrt{2} - \sqrt{3}i$ are all in simplified form.
$\sqrt{-9}$ is not in simplified form; write $\sqrt{-9} = 3i$.
7 is a complex number in simplified form, since it is not necessary to write this as $7 + 0i$.

EXAMPLE 1   **Problem of the Day: Solve an Equation with Nonreal Solutions**

Solve $x^2 + 1 = 0$.

Solution
$$x^2 + 1 = 0 \qquad \textit{Given}$$
$$x^2 = -1 \qquad \textit{Subtract 1 from both sides.}$$
$$x = \pm\sqrt{-1} \qquad \textit{Square root property}$$
$$x = \pm i \qquad \textit{Definition of i}$$

Check: If $x = i$, then

$$x^2 + 1 = 0 \qquad \textit{Given}$$
$$i^2 + 1 = 0 \qquad \textit{Check value x = i.}$$
$$-1 + 1 = 0 \qquad \textit{i}^2 = -1$$
$$0 = 0 \qquad \textit{True equation, so x = i is a solution.}$$

If $x = -i$, $(-i)^2 + 1 = i^2 + 1 = -1 + 1 = 0$, so $x = -i$ also checks.   ■

Our next consideration is to carry out operations with complex numbers.

> **OPERATIONS WITH COMPLEX NUMBERS**
>
> Let $a + bi$ and $c + di$ be any complex numbers.
>
> *Equality* $\quad a + bi = c + di \quad$ if and only if $\quad a = c$ and $b = d$
> *Addition* $\quad (a + bi) + (c + di) = (a + c) + (b + d)i$
> *Subtraction* $\quad (a + bi) - (c + di) = (a - c) + (b - d)i$
> *Distributivity* $\quad c(a + bi) = ca + cbi$

EXAMPLE 2   **Adding and Subtracting Complex Numbers**

Simplify: **a.** $(3 + 4i) + (2 + 3i)$ **b.** $(2 - i) - (3 - 2i)$

c. $(6 + 2i) + (2 - 2i)$ **d.** $(4 + 3i) - (4 - 2i)$ **e.** $\dfrac{2 - \sqrt{-8}}{2}$

Solution   **a.** $(3 + 4i) + (2 + 3i) = 5 + 7i$

b. $(2 - i) - (3 - 2i) = -1 + i$

c. $(6 + 2i) + (2 - 2i) = 8$

d. $(4 + 3i) - (4 - 2i) = 5i$

e. $\dfrac{2 - \sqrt{-8}}{2} = \dfrac{2 - \sqrt{8}\,i}{2}$

$$= \dfrac{2 - 2\sqrt{2}\,i}{2}$$

$$= \dfrac{2(1 - \sqrt{2}\,i)}{2}$$

$$= 1 - \sqrt{2}\,i$$ ∎

Example 2e illustrates the calculations that are necessary to solve quadratic equations over the set of complex numbers.

EXAMPLE 3   **Solving a Quadratic Equation over the Set of Complex Numbers**

Solve $5x^2 - 2x + 3 = 0$.

Solution   We use the quadratic formula:

$$x = \dfrac{-(-2) \pm \sqrt{(-2)^2 - 4(5)(3)}}{2(5)}$$ *If $ax^2 + bx + c = 0$, $a \neq 0$,*
                                                          then $x = \dfrac{-b \pm \sqrt{b^2 - 4ac}}{2a}$.

$$= \dfrac{2 \pm \sqrt{-56}}{10}$$ *Simplify.*

$$= \dfrac{2 \pm \sqrt{4(-14)}}{10}$$ *Factor -56.*

$$= \dfrac{2 \pm 2i\sqrt{14}}{10}$$ *$\sqrt{4} = 2$ and $\sqrt{-1} = i$*

$$= \dfrac{2(1 \pm i\sqrt{14})}{2(5)}$$ *Factor.*

$$= \dfrac{1 \pm i\sqrt{14}}{5}$$ *Simplify.* ∎

Notice that the addition and subtraction of complex numbers conform to the usual way you handle these operations for binomials. Multiplication is defined similarly. Since $i^2 = -1$, multiply in the usual way and replace i^2 by -1 wherever it occurs.

Now consider the complex numbers $a + bi$ and $c + di$. If multiplication of complex numbers is handled in the same manner as binomials, we have

$$\begin{aligned}(a + bi)(c + di) &= ac + adi + bci + bdi^2 \\ &= ac + adi + bci - bd \\ &= (ac - bd) + (ad + bc)i\end{aligned}$$

We use this as the definition of multiplication of complex numbers.

MULTIPLICATION OF COMPLEX NUMBERS

If $a + bi$ and $c + di$ are any two complex numbers, then

$$(a + bi)(c + di) = (ac - bd) + (ad + bc)i$$

It is not necessary (or even desirable) to memorize this definition, since two complex numbers are handled as you would any binomials.

EXAMPLE 4   **Multiplying Complex Numbers**

Simplify:

a. $(2 + 3i)(4 + 2i)$ **b.** $(4 - 3i)(2 - i)$ **c.** $(3 - 2i)(3 + 2i)$

Solution   **a.** $\begin{aligned}(2 + 3i)(4 + 2i) &= 8 + 16i + 6i^2 &&\text{\small \textit{Usual binomial multiplication}}\\ &= 8 + 16i - 6 &&\text{\small Write }i^2 = (-1).\\ &= 2 + 16i\end{aligned}$

b. $\begin{aligned}(4 - 3i)(2 - i) &= 8 - 10i + 3i^2 \\ &= 5 - 10i\end{aligned}$

c. $\begin{aligned}(3 - 2i)(3 + 2i) &= 9 + 6i - 6i - 4i^2 \\ &= 9 - 4i^2 \\ &= 13\end{aligned}$ ■

NOTE

You must be very careful that you use the laws of radicals correctly when you multiply complex numbers in radical form. That is,

if a and b are both positive: $\sqrt{a}\sqrt{b} = \sqrt{ab}$
if a and b are both negative: $\sqrt{a}\sqrt{b} = -\sqrt{ab}$

This means that when working with complex numbers, you should *first* write them in the form $a + bi$ and *then* perform the arithmetic.

EXAMPLE 5

Multiplying Complex Numbers with Radicals

Simplify: **a.** $\sqrt{-4}\sqrt{-4}$ **b.** $(2 - \sqrt{-4})(4 + \sqrt{-9})$ **c.** $(\sqrt{7} - \sqrt{-16})^2$

Solution **a.** $\sqrt{-4}\sqrt{-4} = (\sqrt{4}\,i)(\sqrt{4}\,i)$

$\qquad\qquad\quad = 4i^2$

$\qquad\qquad\quad = -4$

It is WRONG to write $\sqrt{-4}\sqrt{-4} = \sqrt{16} = 4$
because $\sqrt{a}\sqrt{b} = -\sqrt{ab}$ if a and b are both negative.

b. $(2 - \sqrt{-4})(4 + \sqrt{-9}) = (2 - 2i)(4 + 3i)$ *Write in terms of i.*

$\qquad\qquad\qquad\qquad\qquad = 8 - 2i - 6i^2$ *Multiply.*

$\qquad\qquad\qquad\qquad\qquad = 14 - 2i$ *Simplify.*

c. $(\sqrt{7} - \sqrt{-16})^2 = (\sqrt{7} - 4i)^2$

$\qquad\qquad\qquad\quad = 7 - 8\sqrt{7}\,i + 16i^2$

$\qquad\qquad\qquad\quad = 7 - 8\sqrt{7}\,i - 16$

$\qquad\qquad\qquad\quad = -9 - 8\sqrt{7}\,i$ ∎

Example 4c gives a clue for dividing complex numbers. When you multiply a complex number by its conjugate, the result is a real number. The numbers $a + bi$ and $a - bi$ are called **complex conjugates.** In general,

$$(a + bi)(a - bi) = a^2 - b^2i^2 = a^2 + b^2$$

which is a real number since a and b are real numbers. Thus, for division

$$\frac{a + bi}{c + di} = \frac{a + bi}{c + di} \cdot \frac{c - di}{c - di}$$ *Multiply by 1.*

$$= \frac{(ac + bd) + (bc - ad)i}{c^2 - d^2i^2}$$ *Expand.*

$$= \frac{ac + bd}{c^2 + d^2} + \frac{bc - ad}{c^2 + d^2}\,i$$ *Simplify.*

DIVISION OF COMPLEX NUMBERS

If $a + bi$ and $c + di$ are any two complex numbers (c and d are not both zero), then

$$\frac{a + bi}{c + di} = \frac{ac + bd}{c^2 + d^2} + \frac{bc - ad}{c^2 + d^2} i$$

Help Window   ? X

Notice that the result is in the form of a complex number, but instead of memorizing this definition, simply remember: *To divide complex numbers, you multiply both the numerator and the denominator by the conjugate of the denominator.*

EXAMPLE 6   **Division of Complex Numbers**

Simplify: **a.** $\dfrac{7 + i}{2 + i}$ **b.** $\dfrac{15 - \sqrt{-25}}{2 - \sqrt{-1}}$

Solution   **a.** $\dfrac{7 + i}{2 + i} = \dfrac{7 + i}{2 + i} \cdot \dfrac{2 - i}{2 - i}$ *Multiply by 1.*

$\qquad = \dfrac{14 - 5i - i^2}{4 - i^2}$ *Simplify.*

$\qquad = \dfrac{15 - 5i}{5}$ $i^2 = -1$ *and simplify.*

$\qquad = \dfrac{5(3 - i)}{5}$ *Common factor*

$\qquad = 3 - i$ *Simplify.*

b. $\dfrac{15 - \sqrt{-25}}{2 - \sqrt{-1}} = \dfrac{15 - 5i}{2 - i}$

NOTE   *Do not forget to write these in the form $a + bi$ first.*

$\qquad = \dfrac{15 - 5i}{2 - i} \cdot \dfrac{2 + i}{2 + i}$ *Multiply by 1.*

$\qquad = \dfrac{30 + 5i - 5i^2}{4 - i^2}$ *Simplify.*

$\qquad = \dfrac{35 + 5i}{5}$ $i^2 = -1$ *and simplify.*

$\qquad = \dfrac{5(7 + i)}{5}$ *Common factor*

$\qquad = 7 + i$ *Simplify.*

PROBLEM SET 6.1

Essential Concepts

1. What is the number i?

2. What is a complex number?

3. What does it mean to *simplify* a complex number?

4. What are complex conjugates?

5. Is every real number a complex number? Is every complex number a real number?

Problems 6–9: Classify as true or false.

6. $-i^2 = i^2 = -1$

7. $\sqrt{i} = -1$

8. $\sqrt{-2}\sqrt{-2} = \sqrt{(-2)(-2)} = \sqrt{4} = 2$

9. $(2 + \sqrt{-4})(2 - \sqrt{-4}) = 4 - 4 = 0$

Level 1 Drill

Problems 10–26: Simplify each expression.

10. **a.** $\sqrt{-36}$ **b.** $\sqrt{-100}$

11. **a.** $\sqrt{-49}$ **b.** $\sqrt{-8}$

12. **a.** $\sqrt{-20}$ **b.** $\sqrt{-24}$

13. **a.** $\dfrac{-3\sqrt{-144}}{5}$ **b.** $\dfrac{-6\sqrt{-4}}{8}$

14. **a.** $\dfrac{3\sqrt{-49}}{7}$ **b.** $\dfrac{-2\sqrt{-24}}{8}$

15. **a.** $2 + \sqrt{2} - 4 + \sqrt{-2}$
    b. $6 - \sqrt{3} - 8 + \sqrt{-3}$

16. **a.** $(3 + \sqrt{3}) + (5 - \sqrt{-9})$
    b. $(5 - \sqrt{2}) + (2 - \sqrt{-4})$

17. **a.** $i(5 - 2i)$ **b.** $i(2 + 3i)$

18. **a.** $(3 + 3i) + (5 + 4i)$
    b. $(4 - i)(2 + i)$

19. **a.** $(8 - 2i)(8 + 2i)$
    b. $(3 - 4i)(3 + 4i)$

20. **a.** $(5 + 3i)(2 + 6i)$
    b. $(2 - 7i) - (3 - 2i)$

21. **a.** $(5 - 4i) - (5 - 9i)$
    b. $(2 - 3i) - (4 + 5i)$

22. **a.** $-i^2$ **b.** $-i^3$

23. **a.** $-i^5$ **b.** i^7

24. **a.** i^{15} **b.** $-i^{18}$

25. **a.** $-i^9$ **b.** i^{10}

26. **a.** $-i^{1996}$ **b.** i^{1998}

Level 2 Drill

Problems 27–50: Simplify each expression.

27. **a.** $(1 - 3i)^2$ **b.** $(6 - 2i)^2$

28. **a.** $(3 + 4i)^2$ **b.** $(2 - 5i)^2$

29. **a.** $(\sqrt{2} + 3i)^2$ **b.** $(\sqrt{5} - 3i)^2$

30. **a.** $(6 - \sqrt{-1})(2 + \sqrt{-1})$
    b. $(2 - \sqrt{-4})(3 - \sqrt{-4})$

31. **a.** $(3 - \sqrt{-3})(3 + \sqrt{-3})$
    b. $(2 - \sqrt{-3})(2 + \sqrt{-3})$

32. **a.** $(2 - \sqrt{-2})(3 - \sqrt{-3})$
    b. $(4 - \sqrt{-9})(5 + \sqrt{-16})$

33. $\dfrac{4 - 2i}{3 + i}$ 34. $\dfrac{5 + 3i}{4 - i}$

35. $\dfrac{1 + 3i}{1 - 2i}$ 36. $\dfrac{3 - 2i}{5 + i}$

37. $\dfrac{-3}{1 + i}$ 38. $\dfrac{5}{4 - i}$

39. $\dfrac{2}{i}$ 40. $\dfrac{3}{-i}$

41. $\dfrac{-2i}{3 + i}$ 42. $\dfrac{3i}{5 - 2i}$

43. $\dfrac{-i}{2 - i}$ 44. $\dfrac{1 - 6i}{1 + 6i}$

45. $\dfrac{2 + 7i}{2 - 7i}$ 46. $\dfrac{-1 + 4i}{1 + 2i}$

47. $\dfrac{\sqrt{-1} + 1}{\sqrt{-1} - 1}$ 48. $\dfrac{10 - \sqrt{-25}}{2 - \sqrt{-1}}$

49. $\dfrac{4 - \sqrt{-4}}{1 + \sqrt{-1}}$ 50. $\dfrac{2 - \sqrt{-9}}{1 + \sqrt{-1}}$

Level 2 Applications

51. Find the value of $x^2 + 25$ when $x = 5i$.

52. Find the value of $x^2 + 3$ when $x = \sqrt{3}i$.

53. Solve $x^2 + 2 = 0$.

54. Solve $x^2 + 4 = 0$.

55. Find the value of $x^2 - 2x + 5$ when $x = 1 + 2i$.

56. Find the value of $x^2 - 2x + 5$ when $x = 1 - 2i$.

57. Solve $x^2 - 3x + 8 = 0$ by using the quadratic formula.

58. Solve $x^2 + 5x + 10 = 0$ by using the quadratic formula.

59. Evaluate $x^3 - 11x^2 + 40x - 50$ when $x = 3 + i$.

60. Evaluate $2x^3 - 11x^2 + 14x + 10$ when $x = 3 - i$.

61. Solve $x^3 - 1 = 0$. [*Hint:* Find three solutions; begin by factoring.]

62. Solve $x^3 + 1 = 0$. [*Hint:* Find three solutions; begin by factoring.]

Level 3 Drill

63. Simplify $(1.9319 + 0.5176i)(2.5981 + 1.5i)$.

64. Simplify $\dfrac{-3.2253 + 8.4022i}{3.4985 + 1.9392i}$.

Level 3 Theory

Problems 65–67: *Let* $z_1 = a_1 + b_1i$, $z_2 = a_2 + b_2i$, *and* $z_3 = a_3 + b_3i$.

65. Prove the commutative laws for complex numbers. That is, prove

$$z_1 + z_2 = z_2 + z_1, \quad z_1 z_2 = z_2 z_1$$

66. Prove the associative laws for complex numbers. That is, prove

$$z_1 + (z_2 + z_3) = (z_1 + z_2) + z_3, \quad z_1(z_2 z_3) = (z_1 z_2)z_3$$

67. Prove the distributive law for complex numbers. That is, prove

$$z_1(z_2 + z_3) = z_1 z_2 + z_1 z_3$$

SECTION 6.2

DE MOIVRE'S THEOREM

PROBLEM OF THE DAY

What is the solution to the equation $x^3 + 1 = 0$?

How would you answer this question?

In the previous section, we solved a similar Problem of the Day, namely

$$x^2 + 1 = 0$$

and we found that even though there were no real solutions, there were two roots, $x = i$, $x = -i$. In this section, we answer a similar question for higher-degree equations.

You might start by subtracting 1 from both sides and then attempt to "take the cube root of both sides":

$x^3 + 1 = 0$	*Given*
$x^3 = -1$	*Subtract 1 from both sides.*
$x = \sqrt[3]{-1}$	*Cube root of both sides*
$x = -1$	*Simplify.*

This "solution" is not correct because this is a third-degree equation, and consequently by the fundamental theorem of algebra, we should expect three solutions. Let us try again—this time by factoring:

$$x^3 + 1 = 0 \qquad \textit{Given}$$

$$(x + 1)(x^2 - x + 1) = 0 \qquad \textit{Factor a sum of cubes.}$$

$$x + 1 = 0 \qquad x^2 - x + 1 = 0 \qquad \textit{Set each factor equal to 0 (factor theorem).}$$

$$x = -1 \qquad\qquad x = \frac{-(-1) \pm \sqrt{(-1)^2 - 4(1)(1)}}{2(1)}$$

$$= \frac{1 \pm \sqrt{-3}}{2}$$

$$= \frac{1 \pm \sqrt{3}\,i}{2}$$

$$= \frac{1}{2} \pm \frac{\sqrt{3}}{2}\,i$$

The goal of this section is to develop techniques that enable us to find all the roots of such equations (not only the real roots, but those roots that are not real).

We begin by considering a graphical representation of a complex number. To give graphical representations of complex numbers such as

$$2 + 3i, \quad -i, \quad -3 - 4i, \quad 3i, \quad -2 + \sqrt{2}\,i, \quad \tfrac{3}{2} - \tfrac{5}{2}i$$

a two-dimensional coordinate system is used. The horizontal axis represents the **real axis** and the vertical axis the **imaginary axis,** so that the complex number $a + bi$ is represented by the ordered pair (a, b) in the usual manner, as shown in Figure 6.1. The coordinate system in Figure 6.1 is called the **complex plane** or the **Gaussian plane,** in honor of Karl Friedrich Gauss.

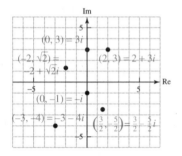

FIGURE 6.1
Complex plane

Historical Note

Karl Gauss (1777–1855) is considered one of the three greatest mathematicians of all time, along with Archimedes and Newton. Gauss graduated from college at the age of 15 and proved what was to become the fundamental theorem of algebra for his doctoral thesis at the age of 22. He published only a small portion of the ideas that seemed to stream from his mind, because he believed that each published result had to be complete, concise, polished, and convincing. His motto was "Few, but ripe." According to Carl B. Boyer in *A History of Mathematics,* Gauss is sometimes described as the last mathematician to know everything in his subject. This is probably only a small exaggeration, but it does emphasize the breadth that Gauss displayed.

The **absolute value** of a complex number z is graphically the distance between z and the origin (just as it is for real numbers). The absolute value of a complex number is also called the **modulus.** The distance formula leads to the following definition.

> ### ABSOLUTE VALUE OR MODULUS OF A COMPLEX NUMBER
>
> If $z = a + bi$ is a complex number, then the **absolute value** or **modulus** of z, denoted by $|z|$, is defined by
>
> $$|z| = \sqrt{a^2 + b^2}$$

EXAMPLE 1   **Finding the Absolute Value of a Complex Number**

Find: **a.** $|3 + 4i|$ **b.** $|-2 + \sqrt{2}\, i|$ **c.** $|-3|$

Solution   **a.** $|3 + 4i| = \sqrt{3^2 + 4^2} = \sqrt{25} = 5$

b. $|-2 + \sqrt{2}\, i| = \sqrt{4 + 2} = \sqrt{6}$

c. You might suggest that -3 is a real number (which it is) and not a complex number (which is not correct). Recall, *all* real numbers are also complex numbers. We can write $-3 = -3 + 0i$.

$$|-3 + 0i| = \sqrt{9 + 0} = 3$$

Note: This is consistent with the absolute value of the real number -3, namely, $|-3| = 3$.   ∎

Trigonometric Form of Complex Numbers

The form $a + bi$ is called the **rectangular form,** but another useful representation uses trigonometry. Consider the graphical representation of a complex number $a + bi$, as shown in Figure 6.2. Let r be the distance from the origin to (a, b) and let θ be the angle the segment makes with the real axis. Then

$$r = \sqrt{a^2 + b^2}$$

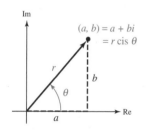

FIGURE 6.2
Trigonometric form and rectangular form of a complex number

and θ, called the **argument,** is chosen so that it is the smallest nonnegative angle the terminal side makes with the positive real axis. From the definition of the trigonometric functions,

$$\cos \theta = \frac{a}{r} \qquad \sin \theta = \frac{b}{r} \qquad \tan \theta = \frac{b}{a}$$

So $a = r \cos \theta \qquad b = r \sin \theta$

Thus

$$a + bi = r \cos \theta + i(r \sin \theta) \qquad \textit{Substitute } a = r \cos \theta \textit{ and } b = r \sin \theta.$$

$$= r(\cos \theta + i \sin \theta) \qquad \textit{Common factor}$$

Sometimes we write $r(\cos \theta + i \sin \theta)$ as r cis θ.

> ### TRIGONOMETRIC FORM OF A COMPLEX NUMBER
>
> The **trigonometric form** of a complex number $z = a + bi$ is
>
> $$r(\cos \theta + i \sin \theta) = r \operatorname{cis} \theta$$
>
> where $r = \sqrt{a^2 + b^2}$; $\tan \theta = \dfrac{b}{a}$ $(a \neq 0)$; $a = r \cos \theta$; $b = r \sin \theta$. This representation is unique for $0° \leq \theta < 360°$ for all z except $0 + 0i$.

STOP

Help Window ? X

This definition tells us how to change forms:

To change from rectangular form to trigonometric form:
   Given $a + bi$. Find r and θ as follows:

$$r = \sqrt{a^2 + b^2} \qquad \text{and} \qquad \tan \theta = \frac{b}{a}$$

To change from trigonometric form to rectangular form:
   Given $r \operatorname{cis} \theta$. Find a and b as follows:

$$a = r \cos \theta \qquad \text{and} \qquad b = r \sin \theta$$

Calculator Window ? X

Many calculators have a built-in conversion for complex numbers. Most calculators call this *rectangular–polar* conversion where

$$(x, y) \leftrightarrow (r, \theta)$$

for

$$x + yi = r \operatorname{cis} \theta$$

Check your owner's manual.

The placement of θ in the proper quadrant is an important consideration because there are two values of $0° \leq \theta < 360°$ that will satisfy the relationship $\tan \theta = b/a$. For example, compare

$-1 + i$, a complex number in Quadrant II

$$\text{where } a = -1, b = 1, \tan \theta = \frac{1}{-1} = -1$$

$1 - i$, a complex number in quadrant IV

$$\text{where } a = 1, b = -1, \tan \theta = \frac{-1}{1} = -1$$

Notice that the trigonometric equation is the same for both complex numbers, even though $-1 + i$ and $1 - i$ are in different quadrants. This consideration is even more important when you are doing the problem on a calculator, because the proper sequence of steps for this example gives the result $-45°$. This is not true for either example because $0° \leq \theta < 360°$ and $-45°$ is not within that domain. The entire process can be dealt with quite simply if we let θ' be the reference angle for θ. Find the reference angle

$$\theta' = \tan^{-1}|-1| = 45°$$

then decide the quadrant for the given complex number, and fix the argument accordingly. For $-1 + i$ in Quadrant II, we see $\theta = 135°$ (reference angle of $45°$ in Quadrant II); for $1 - i$ in Quadrant IV, we find $\theta = 315°$ (reference angle of $45°$ in Quadrant IV).

EXAMPLE 2   **Changing Rectangular Form to Trigonometric Form**

Write each number in trigonometric form:

a. $-1 - \sqrt{3}\,i$ **b.** $6i$ **c.** -5 **d.** $4.310 + 5.516i$

Solution   **a.** $-1 - \sqrt{3}\,i$; $a = -1$, $b = -\sqrt{3}$ is in Quadrant III.

$$r = \sqrt{(-1)^2 + (-\sqrt{3})^2} = \sqrt{4} = 2$$

$$\theta' = \tan^{-1}\left|\frac{-\sqrt{3}}{-1}\right| = 60°; \text{ in Quadrant III, } \theta = 240°$$

$$-1 - \sqrt{3} = 2 \text{ cis } 240°$$

b. $6i$; $a = 0$ and $b = 6$. Notice that $\tan \theta$ is not defined for $\theta = 90°$. By inspection, $6i = 6 \text{ cis } 90°$.

c. -5; $a = -5$, $b = 0$; by inspection $-5 = 5 \text{ cis } 180°$.

d. $4.310 + 5.516i$; $a = 4.310$ and $b = 5.516$ is in Quadrant I.

$$r = \sqrt{(4.310)^2 + (5.516)^2} \approx \sqrt{49} = 7$$

$$\theta' = \tan^{-1}\left|\frac{5.516}{4.310}\right| \approx 52°$$

Thus, $4.310 + 5.516i \approx 7 \text{ cis } 52°$.   ■

EXAMPLE 3   **Changing Trigonometric Form into Rectangular Form**

Write each number in rectangular form.

a. $4 \text{ cis } 330°$ **b.** $5(\cos 38° + i \sin 38°)$

Solution   **a.** $4 \text{ cis } 330° = 4(\cos 330° + i \sin 330°) = 4\left[\dfrac{\sqrt{3}}{2} + i\left(-\dfrac{1}{2}\right)\right] = 2\sqrt{3} - 2i$

b. $5(\cos 38° + i \sin 38°) = 5 \cos 38° + (5 \sin 38°)i \approx 3.94 + 3.08i$ ■

Multiplication and Division of Complex Numbers

The great advantage of the trigonometric form over the rectangular form is the ease with which you can multiply and divide complex numbers.

> **PRODUCTS AND QUOTIENTS OF COMPLEX NUMBERS IN TRIGONOMETRIC FORM**
>
> Let $z_1 = r_1 \operatorname{cis} \theta_1$ and $z_2 = r_2 \operatorname{cis} \theta_2$ be nonzero complex numbers. Then,
>
> $$z_1 z_2 = r_1 r_2 \operatorname{cis}(\theta_1 + \theta_2) \qquad \text{and} \qquad \frac{z_1}{z_2} = \frac{r_1}{r_2} \operatorname{cis}(\theta_1 - \theta_2)$$

YIELD

To prove the product rule, multiply and use the addition laws.

$$
\begin{aligned}
z_1 z_2 &= (r_1 \operatorname{cis} \theta_1)(r_2 \operatorname{cis} \theta_2) \\
&= [r_1(\cos \theta_1 + i \sin \theta_1)][r_2(\cos \theta_2 + i \sin \theta_2)] \\
&= r_1 r_2 (\cos \theta_1 + i \sin \theta_1)(\cos \theta_2 + i \sin \theta_2) \\
&= r_1 r_2 (\cos \theta_1 \cos \theta_2 + i \cos \theta_1 \sin \theta_2 + i \sin \theta_1 \cos \theta_2 + i^2 \sin \theta_1 \sin \theta_2) \\
&= r_1 r_2 [(\cos \theta_1 \cos \theta_2 - \sin \theta_1 \sin \theta_2) + i(\cos \theta_1 \sin \theta_2 + \sin \theta_1 \cos \theta_2)] \\
&= r_1 r_2 [\cos(\theta_1 + \theta_2) + i \sin(\theta_1 + \theta_2)] \\
&= r_1 r_2 \operatorname{cis}(\theta_1 + \theta_2)
\end{aligned}
$$

The proof of the quotient form is similar and is left as a problem.

EXAMPLE 4

Simplifying Expressions in Complex Form

Simplify:

a. $(5 \operatorname{cis} 38°)(4 \operatorname{cis} 75°)$ **b.** $(\sqrt{2} \operatorname{cis} 188°)(2\sqrt{2} \operatorname{cis} 310°)$

c. $(2 \operatorname{cis} 48°)^3$ **d.** $\dfrac{15(\cos 48° + i \sin 48°)}{5(\cos 125° + i \sin 125°)}$

Solution

a. $(5 \operatorname{cis} 38°)(4 \operatorname{cis} 75°) = 5 \cdot 4 \operatorname{cis}(38° + 75°) = 20 \operatorname{cis} 113°$

b. $(\sqrt{2} \operatorname{cis} 188°)(2\sqrt{2} \operatorname{cis} 310°) = 2 \cdot 2 \operatorname{cis}(188° + 310°)$
  $= 4 \operatorname{cis} 498° = 4 \operatorname{cis} 138°$ *Arguments should be between 0° and 360°.*

c. $(2 \operatorname{cis} 48°)^3 = (2 \operatorname{cis} 48°)(2 \operatorname{cis} 48°)^2 = (2 \operatorname{cis} 48°)(4 \operatorname{cis} 96°)$
  $= 8 \operatorname{cis} 144°$

d. $\dfrac{15(\cos 48° + i \sin 48°)}{5(\cos 125° + i \sin 125°)} = \dfrac{15 \operatorname{cis} 48°}{5 \operatorname{cis} 125°} = 3 \operatorname{cis}(48° - 125°)$
  $= 3 \operatorname{cis}(-77°) = 3 \operatorname{cis} 283°$

EXAMPLE 5   **Simplifying a Power**

Simplify (expand) the expression $(1 - \sqrt{3}\, i)^5$.

Solution   First change to trigonometric form:

$$a = 1, b = -\sqrt{3};\ \text{Quadrant IV}$$
$$r = \sqrt{1 + 3} = 2$$
$$\theta' = \tan^{-1}\left| -\frac{\sqrt{3}}{1} \right| = 60°;\ \text{in Quadrant IV},\ \theta = 300°$$

Thus

$$(1 - \sqrt{3}\, i)^5 = (2\ \text{cis}\ 300°)^5$$
$$= (2\ \text{cis}\ 300°)(2\ \text{cis}\ 300°)(2\ \text{cis}\ 300°)(2\ \text{cis}\ 300°)(2\ \text{cis}\ 300°)$$
$$= (2 \cdot 2 \cdot 2 \cdot 2 \cdot 2)\ \text{cis}(300° + 300° + 300° + 300° + 300°)$$
$$= 2^5\ \text{cis}(5 \cdot 300°) = 32\ \text{cis}\ 1{,}500° = 32\ \text{cis}\ 60°$$

If we want the answer in rectangular form, we can now change back:

$$32\ \text{cis}\ 60° = 32\cos 60° + i(32\sin 60°) = 16 + 16\sqrt{3}\, i \qquad \blacksquare$$

Powers of Complex Numbers—De Moivre's Theorem

As you can see from Example 5, multiplication in trigonometric form extends quite nicely to any positive integral power in a result called **De Moivre's theorem.** This theorem is proved by mathematical induction.

DE MOIVRE'S THEOREM

If n is a natural number, then

$$(r\ \text{cis}\ \theta)^n = r^n\ \text{cis}\ n\theta$$

for a complex number $r\ \text{cis}\ \theta = r(\cos\theta + i\sin\theta)$.

Although De Moivre's theorem is useful for powers, as illustrated by Example 5, its real usefulness is in finding the complex roots of numbers. Recall from algebra that $\sqrt[n]{r} = r^{1/n}$ is used to denote the principal nth root of r. However, $r^{1/n}$ is only *one* of the nth roots of r. How do you find *all* the nth roots of r? To find the principal root, you can use a calculator or logarithms along with the following theorem. The following **nth root theorem** follows directly from De Moivre's theorem.

> ### nth ROOT THEOREM
>
> If n is any positive integer, then the nth roots of r cis θ are given by
>
> $$\sqrt[n]{r}\, \text{cis}\left(\frac{\theta + 360°k}{n}\right) \qquad \text{or} \qquad \sqrt[n]{r}\, \text{cis}\left(\frac{\theta + 2\pi k}{n}\right)$$
>
> for $k = 0, 1, 2, 3, \ldots, n - 1$. If $\theta = 0$, the root found when $k = 0$ is called the **principal nth root.** If $\theta = 180°$, the **principal nth root** is the negative real root.

Notice that the principal nth root of a positive or negative real number is the root that you obtain from your calculator. The proof of this theorem is left as a problem.

EXAMPLE 6   **Finding the Square Roots of a Number**

Find the square roots of $-\frac{9}{2} + \frac{9}{2}\sqrt{3}\, i$.

Solution   First, change to trigonometric form.

$$r = \sqrt{\left(-\tfrac{9}{2}\right)^2 + \left(\tfrac{9}{2}\sqrt{3}\right)^2} = 9$$

$$\theta' = \tan^{-1}\left|\frac{\tfrac{9}{2}\sqrt{3}}{-\tfrac{9}{2}}\right| = \tan^{-1}\sqrt{3} = 60°; \text{ in Quadrant II, } \theta = 120°$$

The square roots are

$$9^{1/2}\, \text{cis}\left(\frac{120° + 360°k}{2}\right) = 3\, \text{cis}(60° + 180°k)$$

$$k = 0: \qquad 3\, \text{cis}\, 60° = \tfrac{3}{2} + \tfrac{3}{2}\sqrt{3}\, i$$
$$k = 1: \qquad 3\, \text{cis}\, 240° = -\tfrac{3}{2} - \tfrac{3}{2}\sqrt{3}\, i$$

Check:

$$\left(\tfrac{3}{2} + \tfrac{3}{2}\sqrt{3}\, i\right)^2 = \tfrac{9}{4} + \tfrac{9}{2}\sqrt{3}\, i + \tfrac{9}{4}\cdot 3i^2 = -\tfrac{9}{2} + \tfrac{9}{2}\sqrt{3}\, i$$
$$\left(-\tfrac{3}{2} - \tfrac{3}{2}\sqrt{3}\, i\right)^2 = \tfrac{9}{4} + \tfrac{9}{2}\sqrt{3}\, i + \tfrac{9}{4}\cdot 3i^2 = -\tfrac{9}{2} + \tfrac{9}{2}\sqrt{3}\, i$$

■

EXAMPLE 7   **Finding the Fifth Roots of a Number**

Find the fifth roots of 32.

Solution   Begin by writing 32 in trigonometric form: $32 = 32\, \text{cis}\, 0°$. The fifth roots are found by

$$32^{1/5} \operatorname{cis}\left(\frac{0° + 360°k}{5}\right) = 2 \operatorname{cis} 72°k$$

$k = 0$: $2 \operatorname{cis} 0° = 2$ *The first root, which is located so that its argument is θ/n, is the **principal nth root**.*

$k = 1$: $2 \operatorname{cis} 72° \approx 0.6180 + 1.90211i$

$k = 2$: $2 \operatorname{cis} 144° \approx -1.6180 + 1.1756i$

$k = 3$: $2 \operatorname{cis} 216° \approx -1.6180 - 1.1756i$

$k = 4$: $2 \operatorname{cis} 288° \approx 0.6180 - 1.9021i$

All other integral values for k repeat those listed here.   ■

If we represent the fifth roots of 32 graphically, as shown in Figure 6.3, notice that they all lie on a circle with radius of 2 and are equally spaced.

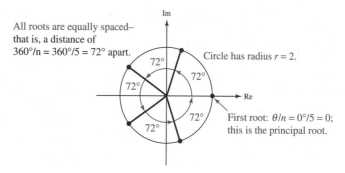

FIGURE 6.3   Graphical representation of the fifth roots of 32

If n is a positive integer, the nth roots of a complex number $a + bi = r \operatorname{cis} \theta$ are equally spaced on the circle of radius r centered at the origin. We return to the Problem of the Day, this time using De Moivre's formula.

EXAMPLE 8

Problem of the Day Using De Moivre's Formula

Solve $x^3 + 1 = 0$.

Solution $x^3 = -1$, so we are looking for the cube roots of -1.
Since $-1 = \operatorname{cis} 180°$, we have

$$1^{1/3} \operatorname{cis} \frac{180° + 360°k}{3} = \operatorname{cis}(60° + 120°k)$$

$k = 0$: $\operatorname{cis} 60° = \frac{1}{2} + \frac{\sqrt{3}}{2} i$

$k = 1$: $\operatorname{cis} 180° = -1$ *This is the principal nth root.*

$k = 2$: $\operatorname{cis} 300° = \frac{1}{2} - \frac{\sqrt{3}}{2} i$

Thus, the roots of $x^3 + 1 = 0$ are $\frac{1}{2} \pm \frac{\sqrt{3}}{2} i$, -1.   ■

PROBLEM SET 6.2

Essential Concepts

1. Let $z = a + bi$ be any complex number.
 a. What is $|z|$?
 b. What is $|x|$ for x a real number?
 c. Since the set of real numbers is a subset of the set of complex numbers, show that your answer to part **b** is a subset of your answer to part **a**.

2. a. What is the rectangular form of a complex number?
 b. What is the trigonometric form of a complex number?
 c. What are the formulas for changing from rectangular form to trigonometric form?
 d. What are the formulas for changing from trigonometric form to rectangular form?

3. IN YOUR OWN WORDS Why would you want to do multiplication in trigonometric form rather than in rectangular form?

4. IN YOUR OWN WORDS Why would you want to do division in trigonometric form rather than in rectangular form?

5. IN YOUR OWN WORDS De Moivre's theorem seems to be about raising to the nth power. How does it relate to the nth root theorem?

6. IN YOUR OWN WORDS Explain the general procedure for finding the nth roots of some given number.

7. What is the meaning of $r \operatorname{cis} \theta$?

Problems **8–13:** *Classify as true or false. If it is false, explain why.*

8. cis 90° does not exist because tan 90° is undefined.

9. All negative real numbers have $\theta = 180°$.

10. If a point is on the coordinate axis, then θ is 0°, 90°, 180°, or 270°.

11. If $\theta = 90°$, then the complex number is plotted on the negative imaginary axis.

12. $2 - 3i$ is a point in Quadrant II.

13. 4 cis 250° is a point in Quadrant IV.

Level 1 Drill

Problems **14–17:** *Plot the complex numbers, and find the modulus of each.*

14. a. $3 + i$ **b.** $7 - i$ **c.** $3 + 2i$

15. a. $-3 - 2i$ **b.** $2 + 4i$ **c.** $5 + 6i$

16. a. $-2 + 5i$ **b.** $-5 + 4i$ **c.** $-3 - 2i$

17. a. $2 - 5i$ **b.** $4 - i$ **c.** $-1 + i$

Problems **18–23:** *Plot the given number, and then change to trigonometric form.*

18. a. $1 + i$ **b.** $1 - i$ **c.** $\sqrt{3} - i$

19. a. $\sqrt{3} + i$ **b.** $1 - \sqrt{3}\,i$ **c.** $-1 - \sqrt{3}\,i$

20. a. -1 **b.** 2 **c.** -3

21. a. i **b.** $-2i$ **c.** $3i$

22. a. $5.7956 - 1.5529i$
 b. $1.5321 - 1.2856i$

23. a. $-0.6946 + 3.9392i$
 b. $-2.0337 - 4.5677i$

Problems **24–27:** *Plot the given number, and then change to rectangular form. Use exact values whenever possible.*

24. a. $2(\cos 45° + i \sin 45°)$ **b.** $\operatorname{cis} \frac{5\pi}{6}$

25. a. $3(\cos 60° + i \sin 60°)$ **b.** $5 \operatorname{cis} \frac{3\pi}{2}$

26. a. $6 \operatorname{cis} 247°$ **b.** $9 \operatorname{cis} 190°$

27. a. $2.5 \operatorname{cis} 300°$ **b.** $4.2 \operatorname{cis} 135°$

Level 2 Drill

Problems **28–39:** *perform the indicated operations.*

28. $(2 \operatorname{cis} 60°)(3 \operatorname{cis} 150°)$ **29.** $(3 \operatorname{cis} 48°)(5 \operatorname{cis} 92°)$

30. $4(\cos 65° + i \sin 65°) \cdot 12(\cos 87° + i \sin 87°)$

31. $\dfrac{5(\cos 315° + i \sin 315°)}{2(\cos 48° + i \sin 48°)}$

32. $\dfrac{8 \operatorname{cis} 30°}{4 \operatorname{cis} 15°}$ **33.** $\dfrac{12 \operatorname{cis} 250°}{4 \operatorname{cis} 120°}$

34. $(2 \operatorname{cis} 50°)^3$ **35.** $(3 \operatorname{cis} 60°)^4$

36. $(\cos 210° + i \sin 210°)^5$ **37.** $(2 \operatorname{cis} 80°)^6$

38. $(1 + i)^6$ **39.** $(\sqrt{3} - i)^8$

Problems 40–47: *Find the indicated roots of the given numbers. Leave your answers in trigonometric form.*

40. square roots of 16 cis 100°

41. fourth roots of 81 cis 88°

42. cube roots of 64 cis 216°

43. cube roots of 27

44. cube roots of 8

45. fourth roots of $1 + i$

46. cube roots of 8 cis 240°

47. fifth roots of 32 cis 200°

Problems 48–53: *Find the indicated roots of the given numbers. Leave your answers in rectangular form. Show the roots graphically.*

48. cube roots of 1

49. cube roots of -8

50. fourth roots of 1

51. cube roots of $4\sqrt{3} - 4i$

52. square roots of $\dfrac{25}{2} - \dfrac{25\sqrt{3}}{2}i$

53. fourth roots of $12.2567 + 10.2846i$

Problems 54–61: *Solve the given equations for all complex roots. Leave your answers in rectangular form and give approximate answers to four decimal places.*

54. $x^3 - 1 = 0$

55. $x^5 - 1 = 0$

56. $x^5 + 1 = 0$

57. $x^6 - 1 = 0$

58. $x^4 + 16 = 0$

59. $x^4 - 16 = 0$

60. $x^4 + x^3 + x^2 + x + 1 = 0$

[*Hint:* Multiply both sides by $(x - 1)$.]

61. $x^5 + x^4 + x^3 + x^2 + x + 1 = 0$

[*Hint:* Multiply both sides by $(x - 1)$.]

Level 2 Applications

62. In calculus, it is shown that

$$e^{i\theta} = \cos\theta + i\sin\theta$$

which is one of the most remarkable formulas in all of mathematics. This means, among other things,

that

$$re^{i\theta} = r\,\text{cis}\,\theta$$

Explain the meaning for the following graffiti:

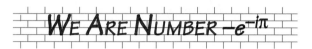

63. Use the result of Problem 62 to plot the following complex numbers.

a. $e^{i\pi/4}$ **b.** $2e^{-i\pi/6}$ **c.** $-e^{-i\pi}$

64. Complex numbers are used in the study of electrical currents. For example, the alternating current, in amperes, in an electric inductor is

$$I = \frac{E}{Z}$$

where E is the voltage and Z is the impedance. Suppose $E = 16$ cis 75° and $Z = 10 + 5i$. Find I, the current in amperes, in rectangular form (rounded to two decimal places).

65. Complex numbers are used in the study of electrical currents. For example, the alternating current, in amperes, in an electric inductor is

$$I = \frac{E}{Z}$$

amperes, where E is the voltage and Z is the impedance. Suppose $E = 12$ cis 27° and $Z = 3 + 2i$. Find I, the current in amperes, in rectangular form (rounded to two decimal places).

66. Two forces $\mathbf{F}_1$ and $\mathbf{F}_2$ are acting on an object at the origin. Find the resultant if

$$\mathbf{F}_1 = 150 \text{ cis } 0° \quad \text{and} \quad \mathbf{F}_2 = 275 \text{ cis } 120°$$

67. Two forces $\mathbf{F}_1$ and $\mathbf{F}_2$ are acting on an object at the origin. Find the resultant (rounded to two decimal places) if

$$\mathbf{F}_1 = 25 \text{ cis } 30° \quad \text{and} \quad \mathbf{F}_2 = 45 \text{ cis } 60°$$

68. Prove that

$$\frac{r_1 \operatorname{cis} \theta_1}{r_2 \operatorname{cis} \theta_2} = \frac{r_1}{r_2} \operatorname{cis}(\theta_1 - \theta_2)$$

69. Prove that

$$\sqrt[n]{r} \operatorname{cis}\left(\frac{\theta + 360^\circ k}{n}\right) = (r \operatorname{cis} \theta)^{1/n}$$

70. Use Problem 62, as well as the laws of exponents, to prove De Moivre's theorem.

71. Prove the *triangle inequality*

$$|z_1 + z_2| \le |z_1| + |z_2|$$

for complex numbers $z_1 = a + bi$ and $z_2 = c + di$.

SECTION 6.3

POLAR COORDINATES

PROBLEM OF THE DAY

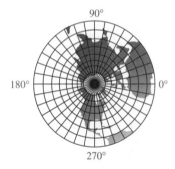

FIGURE 6.4
Paris Observatory

A portion of the floor of the Paris Observatory is reproduced in Figure 6.4. Can you locate Washington, D.C.?

How would you answer this question?

We will first find the approximate location of Washington, D.C.; do you know the location of this city? Can you point to it in Figure 6.4? Can you locate the United States?

Up to now, we have used only the Cartesian coordinate system. This Problem of the Day seems to be using a different system. It is called the *polar coordinate system,* which we introduce in this section. In this system, points are located by specifying their distance from a fixed point and their direction in relation to a fixed line. We use the term **rectangular coordinates** to refer to Cartesian coordinates; this will help you remember that Cartesian coordinates are plotted in a rectangular fashion. If you have completed the first two sections of this chapter, this new system will look a bit familiar to you, since the main idea is taken from the way we plotted complex numbers. Here we go a step further, because we allow r to be either positive or negative. Also, each point in the last section represented a *single* complex number; now we once again consider plotting *ordered pairs.*

Plotting Points in Polar Coordinates

In a **polar coordinate system** we fix a point O, called the **pole,** and draw a horizontal ray (half-line), called the **radial axis,** from the point O. Then each point in the plane is represented by an ordered pair $P(r, \theta)$, where θ (the **polar angle**) measures the angle from the radial axis and r (the **radial distance**)

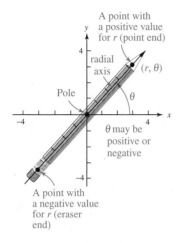

FIGURE 6.5
Polar-form points

measures the directed distance from the pole to the point P. Both r and θ can be any real numbers.

When plotting points, rotate the radial axis through an angle θ, as shown in Figure 6.5. You might find it helpful to rotate your pencil as the axis—the tip points in the positive direction and the eraser, in the negative direction. If θ is measured in a counterclockwise direction, it is positive; if θ is measured in a clockwise direction, it is negative. Next, plot r on the radial axis (the pencil). Notice that any real number can be plotted on this real-number line in the direction of the tip if the number is positive and in the direction of the eraser if it is negative.

Plotting points seems easy, but it is necessary that you completely understand this process. Study each part of Example 1 and make sure you undertand how each point is plotted. We assume an orientation provided by an x-axis and a y-axis with the pole at the origin, and the rotation of the radial axis is measured from the positive x-axis. We consider that points can be plotted using either rectangular or polar coordinates, and that the system we are using in a particular example must be either specified or obvious from the context.

EXAMPLE 1   **Plotting Polar-Form Points**

Plot each of the following polar-form points:

$$A\left(4, \tfrac{\pi}{3}\right), \quad B\left(-4, \tfrac{\pi}{3}\right), \quad C\left(3, -\tfrac{\pi}{6}\right), \quad D\left(-3, -\tfrac{\pi}{6}\right), \quad E(-3, 3),$$

$$F(-3, -3), \quad G(-4, -2), \quad H\left(5, \tfrac{3\pi}{2}\right), \quad I\left(-5, \tfrac{\pi}{2}\right), \quad J\left(5, -\tfrac{\pi}{2}\right)$$

Solution   Points A, B, C, and D illustrate the basic ideas of plotting polar-form points.

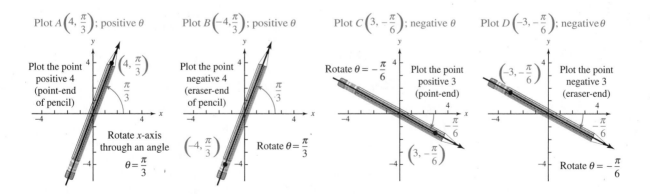

Points E, F, G, H, I, and J illustrate common situations that can sometimes be confusing. Make sure you take time with each example.

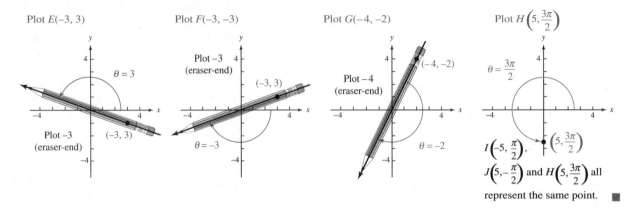

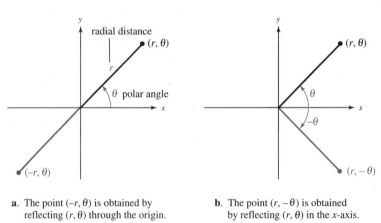

Note that the point $(-r, \theta)$ can be obtained by *reflecting* (r, θ) through the origin, whereas the point $(r, -\theta)$ can be obtained by reflecting (r, θ) in the *x*-axis. These features of the polar coordinate system are illustrated in Figure 6.6.

a. The point $(-r, \theta)$ is obtained by reflecting (r, θ) through the origin.

b. The point $(r, -\theta)$ is obtained by reflecting (r, θ) in the *x*-axis.

FIGURE 6.6 Plotting points using reflections

Primary Representations of a Point

Notice from Example 1 that ordered pairs in polar form are not associated in a one-to-one fashion with points in the plane. Indeed, given any point in the plane, there are infinitely many ordered pairs of polar coordinates associated with that point. If you are given a point $P(r, \theta)$ other than the pole in polar coordinates, then $(-r, \theta + \pi)$ also represents P. In addition, there are infinitely many other representations of P—namely, $(r, \theta + 2n\pi)$ for n an integer. We call (r, θ) and $(-r, \theta + \pi)$ the *primary representations of the point* if the angles θ and $\theta + \pi$ are between 0 and 2π. If the angle is not between 0 and 2π, then you can add or subtract a multiple of 2π so that it is between 0 and 2π. We summarize in the box at the top of page 303.

Remember to simplify so the second component represents a nonnegative angle less than one revolution.

NOTE

> **PRIMARY REPRESENTATIONS OF A POINT IN POLAR FORM**
>
> Every nonzero point in polar form has two **primary representations:**
>
> $$(r, \theta) \qquad \text{and} \qquad (-r, \pi + \theta)$$
>
> where the second component in each case is between 0 and 2π.

EXAMPLE 2

Giving Primary Representations of a Point

Give both primary representations of each point:

a. $\left(3, \frac{\pi}{4}\right)$ **b.** $\left(5, \frac{5\pi}{4}\right)$ **c.** $\left(-6, -\frac{2\pi}{3}\right)$ **d.** $(9, 5)$ **e.** $(9, 7)$

Solution

a. $\left(3, \frac{\pi}{4}\right)$ is primary; the other is $\left(-3, \frac{5\pi}{4}\right)$.   *Change the sign of r and add π.*

b. $\left(5, \frac{5\pi}{4}\right)$ is primary; the other is $\left(-5, \frac{\pi}{4}\right)$.
$\frac{5\pi}{4} + \pi = \frac{9\pi}{4}$, but $\left(-5, \frac{9\pi}{4}\right)$ is not a primary representation because
$\frac{9\pi}{4} > 2\pi$. Thus, we write $\left(-5, \frac{9\pi}{4}\right)$ and then add a multiple of 2π to
obtain the correct primary representation of $\left(-5, \frac{\pi}{4}\right)$.

c. $\left(-6, -\frac{2\pi}{3}\right)$ has primary representations $\left(-6, \frac{4\pi}{3}\right)$ and $\left(6, \frac{\pi}{3}\right)$.

d. $(9, 5)$ is primary; the other is $(-9, 5 - \pi)$; a point like $(-9, 5 - \pi)$ may
be approximated by $(-9, 1.86)$.

e. $(9, 7)$ has primary representations $(9, 7 - 2\pi)$, or $(9, 0.72)$, and
$(-9, 7 - \pi)$, or $(-9, 3.86)$.   ∎

EXAMPLE 3

Problem of the Day: Finding Polar-Form Coordinates

A portion of the floor of the Paris Observatory is reproduced in Figure 6.7.
Can you locate Washington, D.C.?

Solution

There are infinitely many polar forms, but only one actual location. We will
find the primary representations for this city in polar form.

First, we locate the point showing the approximate location of
Washington, D.C. (see arrow). Begin by finding r; we count the concentric
circles and see that Washington, D.C., is located between the 5th and 6th
circles—estimate $r \approx 5.4$. Next, we notice from the rays marked 0°, 90°,
180°, and 270° that each such ray represents 10°. From this, we can estimate
$\theta \approx 280°$.

Now for the two primary representations, we have

$$(r, \theta) = (5.4, 280°) \text{ and } (-r, \theta + 180°) = (-5.4, 460°) \text{ or } (-5.4, 100°)$$

The two primary representations are $(5.4, 280°)$ and $(-5.4, 100°)$.   ∎

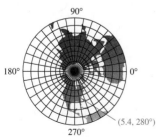

FIGURE 6.7
Map from the floor of the
Paris Observatory

Relationship Between Polar and Rectangular Coordinates

The relationship between polar and rectangular coordinates can be found by using the definition of trigonometric functions (see Figure 6.8).

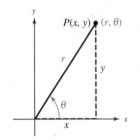

FIGURE 6.8 Relationship between rectangular and polar coordinates

RELATIONSHIP BETWEEN RECTANGULAR AND POLAR COORDINATES

1. To change *from polar to rectangular:*

$$x = r \cos \theta \qquad y = r \sin \theta$$

2. To change *from rectangular to polar:*

$$r = \sqrt{x^2 + y^2} \qquad \theta' = \tan^{-1} \left| \frac{y}{x} \right|, x \neq 0$$

where θ' is the reference angle for θ. Place θ in the proper quadrant by noting the signs of x and y. If $x = 0$, then $\theta' = \frac{\pi}{2}$.

 STOP *You might notice that these formulas are identical to those we stated in the Help Window on page 292.*

EXAMPLE 4

Converting from Polar to Rectangular Coordinates

Change the polar coordinates $\left(-3, \frac{7\pi}{4}\right)$ to rectangular coordinates.

Solution

$$x = -3 \cos \frac{7\pi}{4} = -3\left(\frac{\sqrt{2}}{2}\right) = -\frac{3\sqrt{2}}{2}$$

$$y = -3 \sin \frac{7\pi}{4} = -3\left(-\frac{\sqrt{2}}{2}\right) = \frac{3\sqrt{2}}{2}$$

The rectangular coordinates are $\left(-\frac{3\sqrt{2}}{2}, \frac{3\sqrt{2}}{2}\right)$. ∎

EXAMPLE 5

Converting from Rectangular Coordinates to Polar Coordinates

Write both primary representations of the polar-form coordinates for the point with rectangular coordinates $\left(\frac{5\sqrt{3}}{2}, -\frac{5}{2}\right)$.

Solution

$$r = \sqrt{\left(\frac{5\sqrt{3}}{2}\right)^2 + \left(-\frac{5}{2}\right)^2} = \sqrt{\frac{75}{4} + \frac{25}{4}} = 5$$

Note that θ is in Quadrant IV because x is positive and y is negative.

$$\theta' = \tan^{-1}\left|\frac{-\frac{5}{2}}{\frac{5\sqrt{3}}{2}}\right| = \tan^{-1}\left(\frac{1}{\sqrt{3}}\right) = \frac{\pi}{6}; \text{ thus, } \theta = \frac{11\pi}{6} \quad \text{(Quadrant IV)}$$

The polar-form coordinates are $\left(5, \frac{11\pi}{6}\right)$ and $\left(-5, \frac{5\pi}{6}\right)$. ∎

Points on a Polar Curve

The **graph** of an equation in polar coordinates is the set of all points P having a pair of polar coordinates (r, θ) that satisfy the given equation. Circles, lines through the origin, and rays emanating from the origin have particularly simple equations in polar coordinates.

EXAMPLE 6   **Graphing Circles, Lines, and Rays**

Graph   **a.** $r = 6$ **b.** $\theta = \frac{\pi}{6}, r \geq 0$ **c.** $\theta = \frac{\pi}{6}$

Solution   **a.** The graph is the set of all points (r, θ) such that the first component is 6 for any angle θ. This is a circle with radius 6 centered at the origin.

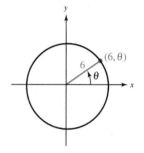

b. The graph is the closed half-line (ray) that emanates from the origin and makes an angle of $\frac{\pi}{6}$ with the positive x-axis.

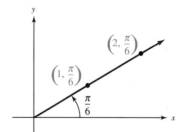

c. This is the line through the origin that makes an angle of $\frac{\pi}{6}$ with the positive x-axis.

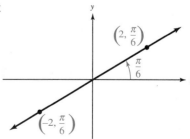

∎

As with other equations, we begin graphing polar-form curves by plotting some points. However, you must first be able to recognize whether a point in polar form satisfies a given equation.

EXAMPLE 7

Verifying That Polar Coordinates Satisfy an Equation

Show that each of the given points lies on the polar graph whose equation is

$$r = \frac{2}{1 - \cos \theta}$$

a. $\left(2, \frac{\pi}{2}\right)$ **b.** $\left(-2, \frac{3\pi}{2}\right)$ **c.** $(-1, 2\pi)$

Solution Begin by substituting the given coordinates into the equation.

a. $2 \overset{?}{=} \dfrac{2}{1 - \cos \frac{\pi}{2}} = \dfrac{2}{1 - 0} = 2;$

Thus, $\left(2, \frac{\pi}{2}\right)$ is on the curve because it satisfies the equation.

b. $-2 \overset{?}{=} \dfrac{2}{1 - \cos \frac{3\pi}{2}} = \dfrac{2}{1 - 0} = 2$ This equation is **not** true.

NOTE

Although the equation is not satisfied, we *cannot* say that the point is not on the curve. Indeed, we see from part **a** that it is on the curve, because $\left(-2, \frac{3\pi}{2}\right)$ and $\left(2, \frac{\pi}{2}\right)$ name the same point! So even if one primary representation of a point does not satisfy the equation, we must still check the other primary representation to see if it satisfies the equation.

c. The primary representation of $(-1, 2\pi)$ is $(-1, 0)$

$$-1 = \frac{2}{1 - \cos 0} \quad \text{is undefined.}$$

Check the other representation—namely, $(1, \pi)$:

$$1 \overset{?}{=} \frac{2}{1 - \cos \pi} = \frac{2}{1 - (-1)} = 1$$

Thus, the point is on the curve.   ■

We will discuss the graphing of polar-form curves in the next section, but sometimes a polar-form equation can be graphed by changing it to rectangular form.

EXAMPLE 8

Polar-Form Graphing by Changing to Rectangular Form

Graph the given polar-form curves by changing to rectangular form.

a. $r = 3 \cos \theta$ **b.** $r \cos \theta = 4$ **c.** $r = \dfrac{6}{2 \sin \theta + \cos \theta}$

Solution   **a.**

$$r = 3\cos\theta \qquad \textit{Given}$$

$$r^2 = 3r\cos\theta \qquad \textit{Multiply by r.}$$

$$x^2 + y^2 = 3x \qquad \textit{Because } x = r\cos\theta \textit{ and } r^2 = x^2 + y^2$$

$$x^2 - 3x + y^2 = 0 \qquad \textit{Subtract 3x from both sides.}$$

$$\left[x^2 - 3x + \left(\tfrac{3}{2}\right)^2\right] + y^2 = \tfrac{9}{4} \qquad \textit{Complete the square.}$$

$$\left(x - \tfrac{3}{2}\right)^2 + y^2 = \tfrac{9}{4} \qquad \textit{Factor.}$$

We see this is a circle with center at $\left(\tfrac{3}{2}, 0\right)$ and radius $\tfrac{3}{2}$. The graph is shown in Figure 6.9a.

b. Because $x = r\cos\theta$, we see that the given equation can be written $x = 4$, which is a vertical line (Figure 6.9b).

c.

$$r = \frac{6}{2\sin\theta + \cos\theta} \qquad \textit{Given}$$

$$2r\sin\theta + r\cos\theta = 6 \qquad \textit{Multiply both sides by } (2\sin\theta + \cos\theta).$$

$$2y + x = 6 \qquad \textit{Substitute } y = \sin\theta \textit{ and } x = \cos\theta.$$

$$y = -\tfrac{1}{2}x + 3$$

We see this is a line with y-intercept 3 and slope $-\tfrac{1}{2}$ (Figure 6.9c).

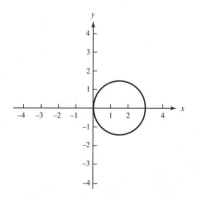

a. Graph of $r = 3\cos\theta$

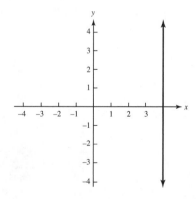

b. Graph of $r\cos\theta = 4$

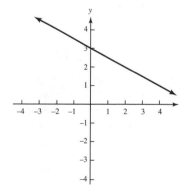

c. Graph of $r = \dfrac{6}{2\sin\theta + \cos\theta}$

FIGURE 6.9   Polar-form graphs

PROBLEM SET 6.3

Essential Concepts

1. **IN YOUR OWN WORDS** Illustrate the procedure for plotting points in polar form.

2. **IN YOUR OWN WORDS** Show the derivation of the polar–rectangular conversion equations.

3. What are the primary representations of a point, and why is it necessary to consider two primary representations?

4. What is the procedure for verifying whether a given ordered pair satisfies an equation?

Level 1 Drill

Problems 5–13: *Plot each polar-form point. Give both primary representations and the rectangular coordinates of the point.*

5. $\left(4, \frac{\pi}{4}\right)$ 6. $\left(6, \frac{\pi}{3}\right)$ 7. $\left(5, \frac{2\pi}{3}\right)$

8. $\left(3, -\frac{\pi}{6}\right)$ 9. $\left(\frac{3}{2}, -\frac{5\pi}{6}\right)$ 10. $(-4, 4)$

11. $(1, 3\pi)$ 12. $\left(-2, -\frac{3\pi}{2}\right)$ 13. $(0, -3)$

Problems 14–22: *Plot each rectangular-form point and give both primary representations in polar form.*

14. $(5, 5)$ 15. $(-1, \sqrt{3})$ 16. $(2, -2\sqrt{3})$

17. $(-2, -2)$ 18. $(3, -3)$ 19. $(3, 7)$

20. $(3, -3\sqrt{3})$ 21. $(\sqrt{3}, -1)$ 22. $(-3, 0)$

Level 2 Drill

Problems 23–30: *Write each equation given in rectangular form.*

23. $r = 4 \sin \theta$ 24. $r = 16$

25. $r = \sec \theta$ 26. $r = 2 \cos \theta$

27. $r^2 = \dfrac{2}{1 + \sin^2 \theta}$ 28. $r = \dfrac{5}{\sin \theta + 4 \cos \theta}$

29. $r = 1 - \sin \theta$ 30. $r = 4 \tan \theta$

Problems 31–38: *Sketch the graph of each equation.*

31. $r = \frac{3}{2}$ 32. $r = \frac{3}{2}, 0 \le \theta < 2$

33. $r = \sqrt{2}, 0 \le \theta < 2$ 34. $r = 4$

35. $\theta = 1$ 36. $\theta = 1, r \ge 0$

37. $\theta = \frac{\pi}{6}, r < 0$ 38. $\theta = -\frac{\pi}{3}$

Problems 39–44: *State whether the given points lie on the curve.*

$$r = \frac{5}{1 - \sin \theta}$$

39. $\left(10, \frac{\pi}{6}\right)$ 40. $\left(5, \frac{\pi}{2}\right)$

41. $\left(-10, \frac{5\pi}{6}\right)$ 42. $\left(-\frac{10}{3}, \frac{5\pi}{6}\right)$

43. $\left(20 + 10\sqrt{3}, \frac{\pi}{3}\right)$ 44. $\left(-10, \frac{\pi}{3}\right)$

Problems 45–52: *State whether the given points lie on the curve $r = 2(1 - \cos \theta)$.*

45. $\left(1, \frac{\pi}{3}\right)$ 46. $\left(1, -\frac{\pi}{3}\right)$

47. $\left(-1, \frac{\pi}{3}\right)$ 48. $\left(-2, \frac{\pi}{2}\right)$

49. $\left(2 + \sqrt{2}, \frac{\pi}{4}\right)$ 50. $\left(-2 - \sqrt{2}, \frac{\pi}{4}\right)$

51. $\left(0, \frac{\pi}{4}\right)$ 52. $\left(0, -\frac{2\pi}{3}\right)$

Problems 53–60: *Find three distinct ordered pairs satisfying each equation. Give both primary representations for each point.*

53. $r^2 = 9 \cos \theta$ 54. $r^2 = 9 \cos 2\theta$

55. $r = 3\theta$ 56. $r = 5\theta$

57. $r = 2 - 3 \sin \theta$ 58. $r = 2(1 + \cos \theta)$

59. $\dfrac{r}{1 - \sin \theta} = 2$ 60. $r = \dfrac{8}{1 - 2 \cos \theta}$

Level 2 Applications

61. Name the country or island in Figure 6.10 (page 309) in which each of the named points is located.
    a. $(5.5, 260°)$ b. $(7.5, 78°)$
    c. $(-2, 140°)$ d. $(-4, 80°)$

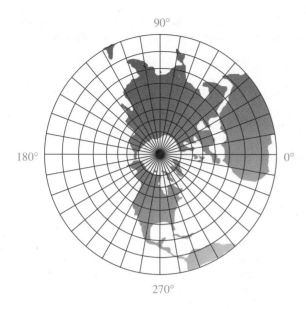

FIGURE 6.10   This map is reproduced from the floor of the Paris Observatory.

62. Name possible coordinates for each city (see Figure 6.10).
  a. Miami, Florida
  b. Los Angeles, California
  c. Mexico City, Mexico
  d. London, England

63. The equation $r = 1 - \sin \theta$ is used by entomologists to describe the "dance" of a certain species of bee. The dance tells other bees the direction and distance of a food source. Plot points to graph the path of this bee's dance. (For more information about this topic, see the Group Research Project 1 at the end of this chapter.)

Level 3 Theory

64. a. What is the distance between the polar-form points $\left(3, \frac{\pi}{3}\right)$ and $\left(7, \frac{\pi}{4}\right)$? Explain why you cannot use the distance formula for these ordered pairs.
  b. What is the distance between the polar-form points (r_1, θ_1) and (r_2, θ_2)?

65. Use the result of Problem 64b to find an equation for a circle of radius a and polar-form center (R, α).

66. Show that the graph of the polar equation $r = a \cos \theta + b \sin \theta$ is a circle. Find its center and radius.

SECTION 6.4

GRAPHING IN POLAR COORDINATES

PROBLEM OF THE DAY

What is an equation for the curve shown in Figure 6.11?

How would you answer this question?

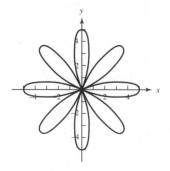

FIGURE 6.11
Problem of the Day

We first note that the curve does not represent a function and also that it does not look like any of the curves we have yet graphed. Since it seems periodic, it appears to be related to some sort of trigonometric curve. The topic of this section is to find equations for graphs such as this, as well as to graph similar equations by simply looking at the equation.

In the previous section, we found that the graph of the polar equation $r = a$ is a circle centered at the origin with radius a, and $\theta = a$ is a line passing through the origin. In this section, we discuss techniques for graphing polar-form equations and categorize several important types of polar-form curves.

Graphing by Plotting Points

We have already examined polar forms for lines and circles; in this section we shall examine curves that are more easily represented in polar coordinates than in rectangular coordinates. We begin with a simple **spiral.**

EXAMPLE 1

Finding a Spiral by Plotting Points

Graph $r = \theta$ for $\theta \geq 0$.

Solution

Set up a table of values.

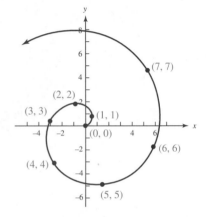

θ	r
0	0
1	1
2	2
3	3
4	4
5	5
6	6

Choose a θ, and then find a corresponding r so that (r, θ) satisfies the equation. Plot each of these points and connect them, as shown in Figure 6.12.*

FIGURE 6.12
Graph of $r = \theta$

Notice that as θ increases, r must also increase.

In Example 1, the polar equation $r = \theta$ has exactly one value of r for each value of θ. Thus, the relationship given by $r = \theta$ is a function of θ, and we can write the polar form $r = f(\theta)$, where $f(\theta) = \theta$. A function of the form $r = f(\theta)$, where θ is a polar angle and r is the corresponding radial distance is called a **polar function.**

Cardioids

Next, we examine a class of polar curves called **cardioids** because of their heartlike shape.

---

* Many books use *polar graph paper*, but such paper is not really necessary. It also obscures the fact that polar curves and rectangular curves are both plotted on a Cartesian coordinate system, only with a different meaning attached to the ordered pairs. You can estimate the angles as necessary without polar graph paper. Remember, just as when graphing rectangular curves, the key is not in plotting many points, but in recognizing the type of curve and then plotting a few key points.

EXAMPLE 2

Finding a Cardioid by Plotting Points

Graph $r = 2(1 - \cos \theta)$.

Solution   Construct a table of values by choosing values for θ and approximating the corresponding values for r. Do not forget that even though we pick θ and find a value for r, we still represent the table values as ordered pairs of the form (r, θ).

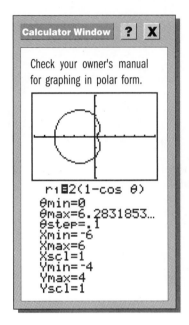

Calculator Window

Check your owner's manual for graphing in polar form.

```
r₁▤2(1-cos θ)
θmin=0
θmax=6.2831853…
θstep=.1
Xmin=-6
Xmax=6
Xscl=1
Ymin=-4
Ymax=4
Yscl=1
```

θ	r
0	0
1	0.9193954
2	2.832294
3	3.979985
4	3.307287
5	1.432676
6	0.079659

The points are connected as shown in Figure 6.13.

FIGURE 6.13
Graph of $r = 2(1 - \cos \theta)$

The general form for a cardioid in standard position is given in the following box.

STANDARD-POSITION CARDIOID

$r = a(1 - \cos \theta)$

In general, a cardioid in standard position can be completely determined by plotting four particular points:

θ	$r = a(1 - \cos \theta)$
0	$r = a(1 - \cos 0) = a(1 - 1) = 0$
$\frac{\pi}{2}$	$r = a\left(1 - \cos \frac{\pi}{2}\right) = a(1 - 0) = a$
π	$r = a(1 - \cos \pi) = a(1 + 1) = 2a$
$\frac{3\pi}{2}$	$r = a\left(1 - \cos \frac{3\pi}{2}\right) = a(1 - 0) = a$

These reference points are all you need when graphing other standard-position cardioids, because they will have the same shape as the one shown in Figure 6.13.

Symmetry and Rotations

When sketching a polar graph, it is often useful to determine whether the graph has been rotated or whether it has any **polar symmetry.**

If an angle α is subtracted from θ in a polar-form equation, it has the effect of rotating the curve (see Problem 73).

> **ROTATION OF POLAR-FORM GRAPHS**
>
> The polar graph of $r = f(\theta - \alpha)$ is the same as the polar graph of $r = f(\theta)$, only rotated through an angle α. If α is positive, the **rotation** is counterclockwise, and if α is negative, then the rotation is clockwise.

EXAMPLE 3   **Rotated Cardioid**

Graph $r = 3 - 3 \cos\left(\theta - \frac{\pi}{6}\right)$.

Solution   Recognize this as a cardioid with $a = 3$ and a rotation of $\frac{\pi}{6}$. Plot the four points shown in Figure 6.14 and draw the cardioid.

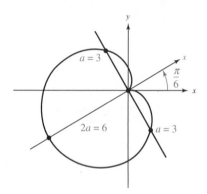

FIGURE 6.14   Graph of $r = 3 - 3 \cos\left(\theta - \frac{\pi}{6}\right)$ ■

If the rotation is $180° = \pi$, the equation simplifies considerably. Consider

$$r = 3 - 3 \cos(\theta - \pi) \qquad \textit{Cardioid with 180° rotation}$$
$$= 3 - 3[\cos\theta \cos\pi + \sin\theta \sin\pi] \qquad \textit{Addition law cosine}$$
$$= 3 - 3[\cos\theta\,(-1) + \sin\theta\,(0)] \qquad \textit{Substitute exact values.}$$
$$= 3 - 3(-1)\cos\theta \qquad \textit{Simplify.}$$
$$= 3 + 3 \cos\theta$$

Compare this with Example 3 and you will see that the only difference is a 180° rotation instead of a 30° rotation. This means that, whenever you graph an equation of the form

$$r = a(1 + \cos \theta)$$ *Note the plus sign.*

instead of $r = a(1 - \cos \theta)$, it is a standard-form cardioid with a 180° rotation. Similarly,

$r = a(1 - \sin \theta)$ is a standard-form cardioid with a 90° rotation.
$r = a(1 + \sin \theta)$ is a standard-form cardioid with a 270° rotation.

These curves, along with other polar-form curves, are graphed and summarized in Table 6.1 on pages 319 and 320.

The cardioid is only one of the polar-form curves we will consider. However, before we sketch other curves, we will consider symmetry. There are three important kinds of polar symmetry, which are described in the following box and shown in Figure 6.15.

SYMMETRY IN THE GRAPH OF THE POLAR FUNCTION $r = f(\theta)$

A polar-form graph $r = f(\theta)$ is symmetric with respect to . . .	. . . if the equation $r = f(\theta)$ is unchanged when (r, θ) is replaced by . . .
x-axis	$(r, -\theta)$
y-axis	$(r, \pi - \theta)$
origin	$(-r, \theta)$

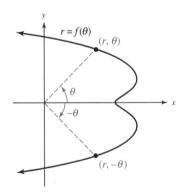

a. Symmetry with respect to the x-axis

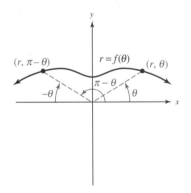

b. Symmetry with respect to the y-axis

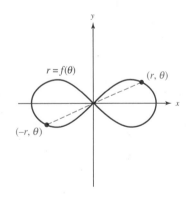

c. Symmetry with respect to the origin

FIGURE 6.15 Symmetry in polar form

Limaçons

We will illustrate symmetry in polar-form curves by graphing a curve called a *limaçon*.

EXAMPLE 4

Graphing a Limaçon Using Symmetry

Graph $r = 3 + 2 \cos \theta$.

Solution   Let $f(\theta) = 3 + 2 \cos \theta$.

Symmetry with respect to the x-axis

$$f(-\theta) = 3 + 2 \cos(-\theta) = 3 + 2 \cos \theta = f(\theta)$$

Yes; it is symmetric, so it is enough to graph f for θ between 0 and π.

Symmetry with respect to the y-axis

$$\begin{aligned} f(\pi - \theta) &= 3 + 2 \cos(\pi - \theta) \\ &= 3 + 2(\cos \pi \cos \theta + \sin \pi \sin \theta) \\ &= 3 - 2 \cos \theta \end{aligned}$$

After checking the other primary representation, we find that it is not symmetric with respect to the y-axis.

Symmetry with respect to the origin

$$-r \neq f(\theta) \quad \text{and} \quad r \neq f(\theta + \pi)$$

Thus, the graph is not symmetric with respect to the origin.

The graph is shown in Figure 6.16. Note that we first sketch the top half of the graph (for $0 \le \theta \le \pi$) by plotting points and then complete the sketch by reflecting the graph in the x-axis. Because $\cos \theta$ steadily decreases from its largest value of 1 at $\theta = 0$ to its smallest value -1 at $\theta = \pi$, the radial distance $r = 3 + 2 \cos \theta$ will also steadily decrease as θ increases from 0 to π. The largest value of r is $r = 3 + 2(1) = 5$ at $\theta = 0$, and its smallest value is $r = 3 + 2(-1) = 1$ at $\theta = \pi$.

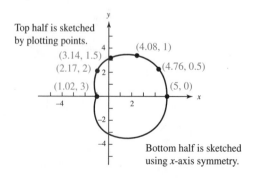

FIGURE 6.16   Graph of $r = 3 + 2 \cos \theta$

The graph of any polar equation of the general form

$$r = b \pm a \cos \theta \qquad \text{or} \qquad r = b \pm a \sin \theta$$

is called a **limaçon** (derived from the Latin word "limax," which means "slug"). The special case where $a = b$ is the *cardioid.* Figure 6.17 shows four different kinds of limaçons that can occur. Note how the appearance of the graph depends on the ratio b/a. We have discussed cases II and III in Examples 2 and 4. Case I (the "inner loop" case) and case IV (the "convex") case are examined in the problem set.

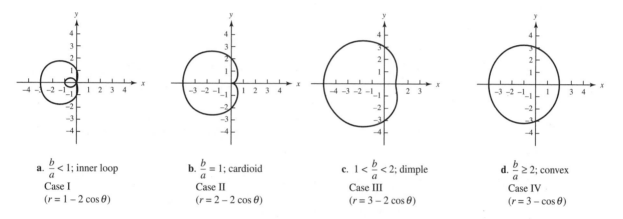

a. $\frac{b}{a} < 1$; inner loop
Case I
$(r = 1 - 2 \cos \theta)$

b. $\frac{b}{a} = 1$; cardioid
Case II
$(r = 2 - 2 \cos \theta)$

c. $1 < \frac{b}{a} < 2$; dimple
Case III
$(r = 3 - 2 \cos \theta)$

d. $\frac{b}{a} \geq 2$; convex
Case IV
$(r = 3 - \cos \theta)$

FIGURE 6.17 Limaçons: $r = b \pm a \cos \theta$ or $r = b \pm a \sin \theta$

Just as with the cardioid, we designate a standard-form limaçon and consider the others as rotations:

$r = b - a \cos \theta$	**Standard form**
$r = b - a \sin \theta$	90° rotation
$r = b + a \cos \theta$	180° rotation
$r = b + a \sin \theta$	270° rotation

STANDARD-POSITION LIMAÇON

$r = b - a \cos \theta$

Rose Curves

There are several polar-form curves known as **rose curves,** which consist of several loops, called *leaves* or *petals.*

EXAMPLE 5   **Graphing a Four-Leaved Rose**

Graph $r = 4 \cos 2\theta$.

Solution   Let $f(\theta) = 4 \cos 2\theta$.

Symmetry with respect to the x-axis

$$f(-\theta) = 4 \cos 2(-\theta) = 4 \cos 2\theta = f(\theta); \text{ Yes, symmetric}$$

Symmetry with respect to the y-axis

$$f(\pi - \theta) = 4 \cos 2(\pi - \theta) = 4 \cos(2\pi - 2\theta) = 4 \cos(-2\theta) = 4 \cos 2\theta$$
$$= f(\theta); \text{ Yes, symmetric}$$

Symmetry with respect to the origin

If $-r = 4 \cos 2\theta$, the the other primary representation is

$$r = 4 \cos 2(\theta + \pi) = 4 \cos(2\theta + 2\pi) = 4 \cos 2\theta; \text{ Yes, symmetric}$$

Because of this symmetry, we shall sketch the graph of $r = f(\theta)$ for θ between 0 and $\frac{\pi}{2}$, and then use symmetry to complete the graph.

θ	$f(\theta)$
0	4
0.2	3.684244
0.4	2.786827
0.6	1.449431
0.8	−0.1167981
1	−1.664587
1.2	−2.949575
1.3	−3.427555
1.4	−3.768889

Keep plotting points until you are satisfied that you have a reasonable representation for the graph. After working through Example 5, you will be able to do other rose curves more easily.

When $\theta = 0$, $r = 4$, and as θ increases from 0 to $\frac{\pi}{4}$, the radial distance r decreases from 4 to 0. Then, as θ increases from $\frac{\pi}{4}$ to $\frac{\pi}{2}$, r becomes negative and decreases from 0 to −4. A table of values is given in the margin.

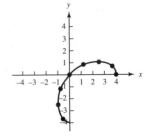

The next part of the graph is obtained by first reflecting in the y-axis.

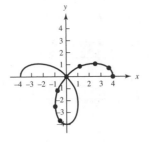

The last part is found by reflecting in the x-axis.

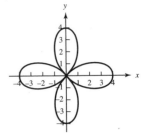

In general, $r = a \cos n\theta$ is the equation of a rose curve in which each petal has length a. If n is an even number, the rose has $2n$ petals; if n is odd, the number of petals is n. The tips of the petals are equally spaced on a circle of radius a. Equations of the form $r = a \sin n\theta$ are handled as rotations.

> **STANDARD-POSITION ROSE CURVE**
>
> $r = a \cos n\theta$

EXAMPLE 6

Graphing a Rose Curve with 8 Petals and a Rotation

Graph $r = 5 \sin 4\theta$.

Solution   We begin by finding the amount of rotation.

$$r = 5 \sin 4\theta$$
$$= 5 \cos\left(\tfrac{\pi}{2} - 4\theta\right) \quad \textit{Cofunctions of complementary angles}$$
$$= 5 \cos\left(4\theta - \tfrac{\pi}{2}\right) \quad \textit{Remember, } \cos(-\theta) = \cos\theta.$$
$$= 5 \cos 4\left(\theta - \tfrac{\pi}{8}\right) \quad \textit{Common factor of } 4$$

Recognize this as a rose curve rotated $\pi/8$. There are $2(4) = 8$ petals of length 5. The petals are a distance of $\pi/4$ (one revolution = 2π divided by the number of petals) apart. The graph is shown in Figure 6.18. ■

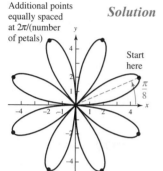

Additional points equally spaced at $2\pi/$(number of petals)

Start here

$\frac{\pi}{8}$

FIGURE 6.18
Graph of $r = 5 \sin 4\theta$

We now return to the Problem of the Day.

EXAMPLE 7

Problem of the Day: Finding the Equation of a Polar-Form Curve

What is the equation of the curve shown in Figure 6.19?

Solution   By comparing this graph with the one we did for Example 6, we see that the curve is a rose curve with 8 leaves, and since 8 is an even number, we know the equation has the form

$$r = a \cos n\theta$$

where $n = 4$. Furthermore, we see that the length of each leaf is 5 units long, so $a = 5$. Thus, the desired equation is

$$r = 5 \cos 4\theta$$ ■

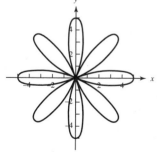

FIGURE 6.19
What is the equation?

Lemniscates

The last general type of polar-form curve we will consider is called a **lemniscate.**

> **STANDARD-POSITION LEMNISCATE**
>
> $r^2 = a^2 \cos 2\theta$

EXAMPLE 8   **Graphing a Lemniscate**

Graph $r^2 = 9 \cos 2\theta$.

Solution   As before, when graphing a curve for the first time, begin by checking symmetry and plotting points. For this example, note that you obtain two values for r when solving this quadratic equation. For example, if $\theta = 0$, then $\cos 2\theta = 1$ and $r^2 = 9$, so $r = 3$ or -3.

Symmetry with respect to the x-axis:

$9 \cos[2(-\theta)] = 9 \cos 2\theta$, so $r^2 = 9 \cos 2\theta$ is not affected when θ is replaced by $-\theta$; yes, symmetric

Symmetry with respect to the y-axis:

$$9 \cos[2(\pi - \theta)] = 9 \cos(2\pi - 2\theta)$$
$$= 9 \cos(-2\theta) = 9 \cos 2\theta; \text{ yes, symmetric}$$

Symmetry with respect to the origin:

$(-r)^2 = r^2$, so $r^2 = 9 \cos 2\theta$ is not affected when r is replaced by $-r$; yes, symmetric

Note that because $r^2 \geq 0$, the function $\cos 2\theta$ is defined only for those values of θ for which $\cos 2\theta \geq 0$; that is,

$$-\frac{\pi}{4} \leq \theta \leq \frac{\pi}{4}; \qquad \frac{3\pi}{4} \leq \theta \leq \frac{5\pi}{4}; \qquad \cdots$$

We begin by restricting our attention to the interval $0 \leq \theta \leq \frac{\pi}{4}$. Note that $\cos 2\theta$ decreases steadily from 3 to 0 as θ varies from 0 to $\frac{\pi}{4}$.

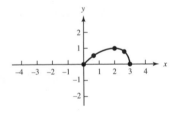

The second step is to use symmetry to reflect the curve in the x-axis.

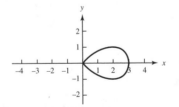

Finally, obtain the rest of the graph by reflecting the curve in the y-axis.

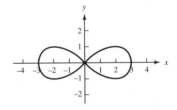

Summary of Polar-form Curves

We conclude this section by summarizing the special types of polar-form curves we have examined. There are many others, some of which are represented in the problem set.

TABLE 6.1 Directory of Polar-form Curves

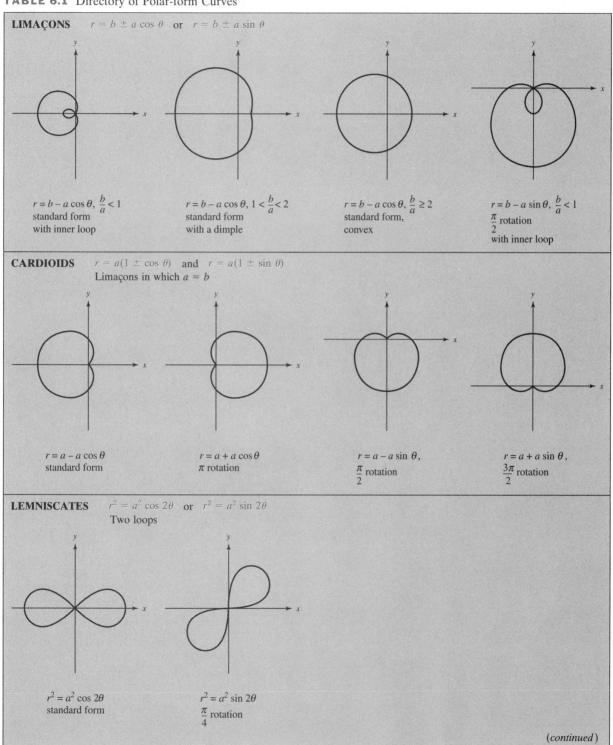

LIMAÇONS $r = b \pm a \cos \theta$ or $r = b \pm a \sin \theta$

$r = b - a \cos \theta, \frac{b}{a} < 1$
standard form
with inner loop

$r = b - a \cos \theta, 1 < \frac{b}{a} < 2$
standard form
with a dimple

$r = b - a \cos \theta, \frac{b}{a} \geq 2$
standard form,
convex

$r = b - a \sin \theta, \frac{b}{a} < 1$
$\frac{\pi}{2}$ rotation
with inner loop

CARDIOIDS $r = a(1 \pm \cos \theta)$ and $r = a(1 \pm \sin \theta)$
Limaçons in which $a = b$

$r = a - a \cos \theta$
standard form

$r = a + a \cos \theta$
π rotation

$r = a - a \sin \theta,$
$\frac{\pi}{2}$ rotation

$r = a + a \sin \theta,$
$\frac{3\pi}{2}$ rotation

LEMNISCATES $r^2 = a^2 \cos 2\theta$ or $r^2 = a^2 \sin 2\theta$
Two loops

$r^2 = a^2 \cos 2\theta$
standard form

$r^2 = a^2 \sin 2\theta$
$\frac{\pi}{4}$ rotation

(continued)

TABLE 6.1 *Continued*

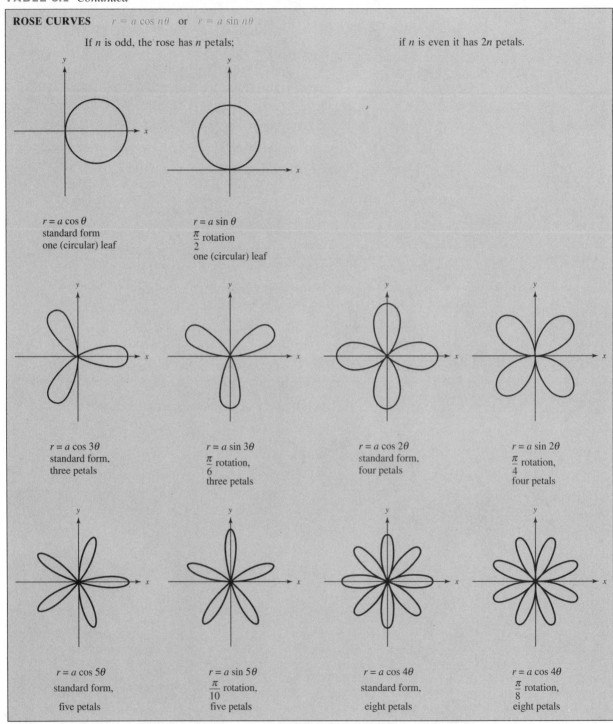

ROSE CURVES $r = a \cos n\theta$ or $r = a \sin n\theta$

If n is odd, the rose has n petals;   if n is even it has $2n$ petals.

$r = a \cos \theta$
standard form
one (circular) leaf

$r = a \sin \theta$
$\dfrac{\pi}{2}$ rotation
one (circular) leaf

$r = a \cos 3\theta$
standard form,
three petals

$r = a \sin 3\theta$
$\dfrac{\pi}{6}$ rotation,
three petals

$r = a \cos 2\theta$
standard form,
four petals

$r = a \sin 2\theta$
$\dfrac{\pi}{4}$ rotation,
four petals

$r = a \cos 5\theta$
standard form,
five petals

$r = a \sin 5\theta$
$\dfrac{\pi}{10}$ rotation,
five petals

$r = a \cos 4\theta$
standard form,
eight petals

$r = a \cos 4\theta$
$\dfrac{\pi}{8}$ rotation,
eight petals

PROBLEM SET 6.4

Essential Concepts

1. **IN YOUR OWN WORDS** Describe a procedure for graphing polar-form curves.

2. **IN YOUR OWN WORDS** Discuss symmetry in the graph of a polar-form function.

3. **IN YOUR OWN WORDS** Compare and contrast the forms of the equation for limaçons, cardioids, rose curves, and lemniscates.

4. Identify each curve as a cardioid, rose curve (state number of petals), lemniscate, limaçon, circle, line, or none of the above.
   a. $r^2 = 9 \cos 2\theta$ **b.** $r = 2 \sin \frac{\pi}{6}$
   c. $r = 3 \sin 3\theta$ **d.** $r = 3\theta$
   e. $r = 2 - 2 \cos \theta$ **f.** $\theta = \frac{\pi}{6}$
   g. $r^2 = \sin 2\theta$ **h.** $r - 2 = 4 \cos \theta$

5. Identify each curve as a cardioid, rose curve (state number of petals), lemniscate, limaçon, circle, line, or none of the above.
   a. $r = 2 \sin 2\theta$ **b.** $r^2 = 2 \cos 2\theta$
   c. $r = 5 \cos 60°$ **d.** $r = 5 \sin 8\theta$
   e. $r\theta = 3$ **f.** $r^2 = 9 \cos\left(2\theta - \frac{\pi}{4}\right)$
   g. $r = \sin 3\left(\theta + \frac{\pi}{6}\right)$ **h.** $\cos \theta = 1 - r$

6. Identify each curve as a cardioid, rose curve (state number of petals), lemniscate, limaçon, circle, line, or none of the above.
   a. $r = 2 \cos 2\theta$ **b.** $r = 4 \sin 30°$
   c. $r + 2 = 3 \sin \theta$ **d.** $r + 3 = 3 \sin \theta$
   e. $\theta = 4$ **f.** $\theta = \tan \frac{\pi}{4}$
   g. $r = 3 \cos 5\theta$ **h.** $r \cos \theta = 2$

Level 1 Drill

Problems 7–38: *Graph the polar-form curves.*

7. $r = 3, 0 \le \theta \le \frac{\pi}{2}$ **8.** $r = -1, 0 \le \theta \le \pi$

9. $\theta = -\frac{\pi}{4}, 0 \le r \le 3$ **10.** $\theta = \frac{\pi}{4}, 1 \le r \le 2$

11. $r = \theta + 1, 0 \le \theta \le \pi$ **12.** $r = \theta - 1, 0 \le \theta \le \pi$

13. $r = 2\theta, \theta \ge 0$ **14.** $r = \frac{\theta}{2}, \theta \ge 0$

15. $r = 2 \cos 2\theta$ **16.** $r = 3 \cos 3\theta$

17. $r = 5 \sin 3\theta$ **18.** $r^2 = 9 \cos 2\theta$

19. $r^2 = 16 \cos 2\theta$ **20.** $r^2 = 9 \sin 2\theta$

21. $r = 3 \cos 3\left(\theta - \frac{\pi}{3}\right)$ **22.** $r = 2 \cos 2\left(\theta + \frac{\pi}{3}\right)$

23. $r = 5 \cos 3\left(\theta - \frac{\pi}{4}\right)$ **24.** $r = \sin 3\left(\theta + \frac{\pi}{6}\right)$

25. $r = \sin\left(2\theta + \frac{2\pi}{3}\right)$ **26.** $r = \cos\left(2\theta + \frac{\pi}{3}\right)$

27. $r^2 = 16 \cos\left(2\theta - \frac{\pi}{3}\right)$ **28.** $r^2 = 9 \cos\left(2\theta - \frac{\pi}{3}\right)$

29. $r = 2 + \cos \theta$ **30.** $r = 3 + \sin \theta$

31. $r = 1 + \sin \theta$ **32.** $r = 1 + \cos \theta$

33. $r \cos \theta = 1$ **34.** $r \sin \theta = 3$

35. $r = 1 + 3 \cos \theta$ **36.** $r = 1 + 2 \sin \theta$

37. $r = -2 \sin \theta$ **38.** $r^2 = \cos 2\theta$

Level 2 Drill

Problems 39–44: *Sketch the graph of the polar functions.*

39. $f(\theta) = \sin 2\theta, 0 \le \theta \le \frac{\pi}{2}$; rose petal

40. $f(\theta) = |\sin 2\theta|, 0 \le \theta \le 2\pi$; four-leaved rose

41. $f(\theta) = 2|\cos \theta|, 0 \le \theta \le 2\pi$

42. $f(\theta) = 4|\sin \theta|, 0 \le \theta \le 2\pi$

43. $f(\theta) = \sqrt{|\cos \theta|}, 0 \le \theta \le 2\pi$; lazy eight

44. $f(\theta) = \sqrt{\cos 2\theta}, 0 \le \theta \le 2\pi$; lemniscate

Problems 45–54: *Graph the set of points (r, θ) so that the inequalities are satisfied.*

45. $0 \le r \le 1, 0 \le \theta \le 2\pi$

46. $2 \le r \le 3, 0 \le \theta < 2\pi$

47. $0 \le r < 4, 0 \le \theta \le \frac{\pi}{2}$

48. $0 \le r \le 4, 0 \le \theta \le \pi$

49. $r > 1, 0 \le \theta < 2\pi$

50. $r \ge 2, \frac{\pi}{2} \le \theta \le \pi$

51. $0 \le \theta \le \frac{\pi}{4}, r \ge 0$

52. $\pi \le \theta \le \frac{5\pi}{4}, r \ge 0$

53. $0 \le \theta \le \frac{\pi}{4}, 1 \le r \le 2$

54. $0 \le \theta \le \frac{\pi}{4}, r > 1$

55. Show that the polar equations
$$r = \cos \theta + 1 \quad \text{and} \quad r = \cos \theta - 1$$
have the same graph in the *xy*-plane.

56. **Spirals** are interesting mathematical curves. There are three special types of spirals:
   a. A *spiral of Archimedes* has the form $r = a\theta$; graph $r = 2\theta$.

b. A *hyperbolic spiral* has the form $r\theta = a$; graph $r\theta = 2$.

c. A *logarithmic spiral* has the form $r = a^{k\theta}$; graph $r = 2^{\theta}$.

57. The **strophoid** is a curve of the form

$$r = a \cos 2\theta \sec \theta$$

Graph this curve where $a = 2$.

58. The **bifolium** has the form

$$r = a \sin \theta \cos^2\theta$$

Graph this curve where $a = 1$.

59. The **folium of Descartes** has the form

$$r = \frac{3a \sin \theta \cos \theta}{\sin^3\theta + \cos^3\theta}$$

Graph this curve where $a = 2$.

60. The curve known as the **ovals of Cassini** has the form

$$r^4 + b^4 - 2b^2r^2 \cos 2\theta = k^4$$

Graph this curve where $b = 2$, $k = 3$.

Problems 61–66: Graph the given pair of curves on the same coordinate axes. The first equation uses (x, y) as rectangular coordinates and the second uses (r, θ) as polar coordinates.

61. $y = \cos x$ and $r = \cos \theta$
62. $y = \sin x$ and $r = \sin \theta$
63. $y = \tan x$ and $r = \tan \theta$
64. $y = \sec x$ and $r = \sec \theta$
65. $y = \csc x$ and $r = \csc \theta$
66. $y = \cot x$ and $r = \cot \theta$

Level 3 Theory

67. **INDIVIDUAL RESEARCH PROJECT** In this problem we investigate the periodicity of graphs of the form

$$r = f(\theta) = \sin m\theta$$

for m a positive integer.

a. Consider $r = f(\theta) = \sin 2\theta$. Since $\sin 2\theta$ has period π, one would think that surely the graph repeats itself for $\theta > \pi$. Graphically, show that this is *not* the case. Verify that it is necessary for $0 \le \theta \le 2\pi$ to obtain the entire graph. Describe the graph.

b. Explain *why* the interval $0 \le \theta \le \pi$ is insufficient.

c. Now systematically study $r = f(\theta) = \sin mx$ for m a positive integer. In particular, describe *why m* being even or odd makes a fundamental difference. Summarize the number of leaves on the roses relative to m.

68. **INDIVIDUAL RESEARCH PROJECT** In this problem we investigate the periodicity of graphs of the form

$$r = f(\theta) = \sin\left(\frac{m\theta}{n}\right)$$

where m and n are positive integers (and relatively prime). Let P denote the period of the sine.

a. First set $m = 1$ and study the behavior of the graphs for $n = 1, 2, 3, \ldots$. Clearly the period of the sine function is $P = 2n\pi$. But what θ interval is required to get the entire graph? Attempt to explain this. Hand in your favorite graph.

b. Now do a study for $m = 1, 2, \ldots$ and $n = 1, 2, \ldots$. Attempt to generalize the situation regarding the period of the sine functions, the necessary θ interval, and the number of loops in the graphs. Hand in your favorite graph.

c. A challenge: Attempt to carefully explain the necessary θ interval in part **b**.

69. Show that the curve $r = f(\theta)$ is symmetric with respect to the x-axis if the equation is unaffected when r is replaced by $-r$ and θ is replced by $\pi - \theta$.

70. Show that the curve $r = f(\theta)$ is symmetric with respect to the y-axis if the equation is unaffected when r is replaced by $-r$ and θ is replaced by $-\theta$.

71. Show that the curve $r = f(\theta)$ is symmetric in the line $y = x$ if the equation is unaffected when θ is replaced by $\frac{\pi}{2} - \theta$.

72. State and prove a result regarding the relationship of the graph of $r = f(\theta)$ to that of $r = f(\theta - \theta_0)$ for $0 < \theta_0 < \frac{\pi}{2}$.

73. **a.** Show that if the polar curve $r = f(\theta)$ is rotated about the pole through an angle α, the equation for the new curve is $r = f(\theta - \alpha)$.

b. Use a rotation to sketch $r = 2 \sec\left(\theta - \frac{\pi}{3}\right)$.

74. Sketch the graph of

$$r = \frac{\theta}{\cos \theta} \quad \text{for } 0 \le \theta \le \frac{\pi}{2}$$

In particular, show that the graph has a vertical asymptote at $\theta = \frac{\pi}{2}$.

6.5 CHAPTER SIX SUMMARY

Objectives

Section 6.1	The Imaginary Unit and Complex Numbers
Objective 6.1	Define the imaginary unit as well as a complex number.
Objective 6.2	Carry out operations (addition, subtraction, multiplication, and division) with complex numbers.

Section 6.2	De Moivre's Theorem
Objective 6.3	Find the absolute value of a complex number and plot that number in the Gaussian plane.
Objective 6.4	Change rectangular-form complex numbers to trigonometric form.
Objective 6.5	Change trigonometric-form complex numbers to rectangular form.
Objective 6.6	Multiply and divide complex numbers.
Objective 6.7	State De Moivre's theorem and the nth root theorem.
Objective 6.8	Find all roots of a complex number.
Objective 6.9	Solve equations for all complex roots.

Section 6.3	Polar Coordinates
Objective 6.10	Plot points in polar form.
Objective 6.11	Write the two primary representations in polar form, given a point in either rectangular form or polar form.
Objective 6.12	Change from rectangular to polar form and from polar to rectangular form.

Section 6.4	Graphing in Polar Coordinates
Objective 6.13	Sketch polar-form curves.
Objective 6.14	Graph cardioids by rotating and comparing to a standard-position cardioid.
Objective 6.15	Recognize and graph limaçons.
Objective 6.16	Recognize and graph rose curves.
Objective 6.17	Recognize and graph lemniscates.
Objective 6.18	Identify cardioids, limaçons, rose curves, and lemniscates by looking at the equation. Also, be able to determine the rotation, if any.

Terms

Absolute value [6.2]
Argument [6.2]
Cardioid [6.4]
Complex conjugates [6.1]
Complex number [6.1]
Complex plane [6.2]
De Moivre's theorem [6.2]
Equality of complex numbers [6.1]
Gaussian plane [6.2]

i [6.1]
Imaginary axis [6.2]
Imaginary unit [6.1]
Lemniscate [6.4]
Limaçon [6.4]
Modulus [6.2]
nth root theorem [6.2]
Polar angle [6.3]
Polar coordinate system [6.3]
Polar function [6.4]

Polar symmetry [6.4]
Pole [6.3]
Primary representation of a point [6.3]
Principal nth root [6.2]
Pure imaginary number [6.1]
Radial axis [6.3]
Radial distance [6.3]
Real axis [6.2]
Real number [6.1]

Rectangular coordinates [6.3]
Rectangular form [6.2]
Rose curve [6.4]
Rotation [6.4]
Simplified complex number [6.1]
Spiral [6.4]
Trigonometric form [6.2]

Sample Test

All of the answers for this sample test are given in the back of the book.

1. Simplify:
  a. $(6 + 3i) + (-4 + 2i)$
  b. $(2 - 5i) - (3 - 2i)$
  c. $2i(-1 + i)(-2 + 2i)$
  d. $(2 - 5i)^2$
  e. $\dfrac{2 + 3i}{1 - i}$

2. Simplify:
  a. $\dfrac{2 \operatorname{cis} 158° \cdot 4 \operatorname{cis} 212°}{(2 \operatorname{cis} 312°)^3}$
  (Give the answer in trigonometric form.)
  b. $(\sqrt{12} - 2i)^4$ (Give the answer in rectangular form.)

3. Change to trigonometric form:
  a. $7 - 7i$ **b.** $-3i$
  Change to rectangular form:
  c. $2 \operatorname{cis} 150°$ **d.** $4\left(\cos \frac{7\pi}{4} + i \sin \frac{7\pi}{4}\right)$

4. Find the indicated roots and represent them in trigonometric form, in rectangular form, and graphically.
  a. Square roots of $\frac{7}{2}\sqrt{3} - \frac{7}{2}i$
  b. Fourth roots of 1

5. Solve $x^3 - 27 = 0$.

6. Give both primary representations in polar form for the given points.
  a. $\left(3, -\frac{2\pi}{3}\right)$ is in polar form.
  b. $\left(-5, -\frac{\pi}{10}\right)$ is in polar form.
  c. $(-3, 3\sqrt{3})$ is in rectangular form.
  d. $(2, 5)$ is in rectangular form.

7. Determine whether the given points lie on the curve defined by the equation
$$r = \frac{10}{2 + \cos \theta}$$
  a. $\left(4, \frac{\pi}{6}\right)$ **b.** $\left(-4, \frac{4\pi}{3}\right)$ **c.** $\left(5, \frac{3\pi}{2}\right)$ **d.** $(5, \pi)$

8. Identify each curve.
  a. $r^2 = 9 \cos 2\theta$ **b.** $r = 2 + 2 \sin \theta$
  c. $r = 4 \cos 5\theta$ **d.** $r = 5 \sin \frac{\pi}{4}$

9. Graph
  a. $r - 3\theta = 0$, for $\theta > 0$ **b.** $\theta - 2 = 0$
  c. $r - 2 \tan \frac{3\pi}{4} = 0$ **d.** $r + 3 \sin \theta = 3$

10. Graph
  a. $r = 6 \cos 3\left(\theta - \frac{\pi}{10}\right)$ **b.** $r^2 = 4 \sin 2\theta$

Discussion Problems

1. What is a number? Discuss different sets of numbers used in mathematics.

2. What is a complex number? Find some examples of complex numbers used in disciplines other than mathematics.

3. Why is De Moivre's theorem important?

4. Write a 500-word paper on trigonometry and astronomy.

5. Describe the procedure you use to graph a polar-form curve.

6. Outline your "method of attack" for graphing polar-form curves.

7. Write a 500-word paper extending the rectangular and polar coordinate systems to three dimensions. Are there other three-dimensional coordinate systems?

Miscellaneous Problems—Chapters 1–6

1. From memory, draw a quick sketch of each of the following. Label axes and scale.
   a. $y = \sin x$
   b. $y = \csc x$

2. From memory, draw a quick sketch of each of the following. Label axes and scale.
   a. $y = \cos x$
   b. $y = \cot x$

3. What is the equation of the graph you would use to draw the graph of $y = \sec \theta$?

4. What is the equation of the graph you would use to draw the graph of $y = \csc \theta$?

5. **IN YOUR OWN WORDS** Draw the graphs of $y = \cos x$; $y = \frac{1}{3} \cos x$; and $y = 3 \cos x$ on the same coordinate axes. Discuss.

6. **IN YOUR OWN WORDS** Draw the graphs of $y = \cos x$; $y = \sin \frac{1}{3} x$; and $y = \sin 3x$ on the same coordinate axes. Discuss.

Problems 7–10: Determine the equation of a standard cosine or standard sine curve.

7.

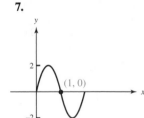

8.

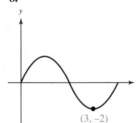

9.

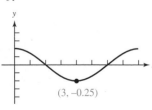

10.

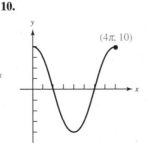

Problems 11–16: Prove that each equation is an identity.

11. $1 + \sin^2\theta = 2 - \cos^2\theta$

12. $\cos^4\theta - \sin^4\theta = \cos 2\theta$

13. $\dfrac{(\sec^2\theta + \tan^2\theta)^2}{\sec^4\theta - \tan^4\theta} = 1 + 2\tan^2\theta$

14. $\dfrac{(\cos^2\theta - \sin^2\theta)^2}{\cos^4\theta - \sin^4\theta} = 2\cos^2\theta - 1$

15. $\sec(-\theta) = \sec\theta$

16. $\cos(\alpha - \beta)\cos\beta - \sin(\alpha - \beta)\sin\beta = \cos\alpha$

Problems 17–20: Solve the given equations for principal values correct to two decimal places.

17. $\cos^2 x - 3\sin x + 3 = 0$ (principal values)

18. $\sin\left(x + \frac{\pi}{6}\right) - \sin\left(x - \frac{\pi}{6}\right) = \frac{1}{2}$ (principal values)

19. $3\cos x - \cos 3x + 2\cos 5x = 1 + 3\cos x - \cos 3x$

20. $\sec^2 x - 1 = 3\tan x$

Problems 21–26: Change the rectangular-form complex numbers to trigonometric form.

21. $-\frac{7}{2}\sqrt{3} + \frac{7}{2}i$ 22. $2 + 3i$ 23. $16i$

24. $-1.5321 - 1.2856i$ 25. $-2 + 3i$ 26. -6

Problems 27–32: Change the trigonometric-form complex numbers to rectangular form.

27. $9\left(\cos\frac{3\pi}{4} + i\sin\frac{3\pi}{4}\right)$ 28. $2(\cos 3 + i\sin 3)$

29. $3\operatorname{cis} 240°$ 30. $5\operatorname{cis} 270°$

31. $5\operatorname{cis} 25°$ 32. $3\operatorname{cis}\frac{11\pi}{6}$

Problems 33–36: Simplify and leave your answer in the requested form.

33. $(\sqrt{3} - i)^6$; rectangular

34. $\dfrac{(3 + 3i)(\sqrt{3} - i)}{1 + i}$; rectangular

35. $\dfrac{2\operatorname{cis} 185° \cdot 4\operatorname{cis} 223°}{(2\operatorname{cis} 300°)^3}$; trigonometric

36. $2i(-1 + i)(-2 + 2i)$; trigonometric

Problems 37–42: Find the indicated roots. Leave your answers in trigonometric form.

37. cube roots of i 38. sixth roots of -64

39. tenth roots of 1   40. tenth roots of i

41. cube roots of $-16 + 16\sqrt{3}\,i$

42. cube roots of $4 \operatorname{cis} \frac{5\pi}{4}$

Problems 43–54: *Graph each curve by comparing it with standard forms.*

43. $r = \cos 2\theta$

44. $r = 2 \cos \theta$

45. $r = \cos 2$

46. $r^2 = \cos 2\theta$

47. $r = 2 - 2 \cos \theta$

48. $r = 2 + 2 \cos \theta$

49. $r = 4 \sin 3\theta$

50. $r = 4 \sin 3$

51. $r = 3 \cos 3\left(\theta - \frac{\pi}{6}\right)$

52. $r^2 = \cos 2\left(\theta - \frac{\pi}{6}\right)$

53. $r = \sin 3\left(\theta - \frac{\pi}{6}\right)$

54. $r^2 = 3 \sin 2\left(\theta - \frac{\pi}{6}\right)$

Problems 55–60: *Graph each curve by plotting points.*

55. $r = \tan 2\theta,\ 0 \le \theta \le \pi$

56. $r - 2\theta = 0,\ 0 \le \theta \le \pi$

57. $r\theta = 1,\ 0 \le \theta \le 2\pi$

58. $r = \dfrac{8}{4 - 2 \cos \theta},\ 0 \le \theta \le 2\pi$

59. $x = 5 \cos \theta,\ y = 3 \sin \theta$

60. $x = 4 \cos \theta,\ y = 4 \sin \theta$

Applications

61. The wave displayed on an oscilloscope for the sound of a tuning fork vibrating at f cycles per second with amplitude a is given by

$$y = a \sin 2\pi f x$$

For a vibrating string (as in a piano) of length L, tension T, and mass m per unit length, the value for f is

$$f = \frac{1}{2L}\sqrt{\frac{T}{m}}$$

For example, the frequency of middle C on a properly tuned piano is 261.626. Using this value for f and an amplitude of 8, we obtain the equation $y = 8 \sin 2\pi f x$. The first overtone for this note is $y = 4 \sin 4\pi f x$, producing the sound wave described by the equation

$$y = 8 \sin 2\pi f x + 4 \sin 4\pi f x$$

Graph this equation where $f = 261.626$.

62. Let R and r be the radii of the circumscribed and inscribed circles of $\triangle ABC$, respectively, as shown in Figure 6.20.

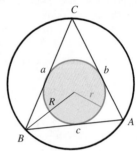

FIGURE 6.20
Given $\triangle ABC$ with inscribed circle (radius r) and circumscribed circle (radius R)

a. Show $R = \dfrac{a}{2 \sin \alpha} = \dfrac{b}{2 \sin \beta} = \dfrac{c}{2 \sin \gamma}$

b. Show $r = \sqrt{\dfrac{(s-a)(s-b)(s-c)}{2}}$

where $s = \frac{1}{2}(a + b + c)$

63. A hot-air balloon is released at noon and rises vertically at the rate of 4 m/s. An observation point A is situated 100 m from a point on the ground directly below the balloon, as shown in Figure 6.21. If t denotes the time (in seconds), write (as a function of time) the distance between the balloon and the observation point.

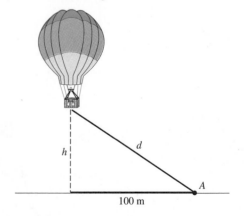

100 m

FIGURE 6.21 Distance as a function of time

Group Research Projects

Working in small groups is typical of most work environments, and this book seeks to develop skills with group activities. At the end of each chapter, we present a list of suggested projects, and even though they could be done as individual projects, we suggest that these projects be done in groups of three or four students.

1. Polar coordinates can be applied to the manner in which bees communicate the distance and source of a food supply. The dance is done on the vertical hanging honeycomb inside the hive (see Figure 6.22). The vertical direction represents the sun's direction, and the angle θ represents the direction of the food source. Write a paper describing how the bees communicate with this polar-form "waggle dance."

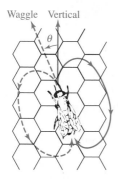

FIGURE 6.22 A bee's waggle dance

2. Suppose a light is attached to the edge of a bike wheel. The path of light is shown in Figure 6.23, and the curve traced by the light is a *cycloid.*

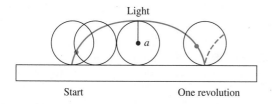

FIGURE 6.23 Side view of a light attached to a wheel

Let $P(x, y)$ be any point of the cycloid, and let θ be the amount of rotation of the circle, as shown in Figure 6.24.

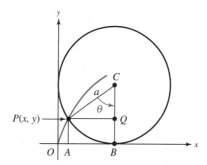

FIGURE 6.24 Figure for equation of a cycloid

Show that the parametric equations for the cycloid are

$$x = a\theta - a \sin \theta = a(\theta - \sin \theta)$$
$$y = a - a \cos \theta = a(1 - \cos \theta)$$

3. A single-arm drive for mounting a camera for time-exposure astrophotography used in astronomy is illustrated in Figure 6.25.

Single-arm drive

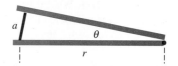

FIGURE 6.25 Single-arm drive

Show $\theta = 2 \arcsin\left(\dfrac{a}{2r}\right)$, and then find θ if $r = 13.519$, $a = 2.000$.

4. A double-arm drive for mounting a camera for time-exposure astrophotography used in astronomy is illustrated in Figure 6.26.

Double-arm drives

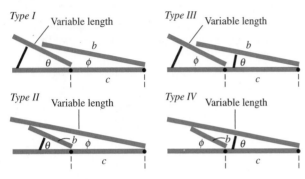

The angle β is opposite side b.

FIGURE 6.26 Double-arm drives

a. Show $\phi = \arcsin\left(\dfrac{\sin\theta}{\beta}\right)$ for type I.

b. Show $\phi = \arctan\left(\dfrac{\beta\sin\theta}{\beta\cos\theta + 1}\right)$ for type II.

c. Show $\phi = \arctan\left(\dfrac{\beta\sin\theta}{\beta\cos\theta - 1}\right)$ for type III.

d. Show $\phi = \theta + \arcsin\left(\dfrac{\sin\theta}{\beta}\right)$ for type IV.

e. If you were to construct a type III mounting so that $r = 13.519$, $\beta = 6.464$, $a = 2.000$, and $b = 12.928$, what is ϕ (measured in degrees)?

CUMULATIVE REVIEW CHAPTERS 4–6

Essential Concepts

1. IN YOUR OWN WORDS Distinguish a trigonometric equation from a trigonometric identity.

2. IN YOUR OWN WORDS Discuss some of the procedures and techniques you might use to solve a trigonometric equation.

3. IN YOUR OWN WORDS Discuss some of the procedures and techniques you might use to prove that a given equation is an identity.

4. IN YOUR OWN WORDS What is an oblique triangle? Draw a correctly labeled triangle. What is the sum of the measures of the angles in any triangle?

5. IN YOUR OWN WORDS What does it mean to "solve a triangle"?

6. IN YOUR OWN WORDS State the law of cosines. Under what conditions will you use the law of cosines to solve a triangle?

7. IN YOUR OWN WORDS State the law of sines. Under what conditions will you use the law of sines to solve a triangle?

8. IN YOUR OWN WORDS Outline a procedure for solving a triangle.

9. IN YOUR OWN WORDS What is a vector?

10. IN YOUR OWN WORDS How do you find a resultant vector?

11. IN YOUR OWN WORDS How do you resolve a vector?

12. IN YOUR OWN WORDS Describe the inclined-plane problem.

13. IN YOUR OWN WORDS What is the algebraic representation of a vector? What is the magnitude of a vector in algebraic form?

14. IN YOUR OWN WORDS What is scalar product and why is it important?

15. IN YOUR OWN WORDS What is the formula for the angle between vectors?

16. IN YOUR OWN WORDS What does it mean for vectors to be orthogonal, and how do you determine when they are orthogonal?

17. IN YOUR OWN WORDS What is the work done by a constant force?

18. IN YOUR OWN WORDS What is a complex number? Discuss both rectangular and trigonometric forms.

19. IN YOUR OWN WORDS Discuss the relationship between De Moivre's theorem and the nth root theorem.

20. IN YOUR OWN WORDS What are the primary representations of a polar-form point?

21. IN YOUR OWN WORDS What is the formula for the rotation of a polar-form curve?

Problems 22–37: *State the requested formula or equation.*

22. Pythagorean theorem

23. Area of a triangle (SAS)

24. Area of a triangle (ASA)

25. Area of a triangle (SSS)

26. Area of a sector

27. Volume of a cone

28. Angle between vectors

29. Orthogonal vectors **v** and **w**

30. Change from rectangular to polar form

31. Change from polar to rectangular form

32. Polar-form circle centered at the origin

33. Polar-form line through the origin

34. Cardioids

35. Limaçons

36. Rose curves

37. Lemniscates

Level 1 Drill

Problems 38–58: *Choose the **best** answer from the choices given.*

38. If $\dfrac{\cos^2\theta(\tan^2\theta + 1)}{\cot^2\theta}$ is changed to sines and cosines and then completely simplified, the result is
A. 1 B. $\cos^4\theta + \sin^2\theta$
C. $\dfrac{\cos^4\theta}{\sin^4\theta}$ D. $\dfrac{\sin^2\theta}{\cos^2\theta}$
E. none of these

39. $\dfrac{1}{1 + \sin 2\theta} + \dfrac{1}{1 - \sin 2\theta}$ is identical to
A. $2\sec^2 2\theta$ B. $2\csc^2 2\theta$
C. 1 D. $\dfrac{2}{\cos^4\theta}$
E. none of these

40. If $\cos^2\theta + \sin\theta = 0$, then a solution is
A. $\dfrac{1 \pm \sqrt{5}}{2}$ B. 3.808
C. 0.666 D. 38.2
E. none of these

41. If $\cos(\alpha + \beta) = \cos\alpha\cos\beta - \sin\alpha\sin\beta$, then an exact value for $\cos 165°$ is
A. $\dfrac{\sqrt{2} - \sqrt{6}}{4}$ B. $\dfrac{\sqrt{6} - \sqrt{2}}{4}$
C. $\dfrac{\sqrt{2} + \sqrt{6}}{4}$ D. -0.9659
E. none of these

42. If $\alpha = \frac{\pi}{2}$ and $\beta = -\theta$, and these values are substituted into the identity given in Problem 41, then the result is
A. $\cos\left(\frac{\pi}{2} - \theta\right) = \sin(-\theta)$ B. $\cos\left(\frac{\pi}{2} - \theta\right) = -\cos\theta$
C. $\cos\left(\frac{\pi}{2} - \theta\right) = -\sin(-\theta)$ D. $\cos\left(\frac{\pi}{2} - \theta\right) = -\sin\theta$
E. none of these

43. If you use the identity in Problem 41 to simplify $\cos 3\theta$, the result is
A. $3\cos\theta$ B. $\cos^3\theta - 3\sin^2\theta\cos\theta$
C. $2\cos^3\theta - 2\sin^2\theta - 1$
D. $2\cos^3\theta - \sin 2\theta\sin\theta - 1$
E. none of these

44. If you are given $a = 38$, $c = 42$, and $\beta = 108°$, the correct method of solution is
A. right triangles B. law of cosines
C. law of sines D. no solution

45. If you are given $a = 55$, $b = 61$, and $\alpha = 57°$, the correct method of solution is
A. right triangles B. law of cosines
C. law of sines D. no solution

46. If you are given $b = 26$, $\beta = 47°$, and $\gamma = 76°$, the correct method of solution is
A. right triangles B. law of cosines
C. law of sines D. no solution

47. Which of the following pairs of vectors are orthogonal?
A. $\mathbf{v} = \cos 20°\mathbf{i} + \sin 20°\mathbf{j}$ B. $\mathbf{v} = \dfrac{1}{\sqrt{3}}\mathbf{i} - \dfrac{2}{\sqrt{5}}\mathbf{j}$
   $\mathbf{w} = \cos 70°\mathbf{i} + \sin 70°\mathbf{j}$ $\mathbf{w} = 2\mathbf{i} - \dfrac{\sqrt{5}}{5\sqrt{3}}\mathbf{j}$
C. $\mathbf{v} = \mathbf{i}$, $\mathbf{w} = \mathbf{j}$ D. all are orthogonal
E. none are orthogonal

48. $(3 + 5i)^2$ simplified is
   A. -16
   B. $-16 + 15i$
   C. $-16 + 30i$
   D. $9 + 30i + 25i^2$
   E. none of these

49. $\dfrac{(5 - 5i)(\sqrt{3} - i)}{1 - i}$ simplified is
   A. $5\sqrt{3} - 5i$
   B. $10(2 - i)$
   C. 20 cis $330°$
   D. $5(1 - i)^2(\sqrt{3} - i)$
   E. none of these

50. $\dfrac{2 \text{ cis } 128° \cdot 9 \text{ cis } 285°}{(3 \text{ cis } 4°)^3}$ simplified is
   A. 6 cis $409°$
   B. 6 cis $49°$
   C. 6 cis $401°$
   D. 6 cis $41°$
   E. none of these

51. One of the cube roots of $\frac{3}{2} - \frac{3}{2}\sqrt{3}\, i$ is
   A. 27 cis $180°$
   B. $\sqrt[3]{3}$ cis $220°$
   C. $\sqrt[3]{3}$ cis $20°$
   D. $1.5 - 2.5981i$
   E. none of these

52. The other primary representation of $\left(3, \frac{\pi}{10}\right)$ is
   A. $\left(-3, \frac{\pi}{10}\right)$
   B. $\left(3, \frac{11\pi}{10}\right)$
   C. $\left(-3, \frac{9\pi}{10}\right)$
   D. $\left(-3, \frac{11\pi}{10}\right)$
   E. none of these

53. A point satisfying the equation $r = \dfrac{3}{1 + \sin\theta}$ is
   A. $\left(\frac{\pi}{3}, 12 - 6\sqrt{3}\right)$
   B. $\left(\frac{1}{2}, \frac{\pi}{3}\right)$
   C. $\left(-6, \frac{\pi}{6}\right)$
   D. $\left(12 - 6\sqrt{3}, \frac{\pi}{6}\right)$
   E. none of these

54. The curve $\theta = $ constant is a
   A. cardioid
   B. rose curve
   C. lemniscate
   D. circle
   E. line

55. The graph of $r = 3 \sin 2\theta$ is a
   A. cardioid
   B. rose curve
   C. lemniscate
   D. circle
   E. line

56. The graph of $r = 3 \sin 60°$ is a
   A. cardioid
   B. rose curve
   C. lemniscate
   D. circle
   E. line

57. What is the rotation of the curve $r^2 = \sin 2\theta$ when compared to the standard-position curve $r^2 = \cos 2\theta$?
   A. $90°$
   B. $60°$
   C. $45°$
   D. $30°$
   E. line

58. In the airport's control tower, Robin observes Batman in the Batplane at a distance of 80 km on a bearing of $120°$ and the Joker in his Learjet 90 km from the airport on a bearing of $140°$. How far apart are Batman and the Joker?
   A. 31 km
   B. 93 km
   C. 147 km
   D. 968 km
   E. none of these

59. Identify each curve.
   a. $r = 3 \sin 2\theta$
   b. $r^2 = 3 \sin 2\theta$
   c. $r = 4 \cos 3\theta$
   d. $r = 3(1 + \sin\theta)$

60. Identify each curve.
   a. $r = 2 \sin\frac{\pi}{2}$
   b. $r - 5 = 5 \cos\theta$
   c. $r = 2 - 3 \cos\theta$
   d. $r^2 = 5 \cos 2\theta$

61. Identify each curve.
   a. $\theta = 3 \sin\frac{\pi}{6}$
   b. $r = 6(2 - \sin\theta)$
   c. $r = 2 + 2 \cos\theta$
   d. $r \sin\theta = 5$

Level 2 Drill

62. Solve $\cos^2\theta = 3 \sin\theta$ for $0 \le \theta < 2\pi$, correct to three decimal places.

63. Solve $2 \cos^2 2\theta = \cos 2\theta + 1$ for $0 \le \theta < 2\pi$.

64. Solve $\cot\theta = \dfrac{-3 \pm \sqrt{9 + 4}}{2}$ for all solutions, correct to the nearest degree.

65. Prove $\dfrac{\sin 3\theta + \cos 3\theta + 1}{\cos 3\theta} = \tan 3\theta + \sec 3\theta + 1$.

66. Prove or disprove: $\tan 2\theta \cos 2\theta = 2 \sin\theta \cos\theta$.

Problems 67–75: *State whether you would use a right-triangle solution, the law of cosines, or the law of sines. Next, solve each triangle.*

67. $a = 38;\ b = 42;\ \beta = 108°$

68. $a = 14.5;\ b = 28.1;\ c = 19.4$

69. $b = 52;\ \alpha = 49°;\ \gamma = 63°$

70. $a = 83;\ \beta = 24°;\ \gamma = 121°$

71. $a = 418; c = 385; \beta = 218.0°$

72. $a = 281; b = 318; c = 412$

73. $a = 3.0; b = 6.0; \alpha = 30°$

74. $a = 18; \alpha = 15°; \beta = 106°$

75. $a = 55; b = 61; \alpha = 57°$

Problems 76–79: *For each polar-form point, give both primary representations in polar form, and then change to rectangular form.*

76. $\left(3, -\frac{\pi}{4}\right)$ **77.** $\left(-5, -\frac{\pi}{6}\right)$

78. $(2, 3\pi)$ **79.** $(-3, 9.4248)$

80. Write the cube roots (in trigonometric form) of $-\frac{5}{2}\sqrt{3} + \frac{5}{2}i$.

81. Write the square roots (in trigonometric form) of $-8i$.

Problems 82–85: *Graph the polar-form curves.*

82. $r = 3 \cos 3\theta$ **83.** $r^2 = 4 \sin 2\theta$

84. $r = 2 + 2 \sin \theta$ **85.** $r = 2 \tan \frac{\pi}{4}$

Level 2 Applications

86. A draftsperson drew to scale (10 ft = 10 in.) a triangularly shaped lot with sides 45, 82, and 60 ft. On the plot plan, what are the measures of the angles?

87. A UFO is sighted by people in two cities 2.300 mi apart. The UFO is between and in the same vertical plane as the two cities. The angle of elevation of the UFO is 10.48° from the first city and 40.79° from the

second. At what altitude is the UFO flying at the instant it is sighted? What is the actual distance of the UFO from each city?

88. A boat (see Figure 6.27) moving at 18.0 knots heads in the direction of S38.2°W across a large river whose current is traveling at 5.00 knots in the direction of S13.1°W. Give the true course of the boat. How fast is it traveling as a result of these forces?

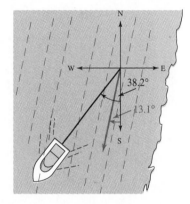

FIGURE 6.27 Navigational application

89. The world's longest deep-water jetty is at Le Havre, France. Since access to the jetty is restricted, it was necessary for me to calculate its length by noting that it forms an angle of 85.0° with the shoreline. After pacing out 1,000 ft along the line making an 85.0° angle with the jetty, I calculated the angle to the end of the jetty to be 83.6°. What is the length of the jetty?

CHAPTER SEVEN

LOGARITHMIC FUNCTIONS

PREVIEW

For years, logarithms were included in a trigonometry course to carry out the complicated calculations necessary to solve triangles. Today, with calculators, we do not need logarithms for those calculations. However, as you will see in this chapter, logarithms are still important in a wide variety of applications.

*The linear, quadratic, polynomial, rational, and radical functions considered in algebra are referred to as **algebraic functions.** An algebraic function is a function that can be expressed in terms of algebraic operations alone. If a function is not algebraic, it is called a **transcendental function.** The trigonometric functions are primary examples of transcendental functions.*

In this chapter, we introduce two transcendental functions, exponential and logarithmic functions. In calculus, exponential and logarithmic functions are important in solving equations called differential equations. The solutions of differential equations often involve both exponential and trigonometric functions. In algebra, they are used in such applications as growth, decay, population studies, and earthquake intensity.

Mathematics is the gate and key of the sciences. . . . Neglect of mathematics works injury to all knowledge, since whoever is ignorant of it cannot know the other sciences or the things of this world.

Roger Bacon (1894)

SECTION 7.1

EXPONENTIAL FUNCTIONS

PROBLEM OF THE DAY

What is $2^{\sqrt{3}}$?

How would you answer this question?

You might say that it is an easy calculator evaluation:

$$2^{\sqrt{3}} \approx 3.321997085$$

But what does this mean? Do you remember the definition of exponents?

$$b^n = \underbrace{b \cdot b \cdot b \cdot \cdots \cdot b}_{n \text{ factors}}$$

According to this definition, does the expression $2^{\sqrt{3}}$ make any sense? The goal of this section is to give meaning to this expression, as well as to consider a general category of functions called *exponential functions*.

We begin by returning to the definition of an exponent so that we can generalize to irrational numbers.

Irrational Exponents

To understand an exponential function, it is necessary to recall the definition of an exponent.

This definition is fundamental to a great deal of mathematics. Spend some time with each line of this definition.

> **RATIONAL EXPONENTS**
>
> If b is any real number and n is any natural number, then
> $$b^n = \underbrace{b \cdot b \cdot b \cdot \cdots \cdot b}_{n \text{ factors}}$$
> If $b \neq 0$, then $b^0 = 1$ and $b^{-n} = \dfrac{1}{b^n}$.
> If $b > 0$, then $b^{1/n} = \sqrt[n]{b}$.
> If m is also a natural number where m/n is reduced, then
> $$b^{m/n} = (b^{1/n})^m = \sqrt[n]{b^m} = (\sqrt[n]{b})^m$$

The number b is called the **base,** n is called the **exponent,** and b^n is called the **nth power** of b. The five laws of exponents can now be summarized.

NOTE

These properties are essential to understand this chapter.

LAWS OF EXPONENTS

Let a and b be real numbers and let m and n be integers. The five rules listed here govern the use of exponents except that the form 0^0 and division by zero are excluded.

Additive law of exponents: $b^m \cdot b^n = b^{m+n}$

Subtractive law of exponents: $\dfrac{b^m}{b^n} = b^{m-n}$

Multiplicative law of exponents: $(b^n)^m = b^{mn}$

Distributive laws of exponents: $(ab)^m = a^m b^m$ and $\left(\dfrac{a}{b}\right)^m = \dfrac{a^m}{b^m}$

Help Window   ? X

In other words, to multiply expressions with the same base, add the exponents. To divide, subtract the exponents. To raise a power to a power, multiply the exponents. To raise a product or quotient to a power, raise each factor to that power.

EXAMPLE 1

Using the Laws of Exponents

Use the laws of exponents to simplify each expression.

a. $2^3 \cdot 2^4$ **b.** $\dfrac{10^5}{10^3}$ **c.** 5^{-4} **d.** $\dfrac{5^7}{5^{10}}$ **e.** $\left(\dfrac{2}{3}\right)^3$ **f.** $(6^3)^{-1}$

g. $(3^2 \cdot 3^{-5})^{-1}$ **h.** $8^{2/3}$

Solution

a. $2^3 \cdot 2^4 = 2^{3+4} = 2^7 = 128$

b. $\dfrac{10^5}{10^3} = 10^{5-3} = 10^2 = 100$

c. $5^{-4} = \dfrac{1}{5^4} = \dfrac{1}{625}$

d. $\dfrac{5^7}{5^{10}} = 5^{7-10} = 5^{-3} = \dfrac{1}{5^3} = \dfrac{1}{125}$

e. $\left(\dfrac{2}{3}\right)^3 = \dfrac{2^3}{3^3} = \dfrac{8}{27}$

f. $(6^3)^{-1} = 6^{-3} = \dfrac{1}{6^3} = \dfrac{1}{216}$

g. $(3^2 \cdot 3^{-5})^{-1} = (3^{2-5})^{-1} = (3^{-3})^{-1} = 3^3 = 27$

h. $8^{2/3} = (2^3)^{2/3} = 2^2 = 4$

We need to define expressions such as $2^{\sqrt{3}}$. To define b^x for x an irrational number (or more generally, for x any real number), the **Squeeze Theorem** (which can be proved using ideas from calculus) will help us to understand the evaluation of an expression with an irrational exponent.

SQUEEZE THEOREM FOR EXPONENTS

Suppose b is a real number greater than 1. Then, for any real number x, there is a unique real number b^x. Moreover, if h and k are any two rational numbers such that $h < x < k$, then

$$b^h < b^x < b^k$$

This theorem gives meaning to expressions such as $2^{\sqrt{3}}$. Consider the graph of the function $f(x) = 2^x$ by plotting points as shown in Figure 7.1. The case where b is a real number $0 < b < 1$ is left for the problem set.

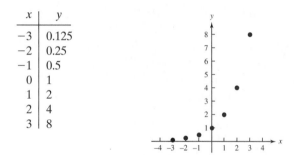

x	y
-3	0.125
-2	0.25
-1	0.5
0	1
1	2
2	4
3	8

FIGURE 7.1   Selected points that satisfy $f(x) = 2^x$

If the points shown in Figure 7.1 are connected with a smooth curve, you can see that $2^{\sqrt{3}}$ is defined and is between 2^1 and 2^2, as shown in Figure 7.2. The number $2^{\sqrt{3}}$ can be approximated to any desired degree of accuracy:

Since $1 < \sqrt{3} < 2$,   we have $2^1 < 2^{\sqrt{3}} < 2^2$

$1.7 < \sqrt{3} < 1.8$ $2^{1.7} < 2^{\sqrt{3}} < 2^{1.8}$

$1.73 < \sqrt{3} < 1.74$ $2^{1.73} < 2^{\sqrt{3}} < 2^{1.74}$

$\vdots$ $\vdots$

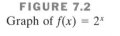

FIGURE 7.2
Graph of $f(x) = 2^x$

We can now assume that b^x is defined for all real numbers x because we have the squeeze theorem for irrational exponents, and a direct definition for rational exponents. We will also accept that the usual laws of exponents hold for all real exponents.

Definition and Graphs of Exponential Functions

We can now define an exponential function.

EXPONENTIAL FUNCTION

The function f is an **exponential function** if

$$f(x) = b^x$$

where b is a positive constant other than 1 and x is any real number. The number x is called the **exponent** and b is called the **base.**

The shape of a particular exponential graph depends on b, as illustrated by the following example.

EXAMPLE 2

Graphing Exponential Functions

a. Graph $f(x) = 10^x$ ($b = 10$: type $b > 1$)

b. Graph $f(x) = 0.1^x$ ($b = 1/10$: type $b < 1$)

Solution   **a.** The function f is an increasing function, with a horizontal asymptote at $y = 0$. The graph, along with a table of values, is shown in Figure 7.3. Notice that it is often necessary to alter the y-scale for exponential functions.

x	$f(x) = 10^x$
-3	0.001
-2	0.01
-1	0.1
0	1
1	10
2	100
3	1,000

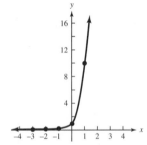

FIGURE 7.3   Graph of $f(x) = 10^x$

b. This function is a decreasing function with a horizontal asymptote at $y = 0$. The graph, along with a table of values, is shown in Figure 7.4.

x	$f(x) = 0.1^x$
-3	1,000
-2	100
-1	10
0	1
1	0.1
2	0.01
3	0.001

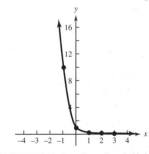

FIGURE 7.4   Graph of $f(x) = 0.1^x$ ■

Example 2 is summarized in Table 7.1.

TABLE 7.1 Directory of Curves   Graphs of Exponential Functions

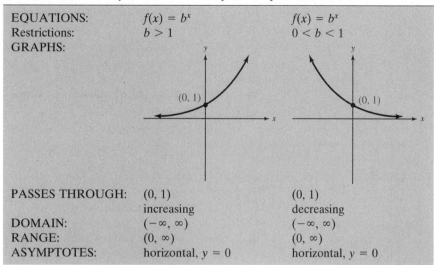

EQUATIONS:	$f(x) = b^x$	$f(x) = b^x$
Restrictions:	$b > 1$	$0 < b < 1$
GRAPHS:		
PASSES THROUGH:	$(0, 1)$	$(0, 1)$
	increasing	decreasing
DOMAIN:	$(-\infty, \infty)$	$(-\infty, \infty)$
RANGE:	$(0, \infty)$	$(0, \infty)$
ASYMPTOTES:	horizontal, $y = 0$	horizontal, $y = 0$

EXAMPLE 3

Graphing Variations of Exponential Graphs

Graph the curves defined by the given equations by comparing with the graph of $y = 2^x$.

a. $y = -2^x$ **b.** $y = 2^{-x}$ **c.** $y = (-2)^x$ **d.** $y = e^{-x}$

Solution   We begin with the graph of $y = 2^x$, as shown in Figure 7.5.

a. Recall that -2^x means "opposite of 2 to the x power," namely $y = -(2^x)$. The graph of $y = -2^x$ is found by reflecting the graph of $y = 2^x$ in the x-axis.

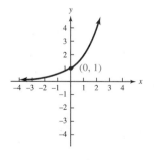

FIGURE 7.5
Graph of $y = 2^x$

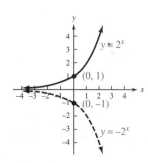

b. $y = 2^{-x}$ is obtained from the graph of $y = 2^x$ by reflecting in the y-axis. Also, note that $y = 2^{-x}$ is the same as $y = \left(\frac{1}{2}\right)^x$, so this reflected graph has the shape of the exponential graph for $0 < b < 1$.

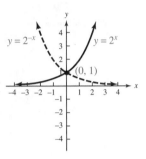

c. $y = (-2)^x$ is not defined because $b = -2$ is not a positive number.

d. Recall from algebra that e is an irrational number (which cannot be represented as a decimal). The graph of $y = e^{-x} = (1/e)^x$ has the general shape of an exponential for $b = \dfrac{1}{e} \approx 0.37 < 1$. (Look for e^x on your calculator; $e \approx 2.71$.) This graph passes through $(0, 1)$, $(1, \ e^{-1})$, $(-1, e)$, $(2, e^{-2})$ as shown in Figure 7.6. ∎

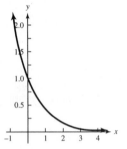

FIGURE 7.6
Graph of $y = e^{-x}$

Harmonic Motion

In this book, we have considered various periodic phenomena such as the motion of a spring or a pendulum, disregarding the effects of friction or air resistance. If we take into account the effects of friction and air resistance, the graphs are often found as the product of two functions. If the product of two functions has the effect of a diminishing amplitude, then the harmonic motion is known as **damped harmonic motion.**

EXAMPLE 4

Graphing a Product That Illustrates a Damped Vibration Curve

The equation of the damped vibration curve is

$$f(x) = e^{-x} \sin x, \quad x \geq 0$$

This equation is derived in calculus and used in the motions of a pendulum, electrical currents, and in the motion of a plucked string.

Solution   This problem is most easily graphed with a graphing calculator, as shown in the calculator window.

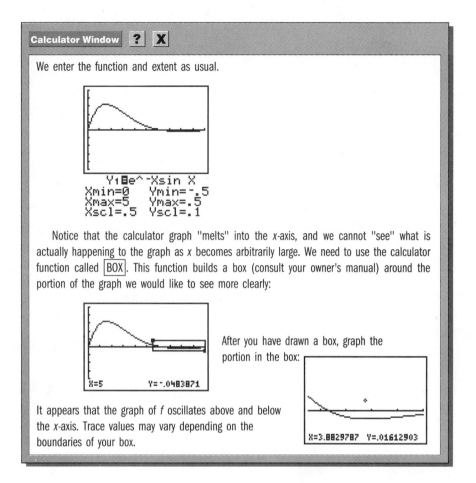

If you use a graphing calculator for this problem, you may not notice the dynamic nature of damped harmonic motion. We can view damped harmonic motion as the interaction between two forces, that of a constant vibrating motion represented by the function $g(x) = a \sin x$ and, secondly, as an ever-increasing constricting force represented by $h(x) = e^{-x}$. To consider the graph without using a calculator, we use the property that $|\sin x| \leq 1$:

$$|e^{-x} \sin x| = |e^{-x}| \cdot |\sin x| \qquad \textit{Property of absolute values: } |ab| = |a| \cdot |b|$$
$$\leq |e^{-x}| \qquad \textit{Since } |\sin x| \leq 1$$
$$= e^{-x} \qquad \textit{For } x \geq 0, e^{-x} > 0, \textit{so } |e^{-x}| = e^{-x}.$$

This tells us that f is between $-e^{-x}$ and e^{-x}:

$$-e^{-x} \leq e^{-x} \sin x \leq e^{-x}$$

We begin by graphing $y_1 = e^{-x}$ and $y_2 = -e^{-x}$ for $x \geq 0$, as shown in Figure 7.7a on page 340. Then we plot values of f to draw the graph in Figure 7.7b.

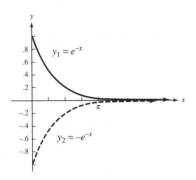 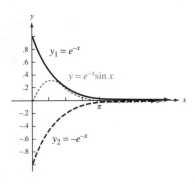

a. Begin by graphing the boundary curve. **b.** Plot points to fill in the desired curves.

FIGURE 7.7 Graph of $f(x) = e^{-x} \sin x$ ■

In Chapter 2, we considered *simple harmonic motion* (amplitude a constant). If the amplitude is decreasing, it is called *damped harmonic motion*. A third possibility is called **resonance** and is characterized by an increasing amplitude.

EXAMPLE 5

Graphing a Product That Illustrates Resonance

Graph $f(x) = x \sin x$.

Solution The function f is the product of $y = x$ and $y = \sin x$. We can graph this product by using a graphing calculator (see the margin), or we can use the fact that $|\sin x| \le 1$.

$$|x \sin x| = |x| \cdot |\sin x| \le |x|(1) = |x|$$

This means

$$-x \le f(x) \le x$$

That is, the graph of the product lies between the lines $y_1 = x$ and $y_2 = -x$. The graph is shown in Figure 7.8.

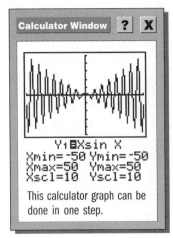

Y₁⊟Xsin X
Xmin=-50 Ymin=-50
Xmax=50 Ymax=50
Xscl=10 Yscl=10

This calculator graph can be done in one step.

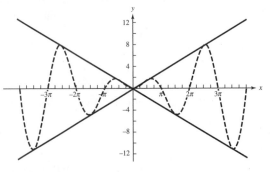

FIGURE 7.8 Graph of $y = x \sin x$ ■

In physics, it is known that the amplitude of vibrations is large compared to the amount of static displacement. For example, it is possible for soldiers marching in step across a suspension bridge to make it vibrate with a very large destructive amplitude, provided that the frequency of their steps happens to be the same as the natural frequency of the bridge. This is why marching soldiers break step when crossing a bridge.

A striking example of the destructive force of resonance is the collapse of the Tacoma Narrows Bridge (see Figure 7.9), which was opened on July 1, 1940.* Early on the morning of November 7, 1940, the wind velocity was 40–45 mi/h—perhaps a greater velocity than any previously encountered by the bridge. Traffic was shut down. By 9:30 A.M., the span was vibrating in eight or nine segments, with a frequenty of 36 vib/min and an amplitude of about 3 ft. While measurements were under way, at about 10:00 A.M. the main span abruptly began to vibrate torsionally in two segments with a frequency of about 14 vib/min. The amplitude of torsional vibrations quickly increased up to 35° in each direction from the horizontal. The main span broke up shortly after 11:00 A.M.

FIGURE 7.9   The Tacoma Narrows Bridge at Puget Sound, Washington

Did we learn anything from the Tacoma Narrows Bridge collapse? We did indeed, but there is still much to learn. For example, engineers reported that the collapse of the elevated freeway in Oakland (1989 San Francisco World Series earthquake), which caused so many fatalities, may have been due as much to resonance as to the quake itself. It was shown that the ground shook with a frequency that matched the natural resonance of the freeway. Dr. Piotr D. Moncarz, principal engineer of Failure Analysis Associates of Palo Alto who investigated the quake, notes "Engineers are only now beginning to understand how to endow structures with resonant frequencies that improve their resistance to quakes.

* A filmstrip *The Tacoma Narrows Bridge Collapse* (#80-218) is available from the Ealing Corporation, Cambridge, MA 02140, which provided the information about the bridge.

PROBLEM SET 7.1

Essential Concepts

1. What is an exponential function?

2. State the requested law of exponents.
   a. Additive law
   b. Subtractive law
   c. Multiplicative law
   d. Distributive law for products
   e. Distributive law for quotients

3. **IN YOUR OWN WORDS** What is the squeeze theorem for exponents? In reference to the definition of irrational exponents, why is a squeeze law necessary?

Level 1 Drill

Problems 4–11: *Use the laws of exponents to simplify each expression.*

4. a. $3^5 \cdot 3^2$ b. $4^2 \cdot 4$ c. $2^5 \cdot 2$

5. a. $\dfrac{3^5}{3^2}$ b. $\dfrac{4^3}{4^2}$ c. $\dfrac{2^5}{2}$

6. a. 2^{-3} b. 3^{-2} c. 4^{-3}

7. a. $\dfrac{2^7}{2^{10}}$ b. $\dfrac{3^4}{3^7}$ c. $\dfrac{4^5}{4^7}$

8. a. $\left(\dfrac{2}{3}\right)^3$ b. $\left(\dfrac{-2}{5}\right)^2$ c. $\left(\dfrac{5}{3}\right)^3$

9. a. $(5^2)^{-1}$ b. $(3^4)^{-1}$ c. $(2^4)^{-1}$

10. a. $(2^5 \cdot 2^{-7})^{-2}$ b. $(3^3 \cdot 3^{-5})^{-2}$

11. a. $27^{2/3}$ b. $81^{3/2}$ c. $25^{3/2}$

Problems 12–15: *Use a calculator to evaluate each expression.*

12. a. $10^{2.3}$ b. $10^{-1/3}$ c. $2^{2/3}$ d. $2^{-4.1}$

13. a. e^3 b. e^{-2} c. $e^{0.12}$ d. $e^{0.08}$

14. a. $\left(1 + \dfrac{0.08}{12}\right)^{24}$ b. $\left(1 + \dfrac{0.05}{6}\right)^{72}$

15. a. $\left(1 + \dfrac{1}{1,000}\right)^{1,000}$ b. $\left(1 + \dfrac{1}{100,000}\right)^{100,000}$

Level 2 Drill

Problems 16–21: *Sketch each graph.*

16. a. $y = 2^x$ b. $y = \left(\dfrac{1}{2}\right)^x$

17. a. $y = 3^x$ b. $y = 3^{-x}$

18. a. $y = 4^x$ b. $y = -4^x$

19. a. $y = \left(\dfrac{3}{2}\right)^x$ b. $y = \left(\dfrac{2}{3}\right)^x$

20. a. $y = \pi^x$ b. $y = \left(\dfrac{1}{\pi}\right)^x$

21. a. $y = e^x$ b. $y = -e^x$

Problems 22–40: *Sketch each graph.*

22. $y = 10^x$ 23. $y = 5^x$

24. $y = 0.1^x$ 25. $y = 0.5^x$

26. $y = x \sin x$ 27. $y = x \cos x$

28. $y = -x \sin x$ 29. $y = -x \cos x$

30. $y = 2 \sin x \cos 6x$ 31. $y = 3 \cos x \sin 6x$

32. $y = 4 \sin x \cos 6x$ 33. $y = \cos x + x \sin x$

34. $y = e^{\sin x}$ 35. $y = e^{\cos x}$

36. $y = \sin e^x$ 37. $y = \cos e^x$

38. $y = e^{x/5}$ for $0 \le x < 2\pi$

39. $y = 2^{x/2} \sin x$ for $0 \le x < 2\pi$

40. $y = e^{x/2} \cos x$ for $0 \le x < 2\pi$

41. Graph for $x \ge 0$ and classify each motion as harmonic motion, damped harmonic motion, or resonance.
    a. $y = 3 \cos 4\pi x$
    b. $y = 3x \cos 4\pi x$
    c. $y = \dfrac{3}{x} \cos 4\pi x$

Problems 42–45: *Graph the functions given in each problem on the same coordinate axes. If you are using a graphing calculator, pay particular attention to the domain and range.*

42. a. $y = x$ b. $y = -x$ c. $y = \dfrac{x}{\sin x}$

43. a. $y = x$ b. $y = -x$ c. $y = x \sin x$

44. a. $y = x^2$ b. $y = -x^2$ c. $y = x^2 \sin x$

45. a. $y = \sqrt{x}$ b. $y = -\sqrt{x}$ c. $y = \sqrt{x} \sin x$

Level 2 Applications

46. Compare the graphs for future value for $1,000 invested at a 5% rate for t years where $0 \leq t \leq 20$.
   a. Simple interest: $A = 1,000(1 + 0.05t)$
   b. Compounded annually: $A = 1,000(1.05)^t$

47. Compare the graphs for future value of $1,000 invested at an 8% rate for t years where $0 \leq t \leq 20$.
   a. Simple interest: $A = 1,000(1 + 0.08t)$
   b. Compounded quarterly:
   $$A = 1,000\left(1 + \frac{0.08}{4}\right)^{4t}$$

48. Use graphical methods to estimate $2^{\sqrt{2}}$.

49. Use graphical methods to estimate $3^{\sqrt{2}}$.

50. Radioactive argon-39 has a half-life of 4 min. This means that the time required for half the argon-39 to decompose is 4 min. If we start with 100 milligrams (mg) of argon-39, the amount A left after t minutes is given by
   $$A = 100\left(\tfrac{1}{2}\right)^{t/4}$$
   Graph this function for $t \geq 0$.

51. Carbon-14 used for archaeological dating has a half-life of 5,700 yr. This means that the time required for half of the carbon-14 to decompose is 5,700 yr. If we start with 100 mg of carbon-14, the amount A left after t years is given by
   $$A = 100\left(\tfrac{1}{2}\right)^{t/5,700}$$
   Graph this function for $t \geq 0$.

52. A graph of world population growth from A.D. 1 to 1990 is shown in Figure 7.10.

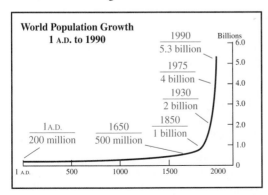

FIGURE 7.10 World population growth

In 1990, the world population was about 5.3 billion. If we assume a growth rate of 1.8%, the world population P (in billions) for t years after 1990 is given by the formula
$$P = 5.3e^{0.018t}$$
Graph this function for 1990–2010.

Problems 53–56: *If P dollars are borrowed for N months at an annual rate of r, then the monthly payment, m, is found by the formula*
$$m = \frac{Pi}{1 - (1 + i)^{-N}}$$
where $i = r/12$.

53. What is the monthly payment for a new car costing $34,560 with a down payment of $2,560? The car is financed for 4 years at 12%.

54. A purchase of $2,650 is financed at 21% for 2 years. What is the monthly payment?

55. A home loan is made for $215,000 at 8.4% for 30 years. What is the monthly payment?

56. A home is financed at $7\frac{1}{2}$% interest for 20 years. If the home cost $240,000 with 20% down, what is the monthly payment?

Level 3 Theory

57. Show that the graph of
   $$f(x) = \left(1 + \frac{1}{x}\right)^x \quad \text{for } x > 0$$
   approaches the horizontal line $y = e$.

58. Find a counterexample to show that the squeeze theorem does not hold for $0 < b < 1$. State an alternative form of the squeeze theorem for this case.

59. Use graphical methods to determine which is larger:
   a. $(\sqrt{3})^{\pi}$ or $\pi^{\sqrt{3}}$
   b. $(\sqrt{5})^{\pi}$ or $\pi^{\sqrt{5}}$
   c. $(\sqrt{6})^{\pi}$ or $\pi^{\sqrt{6}}$
   d. Consider $(\sqrt{N})^{\pi} = \pi^{\sqrt{N}}$, where N is a positive real number. From parts **a–c**, notice that $(\sqrt{N})^{\pi}$ is larger for some values of N, whereas $\pi^{\sqrt{N}}$ is larger for others. For $N = \pi^2$,
   $$(\sqrt{N})^{\pi} = \pi^{\sqrt{N}}$$

is obviously true. Using a graphical method, find another (approximate) value for which the given statement is true.

60. Can an irrational number raised to an irrational power give an answer that is rational?
[*Hint:* Consider $\sqrt{2}^{\sqrt{2}}$. It is either rational or irrational.]

Problems 61–63: PROBLEMS FROM CALCULUS *In calculus, the number e is sometimes introduced using slopes. Problems 61–63 explore this idea.*

61. a. Draw the graph of $y = 2^x$.
   b. Label the points (0, 1) and (2, 4).
   c. Consider the line passing through these points. This line is called a *secant line*. Now, consider the slope of the secant line as the point (2, 4) slides along the curve toward the point (0, 1). Draw the line that you think will result when the point (2, 4) reaches the point (0, 1). This line is called the *tangent line* to the curve $y = 2^x$ at (0, 1).
   d. Using the tangent line, and the fact that the slope of a line is RISE/RUN, estimate (to the nearest tenth) the slope of the tangent line.

62. a. Draw the graph of $y = 3^x$.
   b. Label the points (0, 1) and (2, 9).
   c. Consider the line passing through these points. This line is called a *secant line.* Now, consider the slope of the secant line as the point (2, 9) slides along the curve toward the point (0, 1). Draw the line that you think will result when the point (2, 9) reaches the point (0, 1). This line is called the *tangent line* to the curve $y = 3^x$ at (0, 1).
   d. Using the tangent line, and the fact that the slope of a line is RISE/RUN, estimate (to the nearest tenth) the slope of the tangent line.

63. a. Draw a line passing through (0, 1) with slope 1.
   b. Compare the graphs of $y = 2^x$ (Problem 61) and $y = 3^x$ (Problem 62). It seems reasonable that there exists a number between 2 and 3 such that the slope of the tangent through (0, 1) is 1. Draw such a curve.
   c. On the same coordinate axes, draw the graph of $y = e^x$. How does this curve compare with the curve you drew in part **b**? In calculus, the number between 2 and 3 such that the slope of the tangent through (0, 1) is 1 is used as the number *e*.

SECTION 7.2	# EXPONENTIAL EQUATIONS AND LOGARITHMS

PROBLEM OF THE DAY

What is the solution to the equation $2^x = 14$?

How would you answer this question?

To solve an equation means that you need to find the replacement(s) for the variable that make the equation true. You might try certain values:

$x = 1$:	$2^x = 2^1 = 2$	*Too small*
$x = 2$:	$2^x = 2^2 = 4$	*Too small*
$x = 3$:	$2^x = 2^3 = 8$	*Still too small*
$x = 4$:	$2^x = 2^4 = 16$	*Too big*

It appears that the number you are looking for is between 3 and 4. Our task in this section is to find both an approximate as well as an exact value for *x*, but first we need some definitions.

Historical Note

In his history book, F. Cajori wrote that "The miraculous powers of modern calculation are due to three inventions: the Arabic notation, decimal fractions, and logarithms." Today, we would no doubt add a fourth invention to this list—namely, the hand-held calculator. Nevertheless, the idea of a logarithm was a revolutionary idea that forever changed the face of mathematics. Simply stated, a logarithm is an exponent, and even though logarithms are no longer important as an aid in calculation, the logarithmic and exponential functions are extremely important in advanced mathematics.

Definition of Logarithm

Consider the Problem of the Day; the solution of the equation $2^x = 14$ seeks an x value. What is this x value? We express the idea in words:

x is the exponent on a base 2 that gives the answer 14

This can be abbreviated as

x = exp of 14 to the base 2

We further shorten this notation to

$x = \exp_2 14$

This statement is read, "x is the exponent on a base 2 that gives the answer 14." It appears that the equation is now solved for x, but this is simply a notational change. The expression "exponent of 14 to the base 2" is called, for historical reasons, "the log of 14 to the base 2." That is,

$x = \exp_2 14$ and $x = \log_2 14$

Both equations mean exactly the same thing. This leads us to the following definition of logarithm.

DEFINITION OF LOGARITHM

For positive b and A, $b \neq 1$,

$$x = \log_b A \quad \text{means} \quad b^x = A$$

x is called the **logarithm** and A is called the **argument.**

Help Window   ? X

The statement $x = \log_b A$ should be read as "x is the log (exponent) on a base b that gives the value A." *Do not forget that a logarithm is an exponent.*

EXAMPLE 1

Changing from Exponential Form to Logarithmic Form

Write in logarithmic form: **a.** $5^2 = 25$ **b.** $\frac{1}{8} = 2^{-3}$ **c.** $\sqrt{64} = 8$

Solution

a. In $5^2 = 25$, 5 is the base and 2 is the exponent, so we write

$$2 = \log_5 25$$

Remember, the logarithmic expression "solves" for the exponent.

b. With $\frac{1}{8} = 2^{-3}$, the base is 2 and the exponent is -3:

$$-3 = \log_2 \frac{1}{8}$$

c. With $\sqrt{64} = 8$, the base is 64 and the exponent is $\frac{1}{2}$ (since $\sqrt{64} = 64^{1/2}$):

$$\frac{1}{2} = \log_{64} 8$$ ■

EXAMPLE 2

Changing from Logarithmic Form to Exponential Form

Explain what each expression means, and then rewrite the expression in exponential form.

a. $\log_{10} 100$ **b.** $\log_{10} \frac{1}{1,000}$ **c.** $\log_3 1$

Solution   **a.** $\log_{10} 100$ is the exponent on a base 10 that gives 100. We see that the exponent is 2, so we write $\log_{10} 100 = 2$; the base is 10 and the exponent is 2, so $10^2 = 100$.

b. $\log_{10} \frac{1}{1,000}$ is the exponent on a base 10 that gives $\frac{1}{1,000}$, which is -3. The base is 10 and the exponent is -3, so we write

$$10^{-3} = \frac{1}{1,000}$$

c. $\log_3 1$ is the exponent on a base 3 that gives 1, which is 0. The base is 3 and the exponent is 0, so we write

$$3^0 = 1$$ ■

EXAMPLE 3

Solving Exponential Equations

Solve for x: **a.** $3^x = 5$ **b.** $10^x = 2$ **c.** $e^x = 0.56$

Solution   **a.** $x = \log_3 5$ **b.** $x = \log_{10} 2$ **c.** $x = \log_e 0.56$ ■

In elementary work, the most commonly used base is 10, so we call a logarithm to the base 10 a **common logarithm,** and agree to write it without using a subscript 10. That is, log x is a *common logarithm.* A logarithm to the base e is called a **natural logarithm** and is denoted by ln x. The expression ln x is often pronounced "ell en x" or "lon x."

> **LOGARITHMIC NOTATIONS**
>
> **Common logarithm:**   log x means $\log_{10} x$
> **Natural logarithm:**   ln x means $\log_e x$

The solution for the equation $10^x = 2$ is $x = \log 2$, and the solution for the equation $e^x = 0.56$ is $x = \ln 0.56$.

Evaluating Logarithms

To **evaluate a logarithm** means to find a numerical value for the given logarithm. Calculators have, to a large extent, eliminated the need for logarithm tables. You should find two logarithm keys on your calculator. One is labeled

$\boxed{\text{LOG}}$ for common logarithms, and the other is labeled $\boxed{\text{LN}}$ for natural logarithms.

EXAMPLE 4   **Evaluating Logarithms Using a Calculator**

Use a calculator to evaluate: **a.** log 5.03   **b.** ln 3.49   **c.** log 0.00728

Solution   Be sure you use your own calculator to verify these answers because the number of digits shown may vary. Calculator answers are more accurate than were the old table answers, but it is important to realize that any answer (whether from a table or a calculator) is only as accurate as the input numbers. However, in this book we will not be concerned with significant digits, but instead will use all the accuracy our calculator gives us, rounding only once (if requested) at the end of the problem.

a. log 5.03 $\approx$ 0.7015679851

b. ln 3.49 $\approx$ 1.249901736

c. log 0.00728 $\approx$ -2.137868621 ∎

Example 4 shows fairly straightforward evaluations, since the problems involve common or natural logarithms and because your calculator has both $\boxed{\text{LOG}}$ and $\boxed{\text{LN}}$ keys. However, suppose we wish to evaluate a logarithm to some base *other than* base 10 or base *e*. The first method uses the definition of logarithm (as in Example 3), and the second method uses what is called the **change of base theorem.** Before we state this theorem, we consider its plausibility with the following example.

EXAMPLE 5   **Looking for a Pattern in Evaluating Logarithms**

Evaluate the given expressions.

a. $\log_2 8, \dfrac{\log 8}{\log 2},$ and $\dfrac{\ln 8}{\ln 2}$

b. $\log_3 9, \dfrac{\log 9}{\log 3},$ and $\dfrac{\ln 9}{\ln 3}$

c. $\log_5 625, \dfrac{\log 625}{\log 5},$ and $\dfrac{\ln 625}{\ln 5}$

Solution   **a.** From the definition of logarithm, $\log_2 8 = x$ means

$$2^x = 8$$
$$2^x = 2^3$$
$$x = 3$$

Thus, $\log_2 8 = 3$. By calculator,

$$\frac{\log 8}{\log 2} \approx \frac{0.903089987}{0.3010299957} \approx 3$$

Also, $\dfrac{\ln 8}{\ln 2} \approx \dfrac{2.079441542}{0.6931471806} \approx 3$

b. $\log_3 9 = x$ means $3^x = 3^2$, so that $x = \log_3 9 = 2$. By calculator,

$$\frac{\log 9}{\log 3} \approx \frac{0.9542425094}{0.4771212547} \approx 2$$

Also, $\dfrac{\ln 9}{\ln 3} \approx \dfrac{2.197224577}{1.098612289} \approx 2$

c. $\log_5 625 = x$ means $5^x = 5^4$, so that $\log_5 625 = 4$. By calculator,

$$\frac{\log 625}{\log 5} \approx \frac{2.795880017}{0.6989700043} \approx 4$$

and $\dfrac{\ln 625}{\ln 5} \approx \dfrac{6.43775165}{1.609437912} \approx 4$ ■

You no doubt noticed that the answers to each part of Example 5 are the same. This result is summarized with the following theorem, which is proved on page 368 (Section 7.3).

CHANGE OF BASE THEOREM

$$\log_a x = \frac{\log_b x}{\log_b a}$$

STOP   *Remember this formula so you can evaluate logarithms to bases other than 10 or e.*

EXAMPLE 6   **Evaluating Logarithms with Arbitrary Bases**

Evaluate (round to the nearest hundredth):   **a.** $\log_7 3$ **b.** $\log_3 3.84$

Solution   **a.** $\log_7 3 = \dfrac{\log 3}{\log 7} \approx \underbrace{\dfrac{0.4771212547}{0.84509804} \approx 0.5645750341}_{} \approx 0.56$

This is all done by calculator and not on your paper.

b. $\log_3 3.84 = \dfrac{\log 3.84}{\log 3} \approx \underbrace{\dfrac{0.5843312244}{0.4771212547} \approx 1.224701726}_{} \approx 1.22$

Calculator work   ■

The Problem of the Day can now be solved (sort of). We find

$$2^x = 14$$
$$x = \log_2 14$$

This is called the **exact solution** for the equation, and Example 7 finds an approximate solution.

EXAMPLE 7

Problem of the Day: Solving an Equation

Solve $2^x = 14$ (correct to the nearest hundredth).

Solution

We use the definition of logarithm and the change of base theorem to write

$$x = \log_2 14 \approx 3.807354922 \approx 3.81$$ ■

Exponential Equations

An **exponential equation** is an equation in which one or more variables are included as an exponent or as part of the exponent. Exponential equations will fall into one of three types:

Base:	10 (common log)	e (natural log)	b (arbitrary base)
Example:	$10^x = 5$	$e^{-0.06x} = 3.456$	$8^x = 156.8$

The following example illustrates the procedure for solving each type of exponential equations.

Historical Note

The history of logarithms makes interesting reading. John Napier (1550–1617) is usually credited with the discovery of logarithms because he published the first work on this topic, called *Descriptio*, in 1614. However, similar ideas were developed independently by Jobst Bürgi around 1588. Some sources on the history of logarithms include "Logarithms" by J. W. L. Glaisher (*Encyclopaedia Britannica*, 11th ed. Vol. 16, pp. 868–877) and "History of the Exponential and Logarithmic Concepts," by Florian Cajori [*American Mathematical Monthly*, Vol. 20 (1913)].

In the first article, Glaisher notes "The invention of logarithms and the calculation of the earlier tables form a very striking episode in the history of exact science, and, with the exception of the *Principia* of Newton, there is no mathematical work published in the country which has produced such important consequences, or to which so much interest attaches as to Napier's *Descriptio*."

EXAMPLE 8   **Procedure for Solving Exponential Equations**

Solve the following exponential equations:

a. Base 10 (common log): $10^x = 5$

b. Base e (natural log): $e^{-0.06x} = 3.456$

c. Arbitrary base: $8^x = 156.8$

Solution   Regardless of the base, we use the definition of logarithm to solve an exponential equation.

a. $10^x = 5$

$\qquad x = \log 5$ *Definition of logarithm; this is the exact answer.*

$\qquad \approx 0.6989700043$ *Approximate answer*

b. $e^{-0.06x} = 3.456$

$\qquad -0.06x = \ln 3.456$ *Definition of logarithm*

$\qquad x = \dfrac{\ln 3.456}{-0.06}$ *Exact answer; this can be simplified to $x = -\frac{50}{3} \ln 3.456$.*

$\qquad \approx -20.66853085$ *Approximate answer*

c. $8^x = 156.8$

$\qquad x = \log_8 156.8$ *Definition of logarithm; exact answer*

$\qquad \approx 2.43092725$ *Do not forget the change of base theorem: $\log_8 156.8 = \dfrac{\log 156.8}{\log 8}$.*

Note: Many people will solve this by "taking the log of both sides":

$$8^x = 156.8$$
$$\log 8^x = \log 156.8$$
$$x \log 8 = \log 156.8 \quad \text{\textit{This property of logarithms will be developed in the next section.}}$$
$$x = \frac{\log 156.8}{\log 8}$$

Did you notice that this result is the same? It simply involves several extra steps and some additional properties of logarithms. It is rather like solving quadratic equations by completing the square each time instead of using the quadratic formula. You can see that, before calculators, there were good reasons to avoid representations such as $\log_8 156.8$. Whenever you *see* an expression such as $\log_8 156.8$, you *know* how to calculate it: $\log 156.8/\log 8$.   ■

We now consider some more general exponential equations. We will follow the general procedure illustrated in Example 8. Some algebraic steps will be required to put the equations into the correct form. That is, before we use the definition of logarithm, we first solve for the exponential form.

EXAMPLE 9

Solving Exponential Equations

Solve: **a.** $\dfrac{10^{5x+3}}{5} = 39$ **b.** $1 = 2e^{-0.000425x}$ **c.** $8 \cdot 6^{3x+2} = 1,600$

Solution

Note that, in each case, we use the definition of logarithm.

a. $\dfrac{10^{5x+3}}{5} = 39$ 　　　　　*Given*

$\qquad 10^{5x+3} = 195$ 　　　　*Solve for the exponential (the base, along with its exponent) by multiplying both sides by 5.*

$\qquad 5x + 3 = \log 195$ 　　　*Definition of logarithm:*
$\qquad\qquad\qquad\qquad\qquad\qquad$ $5x + 3$ *is the exponent on a base* 10 *that gives* 195.

$\qquad\qquad 5x = \log 195 - 3$ 　*We now solve the linear equation for x; subtract 3 from both sides.*

$\qquad\qquad\quad x = \dfrac{\log 195 - 3}{5}$ 　*Divide both sides by 5; this is the exact solution.*

$\qquad\qquad\qquad \approx -0.1419930777$ 　*Use a calculator to find an approximate answer.*

b. 　　　$1 = 2e^{-0.000425x}$ 　　　*Given*

$\qquad\quad \dfrac{1}{2} = e^{-0.000425x}$ 　　*Solve for the exponential by dividing both sides by 2.*

$\qquad -0.000425x = \ln \tfrac{1}{2}$ 　　*Definition of logarithm*

$\qquad\qquad\quad x = \dfrac{\ln 0.5}{-0.000425}$ 　*Solve for x by dividing both sides by* −0.000425. *This is the exact solution.*

$\qquad\qquad\qquad \approx 1,630.934542$ 　*Use a calculator to approximate root.*

c. $8 \cdot 6^{3x+2} = 1,600$ 　　　*Given*

$\qquad 6^{3x+2} = 200$ 　　　　*Solve for the exponential by dividing both sides by 8.*

$\qquad 3x + 2 = \log_6 200$ 　　*Definition of logarithm*

$\qquad\qquad 3x = \log_6 200 - 2$ 　*Solve for x; first subtract 2 from both sides.*

$\qquad\qquad\quad x = \dfrac{\log_6 200 - 2}{3}$ 　*Divide both sides by 3; this is the exact solution.*

$\qquad\qquad\qquad \approx 0.3190157417$ 　*Use a calculator to approximate root (you will need the change of base theorem to evaluate).*　■

EXAMPLE 10

Finding the Time Necessary to Compound an Investment

Let P dollars be invested at an annual rate of r for t years. Then the future value A depends on the number of times the money is compounded each year. How long will it take for $1,250 to grow to $2,000 if it is invested at

a. 8% compounded daily? Use the formula $A = P\left(1 + \dfrac{r}{365}\right)^{365t}$.

b. 8% compounded continuously? Use the formula $A = Pe^{rt}$.

Give your answer to the nearest day (assume that one year is 365 days).

Solution   **a.** $P = \$1,250$, $A = \$2,000$, $r = 0.08$, and t is the unknown.

$$A = P\left(1 + \frac{x}{365}\right)^{365t} \qquad \text{\textit{Given}}$$

$$2,000 = 1,250\left(1 + \frac{0.08}{365}\right)^{365t} \qquad \text{\textit{Substitute known values.}}$$

$$1.6 = \left(1 + \frac{0.08}{365}\right)^{365t} \qquad \text{\textit{Divide both sides by 1,250.}}$$

$$365t = \log_{(1+0.08/365)} 1.6 \qquad \text{\textit{Definition of logarithm}}$$

$$t = \frac{\log_{(1+0.08/365)} 1.6}{365} \qquad \text{\textit{Divide both sides by 365.}}$$

$$\text{\textit{Evaluate as }} \frac{\log 1.6}{\log(1 + 0.08/365)} \div 365.$$

$$\approx 5.875689188 \qquad \text{\textit{Approximate solution}}$$

We see that it is almost 6 years, but we need the answer to the nearest day. The time is 5 years + 0.875689188 year. Multiply 0.875689188 by 365 to find 319.6265535. This means that on the 319nd day of the 6th year, we are still a bit short of $2,000, so the time necessary is 5 years 320 days.

b. For continuous compounding, use the formula $A = Pe^{rt}$.

$$A = Pe^{rt} \qquad \text{\textit{Given}}$$

$$2,000 = 1,250e^{0.08t} \qquad \text{\textit{Substitute known values.}}$$

$$1.6 = e^{0.08t} \qquad \text{\textit{Divide both sides by 1,250.}}$$

$$0.08t = \ln 1.6 \qquad \text{\textit{Definition of logarithm}}$$

$$t = \frac{\ln 1.6}{0.08} \qquad \text{\textit{Divide both sides by 0.08; this is the exact solution.}}$$

$$\approx 5.875045366 \qquad \text{\textit{Approximate solution}}$$

To find the number of days, we once again subtract 5 and multiply by 365 to find that the time necessary is 5 years 320 days. (*Note:* 319.39 means that on the 319th day, you do not quite have the $2,000. The necessary time is 5 years 320 days.)

Example 10b illustrates exponential growth. Growth and decay problems are common examples of exponential equations.

GROWTH/DECAY FORMULA

Exponential **growth** or **decay** can be described by the equation

$$A = A_0 e^{rt}$$

where r is the annual growth/decay rate, t is the time (in years), A_0 is the amount present initially (present value), and A is the future value. If r is positive, this formula models growth, and if r is negative, the formula models decay.

Note: You can use this formula as long as the units of time are the same. That is, if the time is measured in days, then the growth/decay rate is a daily growth/decay rate, as illustrated by the next example.

EXAMPLE 11

Determining the Rate of Radioactive Decay

If 100.0 mg of neptunium-239 (^{239}Np) decays to 73.36 mg after 24 hr, find the value of r in the growth/decay formula for t expressed as days.

Solution   Since $A = 73.36$, $A_0 = 100.0$, and $t = 1$ (day), we have

$$A = A_0 e^{rt} \qquad \textit{Growth/decay formula}$$
$$73.36 = 100 e^{r(1)} \qquad \textit{Substitute known values.}$$
$$0.7336 = e^r \qquad \textit{Divide both sides by 100.}$$
$$r = \log_e 0.7336 \qquad \textit{Definition of logarithm}$$
$$r = \ln 0.7336 \approx -0.309791358$$

Thus, the decay rate is approximately 31%.   ■

EXAMPLE 12

Determining the Growth Rate of AIDS-Related Deaths

According to the Centers for Disease Control in Atlanta, Georgia, at the end of January 1992, the total number of AIDS-related deaths for all ages was 209,693. At that time, it was predicted that by January 1996 there would be from 400,000 to 450,000 cumulative deaths from the disease. Assuming these numbers are correct, estimate the cumulative number of AIDS-related deaths at the end of January 2002.

Solution   We assume that the growth rate will remain constant over the years of our study and also that the growth takes place continuously. From the end of January 1992 to January 1996 is 4 years, so $t = 4$. Furthermore, we will work with the more conservative estimate of cumulative deaths.

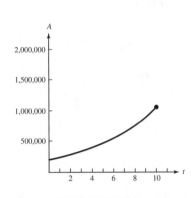

FIGURE 7.11
Cumulative deaths from
AIDS, 1992–2002

$$A = A_0 e^{rt} \qquad \text{\textit{Growth formula}}$$

$$400{,}000 = 209{,}693 e^{4r} \qquad \text{\textit{t = 4; substitute known values.}}$$

$$\frac{400{,}000}{209{,}693} = e^{4r} \qquad \text{\textit{Solve for the exponential.}}$$

$$4r = \ln\left(\tfrac{400{,}000}{209{,}693}\right) \qquad \text{\textit{Definition of logarithm}}$$

$$r = \tfrac{1}{4} \ln\left(\tfrac{400{,}000}{209{,}693}\right) \qquad \text{\textit{Divide both sides by 4.}}$$

$$\approx 0.1614549977 \qquad \text{\textit{Approximate answer by calculator}}$$

Thus, $A = 209{,}693 e^{0.1614549977t}$. At the end of January 2002, we see that $t = 10$ years:

$$A = 209{,}693 e^{0.1614549977(10)} \approx 1{,}053{,}839$$

A graph is shown in Figure 7.11. ■

EXAMPLE 13   **World Population Hits 6 Billion!**

In 1990, the world population was about 5.3 billion. If we assume a growth rate of 1.5%, when will the population reach 6 billion?

Solution

$$6 = 5.3 e^{0.015t} \qquad \text{\textit{Growth formula}}$$

$$\frac{6}{5.3} = e^{0.015t} \qquad \text{\textit{Solve for the exponential.}}$$

$$0.015t = \log_e \frac{6}{5.3} \ \text{ or } \ 0.015t = \ln \frac{6}{5.3} \qquad \text{\textit{Definition of logarithm}}$$

$$t = \frac{\ln 6/5.3}{0.015} \qquad \text{\textit{Divide both sides by 0.018.}}$$

$$t \approx 8.270176578 \qquad \text{\textit{Approximate answer by calculator}}$$

This means that we should pass the 6 billion mark in 1998. ■

EXAMPLE 14   **Calculator Solution of an Exponential Equation**

Solve $e^x \sin x = 5.0$ on $0 \le x < 2\pi$, correct to two significant digits.

Solution   This is an exponential equation, but since it is difficult to isolate the variable x, we seek an approximate solution. We graph two equations:

$$y_1 = e^x \sin x \quad \text{and} \quad y_2 = 5$$

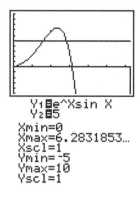

Y₁ᴮe^Xsin X
Y₂ᴮ5
Xmin=0
Xmax=6.2831853...
Xscl=1
Ymin=-5
Ymax=10
Yscl=1

FIGURE 7.12

Graphs of $y = e^x \sin x$ and $y = 5.0$

We graph these equations on the same coordinate axes, as shown in Figure 7.12. We note the points of intersection and obtain the trace values shown in Figure 7.13.

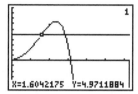
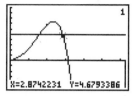

X=1.6042175 Y=4.9711884   X=2.8742231 Y=4.6793386

FIGURE 7.13 Trace values for intersection of $y = e^x \sin x$ and $y = 5.0$

To two significant digits, we find

$$x \approx 1.6 \quad \text{and} \quad x \approx 2.9$$

PROBLEM SET 7.2

Essential Concepts

1. What is the definition of logarithm?

2. What is a common logarithm? What is the notation used for a common logarithm?

3. What is a natural logarithm? What is the notation used for natural logarithm?

4. What does log N mean?

5. What does ln N mean?

6. What does $\log_b N$ mean?

7. How do you use your calculator to evaluate the following?
   a. a common logarithm
   b. a natural logarithm
   c. a logarithm to a base b

8. What is an exponential equation?

9. What are the three types of exponential equations?

10. **IN YOUR OWN WORDS** Outline a procedure for solving exponential equations.

11. What is the growth/decay formula?

Level 1 Drill

Problems 12–16: Use the definition of logarithm to explain what each expression means, and then rewrite it in exponential form.

12. a. $\log_{10} 10$ **b.** $\log_{10} 1{,}000$ **c.** $\log_{10} 10^{-5}$

13. a. $\log_5 5$ **b.** $\log_5 25$ **c.** $\log_5 5^{-3}$

14. a. $\log_e e$ **b.** $\log_e e^2$ **c.** $\log_e e^{-4}$

15. a. $\log_b b$ **b.** $\log_b b^3$ **c.** $\log_b b^{-6}$

16. a. $\log 10^n$ **b.** $\ln e^n$ **c.** $\log_b b^n$

Problems 17–19: Solve each exponential equation. Give the exact value for x.

17. a. $4^x = 8$ **b.** $5^x = 10$ **c.** $6^x = 4.5$

18. a. $10^x = 15$ **b.** $10^x = 2.5$ **c.** $10^x = 45$

19. a. $e^x = 6$ **b.** $e^x = 1.8$ **c.** $e^x = 34.2$

Problems 20–22: Write the equations in logarithmic form.

20. a. $64 = 2^6$ **b.** $100 = 10^2$ **c.** $m = n^p$

21. a. $1{,}000 = 10^3$ **b.** $81 = 9^2$ **c.** $\dfrac{1}{e} = e^{-1}$

22. a. $125 = 5^3$ **b.** $9 = \left(\dfrac{1}{3}\right)^{-2}$ **c.** $a = b^c$

Problems 23–29: *Evaluate the given expressions (to two decimal places).*

23. **a.** $\log 4.27$ **b.** $\log_b b^2$ **c.** $\log_t t^3$

24. **a.** $\log 1.08$ **b.** $\log_e e^4$ **c.** $\log_\pi \sqrt{\pi}$

25. **a.** $\log 71{,}600$ **b.** $\log_3 9$ **c.** $\log_{19} 1$

26. **a.** $\log 0.042$ **b.** $\log 0.321$ **c.** $\log 0.0532$

27. **a.** $\ln 2.27$ **b.** $\ln 16.77$ **c.** $\ln 7.3$

28. **a.** $\log_5 304$ **b.** $\log_4 3.05$ **c.** $\log_2 1{,}513$

29. **a.** $\log_\pi 100$ **b.** $\log_{1.08} 5{,}450$ **c.** $\log_{\sqrt{2}} 8.5$

30. Write each expression in terms of common logarithms and give a calculator approximation.
    a. $\log_3 45$ **b.** $\log_5 91$
    c. $\ln 10$ **d.** $\ln 1{,}000$

31. Write each expression in terms of natural logarithms and give a calculator approximation.
    a. $\log_2 0.0056$ **b.** $\log_{8.3} 105$
    c. $\log e^2$ **d.** $\log e^8$

Level 2 Drill

Problems 32–49: *Solve the exponential equations. Show the approximation you obtain with your calculator without rounding.*

32. **a.** $2^x = 128$ **b.** $3^x = 243$

33. **a.** $125^x = 25$ **b.** $216^x = 36$

34. **a.** $4^x = \frac{1}{16}$ **b.** $27^x = \frac{1}{81}$

35. **a.** $8^{3x} = 2$ **b.** $64^{2x} = 2$

36. **a.** $5^{2x+3} = 4$ **b.** $e^{2x+1} = 25$

37. **a.** $6^{5x-3} = 5$ **b.** $5^{3x-1} = 0.45$

38. **a.** $10^{3x-1} = 42$ **b.** $10^{2x+1} = 0.0234$

39. **a.** $10^{5-3x} = 0.041$ **b.** $10^{2x-1} = 515$

40. **a.** $e^{1-2x} = 3$ **b.** $e^{1-5x} = 15$

41. **a.** $5^{-x} = 8$ **b.** $7^{-x} = 125$

42. $2 \cdot 3^x + 7 = 61$ 43. $3 \cdot 5^x + 30 = 105$

44. $6 \cdot 8^x - 11 = 25$ 45. $8\pi^x - 10 = 102$

46. $\left(1 + \dfrac{0.08}{360}\right)^{360x} = 2$ 47. $\left(1 + \dfrac{0.055}{12}\right)^{12x} = 2$

48. $e^x \cos x = 4.0$ on $0 \le x < 2\pi$
    (2 significant digits)

49. $e^x \sin x = \cos x$ on $0 \le x \le \pi$
    (2 significant digits)

Problems 50–53: *Give the answers correct to two decimal places.*

50. $2^x = x^2$ 51. $3^x = x^3$

52. $x^x = \frac{1}{2}\sqrt{2}$ 53. $x^x = \frac{3}{4}\sqrt{6}$

Level 2 Applications

54. The atmospheric pressure P in pounds per square inch (psi) is approximated by $P = 14.7e^{-0.21a}$, where a is the altitude above sea level in miles. If the atmospheric pressure of Denver is 11.9 psi, estimate Denver's altitude.

55. The atmospheric pressure P in pounds per square inch (psi) is approximated by $P = 14.7e^{-0.21a}$, where a is the altitude above sea level in miles. If the pressure gauge in a small plane is 10.2 psi, estimate the plane's altitude in feet. (Recall 1 mi = 5,280 ft.)

56. A healing law for skin wounds states that $A = A_0 e^{-0.1t}$, where A is the number of square centimeters of unhealed skin after t days when the original area of the wound was A_0. Draw a graph showing the healing of a 100-cm^2 wound. How many days does it take for half the wound to heal?

57. A satellite has an initial radioisotope power supply of 50 watts. The power output in watts is given by the equation $P = 50e^{-t/250}$, where t is the time in days. Draw a graph of the power output. If the satellite will operate if there is at least 10 watts of power, how long would we expect the satellite to operate?

58. The intensity of sound is measured in decibels D and is given by

$$D = \log\left(\frac{I}{I_0}\right)^{10}$$

where I is the power of the sound in watts per cubic centimeter (W/cm^3) and $I_0 = 10^{-16}$ W/cm^3 is the power of sound just below the threshold of hearing. Find the intensity (to the nearest decibel) of
a. a whisper, 10^{-13} W/cm^3
b. normal conversation, 3.16×10^{-10} W/cm^3

59. In Example 12, we used the most conservative estimate for the number of deaths. Now use the higher number, 450,000, to estimate the cumulative number of deaths from AIDS at the end of January in 2002.

60. The half-life of ^{234}U, uranium-234, is 2.52×10^5 yr. If 97.3% of the uranium in the original sample is present, what length of time (to the nearest thousand years) has elapsed?

61. The half-life of ^{22}Na, sodium-22, is 2.6 yr. If 15.5 g of an original 100-g specimen remains, how much time has elapsed (to the nearest year)?

Level 3 Applications

62. In 1986, it was determined that the *Challenger* disaster was caused by failure of the primary O-rings. Linda Tappin gives a formula [in "Analyzing Data Relating to the *Challenger* Disaster," *The Mathematics Teacher*, Vol. 87, No. 6 (Sept. 1994, pp. 423–426)] that relates the temperature x (in degrees Fahrenheit) around the O-rings and the number y of eroded or leaky primary O-rings:

$$y = \frac{6e^{5.085 - 0.1156x}}{1 + e^{5.085 - 0.1156x}}$$

a. What is the predicted number of eroded or leaky O-rings at a temperature of 75°F?

b. What is the predicted number of eroded or leaky O-rings at a temperature of 32°F?

c. Draw the graph of this equation.

d. From the graph, how many O-rings do you think were on the spacecraft?

63. The Arrhenius function is used to relate the viscosity η of a fluid (the fluid's internal friction, which is what makes it resist a tendency to flow) to its absolute temperature T:

$$\frac{1}{\eta} = Ae^{-E/(RT)}$$

where A is a constant specific to that fluid and R is the ideal gas constant. Solve this equation for T. The resulting formula is one you could use to investigate the viscosity of different grades of motor oil at different temperatures.

Problems 64–67: Solve for the indicated variable.

64. $P = P_0 e^{rt}$ for t

65. $I = I_0 e^{-rt}$ for r

66. $A = A_0 \left(\frac{1}{2}\right)^{t/h}$ for t

67. $A = A_0 e^{t/h}$ for h

SECTION 7.3

LOGARITHMIC EQUATIONS

PROBLEM OF THE DAY

In 1989, an earthquake measuring 7.1 on the Richter scale struck San Francisco during the World Series. Over 3,000 people were injured and 67 persons were killed. What was the amount of energy released (in ergs) by this earthquake?

How would you answer this question?

Where would you begin? With a little research, you could find information on the Richter scale, which was developed by Gutenberg and Richter. The formula relating the energy E (in ergs) to the magnitude of the earthquake, M, is given by

$$M = \frac{\log E - 11.8}{1.5}$$

This equation is called a *logarithmic equation*, and the topic of this section is to solve such equations. To answer the question asked in the Problem of the Day, we first solve for log E, and then we use the definition of logarithm to write this as an exponential equation.

$$M = \frac{\log E - 11.8}{1.5} \quad \textit{Given}$$

$$1.5M = \log E - 11.8 \quad \textit{Multiply both sides by 1.5.}$$

$$1.5M + 11.8 = \log E \quad \textit{Add 11.8 to both sides.}$$

$$10^{1.5M+11.8} = E \quad \begin{array}{l}\textit{1.5M + 11.8 is the exponent on a base 10}\\ \textit{that gives E (definition of logarithm).}\end{array}$$

We can now answer the Problem of the Day. Since $M = 7.1$,

$$E = 10^{1.5M+11.8} = 10^{1.5(7.1)+11.8} \approx 2.818 \times 10^{22} \text{ ergs}$$

However, to solve certain logarithmic equations, we must first develop some properties of logarithms.

Logarithmic Functions

The notion of a logarithm can also be considered as a special type of function.

> **LOGARITHMIC FUNCTION**
>
> The function f is a **logarithmic function** if $x > 0$ and
>
> $$f(x) = \log_b x$$
>
> where $b > 0$, $b \neq 1$.

We begin our study of the logarithmic function, as we did the exponential function, by looking at some general properties of graphs of logarithmic functions.

EXAMPLE 1

Graphing a Logarithmic Function for $b > 1$

Sketch $f(x) = \log_2 x$.

Solution

This function is an increasing function with a vertical asymptote $x = 0$. The equation $y = \log_2 x$ is equivalent (by definition) to the equation $x = 2^y$. However, since x is the independent variable, we construct a table of values using the change of base theorem. That is,

$$f(0.1) = \log_2 0.1 = \frac{\log 0.1}{\log 2} \approx -3.321928095$$

A table of values and graph are shown in Figure 7.14; the values are rounded to the nearest hundredth.

x	$f(x) = \log_2 x$
0.1	−3.32
0.5	−1
0.75	−0.42
1	0
2	1
3	1.58
4	2
8	3

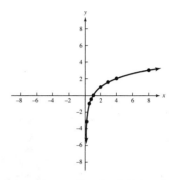

FIGURE 7.14   Graph of $f(x) = \log_2 x$ ■

EXAMPLE 2

Graphing a Logarithmic Function for $0 < b < 1$

Sketch $f(x) = \log_{0.5} x$.

Solution   This function is a decreasing function with a vertical asymptote $x = 0$. The equation $y = \log_{0.5} x$ is equivalent to the equation $x = (0.5)^y$. A table of values and graph are shown in Figure 7.15; the values are, once again, rounded to the nearest hundredth.

x	$f(x) = \log_{0.5} x$
0.1	3.32
0.5	1
0.75	0.42
1	0
2	−1
3	−1.58
4	−2
8	−3

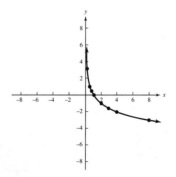

FIGURE 7.15   Graph of $f(x) = \log_{0.5} x$ ■

Examples 1 and 2 are summarized in Table 7.2 (page 360).

TABLE 7.2 Directory of Curves   Graphs of Logarithmic Functions

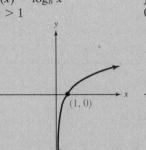

EQUATIONS:	$f(x) = \log_b x$	$f(x) = \log_b x$
Restrictions:	$b > 1$	$0 < b < 1$
GRAPHS:		
PASSES THROUGH:	$(1, 0)$, increasing	$(1, 0)$, decreasing
DOMAIN:	$(0, \infty)$	$(0, \infty)$
RANGE:	$(-\infty, \infty)$	$(-\infty, \infty)$
ASYMPTOTES:	vertical, $x = 0$	vertical, $x = 0$

Exponential and Logarithmic Functions as Inverse Functions

In Section 3.4, we introduced the idea of inverse functions. By looking at the graphs of this section, you may notice a relationship between the graphs of the exponential and logarithmic functions. For example, consider the graphs of $y = 2^x$, $y = \log_2 x$, along with the line $y = x$, as shown in Figure 7.16.

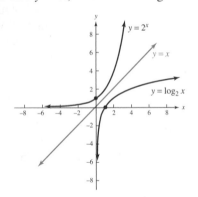

FIGURE 7.16
Graphs of $y = 2^x$, $y = \log_2 x$,
and $y = x$

We need to consider two **fundamental properties of logarithms** that follow from the definition of a logarithm.

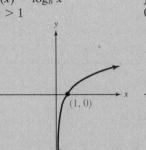

YIELD

Sometimes I call these the Grant's Tomb properties because if you think about them for a moment they become self-evident.

FUNDAMENTAL PROPERTIES OF LOGARITHMS

1. $\log_b b^x = x$
2. $b^{\log_b x} = x$ $x > 0$

To justify property 1, remember the definition of logarithm:

$$b^M = N \qquad \text{means} \qquad \log_b N = M$$

and let $M = x$ and $N = b^x$. Then, since $b^x = b^x$, we have $\log_b b^x = x$. To justify property 2, let $M = \log_b x$ and $N = x$ in the definition. Since $\log_b x = \log_b x$, then using the definition, we have $b^{\log_b x} = x$.

To show that $f(x) = \log_b x$ and $g(x) = b^x$ are inverse functions, we need to show $f[g(x)] = g[f(x)] = x$ for all $x > 0$, $b > 0$, $b \neq 1$:

$$
\begin{aligned}
f[g(x)] &= f(b^x) & g[f(x)] &= g(\log_b x) \\
&= \log_b b^x & &= b^{\log_b x} \\
&= x \quad \textit{By property 1} & &= x \quad \textit{By property 2}
\end{aligned}
$$

This proves the following theorem.

> **EXPONENTIAL AND LOGARITHMIC FUNCTIONS ARE INVERSES**
>
> The exponential and logarithmic functions with base b are inverse functions of one another.

Logarithmic Equations

A **logarithmic equation** is an equation for which there is a logarithm on one or both sides. The key to solving logarithmic equations is the following theorem, which we will call the **log of both sides theorem.**

> **LOG OF BOTH SIDES THEOREM**
>
> If A, B, and b are positive real numbers with $b \neq 1$, then $\log_b A = \log_b B$ is equivalent to $A = B$.

The proof of this theorem is not difficult, and it depends on the two fundamental properties of logarithms given in the previous section.

Basically, all logarithmic equations in this book fall into one of four types:

Type I: The unknown is the logarithm; $\log_2 \sqrt{3} = x$.
Type II: The unknown is the base; $\log_x 6 = 2$.
Type III: The logarithm of an unknown is equal to a number; $\ln x = 5$.
Type IV: The logarithm of an unknown is equal to the logarithm of a number; $\log_5 x = \log_5 72$.

The following example illustrates the procedure for solving each type of logarithmic equation.

EXAMPLE 3

Procedure for Solving Logarithmic Equations

Solve the following logarithmic equations:

a. Type I: $\log_2 \sqrt{3} = x$ **b.** Type II: $\log_x 6 = 2$
c. Type III: $\ln x = 5$ **d.** Type IV: $\log_5 x = \log_5 72$

Solution

a. Type I: $\log_2 \sqrt{3} = x$. If the logarithmic expression does not contain a variable, you can use your calculator to evaluate. Remember, this type was evaluated in Section 7.2. If it is a common logarithm (base 10), use the $\boxed{\text{LOG}}$ key; if it is a natural logarithm (base e), use the $\boxed{\text{LN}}$ key; if it has another base, use the change of base theorem:

$$\log_a N = \frac{\log N}{\log a} \text{ or } \log_a N = \frac{\ln N}{\ln a}$$

For this example, we see

$$x = \log_2 \sqrt{3} = \frac{\log \sqrt{3}}{\log 2} \approx 0.7924812504$$

b. Type II: $\log_x 6 = 2$. If the unknown is the base, then use the definition of logarithm to write an equation that is not a logarithmic equation.

$$\log_x 6 = 2 \qquad \textit{Given}$$
$$x^2 = 6 \qquad \textit{Definition of logarithm}$$
$$x = \pm\sqrt{6} \qquad \textit{Square root property}$$

When solving logarithmic equations, make sure your answers are in the domain of the variable. Remember that, by definition, the base must be positive. For this example, $x = -\sqrt{6}$ is not in the domain, so the solution is $x = \sqrt{6}$.

c. The third and fourth types of logarithmic equations are the most common, and both involve the logarithm of an unknown quantity on one side of an equation. For the third type, use the definition of logarithm (and a calculator for an approximate solution, if necessary).

$$\ln x = 5$$
$$e^5 = x \qquad \textit{5 is the exponent on e that gives x.}$$

This is the exact solution. An approximate solution is $x \approx 148.4131591$.

d. When a log form occurs on both sides, use the log of both sides theorem. *Make sure the log on both sides has the same base:*

$$\log_5 x = \log_5 72$$
$$x = 72$$

Historical Note

For years, logarithms were used as a computational aid, but today they are more important in solving problems. Perhaps we have come full circle: The Babylonians first used logarithms to solve problems, not to do calculations.

The modern basis of logarithms was developed by Tycho Brahe (1546–1601) to disprove the Copernican theory of planetary motion. The name he used for the method was *prostaphaeresis*. In 1590, a storm brought together Brahe and John Craig, who in turn told Napier about Brahe's method. It was John Napier (1550–1617) who was the first to use the word *logarithm*. Napier was the Isaac Asimov of his day, having envisioned the tank, the machine gun, and the submarine. He also predicted that the end of the world would occur between 1688 and 1700. He is best known today as the inventor of logarithms.

Example 3 illustrates the procedures for solving logarithmic equations, but most logarithmic equations are not as easy as those in Example 3. Usually, you must do some algebraic simplification to put the problem into the form of one of the four types of logarithmic equations. You may also have realized that Type IV is a special case of Type III. For example, to solve

$$\log_5 x = \log_3 72$$

which looks like Example 3d, we see that we cannot use the log of both sides theorem because the bases are not the same. We can, however, treat this as a Type III equation by using the definition of logarithm to write

$$x = 5^{\log_3 72}$$

This can be evaluated using a calculator. However, you may find it easier to visualize if you write

$$\log_5 x = \log_3 72 \approx 3.892789261$$

so that 3.892789261 is the exponent on a base 5 that gives x. In other words,

$$x \approx 5^{3.892789261} \approx 525.9481435$$

To simplify logarithmic expressions, remember that a logarithm is an exponent and the laws of exponents correspond to the **laws of logarithms**.

LAWS OF LOGARITHMS

If A, B, and b are positive numbers, p is any real number, and $b \neq 1$:

First Law (Additive)

$$\log_b(AB) = \log_b A + \log_b B$$

The log of the product of two numbers is the sum of the logs of those numbers.

Second Law (Subtractive)

$$\log_b\left(\frac{A}{B}\right) = \log_b A - \log_b B$$

The log of the quotient of two numbers is the log of the numerator minus the log of the denominator.

Third Law (Multiplicative)

$$\log_b A^p = p \log_b A$$

The log of the pth power of a number is p times the log of that number.

The proofs of these laws of logarithms are easy. The first law of logarithms comes from the first law of exponents:

$$b^x b^y = b^{x+y}$$

Let $A = b^x$ and $B = b^y$, so that $AB = b^{x+y}$. Then from the definition of logarithm, these three equations are equivalent to

$$x = \log_b A, \quad y = \log_b B, \quad \text{and} \quad x + y = \log_b(AB)$$

Therefore, by putting these pieces together, we have

$$\log_b(AB) = x + y = \log_b A + \log_b B$$

Similarly, for the second law of logarithms,

$$\frac{A}{B} = \frac{b^x}{b^y} = b^{x-y} \quad \textit{Second law of exponents}$$

Thus $x - y = \log_b\left(\dfrac{A}{B}\right) \quad \textit{Definition of logarithm}$

which means $\log_b\left(\dfrac{A}{B}\right) = \log_b A - \log_b B \quad \textit{Since } x = \log_b A \textit{ and } y = \log_b B$

The proof of the third law of logarithms follows from the third law of exponents, and you are asked to do this in the problem set. We can also prove this third law by using the second law of logarithms for p a positive integer:

$$\log_b A^p = \log_b \underbrace{A \cdot A \cdot A \cdots A}_{p \text{ factors}} \quad \textit{Definition of } A^p$$

$$= \underbrace{\log_b A + \log_b A + \log_b A + \cdots + \log_b A}_{p \text{ terms}} \quad \textit{Second law of logarithms}$$

$$= p \log_b A$$

When logarithms were used for calculations, the laws of logarithms were used to expand an expression such as $\log\left(\dfrac{6 \cdot 45.62^2}{84.2}\right)$. Calculators have made such problems obsolete. Today, logarithms are important in solving equations, and the procedure for solving logarithmic equations requires that we take an algebraic expression involving logarithms and write it as a single logarithm. We might call this *contracting* a logarithmic expression.

EXAMPLE 4

Using the Laws of Logarithms to Contract

Write each statement as a single logarithm.

a. $\log x + 5 \log y - \log z$ **b.** $\log_2 3x - 2 \log_2 x + \log_2(x + 3)$

Solution   **a.** $\log x + 5 \log y - \log z = \log x + \log y^5 - \log z$ *Third law*

$$= \log xy^5 - \log z \qquad \text{\textit{First law}}$$

$$= \log \frac{xy^5}{z} \qquad \text{\textit{Second law}}$$

NOTE

Be sure the bases are the same before you use the laws of logarithms.

b. $\log_2 3x - 2 \log_2 x + \log_2(x + 3) = \log_2 3x - \log_2 x^2 + \log_2(x + 3)$

$$= \log_2 \frac{3x(x + 3)}{x^2}$$

$$= \log_2 \frac{3(x + 3)}{x}$$

■

EXAMPLE 5   **Using the Laws of Logarithms to Solve a Logarithmic Equation**

Solve $\log_8 3 + \frac{1}{2} \log_8 25 = \log_8 x$.

Solution   The goal here is to make this look like a Type IV logarithmic equation so that there is a single log expression on both sides.

$$\log_8 3 + \tfrac{1}{2}\log_8 25 = \log_8 x$$

$$\log_8 3 + \log_8 25^{1/2} = \log_8 x \qquad \text{\textit{Third law of logarithms}}$$

$$\log_8 3 + \log_8 5 = \log_8 x \qquad 25^{1/2} = (5^2)^{1/2} = 5$$

$$\log_8(3 \cdot 5) = \log_8 x \qquad \text{\textit{First law of logarithms}}$$

$$15 = x \qquad \text{\textit{Log of both sides theorem}}$$

The solution is 15. (Check to be sure 15 is in the domain of the variable.)

■

Calculator Window **?** **X**

If you have a graphing calculator, you can solve an equation such as the one given in Example 5 graphically. There are certain conventions you may need to follow when you input the data. On many calculators and computers, it is necessary to enclose the value for a logarithm in parentheses. That is, log 3 is input as LOG(3). Insert parentheses if required by your calculator. Also, to work with a base 8, it is necessary to use the change of base theorem to write $\log_8 3$ as LOG 3/LOG 8 and $\log_8 25$ as LOG 25/LOG 8. Since we have chosen Y1 and Y2 so that Y1 = Y2, we see that the solution is the point of intersection of the graphs. Notice that y_1 is a line because the left side of the equation in Example 5 is constant (no variable).

To find the point of intersection, use the TRACE function of your calculator. For this example, we find X=15.106383, Y=1.3022969. By using the ZOOM, you can find any reasonable degree of accuracy.

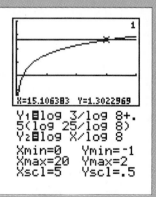

EXAMPLE 6

Using the Laws of Logarithms to Solve an Exponential Equation

Solve: **a.** $5^{x+2} = 9^x$ **b.** $5^{x+2} = 8 \cdot 9^x$

Give both the exact answer and an approximate answer (to the nearest hundredth).

Solution   **a.**

$5^{x+2} = 9^x$	*Given*
$x + 2 = \log_5 9^x$	*Definition of logarithm*
$x + 2 = x \log_5 9$	*Third law of logarithms*
$2 = x \log_5 9 - x$	*Subtract x from both sides; we want all terms involving the variable on one side.*
$2 = x(\log_5 9 - 1)$	*Common factor*
$\dfrac{2}{\log_5 9 - 1} = x$	*Divide both sides by coefficient of x.*

Thus, the exact value is $\dfrac{2}{\log_5 9 - 1}$, and the approximate value is 5.48.

b. This is similar to part **a**, except for the coefficient of 9^x.

$5^{x+2} = 8 \cdot 9^x$	*Given*
$x + 2 = \log_5(8 \cdot 9^x)$	*Definition of logarithm*
$x + 2 = \log_5 8 + \log_5 9^x$	*Second law of logarithms*
$x + 2 = \log_5 8 + x \log_5 9$	*Third law of logarithms*
$x - x \log_5 9 = \log_5 8 - 2$	*Terms involving the variable on one side*
$x(1 - \log_5 9) = \log_5 8 - 2$	*Common factor*
$x = \dfrac{\log_5 8 - 2}{1 - \log_5 9}$	*Divide both sides by coefficient of x.*

Thus, the exact value is $\dfrac{\log_5 8 - 2}{1 - \log_5 9}$ and the approximate value is 1.94. ∎

When solving logarithmic equations, you must be mindful of extraneous solutions. The reason for this is that the logarithm function requires that the arguments be positive, but when solving an equation, we may not know the signs of the arguments. For example, if you solve an equation involving log x and obtain two answers, for example, $x = 3$ and $x = -4$, then the value $x = -4$ must be extraneous because $\log(-4)$ is not defined.

EXAMPLE 7   **Logarithmic Equations**

Solve: **a.** $\log 15 + 2 = \log(x + 250)$ **b.** $\log 5 + \log(2x^2) = \log x + \log 15$

Solution   Use the laws of logarithms to combine the log statements. We have chosen two examples that are quite similar, but whose solutions require slightly different procedures. For part **a**, rewrite the expression so that all parts involving logarithms are on one side. In part **b**, rewrite the expression so that all logarithms involving the variable are on one side.

Type III: *Use definition of a logarithm.*

a. $\log 15 + 2 = \log(x + 250)$

$2 = \log(x + 250) - \log 15$ *Subtract* log 15 *from both sides.*

$2 = \log \dfrac{x + 250}{15}$ *Second law of logarithms*

$10^2 = \dfrac{x + 250}{15}$ *Definition of logarithm*

$1,500 = x + 250$ *Multiply both sides by* 15.

$1,250 = x$ *Subtract* 250 *from both sides.*

The solution is $x = 1,250$.

Type IV: *Use log of both sides theorem.*

b. $\log 5 + \log(2x^2) = \log x + \log 15$

$\log(2x^2) - \log x = \log 15 - \log 5$ *Subtract* log x *and* log 5 *from both sides.*

$\log \dfrac{2x^2}{x} = \log \dfrac{15}{5}$ *Second law of logarithms*

$\log(2x) = \log 3$

$2x = 3$ *Log of both sides theorem*

$x = \dfrac{3}{2}$

The solution is $x = \dfrac{3}{2}$.

EXAMPLE 8

Solving a Logarithmic Equation

Solve: $\ln x - \frac{1}{2} \ln 2 = \frac{1}{2} \ln(x + 4)$

Solution

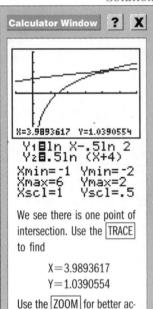

Calculator Window ? X

X=3.9893617 Y=1.0390554
Y₁⊟ln X-.5ln 2
Y₂⊟.5ln (X+4)
Xmin=-1 Ymin=-2
Xmax=6 Ymax=2
Xscl=1 Yscl=.5

We see there is one point of intersection. Use the TRACE to find

X=3.9893617
Y=1.0390554

Use the ZOOM for better accuracy.

$$\ln x - \tfrac{1}{2}\ln 2 = \tfrac{1}{2}\ln(x+4)$$

$2 \ln x - \ln 2 = \ln(x+4)$ *Multiply both sides by 2.*

$2 \ln x - \ln(x+4) = \ln 2$ *Logs with variables on one side*

$\ln x^2 - \ln(x+4) = \ln 2$ *Third law of logarithms*

$$\ln \frac{x^2}{x+4} = \ln 2$$ *Second law of logarithms*

$$\frac{x^2}{x+4} = 2$$ *Log of both sides theorem*

$x^2 = 2(x+4)$

$x^2 - 2x - 8 = 0$

$(x-4)(x+2) = 0$

$x = 4, -2$

Since the domain requires that the argument is positive, we see that the solution is $x = 4$. ■

EXAMPLE 9

Proof of the Change of Base Theorem

Prove: $\log_a x = \dfrac{\log_b x}{\log_b a}$

Proof   Let $y = \log_a x$.

$a^y = x$ *Definition of logarithm*

$\log_b a^y = \log_b x$ *Log of both sides theorem*

$y \log_b a = \log_b x$ *Third law of logarithms*

$$y = \frac{\log_b x}{\log_b a}$$ *Divide both sides by $\log_b a$ ($\log_b a \neq 0$).*

Thus, by substitution, $\log_a x = \dfrac{\log_b x}{\log_b a}$. ■

PROBLEM SET 7.3

Essential Concepts

1. Simplify: **a.** $\log_b b^x$ **b.** $b^{\log_b x}$

2. Simplify:
   a. $e^{\ln 23}$ **b.** $10^{\log 3.4}$ **c.** $4^{\log_4 x}$

3. Simplify:
   a. $\log 10^{4.2}$ **b.** $\ln e^3$ **c.** $\log_6 6^x$

4. What is a logarithmic equation?

5. What are the four types of logarithmic equations?

6. **IN YOUR OWN WORDS** Outline a procedure for solving logarithmic equations.

Problems 7–23: *Classify each as true or false. If it is false, explain why you think it is false.*

7. log 500 is the exponent on 10 that gives 500.

8. A common logarithm is a logarithm in which the base is 2.

9. A natural logarithm is a logarithm in which the base is 10.

10. In $\log_b N$, the exponent is N.

11. To evaluate $\log_5 N$, divide log 5 by log N.

12. If $2 \log_3 81 = 8$, then $\log_3 6{,}561 = 8$.

13. If $2 \log_3 81 = 8$, then $\log_3 81 = 4$.

14. If $\log_{1.5} 8 = x$, then $x^{1.5} = 8$.

15. $\ln \dfrac{x}{2} = \dfrac{\ln x}{2}$

16. $\log_b(A + B) = \log_b A + \log_b B$

17. $\log_b AB = (\log_b A)(\log_b B)$

18. $\dfrac{\log_b A}{\log_b B} = \log_b \dfrac{A}{B}$

19. $\dfrac{\log A}{\log B} = \dfrac{\ln A}{\ln B}$

20. $\dfrac{\log_b A}{\log_b B} = \log_b(A - B)$

21. $\dfrac{\log_b A}{\log_b B} = \log_b A - \log_b B$

22. $\log_b N$ is negative when N is negative.

23. log N is negative when $N > 1$.

Level 1 Drill

Problems 24–29: *Graph the logarithmic functions.*

24. $y = \log_3 x$ 25. $y = \log_{1/3} x$ 26. $y = \log x$

27. $y = \ln x$ 28. $y = \log_\pi x$ 29. $y = \log_6 6^x$

Problems 30–33: *Find a simplified value for x by inspection. Do not use a calculator.*

30. **a.** $\log_5 25 = x$ **b.** $\log_2 128 = x$
    c. $\log_3 81 = x$ **d.** $\log_4 64 = x$

31. **a.** $\log \frac{1}{10} = x$ **b.** $\log 10{,}000 = x$
    c. $\log 1{,}000 = x$ **d.** $\log \frac{1}{1{,}000} = x$

32. **a.** $\log x = 5$ **b.** $\log_x e = 1$
    c. $\ln x = 2$ **d.** $\ln x = 3$

33. **a.** $\ln x = 4$ **b.** $\ln x = \ln 14$
    c. $\ln 9.3 = \ln x$ **d.** $\ln 109 = \ln x$

Level 2 Drill

Problems 34–38: *Contract the given expressions. That is, use the properties of logarithms to write each expression as a single logarithm with a coefficient of 1.*

34. **a.** $\log 2 + \log 3 + \log 4$
    b. $\log 40 - \log 10 - \log 2$
    c. $2 \ln x + 3 \ln y - 4 \ln z$

35. **a.** $3 \ln 4 - 5 \ln 2 + \ln 3$
    b. $3 \ln 4 - 5 \ln(2 + 3)$
    c. $3 \ln 4 - 5(\ln 2 + \ln 3)$

36. **a.** $\ln 3 - 2 \ln 4 + \ln 8$
    b. $\ln 3 - 2 \ln(4 + 8)$
    c. $\ln 3 - 2(\ln 4 + \ln 8)$

37. **a.** $\log(x^2 - 9) - 2 \log(x + 3) + 3 \log x$
    b. $\log(x^2 - 9) - 2[\log(x + 3) + 3 \log x]$
    c. $\ln(x^2 - 4) - \ln(x + 2)$

38. a. $\frac{1}{2}\log(x + 1) - \frac{3}{2}\log(x + 1)$

   b. $\frac{4\pi}{3}\ln x - \frac{\pi}{3}\ln x + 2\pi \ln x$

   c. $-\log x + 8 \log x + \frac{4}{3}\log x - \frac{1}{3}\log x$

***Problems* 39–59:** *Solve the equations by finding the exact solution.*

39. a. $\frac{1}{2}x - 2 = 2$ **b.** $\frac{1}{2}\log x - \log 100 = 2$

40. a. $3 + 2x = 11$ **b.** $\ln e^3 + 2 \log x = 11$

41. a. $\frac{1}{2}x = 5 - x$ **b.** $\frac{1}{2}\log_b x = 3 \log_b 5 - \log_b x$

42. a. $x - 2 = 2$ **b.** $\log 10^x - 2 = \log 100$

43. a. $1 = x - 1$ **b.** $\ln e = \ln \dfrac{\sqrt{2}}{x} - \ln e$

44. a. $1 = \frac{3}{2} - x$ **b.** $\log 10 = \log \sqrt{1{,}000} - \log x$

45. a. $3 - x = 1$ **b.** $\ln e^3 - \ln x = 1$

46. a. $0 + x = 2$ **b.** $\ln 1 + \ln e^x = 2$

47. $\log(\log x) = 1$ **48.** $\ln[\log(\ln x)] = 0$

49. $x^2 5^x = 5^x$ **50.** $x^2 3^x = 9(3^x)$

51. $\log 2 = \frac{1}{4}\log 16 - x$

52. $\log_8 5 + \frac{1}{2}\log_8 9 = \log_8 x$

53. $\ln x + \ln(x - 3) = 2$

54. $5^x = 8^{2x}$ **55.** $10^x = 4^{2x}$

56. $3^x = 8 \cdot 5^{x+3}$ **57.** $6^{x+2} = 123 \cdot 10^x$

58. $10^{5x+1} = e^{2-3x}$ **59.** $5^{2+x} = 6^{3x+2}$

Level 2 Applications

60. An advertising agency conducted a survey and found that the number of units sold, N, is related to the amount a spent on advertising (in dollars) by the following formula:

$$N = 1{,}500 + 300 \ln a \quad (a \geq 1)$$

   a. How many units are sold after spending $1,000?
   b. How many units are sold after spending $50,000?
   c. How much needs to be spent (to the nearest hundred dollars) to sell 4,000 units?

61. The pH of a substance measures its acidity or alkalinity. It is found by the formula

$$pH = -\log[H^+]$$

where $[H^+]$ is the concentration of hydrogen ions in an aqueous solution given in moles per liter.
   a. What is the pH (to the nearest tenth) of a lemon for which $[H^+] = 2.86 \times 10^{-4}$?
   b. What is the pH (to the nearest tenth) of rainwater for which

$$[H^+] = 6.31 \times 10^{-7}?$$

   c. If the pH of a tested substance is 8.1, what is the hydrogen content? Give your answer in scientific notation.

62. The Richter scale for measuring earthquakes was developed by Gutenberg and Richter. It relates the energy E (in ergs) to the magnitude of the earthquake M by the formula

$$M = \frac{\log E - 11.8}{1.5}$$

   a. A small earthquake is one that releases 15^{15} ergs of energy. What is the magnitude (to the nearest hundredth) of such an earthquake on the Richter scale?
   b. A large earthquake is one that releases 10^{25} ergs of energy. What is the magnitude (to the nearest hundredth) of such an earthquake on the Richter scale?
   c. How much energy is released in an 8.0 earthquake?
   d. Solve for E.

63. The "learning curve" describes the rate at which a person learns certain tasks. If a person sets a goal of typing N words per minute (wpm), the length of time t (in days) to achieve this goal is given by

$$t = -62.5 \ln \left(1 - \frac{N}{80}\right)$$

   a. According to this formula, what is the maximum number of words per minute?
   b. Solve for N.

64. The "forgetting curve" for memorizing nonsense syllables is given by

$$R = 80 - 27 \ln t \quad (t \geq 1)$$

where R is the percentage who remember the syllables after t seconds.
a. In how many seconds would only 10% ($R = 10$) of the students remember?
b. Solve for t.

65. Historical Question INDIVIDUAL RESEARCH PROJECT Write an essay on John Napier. Include what he is famous for today, and what he considered to be his crowning achievement. Also include a discussion of "Napier's bones."

66. INDIVIDUAL RESEARCH PROJECT Write an essay on earthquakes. In particular, discuss the Richter scale for measuring earthquakes. What is its relationship to logarithms?

Level 3 Theory

67. Prove the third law of logarithms using the third law of exponents. That is, prove $\log_b A^p = p \log_b A$.

68. A population that grows rapidly at first but then slows its growth can be described by a curve called a *logistic curve*. This equation has the form

$$y = \frac{k}{1 + be^{-cx}}$$

a. Graph this equation for $k = 100$, $b = 5$, and $c = 1.15$
b. Solve this equation for x.

SECTION 7.4

HYPERBOLIC FUNCTIONS

PROBLEM OF THE DAY

In physics, it is shown that a heavy, flexible cable (for example, a power line) that is suspended between two points at the same height assumes a shape described by a curve called a catenary (see Figure 7.17), with an equation of the form

$$y = \frac{a}{2}\left(e^{x/a} + e^{-x/a}\right)$$

Graph this equation for $-20 \leq x \leq 20$ for $a = 50$. Where do you think the poles are located, and how tall is each pole (assume the ground is level and the units are in feet)? Also, what is the minimum distance between the ground and the cable?

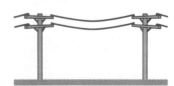

FIGURE 7.17   Hanging cable

How would you answer this question?

You might begin by writing the equation

$$y = 25(e^{x/50} + e^{-x/50})$$

We recognize this as an exponential function. You might use a graphing calculator or manually plot points to find the graph.

x	y
-20	54.054
-10	51.003
0	50
10	51.003
20	54.054

It appears that the poles are located 40 feet apart and the height of the poles is about 54 feet. The minimum distance is at the center, and we see this is 50 feet.

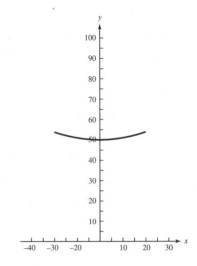

FIGURE 7.18   Graph of hanging cable

Hyperbolic Functions

The catenary is one of several important applications that involve combinations of exponential functions. The equation of the catenary gives rise to a group of transcendental functions that behave very much like the trigonometric functions. These functions are studied in physics and calculus. The goal of this section is to study such combinations and their inverses not only as a prelude to more advanced work, but also as an interesting exercise in handling identities.

In certain ways, the functions we shall study are analogous to the trigonometric functions, and they have essentially the same relationship to the hyperbola that the trigonometric functions have to the circle. For this reason, these functions are called **hyperbolic functions**. Three basic hyperbolic functions are the **hyperbolic sine** (denoted "sinh x" and pronounced "cinch"), the **hyperbolic cosine** (cosh x; pronounced "kosh"), and the **hyperbolic tangent** (tanh x; pronounced "tansh"). They are defined as follows.

HYPERBOLIC FUNCTIONS

$$\sinh x = \frac{e^x - e^{-x}}{2} \qquad \text{for all } x$$

$$\cosh x = \frac{e^x + e^{-x}}{2} \qquad \text{for all } x$$

$$\tanh x = \frac{\sinh x}{\cosh x} = \frac{e^x - e^{-x}}{e^x + e^{-x}} \qquad \text{for all } x$$

We begin with certain properties of these hyperbolic functions.

EXAMPLE 1 **Proving a Hyperbolic Identity**

Prove that $\cosh^2 x - \sinh^2 x = 1$.

Solution We will prove this identity by using the definition.

$$\cosh^2 x - \sinh^2 x = \left(\frac{e^x + e^{-x}}{2}\right)^2 - \left(\frac{e^x - e^{-x}}{2}\right)^2$$

$$= \frac{(e^x + e^{-x})^2}{4} - \frac{(e^x - e^{-x})^2}{4}$$

$$= \frac{e^{2x} + 2e^x e^{-x} + e^{-2x} - e^{2x} + 2e^x e^{-x} - e^{-2x}}{4}$$

$$= \frac{2 + 2}{4} \quad \text{Since } e^x e^{-x} = 1$$

$$= 1$$

There are three additional hyperbolic functions, the **hyperbolic cotangent** (coth x), the **hyperbolic secant** (sech x), and the **hyperbolic cosecant** (csch x). These functions are defined as follows:

$$\coth x = \frac{1}{\tanh x} = \frac{e^x + e^{-x}}{e^x - e^{-x}}$$

$$\operatorname{sech} x = \frac{1}{\cosh x} = \frac{2}{e^x + e^{-x}}$$

$$\operatorname{csch} x = \frac{1}{\sinh x} = \frac{2}{e^x - e^{-x}}$$

We now have eight **hyperbolic identities** that look very much like the eight fundamental identities for the trigonometric functions.

HYPERBOLIC IDENTITIES

$$\operatorname{sech} x = \frac{1}{\cosh x} \qquad \operatorname{csch} x = \frac{1}{\sinh x} \qquad \coth x = \frac{1}{\tanh x}$$

$$\tanh x = \frac{\sinh x}{\cosh x} \qquad \coth x = \frac{\cosh x}{\sinh x}$$

$$\cosh^2 x - \sinh^2 x = 1$$
$$1 - \tanh^2 x = \operatorname{sech}^2 x$$
$$\coth^2 x - 1 = \operatorname{csch}^2 x$$

Help Window ? X

Notice that the only difference between these identities and the eight fundamental identities is the Pythagorean identities have positive signs and the hyperbolic identities have negative signs.

You are asked to prove some of these identities in the problem set at the end of this section.

Graphs of Hyperbolic Functions

A major difference between the trigonometric and hyperbolic functions is that the trigonometric functions are periodic, but the hyperbolic functions are not. Graphs of the three basic hyperbolic functions are shown in Figure 7.19. The hanging cable we graphed for the Problem of the Day is a hyperbolic cosine.

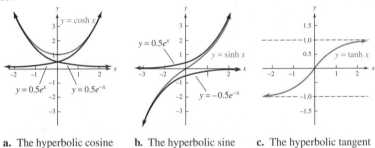

a. The hyperbolic cosine
$y = \cosh x$

b. The hyperbolic sine
$y = \sinh x$

c. The hyperbolic tangent
$y = \tanh x$

FIGURE 7.19   Graphs of the three basic hyperbolic functions

Inverse Hyperbolic Functions

Inverse hyperbolic functions are also of interest in physics and calculus. We define the inverse hyperbolic functions for the stated domains.

			Domain
$y = \cosh^{-1} x$	if and only if	$x = \cosh y$	$x \geq 1$
$y = \sinh^{-1} x$	if and only if	$x = \sinh y$	all x
$y = \tanh^{-1} x$	if and only if	$x = \tanh y$	$-1 < x < 1$
$y = \operatorname{sech}^{-1} x$	if and only if	$x = \operatorname{sech} y$	$0 < x \leq 1$
$y = \operatorname{csch}^{-1} x$	if and only if	$x = \operatorname{csch} y$	$x \neq 0$
$y = \coth^{-1} x$	if and only if	$x = \coth y$	$x < -1$ or $x > 1$

Since the hyperbolic functions are defined in terms of exponential functions, we can express the inverse hyperbolic functions in terms of logarithmic functions. We summarize this relationship in the following example.

EXAMPLE 2

Expressing an Inverse Hyperbolic Sine in Terms of a Logarithmic Function

Write $\sinh^{-1} x$ as a logarithmic function.

Solution   Let $y = \sinh^{-1} x$; then its inverse is

$$x = \sinh y \qquad \textit{Definition of inverse hyperbolic sine}$$

$$= \frac{1}{2}(e^y - e^{-y}) \qquad \textit{Definition of hyperbolic sine}$$

$$2x = e^y - \frac{1}{e^y} \qquad \textit{Multiply both sides by 2.}$$

$$e^{2y} - 2xe^y - 1 = 0 \qquad \textit{Multiply both sides by } e^y.$$

$$e^y = \frac{2x \pm \sqrt{4x^2 + 4}}{2} \qquad \begin{array}{l} \textit{Quadratic formula with} \\ a = 1, b = -2x, c = -1 \end{array}$$

$$= x \pm \sqrt{x^2 + 1}$$

Because $e^y > 0$ for all y, the only solution is $e^y = x + \sqrt{x^2 + 1}$, and from the definition of logarithms,

$$y = \ln(x + \sqrt{x^2 + 1}) \quad \text{so that} \quad \sinh^{-1} x = \ln(x + \sqrt{x^2 + 1})$$ ∎

The graph of the inverse hyperbolic sine function can be obtained by reflecting the graph of $y = \sinh x$ in the line $y = x$, as shown in Figure 7.20. Alternatively, we can graph the logarithmic function, $y = \ln(x + \sqrt{x^2 + 1})$, also shown in Figure 7.20.

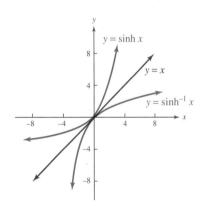

FIGURE 7.20   The graph of $y = \sinh^{-1} x$

PROBLEM SET 7.4

Level 1 Drill

Problems 1–24: Evaluate (correct to four decimal places) the indicated hyperbolic or inverse hyperbolic functions.

1. $\sinh 2$

2. $\cosh 3$

3. $\tanh(-1)$

4. $\sinh 0$

5. $\cosh 1.2$

6. $\tanh 0$

7. $\sinh(-2.3)$

8. $\cosh(-4)$

9. $\tanh(-3)$

10. $\sinh\left(-\frac{\pi}{2}\right)$

11. $\cosh(-\pi)$

12. $\tanh(-2\pi)$

13. $\mathrm{sech}\ 1.5$

14. $\mathrm{csch}\ \frac{\pi}{2}$

15. $\coth 3\pi$

16. $\cosh^{-1} 1.5$

17. $\sinh^{-1} 2.5$

18. $\tanh^{-1} 0$

19. $\cosh(\ln 3)$

20. $\sinh(\ln 2)$

21. $\tanh(\ln 3)$

22. $\mathrm{sech}^{-1} 1.2$

23. $\mathrm{csch}^{-1} 0.3$

24. $\coth^{-1} 0.5$

Level 2 Drill

25. Determine where the graph of $y = \tanh x$ is rising and falling. Sketch the graph and compare with Figure 7.19c.

Problems 26–33: Graph the curves for the given equations.

26. $y = \coth x$ **27.** $y = \operatorname{sech} x$

28. $y = \operatorname{csch} x$ **29.** $y = \cosh^{-1} x$

30. $y = \sinh^{-1} x$ **31.** $y = \tanh^{-1} x$

32. $y = -3 \cosh(x/3)$ **33.** $y = -2 \sinh(x/2)$

34. If $x = a \cosh t$ and $y = b \sinh t$, where a, b are positive constants and t is any number, show that

$$\frac{x^2}{a^2} - \frac{y^2}{b^2} = 1$$

Level 2 Theory

Problems 35–51: Prove the given equations are identities.

35. $\dfrac{1}{\tanh x} = \dfrac{e^x - e^{-x}}{e^x + e^{-x}}$ **36.** $\dfrac{1}{\cosh x} = \dfrac{2}{e^x + e^{-x}}$

37. $\dfrac{1}{\sinh x} = \dfrac{2}{e^x - e^{-x}}$ **38.** $\coth x = \dfrac{\cosh x}{\sinh x}$

39. $\coth x = \dfrac{e^{2x} + 1}{e^{2x} - 1}$ **40.** $\operatorname{sech} x = \dfrac{2e^x}{e^{2x} + 1}$

41. $1 - \tanh^2 x = \operatorname{sech}^2 x$ **42.** $\coth^2 x - 1 = \operatorname{csch}^2 x$

43. $\sinh(-x) = -\sinh x$ **44.** $\cosh(-x) = \cosh x$

45. $\tanh(-x) = -\tanh x$ **46.** $\coth(-x) = -\coth x$

47. $\sinh(x + y) = \sinh x \cosh y + \cosh x \sinh y$

48. $\cosh(x + y) = \cosh x \cosh y + \sinh x \sinh y$

49. $\tanh(x + y) = \dfrac{\tanh x + \tanh y}{1 + \tanh x \tanh y}$

50. $\sinh 2x = 2 \sinh x \cosh x$

51. $\cosh 2x = \cosh^2 x + \sinh^2 x$

Level 2 Applications

52. Graph the hanging cable where the equation

$$y = \frac{a}{2}\left(e^{x/a} + e^{-x/a}\right)$$

for $a = 60$ for $-20 \le x \le 20$. Where do you think the poles are located, and how tall is each pole?

53. Graph the hanging cable where the equation

$$y = \frac{a}{2}\left(e^{x/a} + e^{-x/a}\right)$$

for $a = 40$ for $-20 \le x \le 20$. Where do you think the poles are located, and how tall is each pole?

54. Suppose that the equation of a curve suspended between two poles is given by the equation

$$y = a \cosh \frac{x}{a}$$

where a is a positive constant. Draw the graph of the suspended cable if $a = 46$. Estimate the distance between the 60-ft poles.

55. It can be shown that the sag S of the cable is given by

$$S = a \cosh \frac{b}{a} - a$$

where the x-coordinates of the point of support are $x = b$ and $x = -b$ (for $b > 0$). Estimate the sag for the cable described in Problem 54.

56. It can be shown that the length L of the cable is given by

$$L = 2a \sinh \frac{b}{a}$$

where the x-coordinates of the point of support are $x = b$ and $x = -b$ (for $b > 0$). Estimate the length of the cable described in Problem 54.

57. The Gateway Arch in St. Louis, Missouri, is a catenary whose equation is

$$y = a \cosh(x/a) + 758$$

with $a \approx -128$ ft. Draw this curve on a coordinate grid and then use the photograph shown in Figure 7.21 to indicate where you think the center of the coordinate system is located to produce the Gateway Arch.

© Dennis MacDonald/Photo Edit

FIGURE 7.21   Gateway Arch, St. Louis, Missouri

58. If the Gateway Arch (see Figure 7.21) has equation $y = -128 \cosh(x/128) + 758$ and you know that the Arch is 630 ft tall, how wide is it on the ground?

Level 3 Theory

59. Write $\cosh^{-1} x$ as a logarithmic function for $x \geq 1$.

60. Write $\tanh^{-1} x$ as a logarithmic function for $|x| < 1$.

61. Write $\operatorname{sech}^{-1} x$ as a logarithmic function for $0 < x \leq 1$.

62. Write $\operatorname{csch}^{-1} x$ as a logarithmic function for $x \neq 0$.

63. Write $\coth^{-1} x$ as a logarithmic function for $|x| > 1$.

64. First show that $\cosh x + \sinh x = e^x$, and then use this result to prove that

$$(\cosh x + \sinh x)^n = \cosh nx + \sinh nx$$

for positive integers n.

7.5 CHAPTER SEVEN SUMMARY

Objectives

Section 7.1   Exponential Functions

Objective 7.1   Know the laws of exponents.

Objective 7.2   Know the squeeze theorem for exponents.

Objective 7.3   Define an exponential function.

Objective 7.4   Sketch exponential functions.

Objective 7.5   Define a logarithm, a common logarithm, and a natural logarithm.

Section 7.2   Exponential Equations and Logarithms

Objective 7.6   Evaluate logarithms.

Objective 7.7   State and use the change of base theorem.

Objective 7.8   Solve exponential equations.

Objective 7.9   Solve applied problems involving exponential equations.

Objective 7.10 Know the growth/decay formula.

Section 7.3   Logarithmic Equations

Objective 7.11 Define a logarithmic function.

Objective 7.12 Sketch logarithmic functions.

Objective 7.13 State the fundamental properties of logarithms. Show that the exponential and logarithmic functions are inverse functions.

Objective 7.14 State the log of both sides theorem.

Objective 7.15 State the laws of logarithms—that is, the additive, subtractive, and multiplicative laws.

Objective 7.16 Solve logarithmic equations.

Objective 7.17 Solve exponential equations using the laws of logarithms.

Objective 7.18 Solve applied problems involving logarithmic equations.

Section 7.4   Hyperbolic Functions

Objective 7.19 Know the definition of the hyperbolic functions.

Objective 7.20 Prove identities involving the hyperbolic functions.

Objective 7.21 Draw graphs of the hyperbolic functions.

Objective 7.22 Know the definition of the inverse hyperbolic functions.

Objective 7.23 Solve applied problems involving hyperbolic functions.

Terms

Sample Test

All of the answers for this sample test are given in the back of the book.

Problems 1–4: *Graph the curve defined by each function.*

1. a. $y = 25e^{-x}$ **b.** $y = 25^{-1} \cdot 10^x$

2. a. $y = \log(3x - 1)$ **b.** $y = 100^{-x^2}$

3. a. $y = \log_2 x - 1$ **b.** $y = \log_2(x - 1)$

4. a. $y = 2 \cosh x$ **b.** $y = -3 \sinh x$

Problems 5–7: *Simplify each expression without using a calculator or computer.*

5. $\log 100 + \log \sqrt{10} + 10^{\log 0.5}$

6. $\ln e + \ln 1 + \ln e^{542} + \ln e^{\log 1,000}$

7. $\log_8 4 + \log_8 16 + \log_8 8^{2.3}$

Problems 8–14: *Solve each equation. Give approximate calculator solutions.*

8. $10^x = 85$ **9.** $e^{3x+1} = 45$

10. $435^x = 890$ **11.** $\log_6 x = 4$

12. $\log 2 + 2 \log x = 5$

13. $\log(x + 1) = 2 + \log(x - 1)$

14. $3 \ln \dfrac{e}{\sqrt[3]{5}} = 3 - \ln x$

15. Solve $A = P(1 + i)^x$ for x.

16. Write the equation $y = 14.8(2.5)^x$ as an equation of the form $y = ae^{bx}$.

17. Suppose that \$1,000 is invested at 7% interest compounded monthly. Use the formula

$$A = P\left(1 + \frac{r}{n}\right)^{nt}$$

  a. How long (to the nearest month) before the value is \$1,250?

  b. How long (to the nearest month) before the money doubles?

  c. What is the interest rate (compounded monthly and rounded to the nearest percent) if the money doubles in 5 years?

18. The half-life of arsenic-76 is 26.5 hr.

  a. Find the decay constant.

  b. If 85 mg of arsenic-76 are present at the start, how long would it take (to the nearest hour) for 6 mg to be left?

19. Let $f(x) = \tanh x$:

  a. Find $f(\pi)$.

  b. Graph f.

20. Prove $\tanh 2x = \dfrac{2 \tanh x}{1 + \tanh^2 x}$

Discussion Problems

1. In your own words, explain the procedure you use to solve exponential equations.

2. In your own words, explain the procedure you use to solve logarithmic equations.

3. Discuss graphing using a logarithmic scale rather than a linear scale.

4. Compare and contrast linear growth and exponential growth.

5. Compare and contrast the shapes of a parabola and a catenary.

Miscellaneous Problems—Chapters 1–7

Problems 1–6: *Find* $\dfrac{f(x + h) - f(x)}{h}$.

1. $f(x) = e^x$

2. $f(x) = 10^x$

3. $f(x) = 5^x$

4. $f(x) = \log x$

5. $f(x) = \ln x$

6. $f(x) = \log_2 x$

Problems 7–17: *Evaluate each expression (to two decimal places).*

7. a. $\log_\pi \dfrac{1}{\pi}$ **b.** $\log_2 8$

8. a. $\log 8.43$ **b.** $\log 9{,}760$

9. a. $\ln 2$ **b.** $\ln 0.125$

10. a. $\ln 13$ **b.** $\ln 0.15$

11. a. $\cosh 1.56$ **b.** $\sinh 1.34$

12. a. $\cosh^{-1} 5.3$ **b.** $\sin^{-1} 0.5$

13. a. $\tan^{-1} 0.5$ **b.** $\tanh^{-1} 0.5$

14. $\dfrac{\tan 15° + \tan 30°}{1 - \tan 15° \tan 30°}$ **15.** $\dfrac{\tan 120° - \tan 60°}{1 + \tan 120° \tan 60°}$

16. $\dfrac{2 \tan 30°}{1 - \tan^2 30°}$ **17.** $\dfrac{1 - \cos 120°}{\sin 120°}$

18. Answer the following questions about the standard cosine function.
   a. What is the domain?
   b. What is the range?
   c. What are the amplitude and period?
   d. What is its maximum value?
   e. Where does the graph cross the x-axis?

19. Answer the following questions about the standard hyperbolic cosine function.
   a. What is the domain?
   b. What is the range?
   c. What is the minimum value?
   d. Where does the graph cross the y-axis?

20. Answer the following questions about the standard inverse hyperbolic sine function.
   a. What is the domain?
   b. What is the range?
   c. What is the minimum value?
   d. Where does the graph cross the y-axis?

Problems 21–30: *Match the graph with an equation from this list, and find the values of A and B.*

$$y = A \sin Bx \qquad y = A \tan Bx$$
$$y = A \cos Bx \qquad y = A \cot Bx$$
$$y = A \sec Bx \qquad y = A \sin B(x - h)$$
$$y = A \csc Bx \qquad y = A \cos B(x - h) + k$$

21.

22.

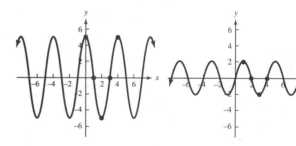

23.

24.

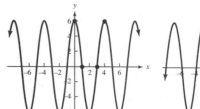
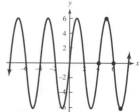

25.

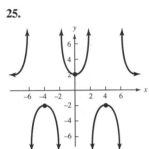

26.

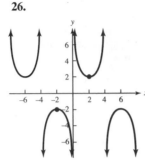

27.

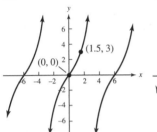

28.

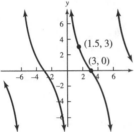

29.

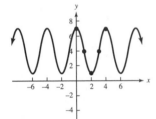

30.

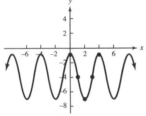

Problems 31–46: Identify each curve.

31. $r = \cos 2\theta$

32. $r = 2 \cos \theta$

33. $r = \cos 2$

34. $r^2 = \cos 2\theta$

35. $r = 2 - 2 \cos \theta$

36. $r = 2 + 2 \cos \theta$

37. $r = 4 \sin 3\theta$

38. $r = 4 \sin 3$

39. $r^2 = 4 \sin 2\theta$

40. $r = 4 + 4 \sin \theta$

41. $r = \cos 3\left(\theta - \frac{\pi}{6}\right)$

42. $r^2 = \cos 2\left(\theta - \frac{\pi}{6}\right)$

43. $r = \sin 3\left(\theta - \frac{\pi}{6}\right)$

44. $r^2 = 3 \sin 2\left(\theta - \frac{\pi}{6}\right)$

45. $r - 2 = 2 \sin(\theta - 3)$

46. $r - 3 = 3 \cos(\theta - 2)$

Problems 47–52: Solve each equation (exact values).

47. $\log 5 = \log x + \log(x + 4)$

48. $2 \ln \dfrac{e}{\sqrt{5}} = 2 - \ln x$

49. $\sin 2x - e^x \sin 2x = 0$

50. $\cos 3x - 2e^{2x} \cos 3x = 0$

51. $3 \log 3 - \frac{1}{2} \log 3 = \log \sqrt{x}$

52. $\frac{1}{2} \log_b x - \log_b \sqrt{2} = \frac{1}{2} \log_b(3x + 1)$

Problems 53–57: Solve each equation (correct to two decimal places).

53. $100 = 6.4(10)^{0.005x^2}$ **54.** $100 = 6.4e^{0.005x^2}$

55. $100 = 6.4(4)^{0.005x^2}$ **56.** $2^{3x-1} = 6$

57. $\ln(x - 2) - 1 = -\ln x$

Problems 58–68: Prove that the given equations are identities.

58. $2^{\sin^2\theta} \cdot 2^{\cos^2\theta} = 2$

59. $\dfrac{10^{\log(\cos x)}}{10^{\log(1 - \sin x)}} = \sec x + \tan x$

60. $\tan 3\theta \sin 3\theta = \sec 3\theta - \cos 3\theta$

61. $2 - \sin^2 3\theta = 1 + \cos^2 3\theta$

62. $2 \sin^2 3\theta - 1 = 1 - 2 \cos^2 3\theta$

63. $(\tan 5\theta - 1)(\tan 5\theta + 1) = \sec^2 5\theta - 2$

64. $\tan^2 7\theta = (\sec 7\theta - 1)(\sec 7\theta + 1)$

65. $(\sin \gamma - \cos \gamma)^2 = 1 - 2 \sin \gamma \cos \gamma$

66. $(\sin \gamma + \cos \gamma)(\sin \gamma - \cos \gamma) = 2 \sin^2 \gamma - 1$

67. $(\sec 2\theta + \csc 2\theta)^2 = \dfrac{1 + 2 \sin 2\theta \cos 2\theta}{\cos^2 2\theta \sin^2 2\theta}$

68. $\dfrac{1}{\sec \theta + \tan \theta} = \sec \theta - \tan \theta$

Problems 69–70: Use the trigonometric substitution $u = a \sin \theta$ for $-\frac{\pi}{2} \le x \le \frac{\pi}{2}$ to rewrite the algebraic expression in terms of θ and simplify.

69. $\dfrac{(4 - u^2)^2}{u^4}$; $a = 2$ **70.** $\dfrac{\sqrt{49 - u^2}}{u}$; $a = 7$

Problems 71–72: Use the trigonometric substitution $u = a \tan \theta$ for $-\frac{\pi}{2} < x < \frac{\pi}{2}$ to rewrite the algebraic expression in terms of θ and simplify.

71. $\dfrac{u^2}{u^2 + 9}$; $a = 3$ **72.** $u^3(4 + u^2)^{-3/2}$; $a = 2$

Problems 73–74: *Use the trigonometric substitution* $u = a \sec \theta$ *for* $0 < x < \frac{\pi}{2}$, $a > 0$, *to rewrite the algebraic expression in terms of* θ *and simplify.*

73. $\dfrac{1}{u\sqrt{u^2 - 1}}$; $a = 1$ **74.** $u^5(u^2 - 4)^{-3/2}$; $a = 2$

75. a. Find and plot the square roots of $\frac{7}{2}\sqrt{3} - \frac{7}{2}i$.

Leave your answer in trigonometric form.
b. Find the cube roots of i. Leave your answer in rectangular form.

76. What are the ninth roots (trigonometric form) of $-1 + i$?

77. What are the fourth roots (rectangular form) of -16?

78. Plot the polar-form points and give both primary representations.

    a. $(5, \sqrt{75})$ **b.** $\left(3, -\frac{2\pi}{3}\right)$ **c.** $(-2, 2)$

Problems 79–84: *Graph the given curves, and state whether there is damping, resonance, neither, or both.*

79. $y = \cos x + x \sin x$ **80.** $y = \sin x - x \cos x$

81. $y = \frac{9}{4} \sin \frac{9}{8} x + \sin \frac{1}{8} x$ **82.** $y = \frac{9}{2} \cos \frac{9}{4} x + \cos \frac{1}{4} x$

83. $y = \frac{13}{5} \cos x - \sin x - \frac{8}{5} \cos \frac{3}{2} x$

84. $y = 2 \sin\left(5x - \frac{\pi}{2}\right) + \sin x$

Applications

85. The half-life of arsenic-76 is 26.5 hours. If 55 mg of arsenic are present initially, how long (to the nearest hour) will it be before there are 30 mg?

86. In 1992, it was reported that the number of teenagers with AIDS doubles every 14 months. Find an equation to model the number of teenagers that may be infected over the next 10 years.

87. The equation of the surface area of a honeycomb (see Figure 7.22) is given by

$$S = 6hs + \frac{3}{2}s^2 \left(\frac{\sqrt{3} - \cos \theta}{\sin \theta}\right) \quad \text{for } 0 \le \theta \le 90°$$

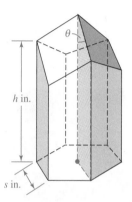

FIGURE 7.22   Surface area of a honeycomb

    a. If $h = 4.5$ in., $s = 1.5$ in., and $\theta = 32°$, what is the surface area?
    b. If $h = 2.5$ in., $s = 0.75$ in., and the surface area is 20.25 in.^{2}, what is θ?
    c. What value of θ gives the minimum surface area for the information in part **b**?

Group Research Projects

Working in small groups is typical of most work environments, and this book seeks to develop skills with group activities. At the end of each chapter, we present a list of suggested projects, and even though they could be done as individual projects, we suggest that these projects be done in groups of three or four students.

1. The entrance of the Aquarium of Americas in New Orleans has a gigantic building-size curve called a *logarithmic spiral*. Find out how to construct a logarithmic spiral, and write a paper about what you learned. Why do you suppose it would appear on the front of an aquarium?

Melissa Lee/Aquarium of the Americas/New Orleans

2. From your local chamber of commerce, obtain the population figures for your city for the years 1970, 1980, and 1990. Find the rate of growth for each period. Forecast the population of your city for the year 2000. Include charts and graphs. List some factors, such as new zoning laws, that could change the growth rate of your city.

3. If we assume that the world population grows exponentially, then it is also reasonable to assume that the use of some nonrenewable resource (such as petroleum) will grow exponentially. In calculus it is shown (under these assumptions) that for some constant k, the formula for the amount of the resource, A, consumed from time $t = 0$ to $t = T$ is given by the formula

$$A = \frac{A_0}{k}(e^{rT} - 1)$$

where r is the relative growth rate of annual consumption.

   a. Solve this equation for T to find a formula for life expectancy of a particular resource.

   b. According to the Energy Information Administration, the annual world production (in millions of barrels per day) of petroleum is shown in the table:

   Year: 1960   1965   1970   1975   1980   1985   1990
   Quantity: 20.96 30.30 45.87 52.78 59.35 53.65 60.07

   Find an exponential equation for these data.

   c. If in 1990, the world petroleum reserves are 933.2 billion barrels, estimate the life expectancy for petroleum.

4. **IN YOUR OWN WORDS** Write an essay on carbon-14 dating. What is its relationship to logarithms?

Problems 5–7: Use the following table, which shows populations for eight cities, to answer the questions. More data are provided than are required to answer these questions, so you will need to make some assumptions to arrive at your prediction. State the assumptions you are making clearly.

City	1960	1970	1980	1990
Cleveland, OH	876,050	750,879	573,822	505,616
Boston, MA	697,197	641,071	574,283	562,994
Pittsburgh, PA	604,332	520,089	423,959	369,879
Milwaukee, WI	741,324	717,372	636,297	628,088
Denver, CO	493,887	514,678	492,686	467,610
San Antonio, TX	587,718	654,153	785,940	935,933
Memphis, TN	497,524	623,988	646,174	610,337
San Jose, CA	203,196	459,913	629,400	782,248

5. Which city seems to have the greatest growth rate for the period 1980–1990? Name the city and predict its population in the year 2000.

6. Which city seems to have had the greatest decline in population for the period 1980–1990? Name the city and predict its population in the year 2000.

7. If the population of Denver, Colorado, had continued to grow at its 1960–1980 rate, what would its 1990 population have been? What statement can you make about the rate of population growth in Denver in the 1980s? In the 1970s?

APPENDICES

CALCULATORS

This book was written with the assumption that you have access to a calculator. Trigonometry is the first course most students take that *requires* the use of a calculator, and it is important that you learn how to use this tool as you learn mathematics. A good calculator can be purchased for under $20, and such a calculator will suffice for this course; but if your goal is to eventually study more advanced mathematics, a graphing calculator is recommended.

The notation used in the book is consistent with that used by graphing calculators. The book is as calculator independent as possible and, for the most part, does not include keystrokes (except in this appendix).

You will also not find "calculator logos" on certain problems. You should consider the calculator a useful *tool* for *any* problem. We have not formulated problems with "ugly answers," set up to specifically fit some artificial calculator logo.

Types of Calculators

Calculators are often used initially to simplify expressions containing basic arithmetic operations. These types of problems provide a good place for you to investigate the logic used by your calculator. The calculator should have **grouping symbols,** usually designated by $($ and $)$; find these keys on your calculator. Grouping symbols are important in algebra and trigonometry and also in properly using your calculator. In algebra, a variety of grouping symbols are used: **parentheses** ( ), **brackets** [ ], and occasionally **braces** { }. These are all entered in your calculator using the $($ and $)$ keys. Grouping symbols are used in mathematics, along with the following **order of operations agreement.**

ORDER OF OPERATIONS AGREEMENT

It is agreed that operations will be carried out in the following order:

1. Carry out all operations within grouping symbols. If there is more than one set of grouping symbols in an expression, begin with the innermost set. A fraction bar is considered to be a grouping symbol that groups the elements in the numerator as well as the elements in the denominator. The overbar on a square (or higher) root symbol also functions as a grouping symbol.
2. Do exponents next.
3. Complete multiplications and divisions, working from left to right.
4. Finally, do additions and subtractions, working from left to right.

To acquaint yourself with your calculator, it is a good idea to begin with a test problem: $2 + 3(4)$; the order of operations agreement tells us that the correct answer is 14. First, look to see whether your calculator has an $\boxed{\text{ENTER}}$ key. If it does not, try this problem on your calculator:

If the output on your calculator is 20, then the logic of the calculator is left to right $(2 + 3) \times 4 = 5 \times 4 = 20$, and is called *arithmetic logic*. If the answer shown is 14, then your calculator used *algebraic logic*.

There are now four types of calculator logic in common use: arithmetic (often found in older, inexpensive calculators), algebraic, RPN, and direct algebraic logic (the type employed by newer calculators, including most graphing calculators). Here is a test problem that you can use to test the type of logic used by your calculator:

$$\sqrt{2 + 3 \times 4}$$

The correct answer is 3.741657387. Obtaining the correct answer for this test problem requires different keystrokes for each type of logic. Determine which one your calculator uses:

Pay attention to the necessary grouping symbols when using direct logic.

Arithmetic logic: $\boxed{3}\boxed{\times}\boxed{4}\boxed{=}\boxed{+}\boxed{2}\boxed{=}\boxed{\sqrt{}}$

Algebraic logic: $\boxed{2}\boxed{+}\boxed{3}\boxed{\times}\boxed{4}\boxed{=}\boxed{\sqrt{}}$

RPN logic: $\boxed{2}\boxed{\text{ENTER}}\boxed{3}\boxed{\text{ENTER}}\boxed{4}\boxed{\times}\boxed{+}\boxed{\sqrt{}}$

Direct (graphing) logic: $\boxed{\sqrt{}}\boxed{(}\boxed{2}\boxed{+}\boxed{3}\boxed{\times}\boxed{4}\boxed{)}\boxed{\text{ENTER or} =}$

Notice that direct (graphing) logic inputs information just as it is written in algebra. One of the advantages of a good graphing calculator is that you can input the mathematics in the same fashion that you would write it on your paper.

We will consider a few other simple problems here to help you become familiar with your calculator, but the goal is to think of the technology NOT as *additional* information, but rather as an extension of your paper and pencil.

EXAMPLE 1

Using a Calculator for Ordinary Operations

Find $\dfrac{2}{3}$ as a decimal on your calculator. From your display, answer the following questions:

a. How many digits are in your calculator display?
b. Does your calculator truncate or round the answers in the display?

Solution   The fraction bar is entered as $\boxed{\div}$.

Graphing: $\boxed{2}$ $\boxed{\div}$ $\boxed{3}$ $\boxed{\text{ENTER}}$
Algebraic: $\boxed{2}$ $\boxed{\div}$ $\boxed{3}$ $\boxed{=}$
RPN: $\boxed{2}$ $\boxed{\text{ENTER}}$ $\boxed{3}$ $\boxed{\div}$
Display: Answers vary, but the display will tell you something about your calculator.

a. Count the number of digits shown. Your owner's manual will tell you how to change the standard display, but most calculators will show only the number of places necessary. (For example, 1 ÷ 2 will display as .5, which is one place.) If the display shows

.6666666667

then your calculator displays 10 digits.
b. Look again at your display; if it shows

.6666666666

then it truncates (cuts off) at a certain decimal location; if it ends in a 7, then it is rounding (see Appendix B). ■

EXAMPLE 2   **Using a Calculator with Mixed Operations**

Find: **a.** $5(2.85 + 0.9) + 2(353)$ **b.** $\dfrac{5 \times 3}{2 \times 8}$ **c.** $\sqrt{5 - 4(3)(-2)}$

Solution   If you are using a graphing calculator or an algebraic-logic calculator, you will need to think about order of operations only when you evaluate expressions involving fractions.

a. Graphing: $\boxed{5}$ $\boxed{(}$ $\boxed{2.85}$ $\boxed{+}$ $\boxed{.9}$ $\boxed{)}$ $\boxed{+}$ $\boxed{2}$ $\boxed{\times}$ $\boxed{353}$ $\boxed{\text{ENTER}}$
Algebraic: $\boxed{5}$ $\boxed{\times}$ $\boxed{(}$ $\boxed{2.85}$ $\boxed{+}$ $\boxed{.9}$ $\boxed{)}$ $\boxed{+}$ $\boxed{2}$ $\boxed{\times}$ $\boxed{353}$ $\boxed{=}$
RPN: $\boxed{5}$ $\boxed{\text{ENTER}}$ $\boxed{2.85}$ $\boxed{\text{ENTER}}$ $\boxed{.9}$ $\boxed{+}$ $\boxed{\times}$
           $\boxed{2}$ $\boxed{\text{ENTER}}$ $\boxed{353}$ $\boxed{\times}$ $\boxed{+}$

The answer is 724.75.
b. There is a special concern when dealing with a fraction. The fractional bar is used as a grouping symbol, so when entering a fraction, you must insert parentheses to group the numbers and operations in the numerator and also in the denominator. (You do not need parentheses for the numerator or denominator if it consists of a single number.)

Graphing: $\boxed{(}$ $\boxed{5}$ $\boxed{\times}$ $\boxed{3}$ $\boxed{)}$ $\boxed{\div}$ $\boxed{(}$ $\boxed{2}$ $\boxed{\times}$ $\boxed{8}$ $\boxed{)}$ $\boxed{\text{ENTER}}$
Algebraic: $\boxed{(}$ $\boxed{5}$ $\boxed{\times}$ $\boxed{3}$ $\boxed{)}$ $\boxed{\div}$ $\boxed{(}$ $\boxed{2}$ $\boxed{\times}$ $\boxed{8}$ $\boxed{)}$ $\boxed{=}$
RPN: $\boxed{5}$ $\boxed{\text{ENTER}}$ $\boxed{3}$ $\boxed{\times}$ $\boxed{2}$ $\boxed{\text{ENTER}}$ $\boxed{8}$ $\boxed{\times}$ $\boxed{\div}$

The answer is 0.9375.

c. When evaluating a square root, you must be sure to group the numbers under the square root symbol.

Graphing: $\boxed{\sqrt{}}\;\boxed{((}\;\boxed{5}\;\boxed{-}\;\boxed{4}\;\boxed{\times}\;\boxed{3}\;\boxed{\times}\;\boxed{(-)}\;\boxed{2}\;\boxed{)}\;\boxed{\text{ENTER}}$

Algebraic: $\boxed{5}\;\boxed{-}\;\boxed{4}\;\boxed{\times}\;\boxed{3}\;\boxed{\times}\;\boxed{2}\;\boxed{+/-}\;\boxed{=}\;\boxed{\sqrt{}}$

RPN: $\boxed{5}\;\boxed{\text{ENTER}}\;\boxed{4}\;\boxed{\text{ENTER}}\;\boxed{3}\;\boxed{\times}\;\boxed{\text{ENTER}}$
$\boxed{2}\;\boxed{\text{CHS}}\;\boxed{\times}\;\boxed{-}$

The answer is 5.385164807.   ■

If the result of a particular operation is very large or very small, many calculators will display the result using a variation of *scientific notation.* A number is in scientific notation if it is written as a power of 10 or as a number between 1 and 10 times a power of 10. For example,

$$4.5036 \times 10^8 \quad \text{is scientific notation for the number } 450{,}360{,}000$$

A calculator representation for this number, called *floating-point notation,* might be

$$4.5036\,\text{E}\,08 \quad \text{or} \quad 4.5036\,\text{E}\,8 \quad \text{or} \quad 4.5036\,08 \quad \text{or} \quad 4.5036\,+08$$

Scientific notation is entered into a calculator using a key labeled $\boxed{\text{EE}}$, $\boxed{\text{SCI}}$, $\boxed{\text{EXP}}$, or $\boxed{\text{EEx}}$. Do not confuse scientific notation with exponents, as illustrated by the next example.

EXAMPLE 3

Using a Calculator for Exponents and for Scientific Notation

Find: **a.** $1.04^{12(15)}$ **b.** 2^{50} **c.** $3{,}800{,}000{,}000{,}000{,}000 \times 0.00000\,00000\,0025$

Solution

a. When carrying out operations with an exponent, you must insert parentheses.

Parentheses are necessary when you enter operations within an exponent.

Graphing: $\boxed{1.04}\;\boxed{\wedge}\;\boxed{((}\;\boxed{12}\;\boxed{\times}\;\boxed{15}\;\boxed{)}\;\boxed{\text{ENTER}}$

Algebraic: $\boxed{1.04}\;\boxed{y^x}\;\boxed{((}\;\boxed{12}\;\boxed{\times}\;\boxed{15}\;\boxed{)}\;\boxed{=}$

RPN: $\boxed{1.04}\;\boxed{\text{ENTER}}\;\boxed{12}\;\boxed{\text{ENTER}}\;\boxed{15}\;\boxed{\times}\;\boxed{y^x}$

The answer is 1,164.128908.

b. Exponents are entered using a $\boxed{y^x}$ or $\boxed{\wedge}$ key:

Graphing: $\boxed{2}\;\boxed{\wedge}\;\boxed{50}\;\boxed{\text{ENTER}}$

Algebraic: $\boxed{2}\;\boxed{y^x}\;\boxed{50}\;\boxed{=}$

RPN: $\boxed{2}\;\boxed{\text{ENTER}}\;\boxed{50}\;\boxed{y^x}$

The answer might be shown as 1.12589990 E 15 or 1.12589 +15 or 1.125899 15. The number of digits in the number between 1 and 10 may vary with brand and model of calculator.

c. Numbers that are too large or too small to enter directly into the calculator should be entered using scientific notation:

$$3{,}800{,}000{,}000{,}000{,}000 = 3.8 \times 10^{15}$$
and $0.00000\ 00000\ 0025 = 2.5 \times 10^{-13}$.

Thus

Graphing: $\boxed{3.8}\ \boxed{\text{EE}}\ \boxed{15}\ \boxed{\text{ENTER}}\ \boxed{\times}\ \boxed{2.5}\ \boxed{\text{EE}}\ \boxed{(-)}\ \boxed{13}$ $\boxed{\text{ENTER}}$

Algebraic: $\boxed{3.8}\ \boxed{\text{EXP}}\ \boxed{15}\ \boxed{\times}\ \boxed{2.5}\ \boxed{\text{EXP}}\ \boxed{13}\ \boxed{+/-}\ \boxed{=}$

RPN: $\boxed{3.8}\ \boxed{\text{EEx}}\ \boxed{15}\ \boxed{\text{ENTER}}\ \boxed{2.5}\ \boxed{\text{EEx}}\ \boxed{13}\ \boxed{\text{CHS}}\ \boxed{\times}$

The answer is 950. Even though you entered these numbers in scientific notation, the answer will generally not be shown in floating-point form unless the output number is very large or very small. ■

Since the technology changes so quickly, we suggest that you check with the owner's manual when you are not sure how to input a particular arithmetic calculation.

Calculator Usage in This Book

Typically, we will show a calculator graph as shown at the left, while your calculator display will look more like the representation at the right:

Textbook representation   **Actual calculator representation**

Y₁■X²−4 ← Input relation
Xmin=-5 Ymin=-5 ⎫
Xmax=5   Ymax=5 ⎬ ← Scaling information (for proper interpretation of graph)
Xscl=1   Yscl=1 ⎭

The representation shown at the left is a compact, calculator-independent representation that will convey all the necessary information you will need to graph this relation using your own calculator.

Different calculators input relations in different formats and with different keystrokes. Because of this variation, we do not show those keystrokes in the text. However, when you input relations into your calculator, there is typically a screen that allows more than one relation to be input at a time. For example, if we wish to graph $y = x^2 - 4$ using a calculator, we show it in the calculator textbook representation as y_1, rather than simply y. A calculator screen for such a relation might actually look like this:

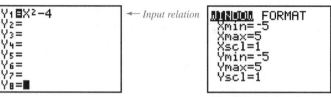

← *Input relation*

*This is the window showing
the dimensions of the
calculator "frame."*

The tick marks shown on the screen must also be identified. This is called the WINDOW or RANGE . Note in the example shown here, Xmin, Xmax, Xscl, Ymin, Ymax, and Yscl values are shown. This means that the minimum value for x is -5, the maximum value is 5, and each tick mark (Xscl) is 1 unit; that is, the x-scale is 1 unit. Similarly, for y:

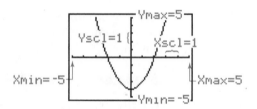

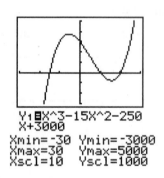

For example, if the text shows a calculator representation such as the one in the margin, you can read the following information:

The relation is $y = x^3 - 15x^2 - 250x + 3{,}000.$

Notice that calculator type wraps to the next line without regard to word breaks or mathematical symbolism. You simply need to read from one line to another. The window shows

$$-30 \le x \le 30 \quad \text{and}$$
$$-3{,}000 \le y \le 5{,}000$$

Each tick mark on the x-axis is 10 units (see Xscl), and each tick mark on the y-axis is 1,000 units (see Yscl).

A good example of calculator usage is illustrated by a simple problem. In this course, you will need to solve quadratic equations. The quadratic formula is stated in Appendix C and is reviewed in the text. We will illustrate solving a quadratic equation using a graphing calculator.

EXAMPLE 4

Computer Solution of a Quadratic Equation

Solve $5x^2 + 2x - 2 = 0$ with a graphing calculator.

Solution

If you have a graphing calculator, you can enter the equation into your calculator: $y = 5x^2 + 2x - 2$. In this book, we show the graph including the function, scaling number, and trace values. This means using the $\boxed{\text{TRACE}}$ key to find the approximate values:

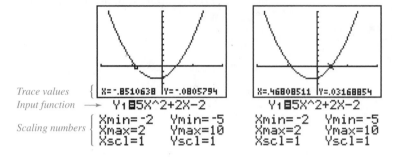

Trace values
Input function →
Scaling numbers

Move the trace to the place where the graph seems to cross the x-axis (that is, where $y = 0$). We identify these as

$$X = -.8510638 \quad \text{and} \quad X = .46808511$$

You can use the $\boxed{\text{ZOOM}}$ to increase the degree of accuracy. ∎

Programming the Quadratic Formula

Most graphing calculators also have the ability to enter programs. Even though you should check your owner's manual, we can illustrate how you might write a simple program to solve quadratic equations. Since you will have occasion to use the quadratic formula over and over again, this is a good time to consider writing a simple program to give the real roots for a quadratic equation. First, write the equation in the form $Ax^2 + Bx + C = 0$, and input the A, B, and C values into the calculator as A, B, and C, respectively. The program will then output the two real values (if they exist). Each brand of graphing calculator is somewhat different, but it is instructive to show the process on the TI-81/82/83/85/86 or 92 model. Press the $\boxed{\text{PRGM}}$ key. You will then be asked to name the program; we call our program QUAD. Next, input the formula

for the two roots (from the quadratic formula). Finally, display the answer. For $5x^2 + 2x - 2 = 0$ input the A, B, and C values as follows:

$\boxed{5}\ \boxed{\text{STO}\rightarrow}\boxed{\text{A}} \quad \boxed{2}\ \boxed{\text{STO}\rightarrow}\boxed{\text{B}} \quad \boxed{-2}\ \boxed{\text{STO}\rightarrow}\boxed{\text{C}} \quad \boxed{\text{PRGM}} \quad \text{QUAD}$

Then run the program for the DISPLAY:

.4633249581
−.8633249581

While most of the calculator references in this book are written for those with a graphing calculator, you may have only an algebraic calculator. We can illustrate how to approximate the roots for $5x^2 + 2x - 2 = 0$ with this type of calculator:

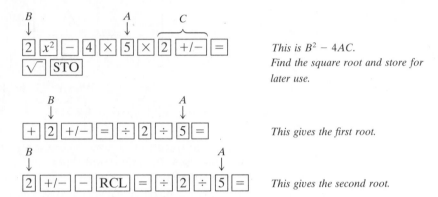

This is $B^2 - 4AC$.
Find the square root and store for later use.

This gives the first root.

This gives the second root.

Some of the steps shown here can be combined because these are simple numbers. These steps give the numerical approximation for a quadratic equation with real roots. For this quadratic equation, these roots are (to four decimal places) 0.4633 and −0.8633.

There are several software programs such as *Mathematica, Mathlab, Maple,* and *Derive* that are particularly easy to use to solve not only quadratic equations, but also most equations (certainly all those equations we will need to solve in this course). To use these programs, consult your owner's manual, but the procedure generally requires entering the equations and then providing a $\boxed{\text{SOLVE}}$ command.

A Personal Note About Calculator Usage from the Author

One of my major criticisms about the calculator-based mathematics material that I have seen published or presented at mathematics conferences is that the author of the material gets bogged down interpreting the precision of the calculator and the inputting of information into the calculator. Consequently, I lose track of the mathematics I'm trying to learn. Knowing about "pixel" size and how my calculator works does not enhance my *mathematical* understanding. Just as I am able to use my car to get to work without understanding the principles of the internal combustion engine, I believe we can use a calculator to advance our mathematical goals without understanding the details of "how" the calculations are accomplished. I do, however, believe it is important to discuss how to interpret calculator output. For example, we need to recognize 2.409 −28 as a number in scientific notation, 2.409×10^{-28}, which for most practical purposes is 0. Similarly, an output of .33333333 or 3.1415927 should be recognized as $\frac{1}{3}$ or π, respectively. Or, at a slightly higher level, if we are tracing a point on a curve and obtain successive points with coordinates (11.982187, −.00239876) and (12.0498731, .0498710), we need to discuss the appropriate mathematics to be able to conclude that the x-intercept is probably (12, 0). On the other hand, many real-world models do not have "nice" or rational intercepts, and these problems are, from a practical standpoint, impossible to solve without the new technology. In such a case, the ZOOM key can be used to approximate the intercept to any reasonable degree of accuracy. In other words, I believe the calculator to be an invaluable *tool* in understanding mathematics, but its use is not the ultimate goal of the material of this book.

APPENDIX B

ACCURACY AND ROUNDING

When we work with numbers that arise from counting, the numbers are **exact.** If we work with measurements, however, the quantities are necessarily **approximations.** The digits known to be correct in a number obtained by a measurement are called **significant digits.** The digits 1, 2, 3, 4, 5, 6, 7, 8, and 9 are always significant, whereas the digit 0 may or may not be significant, as described by the following two rules:

1. Zeros that come between two other digits are significant, as in 203 or 10.04.
2. If the only function of the zero is to place the decimal point, it is not significant, as in

$$\underbrace{0.0000}\,23 \qquad \text{or} \qquad 23,\underbrace{000}$$
$$\quad\ \textit{Placeholders} \qquad\qquad\ \textit{Placeholders}$$

If it does more than fix the decimal point, it is significant, as in

$$0.00230 \qquad \text{or} \qquad 23,\underbrace{000}.01$$
$$\qquad\ \uparrow \qquad\qquad\qquad \textit{These are significant.}$$
$$\textit{This digit is significant.}$$

This second rule can, of course, result in certain ambiguities, as in 23,000 (measured to the *exact* unit). To avoid confusion, we use scientific notation in this case:

2.3×10^4 has two significant digits
2.3000×10^4 has five significant digits

In this book, we do not wish to resort to scientific notation to resolve this ambiguity. We agree to violate this rule and assume the maximum degree of accuracy possible. This means for calculation purposes, we will assume that 23,000 has five significant digits.

Numbers that come about by counting are considered to be exact and are correct to any number of significant digits. Since exponents are usually counting numbers, in this book we will not consider exponents in deciding on the proper number of significant digits used in a particular calculation.

EXAMPLE 1

Recognizing Significant Digits

Give four examples each of two, three, and four significant digits.

Solution Two significant digits: 46,   0.00083, 4.0×10^1,   0.050
Three significant digits: 523,   403, 4.00×10^2,   0.000800
Four significant digits: 600.1, 4.000×10^3,   0.0002345,   8,004   ∎

When we are doing calculations with approximate numbers (particularly when using a calculator), it is often necessary to round off results. In this book, we use the following rounding procedure.

ROUNDING PROCEDURE

To round off a number:

1. Increase the last retained digit by 1 if the remainder is greater than or equal to 5; or
2. Retain the last digit unchanged if the remainder is less than 5.
3. In problems requiring rounding but involving operations with numbers, round only once, at the end. That is, do not work with rounded results, since round-off errors can accumulate.

△ NOTE *Note this agreement.*

Elaborate rules for computation of approximate data can be developed (when it is necessary for some applications, such as in chemistry), but there are four simple rules that will work satisfactorily for the material in this text.

RULES FOR SIGNIFICANT DIGITS

Addition–subtraction: Add or subtract in the usual fashion, and then round the result so that the last digit retained is in the column farthest to the right in which both given numbers have significant digits.

Multiplication–division: Multiply or divide in the usual fashion, and then round the results to the smaller number of significant digits found in either of the given numbers.

Counting numbers: Numbers used to count or whole numbers used as exponents are considered to be correct to any number of significant digits, and will not be considered in deciding on the proper number of significant digits.

Multiple operations: Apply the rounding rules previously given for significant digits only when stating your final answer to avoid errors resulting from repeated rounding.

EXAMPLE 2   **Writing Significant Digits with a Calculation**

a. Calculate $A = \pi r^2$ for $r = 23.4$.   **b.** Calculate $b = 50 \sin 23°$.

Solution   **a.** $A = \pi r^2 = \pi(23.4)^2 \approx 1{,}720.210473$ *Calculator approximation*

Note that this is a multiplication problem, and π is accurate to whatever degree of accuracy of your calculator, but 23.4 is accurate to three significant digits, so the answer should be rounded to three significant digits:

$$C \approx 1{,}720$$

b. $b = 50 \sin 23° \approx 50(0.3907311285)$ *Calculator approximation*
$$\approx 19.53655642$$

Do not work with rounded results, but instead use all the accuracy available in your calculator.

Round your answer to the least accurate of the input numbers. We note that 50 is ambiguous. We have agreed, in this book, to consider a number such as 50 as being accurate to two significant digits. We also note that 23° has two significant digits, so we round our answer to two significant digits:

$$b \approx 20$$ ■

Notice in Example 2**b** we used the degree measure 23° and not the evaluated number (0.3907311285) in determining the number of significant digits. In working with angles in this text, we assume a certain relationship in the accuracy of measurements between lengths and measurements of angles.

SIGNIFICANT DIGITS WITH CALCULATIONS INVOLVING ANGLES	
Accuracy in measuring lengths	Accuracy in measuring angles
2 significant digits	Nearest degree
3 significant digits	Nearest tenth of a degree
4 significant digits	Nearest hundredth of a degree

This chart means that, if the data include one side of a triangle given with two significant digits and another with three significant digits, the angle would be computed to the nearest degree. If one side is given to four significant digits and an angle to the nearest tenth of a degree, then the other side would be given to three significant digits and the angles computed to the nearest tenth of a degree. In general, results computed from this table should not be more accurate than the least accurate item of the given data.

EXAMPLE 3   **Significant Digits in Solving Triangles**

State the number of significant digits of your answers to the indicated problems.

a. $a = 46.6$; $b = 35.3$; $c = 42.75$; $\alpha = 72.7°$; $\beta = 46.3°$; $\gamma = 61°$
b. $a = 68.123$; $b = 95.3$; $c = 128.055678$; $\alpha = 31.4°$; $\beta = 46.7°$, $\gamma = 102°$

Solution
  a. Answer to two significant digits because of the least accurate measurement (γ).
  b. Answer to two significant digits because of the least accurate measurement (γ). ■

One last time: 3°, 33°, and 133° are all considered accurate to two significant digits because they are all measured to the nearest degree; 2.5°, 25.6°, and 112.5° are all considered accurate to three significant digits; and 9.35°, 83.56°, and 108.95° are all considered accurate to four significant digits (they are all measured to the nearest hundredth of a degree).

PROBLEM SET B

Level 1 Drill

Problems 1–22: *Use a calculator to evaluate each of the expressions. Be sure to round each answer to the appropriate number of significant digits.*

1. (14)(351)

2. (218)(263)

3. $(2.00)^4(1,245)(277)$

4. $(3.00)^3(182)$

5. $\dfrac{(1,998)(1,492)}{450}$

6. $\dfrac{(515)(20,600)}{200}$

7. $(990)(1,117)(342) - 96$

8. $[0.14 + (197)(25.08)](19)$

9. $\dfrac{1.00}{0.005 + 0.020}$

10. $\dfrac{1.500 \times 10^4 + 7.000 \times 10^8}{2.00 \times 10^4}$

11. $3,478.06 + 2,256.028 + 1,979.919 + 0.00091$

12. $57.300 + 0.094 + 32.3 + 2.09 + 0.0074$

13. $\sin 85.3°$

14. $\tan 19°$

15. $23.45 \cos 31°$

16. $834 \sin 2°$

17. $834 \sin 2.0°$

18. $834 \sin 2.00°$

19. $1,561 \tan 109°$

20. $1,561 \tan 109.0°$

21. $1,561 \tan 109.00°$

22. $1,561 \tan 9.00°$

Problems 23–28: *State the number of significant digits of the answers for the solution for the triangles.*

23. $a = 803$, $b = 455$, $c = 521.7952$

24. $a = 803$, $b = 455$, $\alpha = 110°$

25. $a = 803$, $\beta = 32.07896°$, $\gamma = 37.52103°$

26. $a = 34.6$, $b = 27.43995$, $c = 61.5$

27. $b = 27.43995$, $c = 61.5$, $\alpha = 8.5°$

28. $b = 27.43995$, $\alpha = 8.5°$, $\gamma = 164.76682°$

Level 2 Drill

Problems 29–50: *Use a calculator to evaluate each of the expressions. Be sure to round each answer to the appropriate number of significant digits.*

29. $6.28^{1/2}(4.85)$

30. $8.23^{1/2}(6.14)$

31. 10π

32. 10.0π

33. 356π

34. $\sqrt{2.00}\,\pi$

35. $\dfrac{1}{\sqrt{4.83} + \sqrt{2.51}}$ where 1 is a counting number

36. $\dfrac{1}{\sqrt{8.48} + \sqrt{21.3}}$ where 1 is a counting number

37. $[(4.083)^2(4.283)^3]^{-2/3}$

38. $[(6.128)^4(3.412)^2]^{-1/2}$

39. $\dfrac{241^2 + 568^2 - (241)(6.48)(0.3462)}{(241)(568)}$

40. $\dfrac{(2.51)^2 + (6.48)^2 - (2.51)(6.48)(0.3462)}{(2.51)(6.48)}$

41. $\dfrac{23.7 \sin 36.2°}{45.1}$

42. $\dfrac{461 \sin 43.8°}{1,215}$

43. $\dfrac{62.8 \sin 81.5°}{\sin 42.3°}$ **44.** $\dfrac{4.381 \sin 49.86°}{\sin 71.32°}$

45. $\dfrac{16 \sin 22°}{25}$ **46.** $\dfrac{42 \sin 52°}{61}$

47. $\dfrac{16^2 + 25^2 - 9.0^2}{2(16)(25)}$

where 2 is a counting number

48. $\dfrac{216^2 + 418^2 - 315^2}{2(216)(418)}$

where 2 is a counting number

49. $\dfrac{4.82^2 + 6.14^2 - 9.13^2}{2(4.82)(6.14)}$

where 2 is a counting number

50. $\dfrac{18.361^2 + 15.215^2 - 13.815^2}{2(18.361)(15.215)}$

where 2 is a counting number

APPENDIX C

REVIEW OF ALGEBRA

We cannot review a complete algebra course; the best we can do here is to review some parts of algebra that are particularly useful in trigonometry. This book assumes that you have completed the equivalent of an intermediate algebra course (that is, two years of high school algebra).

A **numerical expression** is a number, a variable, or numbers and variables connected by defined operations. A **variable expression** is a numerical expression including at least one variable. An **equation** is two expressions (either numerical or variable) connected by an equal sign. Equations are classified as **true** (if the numerical expression on the left represents the same number as the numerical expression on the right), **false** (if the numerical expression on the left does not represent the same number as the numerical expression on the right), or **open** if the equation contains at least one variable. There are three important words, *evaluate, solve,* and *simplify,* that are used in algebra.

*Pay attention to the correct use of the words **evaluate, solve,** and **simplify.***

Evaluate: To *evaluate a numerical expression* is to carry out the order of operations to write the expression as a single number. To *evaluate a variable expression* means to replace the variable or variables by given values, and then to evaluate the resulting numerical expression.

Solve: To *solve* an equation implies that the given equation is an open equation, and we are to find all replacements for the variable or variables that make the equation a true equation.

Simplify: To *simplify* an expression depends on the type of expression that is being considered. We will briefly review the process of simplifying various types of expressions in this section.

Polynomials

A **term** is a constant, a variable, or a product of constants and variables. Thus, $10x$ is a term, but $10 + x$ is not (because the terms 10 and x are connected by addition, not multiplication). A finite sum of terms with *whole-number exponents* on the variables is called a **polynomial.** For example,

POLYNOMIALS: $6x$ $2x^2y + z^4$ $\frac{1}{2}x + 4$ 0 $12\frac{1}{3}$ $\dfrac{x+3}{5}$

NOT POLYNOMIALS: $\dfrac{1}{x}$ $2 + \sqrt{x}$ $\dfrac{\sqrt[5]{x+1}}{5}$ $x^{2/3}$

In a polynomial of the form $a_n x^n$, a_n is called the **coefficient** of x^n. If a_n is the coefficient of the greatest power of x, it is called the **leading coefficient** of the polynomial, and if x is nonzero, the polynomial is said to have **degree** n. Notice that if $n = 2$ (degree 2), then the polynomial is called **quadratic;** if $n = 1$ (degree 1), then the polynomial is said to be **linear;** and if $n = 0$ (degree 0), then it is called a **constant.**

Each $a_k x^k$ of a polynomial is a term. Polynomials are frequently classified according to the number of terms. A **monomial** is a polynomial with one term, a **binomial** has two terms, and a **trinomial** has three terms. It is customary to arrange the terms of a polynomial in order of decreasing powers of the variable. **Similar terms** are terms with the same variable and the same degree. Two polynomials are **equal** if and only if coefficients of similar terms are the same. If some of the numerical coefficients are negative, they are usually written as subtractions of terms. Thus

$$6 + (-3)x + x^3 + (-2)x^2$$

would customarily be written as

$$x^3 - 2x^2 - 3x + 6$$

A polynomial is said to be **simplified** if it is written with all similar terms combined and is expressed in order of decreasing degree.

The distributive property shows you how to add polynomials by adding similar terms.

DISTRIBUTIVE PROPERTY

For real numbers a, b, and c,

$$a(b + c) = ab + ac$$

Since similar terms contain not only the same variables, but also the same exponent (or exponents) on the variables, the distributive property can be applied. For example,

$$5\overset{\downarrow}{x} + 3\overset{\downarrow}{x} = (5 + 3)\overset{\downarrow}{x} = 8x$$

If you also freely use the commutative and associative properties, you can simplify more complicated expressions using the idea of similar terms:

$$
\underbrace{(5x^2 + 2x + 1) + (3x^3 - 4x^2 + 3x - 2) = \underbrace{3x^3}_{\substack{\text{3rd-degree} \\ \text{term}}} + \underbrace{5x^2 + (-4)x^2}_{\substack{\text{2nd-degree} \\ \text{terms}}} + \underbrace{2x + 3x}_{\substack{\text{1st-degree} \\ \text{terms}}} + \underbrace{1 + (-2)}_{\substack{\text{0-degree} \\ \text{terms}}}}
$$

Similar terms   Similar terms   Similar terms

$$
= 3x^3 + x^2 + 5x - 1 \quad \textit{Terms are in order of decreasing degree.}
$$

EXAMPLE 1   Adding Polynomials

Simplify $(4x - 5) + (5x^2 + 2x + 1)$.

Solution
$$
\begin{aligned}
(4x - 5) + (5x^2 + 2x + 1) &= 4x - 5 + 5x^2 + 2x + 1 \\
&= 5x^2 + (4x + 2x) + (-5 + 1) \\
&= 5x^2 + 6x - 4
\end{aligned}
$$

You will usually do the steps for a simplification like this mentally and then generally go directly from the problem to its answer. ∎

When adding polynomials, you simply add similar terms. However, when you subtract polynomials, you first subtract *each term* of the polynomial being subtracted and then combine similar terms, as illustrated by the following example.

EXAMPLE 2   Subtracting Polynomials

Simplify $(4x - 5) - (5x^2 + 2x + 1)$.

Solution
$$
(4x - 5) \underset{\uparrow}{-} (5x^2 + 2x + 1) = 4x - 5 \underset{\uparrow}{-} 5x^2 \underset{\uparrow}{-} 2x \underset{\uparrow}{-} 1
$$

Subtracting polynomials   *First subtract each term of second polynomial.*

$$
= -5x^2 + 2x - 6 \quad \textit{Combine similar terms.}
$$

After you subtract each term, then combine similar terms. ∎

Another way of considering subtraction is to use the *definition of subtraction.*

EXAMPLE 3   Subtracting Polynomials by Adding the Opposite

Simplify $(5x^2 + 2x + 1) - (3x^3 - 4x^2 + 3x - 2)$.

Solution
$$
\begin{aligned}
(5x^2 + 2x + 1) - (3x^3 - 4x^2 + 3x - 2) &= 5x^2 + 2x + 1 + (-1)(3x^3 - 4x^2 + 3x - 2) \\
&= 5x^2 + 2x + 1 + (-1)(3)x^3 + (-1)(-4)x^2 + (-1)(3)x + (-1)(-2) \\
&= -3x^3 + 9x^2 - x + 3
\end{aligned}
$$

Once again, we have shown more steps than you need to show in your work. You can probably do these steps in your head. ∎

EXAMPLE 4 **Multiplying Polynomials**

Simplify $(4x - 5)(3x^3 - 4x^2 + 3x - 2)$.

Solution Use the distributive property (twice):

$$(4x - 5)(3x^3 - 4x^2 + 3x - 2)$$
$$= (4x - 5)(3x^3) + (4x - 5)(-4x^2) + (4x - 5)(3x) + (4x - 5)(-2)$$
$$= 12x^4 - 15x^3 - 16x^3 + 20x^2 + 12x^2 - 15x - 8x + 10$$
$$= 12x^4 - 31x^3 + 32x^2 - 23x + 10$$ ∎

A **factor** of a given algebraic expression (perhaps of some specified type, such as a polynomial) is another algebraic expression that divides evenly (that is, without a remainder) into the given expression. The process of **factoring** involves resolving the given expression into factors. The procedure we will use is to carry out a series of tests for different types of factors.

FACTORING PROCEDURE

To factor a polynomial,

1. Find common factors first.
2. Look for special types.
  Difference of squares: $a^2 - b^2 = (a - b)(a + b)$
  Difference of cubes: $a^3 - b^3 = (a - b)(a^2 + ab + b^2)$
  Sum of cubes: $a^3 + b^3 = (a + b)(a^2 - ab + b^2)$
3. If it is a trinomial, use FOIL (First terms, Outer terms, Inner terms, and then Last terms) to factor it into a pair of binomial factors:

$$acx^2 + (ad + bc)xy + bdy^2 = (ax + by)(cx + dy)$$

EXAMPLE 5 **Common Factoring Monomial Factors**

Factor: **a.** $a^2b + 5a^3b^2 + 7a^2b^3$ **b.** $\cos \theta \sin \theta - \cos^2\theta$

Solution **a.** The common factor is a^2b:

$$a^2b + 5a^3b^2 + 7a^2b^3 = a^2b(1 + 5ab + 7b^2)$$

b. The common factor is $\cos \theta$:

$$\cos \theta \sin \theta - \cos^2\theta = \cos \theta(\sin \theta - \cos \theta)$$ ∎

EXAMPLE 6   **Common Factoring Polynomial Factors**

Factor: **a.** $5x(3a - 5b) + 9y(3a - 5b)$

b. $(\cos \theta - 1)\sin^2\theta + (\cos \theta - 1)\cos^2\theta$

Solution   **a.** The common factor is $(3a - 5b)$:

$$5x(3a - 5b) + 9y(3a - 5b) = (5x + 9y)(3a - 5b)$$

b. The common factor is $(\cos \theta - 1)$:

$$(\cos \theta - 1)\sin^2\theta + (\cos \theta - 1)\cos^2\theta = (\cos \theta - 1)(\sin^2\theta + \cos^2\theta) \quad \blacksquare$$

EXAMPLE 7   **Factoring a Difference of Squares**

Factor: **a.** $3x^2 - 75$ **b.** $\cos^4\theta - \sin^4\theta$

Solution   **a.** First, common factor, then factor as a difference of squares:

$$3x^2 - 75 = 3(x^2 - 25)$$
$$= 3(x - 5)(x + 5)$$

b. $\cos^4\theta - \sin^4\theta = (\cos^2\theta - \sin^2\theta)(\cos^2\theta + \sin^2\theta)$

$$= (\cos \theta - \sin \theta)(\cos \theta + \sin \theta)(\cos^2\theta + \sin^2\theta) \quad \blacksquare$$

EXAMPLE 8   **Factoring a Sum of Cubes**

Factor $(x + 3y)^3 + 8$.

Solution $(x + 3y)^3 + 8 = [(x + 3y) + 2][(x + 3y)^2 - (x + 3y)(2) + (2)^2]$

$$= (x + 3y + 2)(x^2 + 6xy + 9y^2 - 2x - 6y + 4) \quad \blacksquare$$

EXAMPLE 9   **Factoring Trinomials**

Factor **a.** $x^2 - 8x + 15$ **b.** $6x^2 + x - 12$

c. $6w^2 - 9w - 16$ **d.** $6(x + y)^2 - 9(x + y) - 15$

Solution   **a.** $x^2 - 8x + 15 = (x - 5)(x - 3)$

b. $6x^2 + x - 12 = (2x + 3)(3x - 4)$

c. $6w^2 - 9w - 15 = 3(2w^2 - 3w - 5)$

$$= 3(2w - 5)(w + 1)$$

d. This is the same as part **c** where $w = x + y$. Thus

$$6(x + y)^2 - 9(x + y) - 15 = 3[2(x + y) - 5][(x + y) + 1]$$
$$= 3(2x + 2y - 5)(x + y + 1) \quad \blacksquare$$

Rational Expressions

A **rational expression** is a polynomial divided by a nonzero polynomial. This definition allows us to write

$$\frac{x+3}{x-2}$$

without the necessity of saying $x \neq 2$. We will write the general form of a rational expression as

$$\frac{P}{Q}$$

where P is a polynomial and Q is a polynomial with all values of the variable that cause division by 0 eliminated by the domain. The polynomial P is called the **numerator**, and the polynomial Q is called the **denominator**.

A rational expression is said to be **simplified** if all operations are performed according to the order of operations; the numerator and the denominator should be factored, if possible, and all common factors eliminated by using the following **fundamental property of rational expressions.** In the following statement, it is implied that $Q \neq 0$ and $K \neq 0$.

FUNDAMENTAL PROPERTY OF RATIONAL EXPRESSIONS

For any rational expression P/Q,

$$\frac{PK}{QK} = \frac{P}{Q}$$

Help Window ? X

In other words . . . You simplify a rational expression by factoring and eliminating common factors. We also use this property to obtain common denominators (multiply by 1) when adding and subtracting rational expressions.

EXAMPLE 10

Simplifying a Rational Expression by Factoring

Simplify: **a.** $\dfrac{x-2}{x^2-4}$ **b.** $\dfrac{\cos\theta+3}{\cos^2\theta-9}$

Solution **a.** $\dfrac{x-2}{x^2-4} = \dfrac{1(x-2)}{(x+2)(x-2)}$ *Factor.*

$$= \frac{1}{x+2}$$ *Fundamental property*

 NOTE

Note, because of the given expression, it is understood that $x \neq \pm 2$.

b. $\dfrac{\cos\theta + 3}{\cos^2\theta - 9} = \dfrac{\cos\theta + 3}{(\cos\theta - 3)(\cos\theta + 3)}$ *Factor.*

$\qquad\qquad = \dfrac{1}{\cos\theta - 3}$ *Fundamental property*   ∎

Sometimes the factors that are eliminated (as shown in color in Example 10a) are marked off in pairs, as shown in the following example. The slashes should be viewed as replacing the factor K by the number 1, as required by the fundamental property.

Formal

$$\frac{PK}{QK} = \frac{P}{Q}\cdot\frac{K}{K} = \frac{P}{Q}\cdot 1 = \frac{P}{Q}$$

Shortcut notation

$$\frac{P\cancel{K}}{Q\cancel{K}} = \frac{P}{Q}$$

EXAMPLE 11

Simplifying a Rational Expression by Factoring (Shortcut Notation)

Simplify: **a.** $\dfrac{(x-3)(x+2)(x+4)}{(x+4)(x-2)(x-3)}$ **b.** $\dfrac{6x^2 + 2x - 20}{30x^2 - 68x + 30}$ **c.** $\dfrac{x^3 - 1}{x^2 + x + 1}$

Solution

a. We use the first example to illustrate the shortcut notation.

$$\frac{(x-3)(x+2)(x+4)}{(x+4)(x-2)(x-3)} = \frac{\cancel{(x-3)}(x+2)\cancel{(x+4)}}{\cancel{(x+4)}(x-2)\cancel{(x-3)}}$$

$$= \frac{x+2}{x-2}$$

For this problem, it is understood that $x \neq -4$, $x \neq 3$, and $x \neq 2$.

b. To use the fundamental property, make sure that both the numerator and denominator are in factored form:

You can only cancel common factors.

NOTE

$$\frac{6x^2 + 2x - 20}{30x^2 - 68x + 30} = \frac{2(3x^2 + x - 10)}{2(15x^2 - 34x + 15)}$$ *Common factor first.*

$$= \frac{2(3x - 5)(x + 2)}{2(3x - 5)(5x - 3)}$$ *Factor trinomials.*

$$= \frac{\cancel{2}\cancel{(3x-5)}(x + 2)}{\cancel{2}\cancel{(3x-5)}(5x - 3)}$$ *Reduce expression.*

$$= \frac{x + 2}{5x - 3}$$

c. This final example shows the way your work will probably look:

$$\frac{x^3 - 1}{x^2 + x + 1} = \frac{(x - 1)\cancel{(x^2 + x + 1)}}{\cancel{x^2 + x + 1}}$$

$$= x - 1$$ ∎

The procedures for operations on rational expressions are identical to those you learned in arithmetic for fractions. However, with rational expressions, the numerators and denominators are polynomials (division by 0 is excluded) rather than constants. The procedures for handling rational expressions are summarized in the following box.

PROPERTIES OF RATIONAL EXPRESSIONS

Let P, Q, R, S, and K be any polynomials such that all values of the variable that cause division by 0 are excluded from the domain.

Equality: $\dfrac{P}{Q} = \dfrac{R}{S}$ if and only if $PS = QR$

Addition: $\dfrac{P}{Q} + \dfrac{R}{S} = \dfrac{PS + QR}{QS}$

Subtraction: $\dfrac{P}{Q} - \dfrac{R}{S} = \dfrac{PS - QR}{QS}$

Multiplication: $\dfrac{P}{Q} \cdot \dfrac{R}{S} = \dfrac{PR}{QS}$

Division: $\dfrac{P}{Q} \div \dfrac{R}{S} = \dfrac{PS}{QR}$

Some operations on rational expressions are shown in the following examples. The rules for addition and subtraction are simplified if $Q = S$:

$$\frac{P}{Q} + \frac{R}{S} = \frac{P}{Q} + \frac{R}{Q} \qquad \text{\textit{Substitution; } } S = Q \text{ \textit{is given.}}$$

$$= \frac{PQ + RQ}{QQ} \qquad \text{\textit{Definition of addition}}$$

$$= \frac{P + R}{Q} \qquad \text{\textit{Factoring and the fundamental property}}$$

You can similarly show that $\dfrac{P}{Q} - \dfrac{R}{Q} = \dfrac{P - R}{Q}$. Remember, all values of variables that could cause division by 0 are excluded.

EXAMPLE 12

Addition of a Rational Expression with a Common Denominator

Simplify: **a.** $\dfrac{3}{x - y} + \dfrac{2}{x - y}$ **b.** $\dfrac{5}{\cos \theta - \sin \theta} + \dfrac{3}{\cos \theta}$

Solution   **a.** $\dfrac{3}{x-y} + \dfrac{2}{x-y} = \dfrac{5}{x-y}$ *Common denominator*

b. $\dfrac{5}{\cos\theta - \sin\theta} + \dfrac{3}{\cos\theta} = \dfrac{5\cos\theta}{\cos\theta(\cos\theta - \sin\theta)} + \dfrac{3(\cos\theta - \sin\theta)}{\cos\theta(\cos\theta - \sin\theta)}$

$$= \dfrac{5\cos\theta + 3\cos\theta - 3\sin\theta}{\cos\theta(\cos\theta - \sin\theta)}$$

$$= \dfrac{8\cos\theta - 3\sin\theta}{\cos\theta(\cos\theta - \sin\theta)}$$ ∎

EXAMPLE 13   **Subtraction of a Rational Expression by Finding a Common Denominator**

Simplify: **a.** $\dfrac{3}{x-y} - \dfrac{2}{y-x}$ **b.** $\dfrac{5}{\cos\theta + \sin\theta} + \dfrac{3}{\cos\theta - \sin\theta}$

Solution   **a.** $\dfrac{3}{x-y} - \dfrac{2}{y-x} = \dfrac{3}{x-y} - \dfrac{2}{y-x} \cdot \dfrac{-1}{-1}$ *Fundamental property*

$$= \dfrac{3}{x-y} - \dfrac{-2}{x-y}$$ *Multiply.*

$$= \dfrac{3 - (-2)}{x-y}$$ *Add; common denominator*

$$= \dfrac{5}{x-y}$$

b. $\dfrac{5}{\cos\theta + \sin\theta} + \dfrac{3}{\cos\theta - \sin\theta}$

$$= \dfrac{5}{\cos\theta + \sin\theta} \cdot \dfrac{\cos\theta - \sin\theta}{\cos\theta - \sin\theta} + \dfrac{3}{\cos\theta - \sin\theta} \cdot \dfrac{\cos\theta + \sin\theta}{\cos\theta + \sin\theta}$$

$$= \dfrac{5\cos\theta - 5\sin\theta}{\cos^2\theta - \sin^2\theta} + \dfrac{3\cos\theta + 3\sin\theta}{\cos^2\theta - \sin^2\theta}$$

$$= \dfrac{8\cos\theta - 2\sin\theta}{\cos^2\theta - \sin^2\theta}$$

If there are no common factors, it is not necessary to leave your answer in factored form.   ∎

EXAMPLE 14   **Addition of Rational Expressions Using the Definition of Addition**

Simplify $\dfrac{a+b}{a-b} + \dfrac{a-2b}{2a+b}$.

Solution

$$\frac{a+b}{a-b} + \frac{a-2b}{2a+b} = \frac{(a+b)(2a+b) + (a-2b)(a-b)}{(a-b)(2a+b)}$$

$$= \frac{2a^2 + 3ab + b^2 + a^2 - 3ab + 2b^2}{(a-b)(2a+b)}$$

$$= \frac{3a^2 + 3b^2}{(a-b)(2a+b)}$$

$$= \frac{3(a^2 + b^2)}{(a-b)(2a+b)}$$ ∎

With rational expressions such as the answer shown in Example 14, we will write the simplified form as a reduced fraction with both the numerator and denominator factored, if possible.

The key to multiplying and dividing rational expressions is to factor all the numerators and denominators and then use the fundamental property of rational expressions. Divisions are handled using the definition of division to convert the division to a multiplication (sometimes remembered by "invert and multiply").

EXAMPLE 15

Multiplication and Division of a Rational Expression

Simplify: $\left[\dfrac{x^2 + 5x + 6}{2x^2 - x - 1} \cdot \dfrac{2x^2 - 9x - 5}{x^2 + 7x + 12}\right] \div \dfrac{2x^2 - 13x + 15}{x^2 + 3x - 4}$

Solution

$$\left[\frac{x^2 + 5x + 6}{2x^2 - x - 1} \cdot \frac{2x^2 - 9x - 5}{x^2 + 7x + 12}\right] \div \frac{2x^2 - 13x + 15}{x^2 + 3x - 4}$$

$$= \left[\frac{(x+2)(x+3)}{(x-1)(2x+1)} \cdot \frac{(2x+1)(x-5)}{(x+3)(x+4)}\right] \div \frac{(x-5)(2x-3)}{(x-1)(x+4)}$$

$$= \frac{(x+2)\cancel{(x+3)}\cancel{(2x+1)}\cancel{(x-5)}\cancel{(x-1)}\cancel{(x+4)}}{\cancel{(x-1)}\cancel{(2x+1)}\cancel{(x+3)}\cancel{(x+4)}\cancel{(x-5)}(2x-3)}$$

$$= \frac{x+2}{2x-3}$$ ∎

Radical Expressions

The symbol $\sqrt[n]{b}$ is called a **radical**, n is called the **index**, and b is called the **base** or the **radicand**. A **radical expression** is an algebraic expression for which there is a variable under a radical. In trigonometry, we are concerned with radical expressions for which $n = 2$, namely **square roots**. Operations with square roots follow some **laws of square roots**.

> **LAWS OF SQUARE ROOTS**
>
> Let a and b be positive numbers; then
>
> **1.** $\sqrt{0} = 0$ **2.** $\sqrt{x^2} = |x|$
>
> **3.** $\sqrt{ab} = \sqrt{a}\,\sqrt{b}$ **4.** $\sqrt{\dfrac{a}{b}} = \dfrac{\sqrt{a}}{\sqrt{b}}$

A square root is **simplified** when

1. The radicand is written in completely factored form, and there is no factor with an exponent larger than 1.
2. No fraction (or negative exponent) appears within a radical.
3. No radical appears in a denominator.

If an expression involving a square root symbol is in simplest form, then it is irrational. On the other hand, if it is not in simplest form, then it may be rational or irrational. Also, since we frequently work with reciprocals in trigonometry, we sometimes relax the requirement that the denominator cannot be a square root.

EXAMPLE 16

Simplifying a Radical Expression

Simplify: **a.** $\sqrt{a^2b^2 + a^2c^2}$ **b.** $\sqrt{\sin^2\theta\cos\theta + 4\sin^2\theta}$

Solution   **a.** $\sqrt{a^2b^2 + a^2c^2} = \sqrt{a^2(b^2 + c^2)}$ *Factor radicand (we are looking for square factors).*

$$= \sqrt{a^2}\,\sqrt{(b^2 + c^2)}$$ *Law of square root*

$$= |a|\,\sqrt{b^2 + c^2}$$ *There are no square factors in $b^2 + c^2$, only square terms.*

▲ NOTE

b. $\sqrt{\sin^2\theta\cos\theta + 4\sin^2\theta} = \sqrt{\sin^2\theta(\cos\theta + 4)}$ *Common factor*

$$= \sqrt{\sin^2\theta}\,\sqrt{\cos\theta + 4}$$ *Law of square roots*

$$= |\sin\theta|\,\sqrt{\cos\theta + 4}$$ *Law of square roots* ■

In trigonometry, you will need to recognize different forms of the same number. Consider finding the reciprocal of $\sqrt{3}$:

The reciprocal of $\sqrt{3}$ is $\dfrac{1}{\sqrt{3}}$; this can also be written as $\dfrac{1}{\sqrt{3}} \cdot \dfrac{\sqrt{3}}{\sqrt{3}} = \dfrac{\sqrt{3}}{3}$.

This is also written as $\frac{1}{3}\sqrt{3}$.

The following forms are equivalent:

$$\frac{1}{\sqrt{3}} = \frac{\sqrt{3}}{3} = \frac{1}{3}\sqrt{3}$$

NOTE

EXAMPLE 17   **Simplifying Radical Expressions**

a. $\dfrac{6 + 3\sqrt{5}}{3}$ b. $\dfrac{-(-2) + \sqrt{(-2)^2 - 4(2)(-1)}}{2(2)}$

c. $\dfrac{x}{\sqrt{x - 3}}$ d. $\dfrac{x}{\sqrt{x} - 3}$

Solution   a. $\dfrac{6 + 3\sqrt{5}}{3} = \dfrac{3(2 + \sqrt{5})}{3}$ *Do not try to "cancel"; factor first, then use the fundamental property of rational expressions.*

$= 2 + \sqrt{5}$

b. $\dfrac{-(-2) + \sqrt{(-2)^2 - 4(2)(-1)}}{2(2)} = \dfrac{2 + \sqrt{4 + 4(2)}}{4}$

$= \dfrac{2 + \sqrt{4}\sqrt{1 + 2}}{4}$

$= \dfrac{2 + 2\sqrt{3}}{4}$

$= \dfrac{2(1 + \sqrt{3})}{4}$

$= \dfrac{1 + \sqrt{3}}{2}$

c. $\dfrac{x}{\sqrt{x - 3}} = \dfrac{x}{\sqrt{x - 3}} \cdot \dfrac{\sqrt{x - 3}}{\sqrt{x - 3}}$ *Multiply by 1 to simplify.*

$= \dfrac{x\sqrt{x - 3}}{x - 3}$

d. $\dfrac{x}{\sqrt{x} - 3} = \dfrac{x}{\sqrt{x} - 3} \cdot \dfrac{\sqrt{x} + 3}{\sqrt{x} + 3}$ *Multiply by 1 to simplify.*

$= \dfrac{x(\sqrt{x} + 3)}{x - 9}$ *Multiply binomials.*   ■

Solving Equations

Linear Equations

To solve a first-degree (or linear) equation, isolate the variable on one side. That is, the solution of $ax + b = 0$ is

$$x = -\dfrac{b}{a}, \quad \text{where } a \neq 0$$

To solve a linear equation, use the following steps:

Step 1. Use the distributive property to clear the equation of parentheses. If it is a rational expression, multiply both sides by the appropriate expression to eliminate the denominator in the problem. Be sure to check the solutions obtained back in the original.

Step 2. Add the same number to both sides of the equality to obtain an equation in which all of the terms involving the variable are on one side and all the other terms are on the other side.

Step 3. Multiply (or divide) both sides of the equation by the same nonzero number to isolate the variable on one side.

EXAMPLE 18   **Solving a Linear Equation**

Solve $4(x - 3) + 5x = 5(8 + x)$.

Solution

$$4(x - 3) + 5x = 5(8 + x)$$
$$4x - 12 + 5x = 40 + 5x$$
$$9x - 12 = 40 + 5x$$
$$4x - 12 = 40$$
$$4x = 52$$
$$x = 13$$

■

EXAMPLE 19   **Solving a Rational Equation**

Solve $\dfrac{x + 1}{x - 2} = \dfrac{x + 2}{x - 2}$.

Solution

$$(x + 1)(x - 2) = (x + 2)(x - 2), \quad x \neq 2$$
$$x^2 - 2x + x - 2 = x^2 - 4$$
$$-x - 2 = -4$$
$$-x = -2$$
$$x = 2$$

Notice that $x = 2$ causes division by 0, so the solution set is empty.   ■

EXAMPLE 20   **Solving a Literal Equation**

Solve $4x + 5xy + 3y^2 = 10$ for x.

Solution

$$4x + 5xy = 10 - 3y^2$$
$$(4 + 5y)x = 10 - 3y^2$$
$$x = \frac{10 - 3y^2}{4 + 5y}, \quad y \neq -\frac{4}{5}$$

■

Quadratic Equations

To solve a second-degree (or quadratic) equation, first obtain a 0 on one side. Next, try to factor the quadratic. If it is factorable, set each factor equal to 0 and solve. If it is not factorable, use the quadratic formula:

QUADRATIC FORMULA

If $ax^2 + bx + c = 0$, $a \neq 0$, then

$$x = \frac{-b \pm \sqrt{b^2 - 4ac}}{2a}$$

EXAMPLE 21 **Solving a Quadratic Equation by Factoring**

Solve $2x^2 - x - 3 = 0$.

Solution Factoring, $(2x - 3)(x + 1) = 0$. If $2x - 3 = 0$, then $x = \frac{3}{2}$; if $x + 1 = 0$, then $x = -1$. The solution is $x = \frac{3}{2}, -1$. ∎

EXAMPLE 22 **Solving a Quadratic Equation Using the Quadratic Formula**

Solve $5x^2 - 3x - 4 = 0$.

Solution $a = 5$, $b = -3$, and $c = -4$; thus (since it does not factor)

$$x = \frac{3 \pm \sqrt{9 - 4(5)(-4)}}{2(5)} = \frac{3 \pm \sqrt{89}}{10}$$ ∎

EXAMPLE 23 **Solving a Literal Equation Using the Quadratic Formula**

Solve $x^2 + 2xy + 3y^2 - 4 = 0$ for x.

Solution $a = 1$, $b = 2y$, $c = 3y^2 - 4$; thus

$$x = \frac{-2y \pm \sqrt{4y^2 - 4(1)(3y^2 - 4)}}{2(1)}$$
$$= \frac{-2y \pm \sqrt{16 - 8y^2}}{2}$$
$$= \frac{-2y \pm \sqrt{4}\sqrt{4 - 2y^2}}{2}$$
$$= -y \pm \sqrt{4 - 2y^2}$$ ∎

PROBLEM SET C

Level 1 Drill

Problems 1–12: *Simplify each expression.*

1. a. $(x + 2)(x + 1)$ **b.** $(y - 2)(y + 3)$
    c. $(x + 1)(x - 2)$ **d.** $(y - 3)(y + 2)$

2. a. $(a - 5)(a - 3)$ **b.** $(b + 3)(b - 4)$
    c. $(c + 1)(c - 7)$ **d.** $(d - 3)(d + 5)$

3. a. $(2x + 1)(x - 1)$ **b.** $(2x - 3)(x - 1)$
    c. $(x + 1)(3x + 1)$ **d.** $(x + 1)(3x + 2)$

4. a. $(x + y)(x - y)$ **b.** $(a + b)(a - b)$
    c. $(5x - 4)(5x + 4)$ **d.** $(3y - 2)(3y + 2)$

5. a. $(a + 2)^2$ **b.** $(b - 2)^2$
    c. $(x + 4)^2$ **d.** $(y - 3)^2$

6. a. $(3x - 1)(x^2 + 3x - 2)$ **b.** $(2x + 1)(x^2 + 2x - 5)$

7. a. $\dfrac{x^2 - y^2}{3x - 3y}$ **b.** $\dfrac{x^2 - y^2}{2x + 2y}$

8. a. $(x^2 - 36) \cdot \dfrac{3x + 1}{x + 6}$ **b.** $(y^2 - 9) \div \dfrac{y + 3}{y - 3}$

9. a. $\dfrac{x + 3}{x} + \dfrac{3 - x}{2}$ **b.** $\dfrac{x - 2}{x^2} - \dfrac{1 - x}{x}$

10. a. $\dfrac{2}{x - y} + \dfrac{5}{y - x}$ **b.** $\dfrac{a}{a - 1} + \dfrac{a - 3}{1 - a}$

11. $[7^3 + 2^6(3^3 + 4^4)]^0$ **12.** $[9^3 + 3^5(5^3 + 8^3)]^0$

Problems 13–21: *Factor completely, if possible.*

13. a. $me + mi + my$ **b.** $a^2 - b^2$
    c. $a^2 + b^2$ **d.** $a^3 - b^3$

14. a. $a^3 + b^3$ **b.** $x^2 + 2xy + y^2$
    c. $x^2 - 2xy + y^2$ **d.** $u^2 + 2uv + v^2$

15. a. $3x^2 - 5x - 2$ **b.** $6y^2 - 7y + 2$
    c. $8a^2b + 10ab - 3b$ **d.** $2s^2 - 10s - 48$

16. a. $4y^3 + y^2 - 21y$ **b.** $12m^2 - 7m - 12$
    c. $x^2 - 2x - 35$ **d.** $2x^2 + 7x - 15$

17. a. $(x - y)^2 - 1$ **b.** $(2x + 3)^2 - 1$

18. a. $(5a - 2)^2 - 9$ **b.** $(3p - 2)^2 - 16$

19. a. $2x^2 + x - 6$ **b.** $3x^2 - 11x - 4$

20. a. $6x^2 + 47x - 8$ **b.** $6x^2 - 47x - 8$

21. a. $6x^2 + 49x + 8$ **b.** $6x^2 - 49x + 8$

Level 2 Drill

Problems 22–34: *Classify as true or false; if it is false, explain why you think it is false.*

22. $(5x)^3 = 5x^3$ **23.** $(2x^2y^3)^2 = 4x^4y^5$

24. $(x + y)^3 = x^3 + y^3$ **25.** $(x^2 + y^2) = (x + y)^2$

26. $\dfrac{x - y}{y - x} = 1$ **27.** $\dfrac{x}{5} + \dfrac{y}{3} = \dfrac{x + y}{8}$

28. $\dfrac{a}{b} + \dfrac{c}{b} = \dfrac{a + c}{b}$ **29.** $\dfrac{\cancel{a}}{b} + \dfrac{c}{\cancel{a}} = \dfrac{c}{b}$

30. x^3 means $x + x + x$ **31.** $\dfrac{\cancel{x} + y}{3\cancel{x}} = \dfrac{1 + y}{3}$

32. $b^n = \underbrace{b \cdot b \cdot \cdots \cdot b}_{n \text{ factors}}$ for all n

33. $\sqrt{x^2 + y^2} = x + y$

34. $(x + y)^{-2} = x^{-2} + y^{-2}$

Problems 35–49: *Simplify each expression.*

35. $\dfrac{5}{x - 4} + \dfrac{x}{x + 1}$ **36.** $\dfrac{2}{x + 2} - \dfrac{x + 3}{x + 1}$

37. $\dfrac{x}{x - 2} + \dfrac{5}{x + 2}$ **38.** $\dfrac{3}{x - y} + 5$

39. $6 - \dfrac{2}{x + y}$ **40.** $\dfrac{3x^2 - 4x - 4}{x^2 - 4}$

41. $\dfrac{y^2 + 3y - 18}{y^2 - 9}$ **42.** $\dfrac{3}{x + y} + \dfrac{5}{2x + 2y}$

43. $\dfrac{4x - 12}{x^2 - 49} \div \dfrac{18 - 2x^2}{x^2 - 4x - 21}$

44. $\dfrac{36 - 9y}{3y^2 - 48} \div \dfrac{15 + 13y + 2y^2}{12 + 11y + 2y^2}$

45. $\dfrac{x + 1}{x^2 + 7x + 12} + \dfrac{x}{x + 4}$

46. $\dfrac{x - 2}{x^2 + 2x - 15} + \dfrac{x}{x - 3}$

47. $\dfrac{1}{x - 2} + \dfrac{4}{x + 3} + \dfrac{2x + 5}{x^2 + x - 6}$

48. $\dfrac{x + 1}{x - 1} + \dfrac{3}{x + 2} - \dfrac{2x - 1}{x^2 + x - 2}$

49. $\dfrac{2x}{x - 4} - \dfrac{x}{x + 4} + \dfrac{2x + 1}{x^2 - 16}$

Problems 50–71: Solve each equation for x.

50. $(x - 3)(2x + 1) = 0$ **51.** $(3x - 4)(x + 2) = 0$

52. $(4x + 1)(4x + 1) = 25$ **53.** $(2x - 3)(2x - 3) = 4$

54. $x^2 - 144 = 0$ **55.** $5x^2 - 45 = 0$

56. $4x^2 = 9$ **57.** $16x^2 = 1$

58. $(2x - 1)^2 = 0$ **59.** $(6x + 1)^2 = 0$

60. $12x^2 - 5x - 3 = 0$ **61.** $30x^2 + 13x - 10 = 0$

62. $x^2 - 4x + 1 = 0$ **63.** $x^2 - 4x - 1 = 0$

64. $x^2 + 10x + 21 = 0$ **65.** $x^2 - 6x - 43 = 0$

66. $6x^2 - x - 1 = 0$ **67.** $x^2 - 14x + 39 = 0$

68. $x^2 - 8x + 17 = 0$ **69.** $x^2 - 2x + 10 = 0$

70. $(x - 4)(x + 4) = 4x$ **71.** $(x - 3)(x + 3) = 6(x + 1)$

Level 3 Theory

72. PROBLEM FROM CALCULUS The expression $\dfrac{x^2 - 9}{x - 3}$ is not defined when $x = 3$. In this problem, we investigate the behavior of this expression for values of x close to 3. Suppose we choose x so that $|x - 3| < 0.1$ (but $x \neq 3$).
a. Find some x_0 satisfying this inequality.
b. Calculate the value of the expression for x_0:
$$\frac{x_0^2 - 9}{x_0 - 3}$$
c. Repeat parts **a** and **b** for two other values. Can you form a conclusion about the value of the given expression?
d. How could factoring have been used to simplify your work in part **b**?

73. PROBLEM FROM CALCULUS The expression $\dfrac{x^3 - 27}{x - 3}$ is not defined when $x = 3$. In this problem, we investigate the behavior of this expression for values of x close to 3. Suppose we choose x so that $|x - 3| < 0.01$ (but $x \neq 3$).
a. Find some x_0 satisfying this inequality.
b. Calculate the value of the expression for x_0:
$$\frac{x_0^3 - 27}{x_0 - 3}$$
c. Repeat parts **a** and **b** for two other values. Can you form a conclusion about the value of the given expression?
d. How could factoring have been used to simplify your work in part **b**?

74. JOURNAL PROBLEM *What is wrong,* if anything, *with the following problem? It was published in* The College Mathematics Journal (FFF #39). Simplify

$$\frac{bx(a^2x^2 + 2a^2y^2 + b^2y^2) + ay(a^2x^2 + 2b^2x^2 + b^2y^2)}{bx + ay}$$

Solution:

$$\frac{bx(a^2x^2 + 2a^2y^2 + b^2y^2) + ay(a^2x^2 + 2b^2x^2 + b^2y^2)}{bx + ay}$$
$$= \frac{bx(ax + by)^2 + ay(ax + by)^2}{bx + ay}$$
$$= \frac{(bx + ay)(ax + by)^2}{bx + ay}$$
$$= (ax + by)^2$$

The 1988 American High School Mathematics Examination included this problem (#19). Many students obtained the correct answer, $(ax + by)^2$, but included the preceding incorrect solution.

APPENDIX D

REVIEW OF GEOMETRY

In this section, we use the following variables when stating formulas: K = area, P = perimeter, C = circumference, S = surface area, and V = volume. Also, r denotes radius, h altitude, l slant height, b base, B area of base, and θ central angle expressed in radians.

Polygons

Type	Number of sides
triangle	3
quadrilateral	4
pentagon	5
hexagon	6
heptagon	7
octagon	8
nonagon	9
decagon	10
undecagon	11
dodecagon	12

Polygons are classified according to the number of sides. These classifications are shown in the margin.

Triangles

The sum of the measures of the angles of a triangle is 180°.

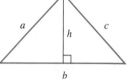

Triangle

$$K = \tfrac{1}{2}bh$$
$$P = a + b + c$$

Pythagorean theorem: If angle C is a right angle, then $c^2 = a^2 + b^2$.

45°–45°–90° triangle theorem: For any right triangle with acute angles measuring 45°, the legs are the same length, and the hypotenuse has a length equal to $\sqrt{2}$ times the length of one of those legs.

30°–60°–90° triangle theorem: For any right triangle with acute angles measuring 30° and 60°:

1. The hypotenuse is twice as long as the leg opposite the 30° angle (the shorter leg).
2. The leg opposite the 30° angle (the shorter leg) is $\tfrac{1}{2}$ as long as the hypotenuse.
3. The leg opposite the 60° angle (the longer leg) equals the length of the other (shorter) leg times $\sqrt{3}$.
4. The leg opposite the 30° angle equals the length of the other leg divided by $\sqrt{3}$.

Equilateral triangle: For any equilateral triangle,

$$\alpha = \beta = \gamma = 60°; \quad K = \tfrac{1}{2}bh = \tfrac{1}{2}b\left(\frac{\sqrt{3}}{2}b\right) = \frac{\sqrt{3}}{4}b^2$$

Quadrilaterals

Rectangle

$$K = \ell w$$
$$P = 2\ell + 2w$$
$$\text{Diagonal} = \sqrt{\ell^2 + w^2}$$

Square

$$K = s^2$$
$$P = 4s$$
$$\text{Diagonal} = s\sqrt{2}$$

Parallelogram

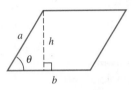

Trapezoid

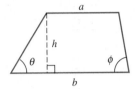

$$K = bh = ab \sin \theta$$

$$P = 2a + 2b$$

$$K = \tfrac{1}{2}h(a + b)$$

$$P = a + b + h(\csc \theta + \csc \phi)$$

Regular Polygon of *n* Sides

central angle: $\dfrac{2\pi}{n}$

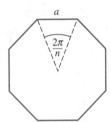

Basic Formulas

$$K = \tfrac{1}{4}na^2 \cot \tfrac{\pi}{n} \qquad P = an$$

Sector

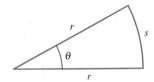

$$K = \tfrac{1}{2}r^2\theta$$

$$s = r\theta$$

Segment

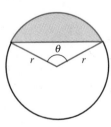

$$K = \tfrac{1}{2}r^2(\theta - \sin \theta)$$

Circles

Terminology

Circumference: Distance around a circle
Chord: A line joining two points of a circle
Diameter: A chord through the center—
  AB in Figure D.1
Arc: Part of a circle—*BC*, *AC*, or *ACB* in
  Figure D.1

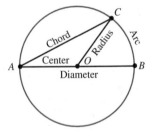

FIGURE D.1   Circle
$$K = \pi r^2$$
$$C = 2\pi r = \pi d$$

The length *s* of an arc of a circle of radius *r* with central angle θ (measured in radians) is $s = r\theta$.

To *intercept an arc* is to cut off the arc; in Figure D.1, $\angle COB$ intercepts *BC*.
Tangent of a circle is a line that intersects the circle at one and only one point.
Secant of a circle is a line that intersects the circle at exactly two points.
Inscribed polygon is a polygon all of whose sides are chords of a circle. A regular inscribed polygon is a polygon all of whose sides are the same length.
Inscribed circle is a circle to which all the sides of a polygon are tangents.

Circumscribed polygon is a polygon all of whose sides are tangents to a circle.

Circumscribed circle is a circle passing through each vertex of a polygon.

Congruent Triangles

We say that two figures are **congruent** if they have the same size and shape. For **congruent triangles** ABC and DEF, denoted by $\triangle ABC \simeq \triangle DEF$, we may conclude that all six corresponding parts (three angles and three sides) are congruent. We call these **corresponding parts.**

EXAMPLE 1   **Corresponding Parts of a Triangle**

Name the corresponding parts of the given triangles.

a. $\triangle ABC \simeq \triangle A'B'C'$ **b.** $\triangle RST \simeq \triangle UST$

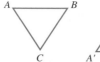 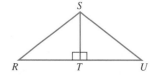

Solution   **a.** $\overline{AB}$ corresponds to $\overline{A'B'}$
$\overline{AC}$ corresponds to $\overline{A'C'}$
$\overline{BC}$ corresponds to $\overline{B'C'}$
$\angle A$ corresponds to $\angle A'$
$\angle B$ corresponds to $\angle B'$
$\angle C$ corresponds to $\angle C'$

b. $\overline{RS}$ corresponds to $\overline{US}$
$\overline{RT}$ corresponds to $\overline{UT}$
$\overline{ST}$ corresponds to $\overline{ST}$
$\angle R$ corresponds to $\angle U$
$\angle RTS$ corresponds to $\angle UTS$
$\angle RST$ corresponds to $\angle UST$ ■

Line segments, angles, triangles, or other geometric figures are *congruent* if they have the same size and shape. In this section, we focus on triangles.

CONGRUENT TRIANGLES

Two triangles are **congruent** if their corresponding sides have the same length and their corresponding angles have the same measure.

To prove that two triangles are congruent, you must show that they have the same size and shape. It is not necessary to show that all six parts (three sides and three angles) are congruent; if some of these six parts are congruent, it necessarily follows that the other parts are congruent. Three important properties are used to show triangle congruence:

> ### CONGRUENT TRIANGLE PROPERTIES
>
> ### SIDE–SIDE–SIDE (SSS)
> If three sides of one triangle are congruent to three sides of another triangle, then the two triangles are congruent.
>
> ### SIDE–ANGLE–SIDE (SAS)
> If two sides of one triangle and the angle between those sides are congruent to the corresponding sides and angle of another triangle, then the two triangles are congruent.
>
> ### ANGLE–SIDE–ANGLE (ASA)
> If two angles and the side that connects them on one triangle are congruent to the corresponding angles and side of another triangle, then the two triangles are congruent.

EXAMPLE 2   **Finding Congruent Triangles**

Determine whether each pair of triangles is congruent. If so, cite one of the congruent triangle properties.

Solution

a.

Congruent; SAS

b.

Not congruent

c.

Congruent; ASA

d.

Congruent; SSS

A side that is in common to two triangles obviously is equal in length to itself, and does not need to be marked.   ■

In geometry, we use congruent triangles mainly when we want to know whether an angle from one triangle is congruent to an angle from a different triangle, or when we want to know whether a side from one triangle is the same length as the side from another triangle. To do this, we often prove that one triangle is congruent to the other (using one of the three congruent triangle properties) and then use the following property:

> ### CONGRUENT TRIANGLE PROPERTY
>
> Corresponding parts of congruent triangles are congruent.

Problem Set 1.1, page 8

1. a. theta   **b.** alpha   **c.** phi   **d.** omega   **e.** mu
3. a. δ **b.** ϕ **c.** θ **d.** ω **5. a.** 360°   **b.** 180°
7. a. 90°   **b.** 135°   **9. a.** −90°   **b.** −45°
11. a. acute

b. straight

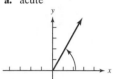

c. acute

d. none

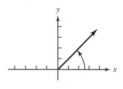

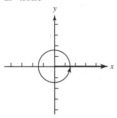

e. none

13. a. none

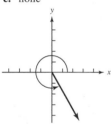

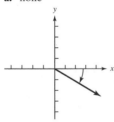

b. obtuse

c. none

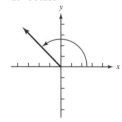

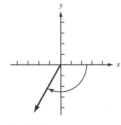

d. none

e. none

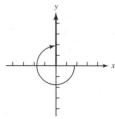

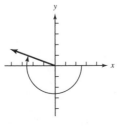

15. a. 65.67°   **b.** 146.83°   **c.** 85.33°
17. a. 62.92°   **b.** 315.33°   **c.** 25.42°
19. a. 128.178°   **b.** 13.514°   **c.** 48.469°
21. a. 16.012°   **b.** 29.005°   **c.** 143.006°   **23.** $\beta = 50°$
25. $\gamma = 45°$ **27.** $\alpha = 180° - \beta - \gamma$
29. $\gamma = 180° - \alpha - \beta$ **31.** $\beta = 40°$ **33.** $c = 2\sqrt{2}$
35. $b = 4$ **37.** $x = 5$, so $a = 5$, $b = 12$, $c = 13$
39. $x = 3$, so $a = 2$, $b = 3$, $c = \sqrt{13}$
41. The integer part (to the left of the decimal point) is the number of degrees. Subtract the integer part and multiply by 60; the whole number part of this result is the number of minutes. Subtract this number and multiply by 60; round this number to the nearest whole number, which is the number of seconds.
43. The length of one side of the roof is 30 ft.

Problem Set 1.2, page 13

1. The vertices are labeled A, B, and C. The sides opposite those vertices are labeled a, b, and c, respectively, and the angles are labeled α, β, and γ, respectively.
3. $\overline{AB} \simeq \overline{ED}$; $\overline{AC} \simeq \overline{EF}$; $\overline{CB} \simeq \overline{FD}$; $\angle A \simeq \angle E$; $\angle B \simeq \angle D$; $\angle C \simeq \angle F$
5. $\overline{RS} \simeq \overline{TU}$; $\overline{RT} \simeq \overline{TR}$; $\overline{ST} \simeq \overline{UR}$; $\angle SRT \simeq \angle UTR$; $\angle S \simeq \angle U$; $\angle STR \simeq \angle URT$
7. similar   **9.** not similar   **11.** similar   **13.** 4
15. $\frac{16}{3}$ **17.** $\frac{8}{3}$ **19.** 24 ft   **21.** 10 ft   **23.** $3\sqrt{2}$
25. $5\sqrt{13}$ ft; 18 ft; 55 ft   **27.** $\frac{20}{3}$ **29.** 125   **31.** 29 ft

Problem Set 1.3, page 22

5. a. $\cos \alpha = \frac{1}{2}$; $\sin \alpha = \frac{1}{2}\sqrt{3}$; $\tan \alpha = \sqrt{3}$
b. $\cos \beta = \frac{1}{2}\sqrt{3}$; $\sin \beta = \frac{1}{2}$; $\tan \beta = \dfrac{1}{\sqrt{3}}$ or $\frac{1}{3}\sqrt{3}$
7. a. 0.9903   **b.** 0.4067   **c.** 1.8807
9. a. 0.5774   **b.** 0.3420   **c.** 0.7660
11. a. 0.6428   **b.** 1.5557   **c.** 0.4132
13. a. 0.6561   **b.** 0.6561   **15. a.** 0.9063   **b.** −0.9063
17. a. 1.0000   **b.** 1.0000   **c.** 1.0000   **19.** 53°   **21.** 31°
23. 68°   **25.** 24°   **27. a.** 60°   **b.** 1.003819838
29. a. 50.19442891°   **b.** 47.73950141   **31.** 0.9913863805
33. 33.70463602
35. $\alpha = 69°$; $\beta = 21°$; $\gamma = 90°$; $a = 96$; $b = 37$; $c = 100$
37. $\alpha = 77°$; $\beta = 13°$; $\gamma = 90°$; $a = 390$; $b = 90$; $c = 400$
39. $\alpha = 50°$; $\beta = 40°$; $\gamma = 90°$; $a = 98$; $b = 82$; $c = 130$
41. $\alpha = 67.8°$; $\beta = 22.2°$; $\gamma = 90.0°$; $a = 26.6$; $b = 10.8$; $c = 28.7$
43. $\alpha = 28.95°$; $\beta = 61.05°$; $\gamma = 90.00°$; $a = 202.7$; $b = 366.4$; $c = 418.7$
45. $\alpha = 56.00°$; $\beta = 34.00°$; $\gamma = 90.00°$; $a = 3{,}484$; $b = 2{,}350$; $c = 4{,}202$
47. $\alpha = 57.83°$; $\beta = 32.17°$; $\gamma = 90.00°$; $a = 290.8$; $b = 182.9$; **c.** 343.6

49. $\alpha = 28°$; $\beta = 62°$; $\gamma = 90°$; $a = 1,600$; $b = 3,100$; $c = 3,500$
51. $\alpha = 34.2°$; $\beta = 55.8°$; $\gamma = 90.0°$; $a = 85.3$; $b = 126$; $c = 152$
53. $\alpha = 15.3°$; $\beta = 74.7°$; $\gamma = 90.0°$; $a = 12.5$; $b = 45.6$; $c = 47.3$
55. $\alpha = 27.8°$; $\beta = 62.2°$; $\gamma = 90.0°$; $a = 135$; $b = 256$; $c = 289$
57. $\alpha = 30.54°$; $\beta = 59.46°$; $\gamma = 90.00°$; $a = 111.7$; $b = 189.4$; $c = 219.9$
59. The distance across the river is 32 m.
61. The top of the ladder is 13 ft above the ground.
63. The fences from an angle of 44°.
65. The chimney is about 1,251 ft.
67. The sun is 35.5° above the horizon.

Problem Set 1.4, page 28

5. The height of the building is 120 ft.
7. The car is 928 ft from the point directly below the helicopter.
9. The perimeter of the fence is 83 ft.
11. The height of the solar panel should be 8 ft.
13. The height of the rafter should be 11 ft.
15. The proper angle for the rafters is 81°.
17. The distance $|\overline{CB}|$ is 350 ft.
19. The bottom of the wheel is about 4.07 ft above the base of the incline.
21. The Sears Tower is 1,454 ft tall, and the Empire State Building is 1,250 ft.
23. It is 173 m across the river.
25. The height of Devil's Tower is about 263 m.
27. The distance from the earth to Venus is about 6.34×10^7 mi.
31. The sun will be 30° above the horizon at 8:08 A.M.
33. a. The angle of the sun is about 55°.
   b. It is about 10:00 A.M. (9:55 A.M.)
35. The lengths are 14 ft, 2 in., 10 ft, 0 in.; 12 ft, 0 in., and 15 ft, 7 in.

Problem Set 1.5, page 38

5. a. II   **b.** 45°   **c.** 135°   **d.** −225°
7. a. III   **b.** 20°   **c.** 200°   **d.** −160°
9. a. III   **b.** 45°   **c.** 225°   **d.** −135°
11. a. II   **b.** 20°   **c.** 160°   **d.** −200°
13. a. no quadrant   **b.** no reference angle
   c. 90°   **d.** −270°
15. a. IV   **b.** 30°   **c.** 330°   **d.** −30°
17.

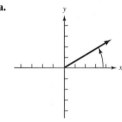

   a. reference angle
   b. coterminal
   c. coterminal
   d. none
   e. coterminal

19. a. θ is in Quadrant I   **b.** θ is in Quadrant I or II (or 90°)
   c. θ is in Quadrant I or IV (or 0°)   **d.** θ is in Quadrant III
21. a. $0° < \alpha < 90°$ **b.** $90° < \beta < 180°$
   c. $180° < \gamma < 270°$ **d.** $270° < \delta < 360°$
23. 18.85 in.   **25.** 15.71 ft   **27.** 14.07 cm   **29.** 118.96 ft
31. The tip moves 3.14 cm.
33. The larger pulley rotates 1.59 revolutions.
35. Columbia is 2,600 mi from Disneyland.
37. The cities are 890 km apart.
39. The diameter of the moon is 5,200 km.

Chapter 1 Sample Test, page 41

1. a. alpha   **b.** theta   **c.** gamma   **d.** lambda   **e.** beta
   f. delta
2. a.

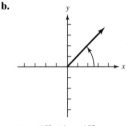

b.

$\theta = 30°$; $\theta' = 30°$ $\theta = 45°$; $\theta' = 45°$

c.

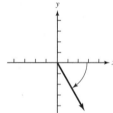

d.

$\theta = -60°$; $\theta' = 60°$ $\theta = 180°$; $\theta' = 0°$

e.

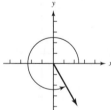

$\theta = 300°$; $\theta' = 60°$

3. 50.60°
4. a. If θ is an acute angle in a right triangle, then
   $\cos \theta = \dfrac{\text{ADJACENT SIDE}}{\text{HYPOTENUSE}}$, $\sin \theta = \dfrac{\text{OPPOSITE SIDE}}{\text{HYPOTENUSE}}$,
   $\tan \theta = \dfrac{\text{OPPOSITE SIDE}}{\text{ADJACENT SIDE}}$
   b. $\alpha = 41°$; $\beta = 49°$; $\gamma = 90°$; $a = 2.6$; $b = 3.0$; $c = 4.0$
5. a. 0.9397   **b.** 0.6428   **c.** 0.2679

6. a. $a = 59.8; b = 77.9; c = 112; \alpha = 30.7°; \beta = 41.7°;$
$\gamma = 107.6°$

b. $a = 3.5; b = 0.89; c = 3.6; \alpha = 79°; \beta = 87°; \gamma = 14°$

c. $a = 0.60; b = 7.7; c = 8.0; \alpha = 4°; \beta = 58°; \gamma = 118°$

7. The tower is 2,063 ft.

8. a. $\cos \beta = \dfrac{a}{c}$ **b.** $\sin \beta = \dfrac{b}{c}$ **c.** $\tan \alpha = \dfrac{a}{b}$

d. $\alpha = 23.5°; \beta = 66.5°; \gamma = 90.0°; a = 5.30; b = 12.2; c = 13.3$

9. It is 278 ft across the bridge. **10.** 50π m; 157 m

Miscellaneous Problems, Chapter 1, page 42

1. 17 in. **3.** 7 ft **5.** I; 15° **7.** II; 95° **9.** 0.9965
11. 0.9994 **13.** 0.8192 **15.** 1.7321 **17.** −0.6873
19. 0.7720
21. $\alpha = 27°; \beta = 63°; \gamma = 90°; a = 1.8; b = 3.6; c = 4.0$
23. $\alpha = 71.2°; \beta = 18.8°; \gamma = 90.0°; a = 22.0; b = 7.48; c = 23.2$
25. $\alpha = 23°; \beta = 67°; \gamma = 90°; a = 2.7; b = 6.5; c = 7.0$
27. $\alpha = 56.0°; \beta = 34.0°; \gamma = 90.0°; a = 5.81; b = 3.91; c = 7.00$
29. $\alpha = 14°; \beta = 76°; \gamma = 90°; a = 2.1; b = 8.2; c = 8.5$
31. $\alpha = 66.3°; \beta = 23.7°; \gamma = 90.0°; a = 45.8; b = 20.1; c = 50.1$
33. A nautical mile is 1.149 statute miles. **35.** $\sqrt{7}$
37. 2,900 miles
39. A person in Sacramento would see the sun about 5
minutes earlier than a person in Santa Rosa.
41. The sign is 16.3 ft.

Problem Set 2.1, page 54

1. $30° = \frac{\pi}{6}$, not $\frac{\pi}{3}$

3. A positive number cannot be equal to a negative number.

5. The circumference of a circle with radius 10π is
$C = 2\pi r = 20\pi$.

7. a. $30° = \frac{\pi}{6}$

b. $90° = \frac{\pi}{2}$

c. $45° = \frac{\pi}{4}$

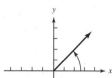

d. $60° = \frac{\pi}{3}$

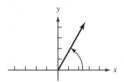

e. $180° = \pi$

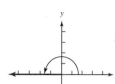

9. a. $\frac{\pi}{3} = 60°$

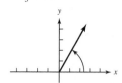

b. $\frac{\pi}{6} = 30°$

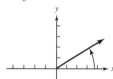

c. $\frac{\pi}{2} = 90°$

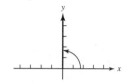

d. $\frac{\pi}{4} = 45°$

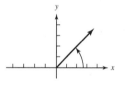

e. $2\pi = 360°$

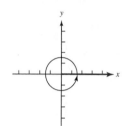

11. a. F **b.** D **c.** A **d.** E **e.** C **f.** B

13.

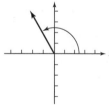

a. reference angle **b.** none
c. coterminal **d.** coterminal
e. equal

15.

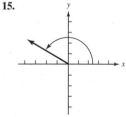

a. reference angle **b.** equal
c. none **d.** none
e. coterminal

17. a. $150° = \frac{5\pi}{6}$

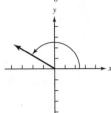

b. $135° = \frac{3\pi}{4}$

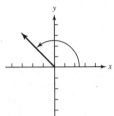

b. $-210° \approx -3.67$

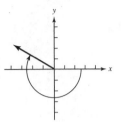

c. $-825° \approx -14.40$

c. $20° = \frac{\pi}{9}$

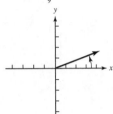

19. a. $225° = \frac{5\pi}{4}$

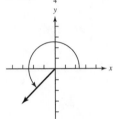

25. a. $-\frac{5\pi}{3} = -300°$

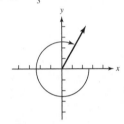

b. $\frac{5\pi}{3} = 300°$

b. $-240° = -\frac{4\pi}{3}$

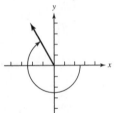

c. $250° = \frac{25\pi}{18}$

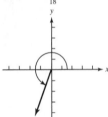

c. $-2\pi = -360°$

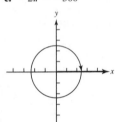

27. a. $-\frac{\pi}{4} = -45°$

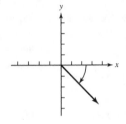

21. a. $-60° \approx -1.05$

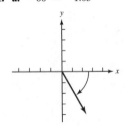

b. $400° \approx 6.98$

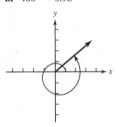

b. $\frac{5\pi}{4} = 225°$

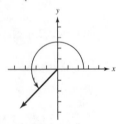

c. $-\frac{11\pi}{4} = -495°$

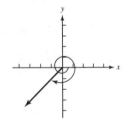

c. $23.7° \approx 0.41$

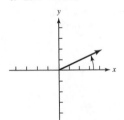

23. a. $38.4° \approx 0.67$

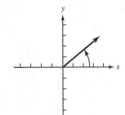

29. a. $2 \approx 115°$

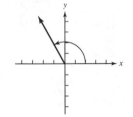

b. $-3 \approx -172°$

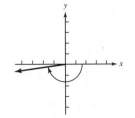

c. $0.5 \approx 29°$

31. a. $12 \approx 688°$

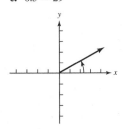

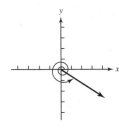

b. $-1.5 \approx -86°$

c. $4.712389 \approx 270°$

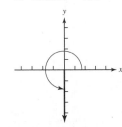

33. a. $\frac{7\pi}{4}$ **b.** $\frac{\pi}{4}$ **c.** $\frac{5\pi}{3}$ **d.** $2\pi - 2$ **e.** $8 - 2\pi$ **f.** $\frac{11\pi}{6}$

35. a. $2\pi - 6 \approx 0.2832$ **b.** 0.0000 **c.** $3\sqrt{5} - 2\pi \approx 0.4250$

37. a. $2\pi - 0.7854 = 5.4978$ **b.** $6.8068 - 2\pi = 0.5236$

c. $\frac{9\pi}{4} - 2\pi = 0.7854$

39. a. $s = 14.04$ **b.** $s = 31.42$

41. $\omega = \frac{18\pi}{5}$ rad/min; $v = 18\pi$ ft/min

43. 365 days $\approx 8{,}760$ hr; $7.172585967 \times 10^{-4}$ rev/h; $v \approx 66{,}700$ mi/h

45. $s = 141.4$ in.

47. The angular velocity is 28 rad/s.

49. The arc length is 4 cm.

51. The linear velocity is about 17 ft/s.

53. The wheel travels 2.5 revolutions for one revolution of the drive sprocket.

55. The angular velocity is 35.2 rad/s.

57. The angular velocity is $4{,}800\pi$ rad/h and the linear velocity is about 357 mi/h.

59. $\theta_2 = \frac{r_1 \theta_1}{r_2}$

Problem Set 2.2, page 63

3. $(\cos \beta, \sin \beta)$

5. $[\cos(\pi - \beta), \sin(\pi - \beta)]$ or $[(\cos(\beta - \pi), \sin(\beta - \pi)]$

7. $\sin 30° = 0.5$, not -0.9880316241

9. $\cos^{-1}\theta$ is the inverse cosine, whereas $\frac{1}{\cos \theta}$ is the reciprocal of cosine.

11. The sine function is positive in Quadrants I and II, and is negative in Quadrant IV.

13. In Quadrant II, sine is positive, not negative.

15. a. 0.6 **b.** 0.9 **c.** -0.5

17. a. -0.4 **b.** -0.8 **c.** 0.3

19. a. 0.6427876097 **b.** 0.3420201433 **c.** 2.9238044
d. 0.8290375726 **e.** -0.9961946981 **f.** 0.3639702343

21. a. 0.6225146366 **b.** -0.1124755272 **c.** -0.1154340956
d. -0.7071067812 **e.** 0.7071067812 **f.** 1

23. a. 0.9092974268 **b.** -0.9364566873 **c.** 4.637332055

25. a. 1.191753593 **b.** -0.2958129155 **c.** 2.7474774195

27. a. positive **b.** positive **c.** negative **d.** positive

29. a. negative **b.** negative **c.** negative

31. a. negative **b.** negative **c.** negative

33. a. I, IV **b.** I, II **35. a.** II, III **b.** II, IV

37. a. II **b.** III

39. $\cot \theta = \frac{a}{b}$ *Definition*

$\quad = \frac{\cos \theta}{\sin \theta}$ *Definition and substitution*

41. As $\theta \to 0$, $\frac{\sin \theta}{\theta} = 1$

Problem Set 2.3, page 73

1. $\sec \theta = \frac{1}{\cos \theta}$; $\csc \theta = \frac{1}{\sec \theta}$; $\cot \theta = \frac{1}{\tan \theta}$

3. $\cos^2\theta + \sin^2\theta = 1$; $1 + \tan^2\theta = \sec^2\theta$; $\cot^2\theta + 1 = \csc^2\theta$

5. $\sin \theta = \pm\sqrt{1 - \cos^2\theta}$ **7.** $\cot \theta = \pm\sqrt{\csc^2\theta - 1}$

9. $\cos \theta = \frac{a}{u}$; $\sin \theta = \frac{\sqrt{u^2 - a^2}}{u}$; $\tan \theta = \frac{\sqrt{u^2 - a^2}}{a}$;

$\sec \theta = \frac{u}{a}$; $\csc \theta = \frac{u}{\sqrt{u^2 - a^2}}$; $\cot \theta = \frac{a}{\sqrt{u^2 - a^2}}$

11. $\frac{1}{\sec 352°} = \cos 352° \approx 0.9902680687$

13. $\frac{1}{\sin 2.6} \approx 1.939859047$

15. $\frac{\sin 0.25}{\cos 0.25} = \tan 0.25 \approx 0.2553419212$

17. $\frac{\cos 128°}{\sin 128°} = \frac{1}{\tan 128°} \approx -0.7812856265$

19. $1 - \sin^2 0.5 = \cos^2 0.5 = 0.7701511529$

21. $-\sqrt{1 - \sin^2 100°} = \cos 100° \approx -0.1736481777$

23. $-\sqrt{\sec^2 135° - 1} = \tan 135° \approx -1$

25. $\tan\frac{5\pi}{6} \cos\frac{5\pi}{6} = \sin\frac{5\pi}{6} = 0.5$

27. $\sin\left(-\frac{2\pi}{3}\right)\cot\left(-\frac{2\pi}{3}\right) = \cos\left(-\frac{2\pi}{3}\right) = -0.5$

29. $\cos \theta = \frac{\pm 1}{\sqrt{1 + \tan^2\theta}}$; $\sin \theta = \frac{\pm\tan \theta}{\sqrt{1 + \tan^2\theta}}$; $\tan \theta = \tan \theta$;

$\sec \theta = \pm\sqrt{1 + \tan^2\theta}$; $\csc \theta = \frac{\pm\sqrt{1 + \tan^2\theta}}{\tan \theta}$; $\cot \theta = \frac{1}{\tan \theta}$

31. $\cos \theta = \frac{1}{\sec \theta}$; $\sin \theta = \frac{\pm\sqrt{\sec^2\theta - 1}}{\sec \theta}$; $\tan \theta = \pm\sqrt{\sec^2\theta - 1}$;

$\sec \theta = \sec \theta$; $\csc \theta = \frac{\pm\sec \theta}{\sqrt{\sec^2\theta - 1}}$; $\cot \theta = \frac{\pm 1}{\sqrt{\sec^2\theta - 1}}$

33. $\sin \theta = -\frac{12}{13} \approx -0.92$; $\tan \theta = -\frac{12}{5} = -2.40$; $\sec \theta = \frac{13}{5} = 2.60$;

$\csc \theta = -\frac{13}{12} \approx -1.08$; $\cot \theta = -\frac{5}{12} \approx -0.42$

35. $\cos\theta \approx 0.76$; $\tan\theta \approx 0.86$; $\csc\theta \approx 1.54$; $\sec\theta \approx 1.32$;
$\cot\theta \approx 1.17$

37. Let $u = a\tan\theta$; $\dfrac{\sqrt{u^2 + a^2}}{a}$

39. Let $u = \sqrt{5}\tan\theta$; $\dfrac{u^2\sqrt{u^2 + 5}}{\sqrt{5}}$

41.

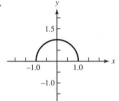

43.

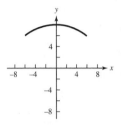

45.

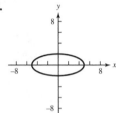

47.

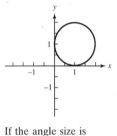

49.

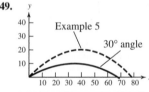

If the angle size is decreased, the height is decreased and the range is also decreased. The time of flight is decreased.

51.

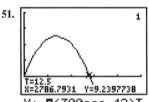

```
T=12.5
X=2786.7931   Y=9.2397738

X₁ᴛ◼(300cos 42)T

Y₁ᴛ◼(300sin 42)T
  -16T^2
    Tmin=0
    Tmax=15
    Tstep=.5
    Xmin=0
    Xmax=5000
    Xscl=1000
    Ymin=-250
    Ymax=1000
    Yscl=250
```

The calculator graph is shown. Notice that we used the cursor to find the values of x and t that most closely approximate $y = 0$. We see that it looks like the range is about 2,800 ft with a time of impact in about 12.5 sec.

Problem Set 2.4, page 80

3. a. 1   **b.** 1   **c.** $\frac{\sqrt{3}}{2}$ **d.** $\frac{\sqrt{3}}{2}$ **e.** 0   **f.** $\frac{\sqrt{3}}{3}$

5. a. $\frac{1}{2}$ **b.** $\frac{1}{2}$ **c.** 0   **d.** -1 **e.** $\sqrt{3}$ **f.** 0

7. a. 2   **b.** $\sqrt{3}$ **c.** 0   **d.** $\frac{2\sqrt{3}}{3}$ **e.** $\frac{\sqrt{3}}{3}$ **f.** undefined

9. a. undefined   **b.** -1 **c.** undefined   **d.** 1   **e.** $\sqrt{2}$
f. -1

11. a. 0   **b.** $-\frac{\sqrt{2}}{2}$ **c.** $-\frac{\sqrt{2}}{2}$ **d.** $\frac{1}{2}$ **e.** $\frac{\sqrt{3}}{2}$ **f.** $-\sqrt{3}$

13. a. $-\frac{1}{2}$ **b.** -1 **c.** $-\frac{\sqrt{3}}{2}$

15.

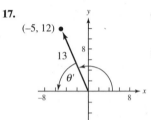

$\cos\theta = \frac{3}{5}$ $\sec\theta = \frac{5}{3}$
$\sin\theta = \frac{-4}{5}$ $\csc\theta = \frac{-5}{4}$
$\tan\theta = \frac{-4}{3}$ $\cot\theta = \frac{-3}{4}$

17.

$\cos\theta = \frac{-5}{13}$ $\sec\theta = \frac{-13}{5}$
$\sin\theta = \frac{12}{13}$ $\csc\theta = \frac{13}{12}$
$\tan\theta = \frac{-12}{5}$ $\cot\theta = \frac{-5}{12}$

19.

$\cos\theta = \frac{-2}{\sqrt{13}}$ or $-\frac{2}{13}\sqrt{13}$
$\sin\theta = \frac{-3}{\sqrt{13}}$ or $-\frac{3}{13}\sqrt{13}$
$\tan\theta = \frac{3}{2}$
$\sec\theta = \frac{-\sqrt{13}}{2}$
$\sec\theta = \frac{-\sqrt{13}}{3}$
$\cot\theta = \frac{2}{3}$

21. Choose a value, say $\theta = 10°$, to show it is false.

23. Choose a value, say $\theta = 20°$, to show it is false.

25. The statement $\cos\theta = \dfrac{\sin\theta}{\tan\theta}$ is equivalent to $\tan\theta = \dfrac{\sin\theta}{\cos\theta}$, which is true because it is a fundamental identity.

27. Choose a value, say $\theta = 40°$, to show it is false.

29. Choose a value, say $\theta = 60°$, to show it is false.

31. Choose a value, say $\theta = 80°$, to show it is false.

33. If $\sin\frac{1}{2}\theta$ is in the first quadrant, it is positive, so it cannot be equal to the opposite of a square root (which is negative).

35. If $\cos\theta$ is in the first or fourth quadrant, then it is positive so it cannot be equal to the opposite of a square root (which is negative).

37. For an angle of $\frac{5\pi}{4}$, we see $x = y$ (see figure at the right; both x and y are negative).

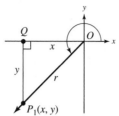

$$\cos \frac{5\pi}{4} = \frac{x}{r}$$

From the ratio definition (this is true for any angle)

$$= \frac{x}{\sqrt{x^2 + y^2}}$$

$r = \sqrt{x^2 + y^2}$ *for any angle*

$$= \frac{x}{\sqrt{x^2 + x^2}}$$

If $\theta = \frac{5\pi}{4}$, then θ bisects Quadrant III so that $x = y$.

$$= \frac{x}{\sqrt{2x^2}}$$

$$= \frac{x}{-x\sqrt{2}}$$

$\sqrt{x^2} = |x| = -x$ *since x is negative in Quadrant III*

$$= -\frac{1}{\sqrt{2}} \text{ or } -\frac{1}{2}\sqrt{2}$$

39. For $\theta = -\frac{\pi}{4}$, x is positive, y is negative, and $x = -y$.

$$\sin\left(-\frac{\pi}{4}\right) = \frac{y}{r} = \frac{y}{\sqrt{x^2 + y^2}} = \frac{y}{\sqrt{(-y)^2 + y^2}} = \frac{y}{\sqrt{2y^2}}$$

$$= \frac{y}{-y\sqrt{2}} = -\frac{1}{\sqrt{2}} \text{ or } -\frac{1}{2}\sqrt{2}$$

41. For $\theta = 210°$, we see that the reference angle is $30°$, so x and y are both negative and $-2y = r$.

$$r^2 = x^2 + y^2$$
$$(-2y)^2 = x^2 + y^2 \qquad \text{Since } -2y = r$$
$$3y^2 = x^2$$
$$\sqrt{3}\,|y| = |x|$$

Since x and y are both negative when $\theta = 210°$, we see $x = \sqrt{3}y$.

$$\cos 210° = \frac{x}{r} = \frac{\sqrt{3}y}{-2y} = -\frac{\sqrt{3}}{2}$$

43. a. 0   **b.** -1 **45. a.** 0.75   **b.** 0.5
47. a. 1   **b.** 0.5   **49. a.** $\frac{\sqrt{2}}{2}$ **b.** $\frac{\sqrt{2}}{2}$
51. a. 2   **b.** $\frac{\sqrt{3}}{3}$ **53. a.** $-\sqrt{3}$ **b.** $-\sqrt{3}$
55. Approximate value (for both parts): 0.0523359562

57. $m = \sqrt{3}$

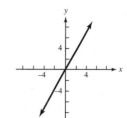

59. a. $y = \frac{1}{3}\sqrt{3}(x - 2) + 3$ **b.** $y = -\sqrt{3}(x - 1) + 4$
   c. $y = (x - 9) - 5$ or $y = x - 14$
   d. $y = -(x + 3) - 8$ or $y = -x - 11$
61. a. $0° < \theta < 90°$ **b.** $90° < \theta < 180°$ **c.** $\theta = 0°$
   d. $\theta = 90°$
63. $d \approx 9.3358$ cm

Chapter 2 Sample Test, page 84

1. Let θ be any angle in standard position with a point $P(x, y)$ on the terminal side a distance of r from the origin ($r \neq 0$). Then, $\cos \theta = \frac{x}{r}$; $\sin \theta = \frac{y}{r}$; $\tan \theta = \frac{y}{x}$; $x \neq 0$;

$\sec \theta = \frac{r}{x}$, $x \neq 0$; $\csc \theta = \frac{r}{y}$, $y \neq 0$; $\cot \theta = \frac{x}{y}$, $y \neq 0$.

2. 8.73   **3.** $-218°$
4. $\cos \alpha = \frac{3}{4}$; $\sin \alpha = \frac{\sqrt{7}}{4}$; $\tan \alpha = \frac{\sqrt{7}}{3}$; $\sec \alpha = \frac{4}{3}$;
   $\csc \alpha = \frac{4}{\sqrt{7}}$; $\cot \alpha = \frac{3}{\sqrt{7}}$
5. a. -0.4161 **b.** -0.7568 **c.** 1.1383   **d.** -1.0025
   e. -1.6611
6. $(\cos \alpha, \sin \alpha)$
7. a. $\cos \frac{\pi}{3}$ **b.** $-\sin \frac{\pi}{4}$ **c.** $\tan \frac{\pi}{4}$ **d.** $\cot \frac{\pi}{4}$
   e. $\csc(\pi - 2)$ **f.** $-\sec(4 - \pi)$
8. a. $\sec \theta = \frac{1}{\cos \theta}$; $\csc \theta = \frac{1}{\sin \theta}$; $\cot \theta = \frac{1}{\csc \theta}$

$$\frac{1}{\cos \theta} = \frac{1}{a} \qquad \text{\textit{Definition of cosine}}$$

$$= \sec \theta \qquad \text{\textit{Definition of secant}}$$

The other two proofs are similar.

b. $\tan \theta = \frac{\sin \theta}{\cos \theta}$; $\cot \theta = \frac{\cos \theta}{\sin \theta}$

$$\frac{\sin \theta}{\cos \theta} = \frac{b}{a} \qquad \text{\textit{Definition of sine and cosine}}$$

$$= \tan \theta \qquad \text{\textit{Definition of tangent}}$$

The other proof is similar.

c. $\cos^2 \theta + \sin^2 \theta = 1$; $\tan^2 \theta + 1 = \sec^2 \theta$; $1 + \cot^2 \theta = \csc^2 \theta$

$$a^2 + b^2 = 1 \qquad \text{\textit{Pythagorean theorem}}$$

$$\frac{a^2}{a^2} + \frac{b^2}{a^2} = \frac{1}{a^2} \qquad \text{\textit{Divide both sides by } } a^2.$$

$$1 + \left(\frac{b}{a}\right)^2 = \left(\frac{1}{a}\right)^2 \qquad \textit{Simplify.}$$

$1 + (\tan\theta)^2 = (\sec\theta)^2$ *Definition of tangent and secant*

$1 + \tan^2\theta = \sec^2\theta$ *Squared notation*

     The other proofs are similar.

9. a. 180°   **b.** −240°   **c.** $-\frac{\pi}{4}$ **d.** $\frac{7\pi}{6}$ **e.** $\frac{\pi}{2}$ **f.** $-\frac{\sqrt{2}}{2}$
g. 0   **h.** $\frac{\sqrt{3}}{2}$ **i.** $-\frac{1}{2}$ **j.** 1   **k.** $\frac{\sqrt{2}}{2}$ **l.** −1   **m.** $-\frac{1}{2}$
n. $-\frac{\sqrt{3}}{2}$ **o.** 0   **p.** −1   **q.** 0   **r.** $-\sqrt{3}$ **s.** $\frac{\sqrt{3}}{3}$
t. undefined
10. 3.6 mi/h

Miscellaneous Problems, page 85

1. $(2\cos\alpha,\, 2\sin\alpha)$ **3.** 2.62   **5.** −1.75   **7.** 120°
9. −143°   **11.** 30 in.   **13.** 16 ft   **15.** II; $9 - 2\pi$
17. II; $\frac{4\pi}{3}$ **19.** 0.9965   **21.** 0.9131   **23.** 0.0831
25. −3.4364   **27.** 2.3048
29. $\cos\theta = \frac{4}{\sqrt{41}}$; $\sin\theta = \frac{5}{\sqrt{41}}$; $\tan\theta = \frac{5}{4}$;
    $\sec\theta = \frac{\sqrt{41}}{4}$; $\csc\theta = \frac{\sqrt{41}}{5}$; $\cot\theta = \frac{4}{5}$
31. $\cos\theta = \frac{5}{6}$; $\sin\theta = \frac{\sqrt{11}}{6}$; $\tan\theta = \frac{\sqrt{11}}{5}$;
    $\sec\theta = \frac{6}{5}$; $\csc\theta = \frac{6}{\sqrt{11}}$; $\cot\theta = \frac{5}{\sqrt{11}}$
33. $\frac{\sqrt{3}}{2}$ **35.** $-\frac{1}{2}$ **37.** $-\frac{2}{3}\sqrt{3}$ **39.** $\sqrt{3}$ **41.** $-\sqrt{3}$
43. undefined   **45.** $-\sqrt{3}$ **47.** $-\sqrt{2}$
49. a. 200π rad/min
    b. The larger wheel is traveling at 5.35 mi/h
    c. The angular speed of the smaller wheel is about 300 rev/min.

Problem Set 3.1, page 100

1. Cosine: amplitude is 1, period is 2π; sine: amplitude is 1, period is 2π; tangent: no amplitude, period is π.

3.   **5.**

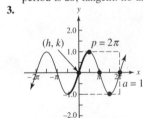 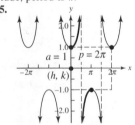

7.

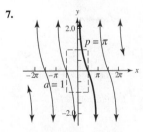

9. x = angle:

x = angle:	$\frac{2\pi}{3}$	$\frac{3\pi}{4}$	$\frac{5\pi}{6}$	$\frac{7\pi}{6}$	$\frac{5\pi}{4}$	$\frac{4\pi}{3}$	$\frac{7\pi}{4}$	$\frac{11\pi}{6}$
Quadrant:	II	II	II	III	III	III	IV	IV
y = tan x:	$-\sqrt{3}$	−1	$-\frac{\sqrt{3}}{3}$	$\frac{\sqrt{3}}{3}$	1	$\sqrt{3}$	−1	$-\frac{\sqrt{3}}{3}$
y approx.:	−1.73	−1.00	−0.58	0.58	1.00	1.73	−1.00	−0.58

For the graph, see Figure 3.11.
11. $a = 6,\ p = 4$ **13.** $a = 6,\ p = 2$
15. $a = 2,\ p = \frac{1}{3},\ f = 3$ **17.** $a = 0.3,\ p = \frac{1}{6},\ f = 6$

23.   **25.**

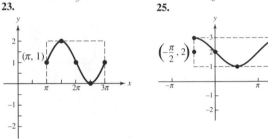

27.   **29.**

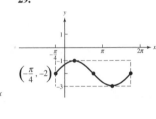

31.   **33.**

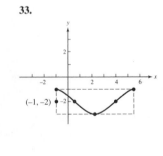

35.

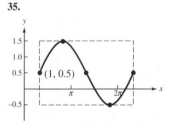

37. no vertical asymptotes
39. vertical asymptotes at $x = \frac{\pi}{2} + n\pi$
41. vertical asymptotes at $x = n\pi$

43. not defined at $x = 0$, (1, 1.19), (2, 1.10), (3, 7.09), (4, −1.32), (5, −1.04), (6, −3.58), (7, 1.52)

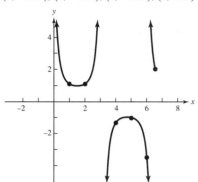

For Problems 45–47, the following window of values was used:

45. a.

b.

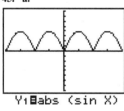

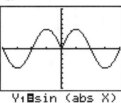

47. a.

b.

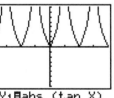

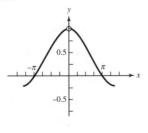

49.

51.

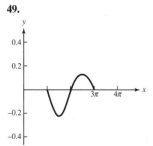

53.

55.

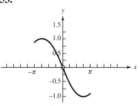

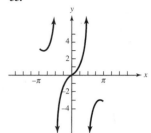

59.

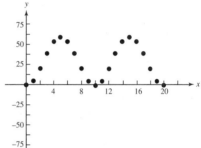

Problem Set 3.2, page 117

7. a.

b.

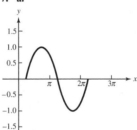

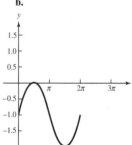

c. The starting point changes. In part **a**, $(h, k) = (1, 0)$, and in part **b**, $(h, k) = (0, −1)$.

9. a.

b.

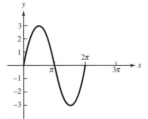

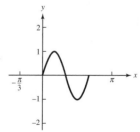

c. The amplitude and period change. In part **a**, $a = 3$, $p = 2\pi$, and in part **b**, $a = 1$ and $p = 2\pi/3$.

11. a.

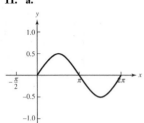

b.

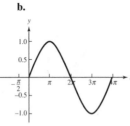

c. The amplitude and period change. In part **a**, $a = 1/2$, $p = 2\pi$, and in part **b**, $a = 1$ and $p = 2\pi/(1/2) = 4\pi$.

13.

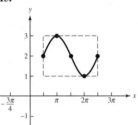

15.

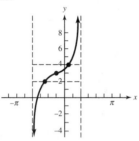

17.

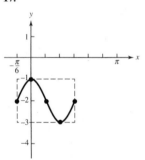

19. $a = 4$, $p = 2\pi$; $y = 4 \sin x$

21. $a = 2$, $p = 2\pi$; $y = 2 \cos(x - \pi)$

23. $a = 10$, $p = 2$; $y - 20 = 10 \sin \pi x$

25. $(h, k) = \left(-\frac{\pi}{6}, 0\right)$, $a = \frac{1}{2}$, $b = 1$, $p = 2\pi$

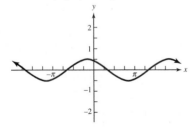

27. $(h, k) = (0, 0)$, $a = 2$, $b = 2\pi$, $p = 1$

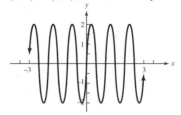

29. $(h, k) = (0, 0)$, $a = 4$, $b = \frac{\pi}{5}$, $p = 5$

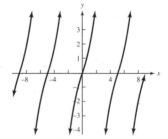

31. $y = \sin 4\left(x + \frac{\pi}{4}\right)$; $(h, k) = \left(-\frac{\pi}{4}, 0\right)$, $a = 1$, $b = 4$, $p = \frac{\pi}{2}$

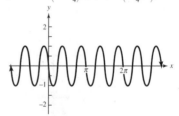

33. $y = \tan 2\left(x - \frac{\pi}{4}\right)$; $(h, k) = \left(\frac{\pi}{4}, 0\right)$, $a = 2$, $b = 2$, $p = \frac{\pi}{2}$

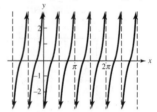

35. $y + 2 = 2 \cos 3\left(x + \frac{2\pi}{3}\right)$; $(h, k) = \left(-\frac{2\pi}{3}, -2\right)$, $a = 2$, $b = 3$, $p = \frac{2\pi}{3}$

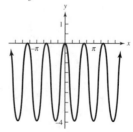

37. $(h, k) = (\sqrt{2}, 1)$, $a = \sqrt{2}$, $b = 1$, $p = 2\pi$

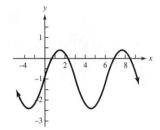

39. $(h, k) = (0, 0)$, $a = 2$, $b = 1$, $p = 2\pi$

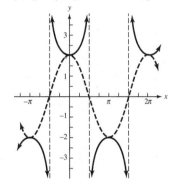

41. $(h, k) = (0, 0)$, $a = \frac{1}{2}$, $b = 1$, $p = \pi$

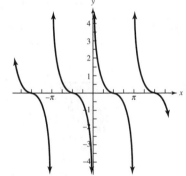

43. $(h, k) = (0, 1)$, $a = 1$, $b = 2$, $p = \pi$

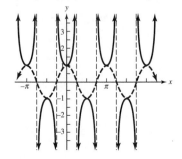

45. $(h, k) = \left(-\frac{\pi}{3}, 0\right)$, $a = 2$, $b = 2$, $p = 2\pi$

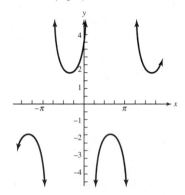

47. a. 20 mm of mercury   **b.** $\dfrac{60}{0.6} = 100$ beats/min

49. $y = 0.02 \cos 660\pi x$

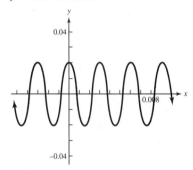

51.

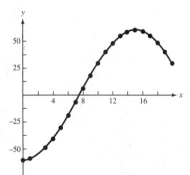

Answers vary; it looks like $(h, k) = (7.5, 0)$, $a = 60$ and the period is found by solving $30 = \dfrac{2\pi}{b}$ so that $b = \dfrac{\pi}{15}$. A possible equation is $y = 60 \sin \frac{\pi}{15}(x - 7.5)$.

53.

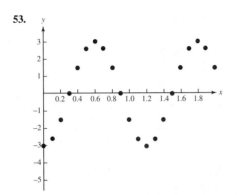

Answers vary; it looks like $(h, k) = (0.6, 0)$, $a = 3$, and $p = 1.2$ so that $b = \dfrac{2\pi}{1.2}$ or $b = \dfrac{5\pi}{3}$. A possible equation is $y = 3 \cos \dfrac{5\pi}{3}(x - 0.6)$.

55. Answers vary; the 1.047 is related to the length of the rod. Yes, the period is, by definition, $\dfrac{2\pi}{\sqrt{g/L}}$, which shows that it depends (in part) on L.

57. a.

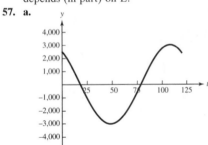

b. 3,000 mi **c.** 2 h $\left(2\pi \div \dfrac{\pi}{60} = 120 \min\right)$

59. The period of the curve is $\frac{1}{440}$, so on the interval $0 \le x \le 0.1$, there are 44 periods; the amplitude of the curve is 0.04. The "beats" that are seen with the calculator display are a function of the selection of the points (pixels) plotted by the calculator. For example, using a computer program, we see more refinement of the "beats":

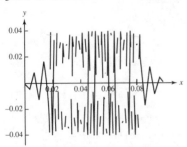

Plot 30 points per unit.

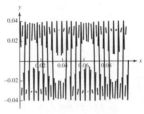

Plot 100 points per unit.

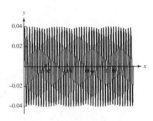

Plot 400 points per unit.

Problem Set 3.3, page 134

3. $x \approx 0$ and $x \approx 2.5$

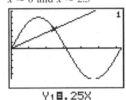

```
Y₁◻.25X
Y₂◻sin X
Xmin=0
Xmax=6.2831853...
Xscl=1
Ymin=-1.2
Ymax=1.2
Yscl=.2
```

5. $x \approx 4.4$

```
Y₁◻X-1
Y₂◻tan X
Xmin=0
Xmax=6.2831853...
Xscl=1
Ymin=-5
Ymax=5
Yscl=1
```

7. $x \approx 1.0$, and $x \approx 5.3$

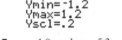
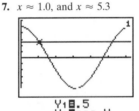

```
Y₁◻.5
Y₂◻cos X
Xmin=0
Xmax=6.2831853...
Xscl=1
Ymin=-1.2
Ymax=1.2
Yscl=.2
```

```
Y₁◻.5
Y₂◻cos X
Xmin=0
Xmax=6.2831853...
Xscl=1
Ymin=-1.2
Ymax=1.2
Yscl=.2
```

9. $x \approx 0.9$

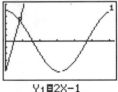

```
Y₁◻2X-1
Y₂◻cos X
Xmin=0
Xmax=6.2831853...
Xscl=1
Ymin=-1.2
Ymax=1.2
Yscl=.2
```

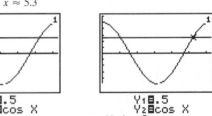

11. $x \approx 0.8$ and $x \approx 2.4$

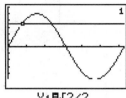

```
Y₁◼√2/2
Y₂◼sin X
Xmin=0
Xmax=6.2831853...
Xscl=1
Ymin=⁻1.2
Ymax=1.2
Yscl=.2
```

```
Y₁◼√2/2
Y₂◼sin X
Xmin=0
Xmax=6.2831853...
Xscl=1
Ymin=⁻1.2
Ymax=1.2
Yscl=.2
```

13. $x \approx 0.5$ and $x \approx 3.7$

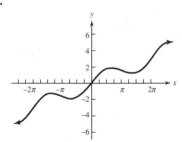

```
Y₁◼√3/3
Y₂◼tan X
Xmin=0
Xmax=6.2831853...
Xscl=1
Ymin=⁻1.2
Ymax=1.2
Yscl=.2
```

```
Y₁◼√3/3
Y₂◼tan X
Xmin=0
Xmax=6.2831853...
Xscl=1
Ymin=⁻1.2
Ymax=1.2
Yscl=.2
```

15.

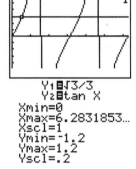

17.

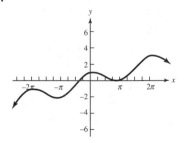

19.

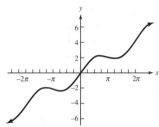

21.

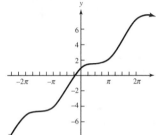

25.

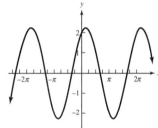

27.

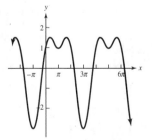

29.

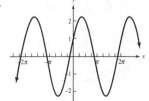

31. The graphs appear to be the same.

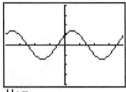

Y₁=
Y₂🔲√2sin (X+π/4)
Xmin=⁻6.152285…
Xmax=6.1522856…
Xscl=1.5707963…
Ymin=⁻4
Ymax=4
Yscl=1

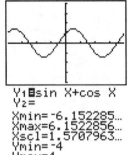

Y₁🔲sin X+cos X
Y₂=
Xmin=⁻6.152285…
Xmax=6.1522856…
Xscl=1.5707963…
Ymin=⁻4
Ymax=4
Yscl=1

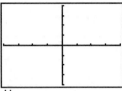

Y₁
Y₂
Y₃🔲sin X+cos (X+
π/2)
Xmin=⁻6.283185…
Xmax=6.2831853…
Xscl=1.5707963…
Ymin=⁻4
Ymax=4
Yscl=1

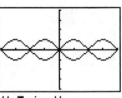

Y₁🔲sin X
Y₂🔲cos (X+π/2)
Y₃🔲Y₁+Y₂
Xmin=⁻6.283185…
Xmax=6.2831853…
Xscl=1.5707963…
Ymin=⁻4
Ymax=4
Yscl=1

33.

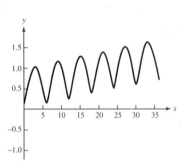

35. a. sun curve: $y = \cos\frac{\pi}{6}x$ **b.** moon curve: $y = 4\cos\frac{\pi}{6}x$
c. combined curve: $y = \cos\frac{\pi}{6}x + 4\cos\frac{\pi}{6}x = 5\cos\frac{\pi}{6}x$

37. Graph $y_1 = \sin x$ and $y_2 = \cos\left(x + \frac{\pi}{2}\right)$; then add ordinates to find $y_1 + y_2$ and note that the sum is the x-axis. We show these graphs individually, and then show how they would look on one calculator screen.

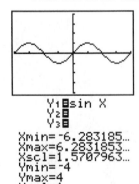

Y₁🔲sin X
Y₂🔲
Y₃🔲
Xmin=⁻6.283185…
Xmax=6.2831853…
Xscl=1.5707963…
Ymin=⁻4
Ymax=4
Yscl=1

Y₁🔲
Y₂🔲cos (X+π/2)
Y₃🔲
Xmin=⁻6.283185…
Xmax=6.2831853…
Xscl=1.5707963…
Ymin=⁻4
Ymax=4
Yscl=1

39. $y = 2 + \cos 588\pi x$

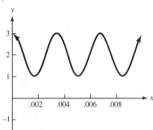

41. $a = 3$; $n = \frac{1}{6}$, $p = 6$; $\lambda \approx 5.09(6^2) \approx 183.24$; $v = \dfrac{\lambda}{p} \approx 30.54$ ft/sec ≈ 20.82 mi/h; $y = 3 \sin 0.03(x - 30.54t)$

43. 24 ft from trough to crest; wavelength $\lambda \approx \dfrac{2\pi}{1} \approx 6.28$; $2\pi n = 37.7$, so $n = 6$; phase velocity is $v \approx 37.7$ ft/s ≈ 25.7 mi/h.

45.

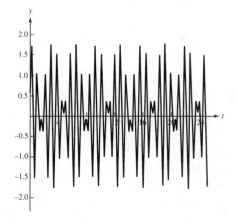

47.

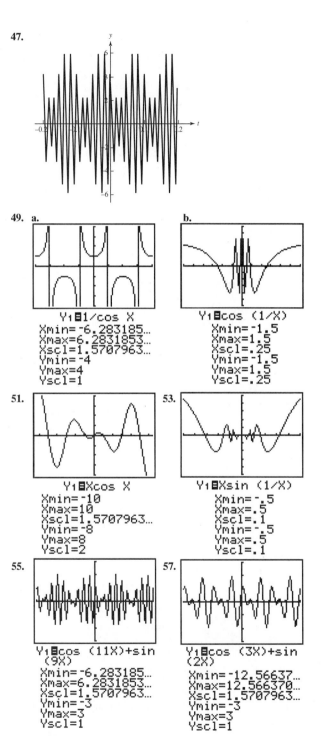

49. a. **b.**

Y₁**目**1/cos X
Xmin=-6.283185...
Xmax=6.2831853...
Xscl=1.5707963...
Ymin=-4
Ymax=4
Yscl=1

Y₁**目**cos (1/X)
Xmin=-1.5
Xmax=1.5
Xscl=.25
Ymin=-1.5
Ymax=1.5
Yscl=.25

51. **53.**

Y₁**目**Xcos X
Xmin=-10
Xmax=10
Xscl=1.5707963...
Ymin=-8
Ymax=8
Yscl=2

Y₁**目**Xsin (1/X)
Xmin=-.5
Xmax=.5
Xscl=.1
Ymin=-.5
Ymax=.5
Yscl=.1

55. **57.**

Y₁**目**cos (11X)+sin (9X)
Xmin=-6.283185...
Xmax=6.2831853...
Xscl=1.5707963...
Ymin=-3
Ymax=3
Yscl=1

Y₁**目**cos (3X)+sin (2X)
Xmin=-12.56637...
Xmax=12.566370...
Xscl=1.5707963...
Ymin=-3
Ymax=3
Yscl=1

59. **61.**

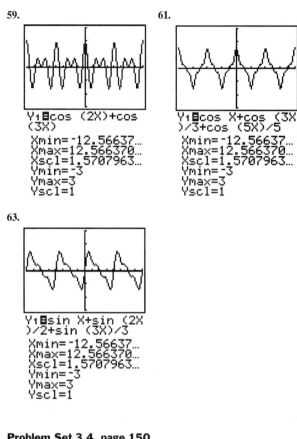

Y₁**目**cos (2X)+cos (3X)
Xmin=-12.56637...
Xmax=12.566370...
Xscl=1.5707963...
Ymin=-3
Ymax=3
Yscl=1

Y₁**目**cos X+cos (3X)/3+cos (5X)/5
Xmin=-12.56637...
Xmax=12.566370...
Xscl=1.5707963...
Ymin=-3
Ymax=3
Yscl=1

63.

Y₁**目**sin X+sin (2X)/2+sin (3X)/3
Xmin=-12.56637...
Xmax=12.566370...
Xscl=1.5707963...
Ymin=-3
Ymax=3
Yscl=1

Problem Set 3.4, page 150

1. $0 \le y \le \pi$ **3.** I, IV   **5.** $-\frac{\pi}{2} < y < \frac{\pi}{2}$

7. $\cos^{-1} x$; $\boxed{x}$ $\boxed{\text{inv}}$ $\boxed{\cos}$ or $\boxed{\cos^{-1}}$ $\boxed{x}$ $\boxed{\text{ENTER}}$

9. $\tan^{-1} x$; $\boxed{x}$ $\boxed{\text{inv}}$ $\boxed{\tan}$ or $\boxed{\tan^{-1}}$ $\boxed{x}$ $\boxed{\text{ENTER}}$

11. $\sec^{-1} x$; $\boxed{x}$ $\boxed{1/x}$ $\boxed{\text{inv}}$ $\boxed{\cos}$ or $\boxed{\cos^{-1}}$ $\boxed{x}$ $\boxed{x^{-1}}$ $\boxed{\text{ENTER}}$

13. a. 0.75 **b.** 0.0 **c.** -1.0 **d.** -0.75 **e.** 0.0 **f.** $\frac{\pi}{2}$
g. $\frac{3\pi}{8}$ **h.** π **i.**

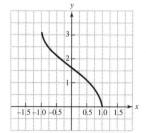

15. a. 1.1 **b.** 0.4 **c.** 0.0 **d.** −1.0 **e.** $\frac{\pi}{4}$ **f.** 1.1
g. 0.5 **h.** −$\frac{\pi}{4}$ **i.**

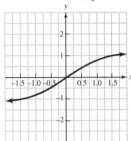

17. a. $\frac{\pi}{4}$ **b.** $\frac{\pi}{4}$ **c.** −$\frac{\pi}{2}$ **d.** 0
19. a. $\frac{\pi}{3}$ **b.** −$\frac{\pi}{2}$ **c.** −$\frac{\pi}{3}$ **d.** $\frac{3\pi}{4}$
21. a. −$\frac{\pi}{6}$ **b.** 0 **c.** $\frac{\pi}{3}$ **d.** $\frac{\pi}{3}$
23. a. −1.31 **b.** 2.81 **c.** 1.11 **d.** 0.98
25. a. 21° **b.** 19° **c.** 69° **d.** −28°
27. a. 153° **b.** 15° **c.** 100° **d.** −72° **29.** true
31. False; the inverse cotangent is negative in Quadrant II.
33. true **35.** true **37.** false; pick a counterexample
39. False, it is cosine and arccosine that are inverse functions;
no such relationship exists for reciprocal functions.
41. a. a **b.** $\frac{\pi}{6}$ **43. a.** $\frac{2}{3}$ **b.** $\frac{2\pi}{15}$ **45. a.** $\frac{1}{2}$ **b.** $\frac{\sqrt{5}}{3}$
47. $\cos(\sin^{-1} x) = \pm\sqrt{1-x^2}$
49. $\cos(\sec^{-1} x) = \dfrac{1}{x}$ **51.** $\tan(\sec^{-1} x) = \pm\sqrt{x^2-1}$
53. Exact value is $\tan^{-1}\left(\frac{5}{8}\right)$, and an approximate value is
32.005°.
55. 1.50 **57.** 62 in.
59.

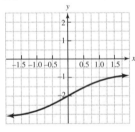

61.

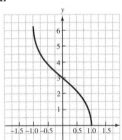

63.

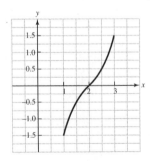

65. Let $\theta = \sec^{-1} x$; then by definition of inverse secant,
$\sec \theta = x. \cos \theta = \dfrac{1}{\sec \theta} = \dfrac{1}{x}$ so that (by definition of
inverse cosine), $\theta = \cos^{-1}\dfrac{1}{x}$. Thus, $\sec^{-1} x = \cos^{-1}\dfrac{1}{x}$.
67. Let $\theta = \cot^{-1} x$; then by definition of inverse cotangent,
$\cot \theta = x$ where θ is in Quadrant II. $\tan \theta = \dfrac{1}{\cot \theta} = \dfrac{1}{x}$ so
that (by definition of inverse tangent) $\theta' = \tan^{-1}\dfrac{1}{x}$ where
$-\frac{\pi}{2} < \theta' < 0$ and $\theta = \theta' + \pi$ (same reference angles in
Quadrants II and IV). By substitution, $\theta = \tan^{-1}\dfrac{1}{x} + \pi$.
Thus, $\cot^{-1} x = \tan^{-1}\dfrac{1}{x} + \pi$.

Chapter 3 Sample Test, page 154

1. a. $y = \tan x$ **b.** $y = \sec x$ **c.** $y = \cos x$
d. $y = \cot x$ **e.** $y = \csc x$ **f.** $y = \sin x$
2. a.

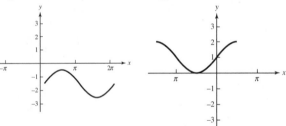

b.

c.

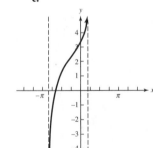

d.

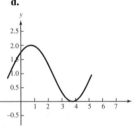

3. a. −$\frac{\pi}{4}$ **b.** $\frac{2\pi}{3}$ **c.** $\frac{3\pi}{4}$ **d.** $\frac{2\pi}{3}$ **e.** 0.85 **f.** 1.32
g. 1.117 **h.** −1.3090
4. a. 0.4 **b.** 0.4 **c.** 1.28 **d.** not defined **e.** $\frac{3}{4}$ **f.** $\frac{2}{7}$
g. 0.29 **h.** −0.64
5. a. (h, k) **b.** $2|a|$ **c.** $\frac{2\pi}{b}$ **d.** $\frac{\pi}{b}$

6.

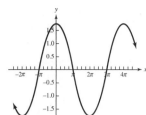

7. $y + 2 = 2 \sin 3\left(x + \frac{\pi}{3}\right)$

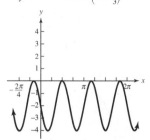

Miscellaneous Problems, Chapters 1–3, pages 155

1.

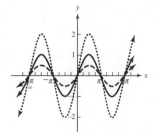

These graphs are all of the form $y = a \sin x$, where $a = 1$, $a = \frac{1}{2}$, $a = 2$. We see all the curves have the same period, and it is the amplitude that changes. The amplitude is a.

3. a. 0; 1   **b.** decreases from 1 to 0
c. decreases from 0 to -1 **d.** increases from -1 to 0

5. a. $\frac{2}{\sqrt{13}}$ **b.** $\frac{\sqrt{21}}{5}$

7. $A: (-4\pi, 0)$ $B: (-\pi, 0)$ $C: \left(\frac{\pi}{2}, 0\right)$ $D: (2\pi, 0)$ $E: \left(\frac{7\pi}{2}, -1\right)$

9. $(5 \cos \beta, 5 \sin \beta)$ **11.** $y = 0.5 \sin \frac{x}{2}$ **13.** $y = 2 \cos \frac{\pi}{5} x$

15.

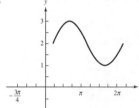

8.

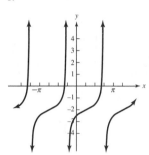

9.

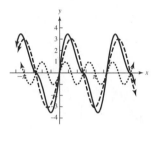

17.

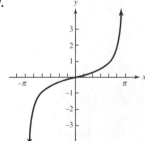

10. a.

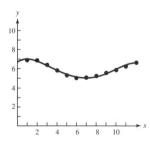

b. $y = 0.97 \cos\left(\frac{\pi}{6}x - \frac{1}{2}\right) + 5.9$

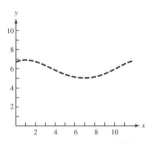

c. The graphs are virtually identical:

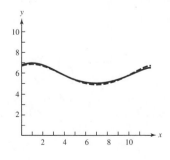

19.

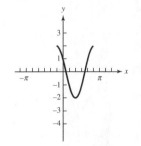

21.

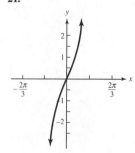

23.

25.

41.

43.

27.

29.

45.

47.

31. a.

b.

49.

51.

33.

35.

37.

39.

53. $x \approx 0.7$ and $x \approx 2.5$

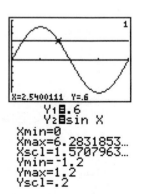

55. $x \approx 0.0$

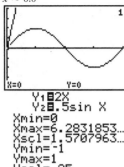

```
Y₁■2X
Y₂■.5sin X
Xmin=0
Xmax=6.2831853…
Xscl=1.5707963…
Ymin=-1
Ymax=1
Yscl=.25
```

57. The graphs appear to be the same

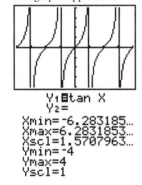   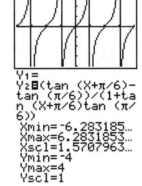

```
Y₁■tan X
Y₂=
Xmin=-6.283185…
Xmax=6.2831853…
Xscl=1.5707963…
Ymin=-4
Ymax=4
Yscl=1
```

```
Y₁=
Y₂■(tan (X+π/6)-
tan (π/6))/(1+ta
n (X+π/6)tan (π/
6))
Xmin=-6.283185…
Xmax=6.2831853…
Xscl=1.5707963…
Ymin=-4
Ymax=4
Yscl=1
```

59.

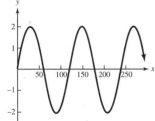

61. It will take about 1,000 years for the north pole to drift about 1°.

63.

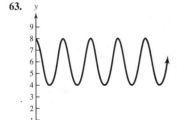

Cumulative Review for Chapters 1–3, page 159

3. a. $\frac{\pi}{6}$ **b.** $\frac{\pi}{4}$ **c.** $\frac{\pi}{3}$ **d.** $\frac{\pi}{2}$ **e.** π **f.** $\frac{3\pi}{2}$ **g.** 2π
h. $\frac{2\pi}{3}$ **i.** $\frac{3\pi}{4}$ **j.** $\frac{5\pi}{6}$ **k.** $-\frac{\pi}{4}$ **ℓ.** $-\frac{5\pi}{3}$ **m.** $-\frac{4\pi}{3}$
n. $-\frac{5\pi}{4}$

5. Right-triangle definition: If θ is an acute angle in a right triangle, then $\cos \theta = \frac{\text{ADJACENT SIDE}}{\text{HYPOTENUSE}}$; $\sin \theta = \frac{\text{OPPOSITE SIDE}}{\text{HYPOTENUSE}}$; $\tan \theta = \frac{\text{OPPOSITE SIDE}}{\text{ADJACENT SIDE}}$

7. Ratio definition: Let θ be an angle in standard position with a point (x, y) the point of intersection of the terminal side of θ a distance of r from the origin $(r \pm 0)$. Then $\cos \theta = \frac{x}{r}$; $\sin \theta = \frac{y}{r}$; $\tan \theta = \frac{y}{x}$, $x \neq 0$; $\sec \theta = \frac{r}{x}$, $x \neq 0$; $\csc \theta = \frac{r}{y}$, $y \neq 0$; $\cot \theta = \frac{x}{y}$, $y \neq 0$

9. a. secant **b.** tangent **c.** cosecant **d.** cosine **e.** sine
f. cotangent

11. $\cos^2\theta + \sin^2\theta = 1$; $1 + \tan^2\theta = \sec^2\theta$; $\cot^2\theta + 1 = \csc^2\theta$

13. a. 1 **b.** 1 **c.** $\frac{\sqrt{3}}{2}$ **d.** 1 **e.** 0 **f.** $\frac{\sqrt{2}}{2}$

15. a. $\frac{1}{2}$ **b.** $\frac{1}{2}$ **c.** 0 **d.** 1 **e.** undefined **f.** undefined

17. a. 2 **b.** $\sqrt{3}$ **c.** 0 **d.** -1 **e.** $\sqrt{3}$ **f.** 0

19. a. undefined **b.** -1 **c.** undefined **d.** $\frac{2\sqrt{3}}{3}$ **e.** $\frac{\sqrt{3}}{3}$
f. undefined

21. a. $\frac{\pi}{6}$ **b.** $\frac{\pi}{6}$ **c.** 0 **d.** $\frac{\pi}{4}$ **e.** $\frac{\pi}{2}$ **f.** $\frac{\pi}{3}$

23.   **25.**

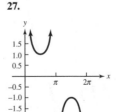   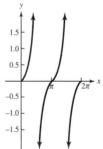

27.   **29.**

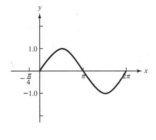   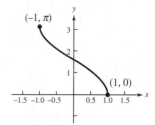

31.

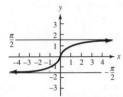

33. $y - k = a \cos b(x - h)$; (h, k) is the starting point for the frame, $2|a|$ is the height of the frame, $2\pi/b$ is the length of the frame, and (x, y) is any point on the curve.

35. $y - k = a \tan b(x - h)$; (h, k) is the starting point for the frame, $2|a|$ is the height of the frame, π/b is the length of the frame, and (x, y) is any point on the curve.

37. a. I **b.** θ between -1.57 and 0 **c.** I
d. θ between 0 and 1.57 **e.** II
f. θ between 0 and 1.57 **g.** IV
h. θ between -1.57 and 0 **i.** II
j. θ between 1.57 and 3.14

39. A **41.** C **43.** A **45.** C **47.** A **49.** C
51. D **53.** B **55.** E

57. $\cos \alpha = \sqrt{7}/4$; $\sin \alpha = 3/4$; $\tan \alpha = 3/\sqrt{7}$; $\sec \alpha = 4/\sqrt{7}$; $\csc \alpha = 4/3$; $\cot \alpha = \sqrt{7}/3$

59. $\alpha = 33°$; $\beta = 57°$; $\gamma = 90°$; $a = 15$; $b = 23$; $c = 28$

61. $\alpha = 35.9°$; $\beta = 54.1°$; $\gamma = 90.0°$; $a = 5.37$; $b = 7.41$; $c = 9.15$

63. a.

b.

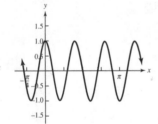

65. a.

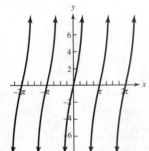

b.

67.

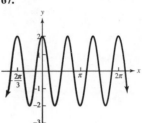

69.

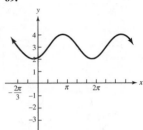

71.

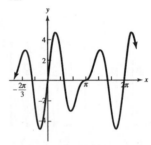

73. The UFO is flying at 10,800 ft.

75. middle gear is turning at $33\frac{1}{3}$ rev/s; small gear is turning at 100 rev/s

77. a.

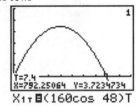

X₁ᴛ◻(160cos 48)T
Y₁ᴛ◻(160sin 48)T
−16T²
Tmin=0
Tmax=9
Tstep=.1
Xmin=0   Ymin=-50
Xmax=1000 Ymax=300
Xscl=100 Yscl=50

b. The time of impact is approximately 7.4 s.
c. The horizontal distance is approximately 792.3 ft.

Problem Set 4.1, page 172

5. a. $60°$ **b.** $60°$ **c.** $60° + 360°k$; $300° + 360°k$

7. a. 2.09 **b.** $\frac{2\pi}{3}, \frac{4\pi}{3}$ **c.** $\frac{2\pi}{3} + 2k\pi$; $\frac{4\pi}{3} + 2k\pi$

9. a. $45°$ **b.** $45°, 225°$ **c.** $45° + 180°k$

11. a. 2.09 **b.** $\frac{2\pi}{3}$ **c.** $\frac{2\pi}{3} + 2k\pi$; $\frac{4\pi}{3} + 2k\pi$

13. a. 0.79 **b.** $\frac{\pi}{4}, \frac{3\pi}{4}$ **c.** $\frac{\pi}{4} + 2k\pi$; $\frac{3\pi}{4} + 2k\pi$

15. a. $x = -0.52$ **b.** $\frac{2\pi}{3}, \frac{5\pi}{6}, \frac{5\pi}{3}, \frac{11\pi}{6}$

c. $\frac{2\pi}{3} + 2k\pi$; $\frac{5\pi}{6} + 2k\pi$; $\frac{5\pi}{3} + 2k\pi$; $\frac{11\pi}{6} + 2k\pi$;

17. a. 0.26 **b.** $\frac{\pi}{12}, \frac{\pi}{4}, \frac{3\pi}{4}, \frac{11\pi}{12}, \frac{17\pi}{12}, \frac{19\pi}{12}$

c. $\frac{\pi}{12} + 2k\pi$; $\frac{\pi}{4} + 2k\pi$; $\frac{3\pi}{4} + 2k\pi$; $\frac{11\pi}{12} + 2k\pi$; $\frac{17\pi}{12} + 2k\pi$;

$\frac{19\pi}{12} + 2k\pi$

19. a. -0.39 **b.** $\frac{5\pi}{8}, \frac{7\pi}{8}, \frac{13\pi}{8}, \frac{15\pi}{8}$

c. $\frac{5\pi}{8} + \pi k$; $\frac{7\pi}{8} + k\pi$ or $\frac{5\pi}{8} + 2k\pi$; $\frac{7\pi}{8} + 2k\pi$; $\frac{13\pi}{8} + 2k\pi$;

$\frac{15\pi}{8} + 2k\pi$

21. a. -0.26 **b.** $\frac{\pi}{4}, \frac{7\pi}{12}, \frac{11\pi}{12}, \frac{5\pi}{4}, \frac{19\pi}{12}, \frac{23\pi}{12}$

c. $\frac{\pi}{4} + k\pi$; $\frac{7\pi}{12} + k\pi$; $\frac{11\pi}{12} + k\pi$

23. a. -0.39 **b.** $\frac{3\pi}{8}, \frac{7\pi}{8}, \frac{11\pi}{8}, \frac{15\pi}{8}$

c. $\frac{3\pi}{8} + k\pi, \frac{7\pi}{8} + k\pi$ or $\frac{3\pi}{8} + 2k\pi, \frac{7\pi}{8} + 2k\pi, \frac{11\pi}{8} + 2k\pi,$

$\frac{15\pi}{8} + 2k\pi$

25. a. 0 **b.** $0, \pi$ **c.** $k\pi$

27. a. 0, 1.57 **b.** $0, \pi, \frac{\pi}{2}, \frac{3\pi}{2}$ **c.** $\frac{k\pi}{2}$

29. a. 1.05, 0.52 **b.** $\frac{\pi}{3}, \frac{5\pi}{3}, \frac{\pi}{6}, \frac{5\pi}{6}$

c. $\frac{\pi}{3} + 2k\pi$; $\frac{5\pi}{3} + 2k\pi$; $\frac{\pi}{6} + 2k\pi$; $\frac{5\pi}{6} + 2k\pi$

31. $0, \pi, \frac{\pi}{3}, \frac{4\pi}{3}$ **33.** $\frac{\pi}{4}, \frac{3\pi}{4}, \frac{5\pi}{4}, \frac{7\pi}{4}$ **35.** $0, \pi, \frac{\pi}{4}, \frac{7\pi}{4}$

37. $\frac{5\pi}{6} + 2k\pi$; $\frac{7\pi}{6} + 2k\pi$

39. $0.4014 + 6.2832k$; $2.7402 + 6.2832k$

41. $0.9423 + 3.1416k$ **43.** 4.71 **45.** 0.67

47. 0.00, 0.35 **49.** 0.00, 2.09 **51.** 0.38 **53.** 0.32

55. 0.52, -1.57 **57.** 1.57, -0.52 **59.** 1.57, 0.52

61. 0.30 **63.** 0.185 **65.** 0.0, 2.4, 3.1, 5.5 **67.** none

$(-0.7$ is not in the domain) **69.** 1.9 **71.** 0.5, 4.2

Problem Set 4.2, page 184

7. $\cos \theta = \sqrt{1 - 9u^2}$; $\sin \theta = -3u$; $\tan \theta = \dfrac{-3u\sqrt{1 - 9u^2}}{1 - 9u^2}$;

$\sec \theta = \dfrac{\sqrt{1 - 9u^2}}{1 - 9u^2}$; $\csc \theta = -\dfrac{1}{3u}$; $\cot \theta = \dfrac{-\sqrt{1 - 9u^2}}{3u}$

9. $\cos \theta = -\dfrac{2u}{5}$; $\sin \theta = \dfrac{\sqrt{25 - 4u^2}}{5}$; $\tan \theta = \dfrac{-\sqrt{25 - 4u^2}}{2u}$;

$\sec \theta = -\dfrac{5}{2u}$; $\csc \theta = \dfrac{5\sqrt{25 - 4u^2}}{25 - 4u^2}$; $\cot \theta = \dfrac{-2u\sqrt{25 - 4u^2}}{25 - 4u^2}$

11. $\cot \theta \tan^2\theta = (\cot \theta \tan \theta)\tan \theta$

$\qquad = \tan \theta$

13. $\tan^2\theta \sin^2\theta = \tan^2\theta(1 - \cos^2\theta)$

$\qquad = \tan^2\theta - \tan^2\theta \cos^2\theta$

$\qquad = \tan^2\theta - \dfrac{\sin^2\theta}{\cos^2\theta}\cos^2\theta$

$\qquad = \tan^2\theta - \sin^2\theta$

15. $\tan \theta + \cot \theta = \dfrac{\sin \theta}{\cos \theta} + \dfrac{\cos \theta}{\sin \theta}$

$\qquad = \dfrac{\sin^2\theta + \cos^2\theta}{\cos \theta \sin \theta}$

$\qquad = \dfrac{1}{\cos \theta \sin \theta}$

$\qquad = \sec \theta \csc \theta$

17. $\dfrac{1 - \sec^2\beta}{\sec^2\beta} = \dfrac{1}{\sec^2\beta} - 1$

$\qquad = \cos^2\beta - 1$

$\qquad = -\sin^2\beta$

19. $\dfrac{1 - \sin^2 2\theta}{1 + \sin 2\theta} = \dfrac{(1 - \sin 2\theta)(1 + \sin 2\theta)}{1 + \sin 2\theta}$

$\qquad = 1 - \sin 2\theta$

21. $\dfrac{1 + \cos 2\lambda \sec 2\lambda}{\tan 2\lambda + \sec 2\lambda} = \dfrac{1 + \cos 2\lambda \cdot \dfrac{1}{\cos 2\lambda}}{\dfrac{\sin 2\lambda}{\cos 2\lambda} + \dfrac{1}{\cos 2\lambda}}$

$\qquad = \dfrac{2}{\dfrac{\sin 2\lambda + 1}{\cos 2\lambda}}$

$\qquad = \dfrac{2\cos 2\lambda}{\sin 2\lambda + 1}$

23. $(\sin \alpha + \cos \alpha)^2 + (\sin \alpha - \cos \alpha)^2$

$\qquad = \sin^2\alpha + 2\sin \alpha \cos \alpha + \cos^2\alpha + \sin^2\alpha$

$\qquad \quad - 2\sin \alpha \cos \alpha + \cos^2\alpha$

$\qquad = 2$

25. $\dfrac{1 + \cot^2\gamma}{1 + \tan^2\gamma} = \dfrac{\csc^2\gamma}{\sec^2\gamma}$

$\qquad = \dfrac{\dfrac{1}{\sin^2\gamma}}{\dfrac{1}{\cos^2\gamma}}$

$\qquad = \dfrac{\cos^2\gamma}{\sin^2\gamma}$

$\qquad = \cot^2\gamma$

27. $\tan^2\phi + \sin^2\phi + \cos^2\phi = \tan^2\phi + 1$

$\qquad = \sec^2\phi$

29. $1 + \sin^2\theta = 1 + (1 - \cos^2\theta)$

$\qquad = 2 - \cos^2\theta$

31. $\dfrac{1}{1 + \cos 2\theta} + \dfrac{1}{1 - \cos 2\theta} = \dfrac{1 - \cos 2\theta + 1 + \cos 2\theta}{(1 + \cos 2\theta)(1 - \cos 2\theta)}$

$\qquad = \dfrac{2}{1 - \cos^2 2\theta}$

$\qquad = \dfrac{2}{\sin^2 2\theta}$

$\qquad = 2\csc^2 2\theta$

33. $\dfrac{1}{\tan 2\beta} + \dfrac{\cos 2\beta}{\cot 2\beta} = \cot 2\beta + \dfrac{\cos 2\beta}{\dfrac{\cos 2\beta}{\sin 2\beta}}$

$\qquad\qquad = \cot 2\beta + \sin 2\beta$

41. The graphs are the same.   **43.** The graphs are the same.

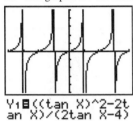

Y₁■((tan X)^2-2t
an X)/(2tan X-4)

Y₂■(1/2)tan X
Xmin=-6.283185...
Xmax=6.2831853...
Xscl=1.5707963...
Ymin=-5
Ymax=5
Yscl=1

Y₁■sin X/(1/sin
X)+cos X/(1/cos
X)
Y₂■1
Xmin=-3.141592...
Xmax=6.2831853...
Xscl=1.5707963...
Ymin=-2
Ymax=3
Yscl=1

45. $\dfrac{\sin 2\theta}{2} \neq \sin \theta$; to show this, find a counterexample,

say $\theta = 30°$.

47. $\dfrac{1 + \tan \alpha}{1 - \tan \alpha} = \dfrac{1 + \tan \alpha}{1 - \tan \alpha} \cdot \dfrac{1 + \tan \alpha}{1 + \tan \alpha}$

$\qquad = \dfrac{1 + 2 \tan \alpha + \tan^2 \alpha}{1 - (\sec^2 \alpha - 1)}$

$\qquad = \dfrac{\sec^2 \alpha + 2 \tan \alpha}{2 - \sec^2 \alpha}$

49. $\dfrac{\sin^3 x - \cos^3 x}{\sin x - \cos x} = \dfrac{(\sin x - \cos x)(\sin^2 x + \sin x \cos x + \cos^2 x)}{\sin x - \cos x}$

$\qquad\qquad = \sin^2 x + \sin x \cos x + \cos^2 x$

$\qquad\qquad = 1 + \sin x \cos x$

51. $\dfrac{(\sec^2 \gamma + \tan^2 \gamma)^2}{\sec^4 \gamma - \tan^4 \gamma} = \dfrac{(\sec^2 \gamma + \tan^2 \gamma)^2}{(\sec^2 \gamma - \tan^2 \gamma)(\sec^2 \gamma + \tan^2 \gamma)}$

$\qquad\qquad = \dfrac{\sec^2 \gamma + \tan^2 \gamma}{\sec^2 \gamma - \tan^2 \gamma}$

$\qquad\qquad = \dfrac{\sec^2 \gamma + \tan^2 \gamma}{(1 + \tan^2 \gamma) - \tan^2 \gamma}$

$\qquad\qquad = \sec^2 \gamma + \tan^2 \gamma$

$\qquad\qquad = (1 + \tan^2 \gamma) + \tan^2 \gamma$

$\qquad\qquad = 1 + 2 \tan^2 \gamma$

53. $\dfrac{1}{\sec \theta + \tan \theta} = \dfrac{1}{\sec \theta + \tan \theta} \cdot \dfrac{\sec \theta - \tan \theta}{\sec \theta - \tan \theta}$

$\qquad = \dfrac{\sec \theta - \tan \theta}{\sec^2 \theta - \tan^2 \theta}$

$\qquad = \dfrac{\sec \theta - \tan \theta}{1 + \tan^2 \theta - \tan^2 \theta}$

$\qquad\qquad = \dfrac{\sec \theta - \tan \theta}{1}$

$\qquad\qquad = \sec \theta - \tan \theta$

55. $\dfrac{1 - \sec^3 \theta}{1 - \sec \theta} = \dfrac{(1 - \sec \theta)(1 + \sec \theta + \sec^2 \theta)}{1 - \sec \theta}$

$\qquad\qquad = 1 + \sec \theta + \tan^2 \theta + 1$

$\qquad\qquad = \tan^2 \theta + \sec \theta + 2$

57. $\sqrt{(3 \cos \theta - 4 \sin \theta)^2 + (3 \sin \theta + 4 \cos \theta)^2}$

$\qquad = \sqrt{9(\cos^2 \theta + \sin^2 \theta) + 16(\sin^2 \theta + \cos^2 \theta)}$

$\qquad = \sqrt{9 + 16}$

$\qquad = 5$

59. $\dfrac{\cos^4 \theta - \sin^4 \theta}{(\cos^2 \theta - \sin^2 \theta)^2} = \dfrac{(\cos^2 \theta - \sin^2 \theta)(\cos^2 \theta + \sin^2 \theta)}{(\cos^2 \theta - \sin^2 \theta)^2}$

$\qquad\qquad = \dfrac{1}{\cos^2 \theta - \sin^2 \theta}$

Also,

$\qquad \dfrac{\cos \theta}{\cos \theta + \sin \theta} + \dfrac{\sin \theta}{\cos \theta - \sin \theta}$

$\qquad\qquad = \dfrac{\cos^2 \theta - \cos \theta \sin \theta + \sin \theta \cos \theta + \sin^2 \theta}{\cos^2 \theta - \sin^2 \theta}$

$\qquad\qquad = \dfrac{1}{\cos^2 \theta - \sin^2 \theta}$

61. $(\sec \alpha + \sec \beta)^2 - (\tan \alpha - \tan \beta)^2$

$\qquad = \sec^2 \alpha + 2 \sec \alpha \sec \beta + \sec^2 \beta - \tan^2 \alpha$

$\qquad\quad + 2 \tan \alpha \tan \beta - \tan^2 \beta$

$\qquad = \tan^2 \alpha + 1 + 2 \sec \alpha \sec \beta + \tan^2 \beta + 1 - \tan^2 \alpha$

$\qquad\quad + 2 \tan \alpha \tan \beta - \tan^2 \beta$

$\qquad = 2 + 2(\sec \alpha \sec \beta + \tan \alpha \tan \beta)$

63. $\tan \theta + \sec \theta + 1 = (\tan \theta + \sec \theta + 1)\left(\dfrac{\tan \theta + \sec \theta - 1}{\tan \theta + \sec \theta - 1}\right)$

$\qquad = (\tan^2 \theta + \tan \theta \sec \theta - \tan \theta + \sec \theta \tan \theta + \sec^2 \theta$

$\qquad\quad - \sec \theta + \tan \theta + \sec \theta - 1)\left(\dfrac{1}{\tan \theta + \sec \theta - 1}\right)$

$\qquad = \dfrac{\tan^2 \theta + 2 \tan \theta \sec \theta + (\sec^2 \theta - 1)}{\tan \theta + \sec \theta - 1}$

$\qquad = \dfrac{2\tan^2 \theta + 2 \tan \theta \sec \theta}{\tan \theta + \sec \theta - 1}$

65. $\sqrt{4 - u^2} = \sqrt{4 - (2 \sin x)^2}$

$\qquad = \sqrt{4 - 4 \sin^2 x}$

$\qquad = \sqrt{4(1 - \sin^2 x)}$

$\qquad = \sqrt{4 \cos^2 x}$

$\qquad = 2 \cos x$

67. $\sqrt{1 + u^2} = \sqrt{1 + \tan^2 x}$

$\qquad = \sqrt{\sec^2 x}$

$\qquad = \sec x$

Problem Set 4.3, page 195

9. $(\sin\theta + \cos\theta)\frac{\sqrt{2}}{2}$ **11.** $(\cos\theta + \sin\theta)\frac{\sqrt{2}}{2}$ **13.** $\frac{1+\tan\theta}{1-\tan\theta}$

15. $2\sin\theta\cos\theta$ **17.** $\cos 75°$ **19.** $\cot 28°$ **21.** $-\sin\frac{\pi}{3}$

23. $-\sin 23°$ **23.** $\cos 57°$ **27.** $-\tan 29°$ **29.** 0.9135

31. 0.9511 **33.** 1.1918

35. $\cos 15° = \frac{\sqrt{6}+\sqrt{2}}{4}$; $\sin 15° = \frac{\sqrt{6}-\sqrt{2}}{4}$; $\tan 15° = 2-\sqrt{3}$

37. $\cos 75° = \frac{\sqrt{6}-\sqrt{2}}{4}$; $\sin 75° = \frac{\sqrt{6}+\sqrt{2}}{4}$; $\tan 75° = 2+\sqrt{3}$

39. $\cos 165° = \frac{-\sqrt{6}-\sqrt{2}}{4}$; $\sin 165° = \frac{\sqrt{6}-\sqrt{2}}{4}$;

$\tan 165° = \sqrt{3}-2$

41. **a.** $\cos\left(\theta-\frac{\pi}{3}\right)$ **b.** $-\sin\left(\theta-\frac{\pi}{3}\right)$ **c.** $-\tan\left(\theta-\frac{\pi}{3}\right)$

43. 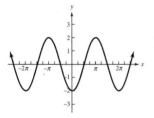 **45.**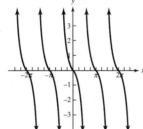

47.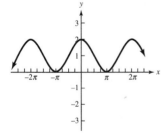

49. **a.** $f(-x) = \sec(-x) = \dfrac{1}{\cos(-x)} = \sec x = f(x)$; even

b. $g(-x) = \dfrac{\sin(-x)}{-x} = \dfrac{-\sin x}{-x} = \dfrac{\sin x}{x} = g(x)$; even

c. $h(-x) = \dfrac{-x}{\cos(-x)} = \dfrac{-x}{\cos x} = -h(x)$; odd

d. $F(-x) = \sin(-x)\tan(-x) = \sin x\tan x = F(x)$; even

51. $\tan(\alpha-\beta) = \tan[\alpha+(-\beta)]$

$= \dfrac{\tan\alpha + \tan(-\beta)}{1-\tan\alpha\tan(-\beta)}$

$= \dfrac{\tan\alpha - \tan\beta}{1+\tan\alpha\tan\beta}$

53. $\dfrac{\sin 6\theta}{\sin 3\theta} - \dfrac{\cos 6\theta}{\cos 3\theta} = \dfrac{\sin 6\theta\cos 3\theta - \cos 6\theta\sin 3\theta}{\sin 3\theta\cos 3\theta}$

$= \dfrac{\sin(6\theta - 3\theta)}{\sin 3\theta\cos 3\theta}$

$= \dfrac{1}{\cos 3\theta}$

$= \sec 3\theta$

55. $\cos(\alpha-\beta)\cos\beta - \sin(\alpha-\beta)\sin\beta$

$= \cos\alpha\cos^2\beta + \sin\alpha\sin\beta\cos\beta$

$\quad - \sin\alpha\cos\beta\sin\beta + \cos\alpha\sin^2\beta$

$= \cos\alpha(\cos^2\beta + \sin^2\beta)$

$= \cos\alpha$

57. $\dfrac{7}{25}$ **59.** $\dfrac{\sqrt{15}+2}{6}$

61. In a correctly labeled triangle, $\alpha + \beta + \gamma = 180°$;

$\tan(\alpha+\beta) = \tan(180° - \gamma)$

$= \dfrac{\tan 180° - \tan\gamma}{1+\tan 180°\tan\gamma}$

$= \dfrac{0 - \tan\gamma}{1+0}$

$= -\tan\gamma$

63. In a correctly labeled triangle, $\alpha + \beta + \gamma = 180°$;

$\sin\left(\dfrac{\alpha+\beta}{2}\right) = \cos\left(90° - \dfrac{\alpha+\beta}{2}\right)$

$= \cos\left(\dfrac{180° - \alpha - \beta}{2}\right)$

$= \cos\dfrac{\gamma}{2}$

65. Label α, β, and γ at $\angle A$, $\angle B$, and $\angle C$, respectively. Since $\triangle ABC$ is isosceles, $\alpha = \beta$ and from the given figure $\gamma = 150°$; since $\alpha + \beta + \gamma = 180°$, we have $\alpha = \beta = 15°$.

$$\cos 15° = \dfrac{|\overline{AD}|}{|\overline{AB}|} = \dfrac{|\overline{AC}| + |\overline{CD}|}{|\overline{AB}|}$$

Now, $|\overline{AC}| = 2$; $\sin 30° = \dfrac{|\overline{BD}|}{2}$, so $|\overline{BD}| = 1$;

$\cos 30° = \dfrac{|\overline{CD}|}{2}$, so $|\overline{CD}| = \sqrt{3}$; Finally,

$|\overline{AB}|^2 = (|\overline{AC}| + |\overline{CD}|)^2 + |\overline{BD}|^2$

$= (2+\sqrt{3})^2 + 1 = 8 + 4\sqrt{3}$ so that

$|AB| = \sqrt{8+4\sqrt{3}} = 2\sqrt{2+\sqrt{3}}$. We now substitute these values:

$$\cos 15° = \dfrac{|\overline{AC}| + |\overline{CD}|}{|\overline{AB}|} = \dfrac{2+\sqrt{3}}{2\sqrt{2+\sqrt{3}}}$$

$$= \dfrac{2+\sqrt{3}}{2\sqrt{2+\sqrt{3}}} \cdot \dfrac{\sqrt{2+\sqrt{3}}}{\sqrt{2+\sqrt{3}}} = \dfrac{\sqrt{2+\sqrt{3}}}{2}$$

67. $\dfrac{\cos(x + h) - \cos x}{h} = \dfrac{\cos x \cos h - \sin x \sin h - \cos x}{h}$

$$= -\sin x \left(\dfrac{\sin h}{h}\right) + \dfrac{-\cos x + \cos x \cos h}{h}$$

$$= -\sin x \left(\dfrac{\sin h}{h}\right) - \cos x \left(\dfrac{1 - \cos h}{h}\right)$$

Problem Set 4.4, page 205

1. $\cos 2\theta = \cos^2\theta - \sin^2\theta$, not $\cos^2\theta + \sin^2\theta$.

3. The number $\frac{1}{2}$ is not a factor of $\cos \frac{1}{2}\theta$. It can be shown that they are not equal by finding a counterexample; say $\theta = 60°$.

5. $\tan(45° + \theta)$; cannot distribute "tan."

7. In $\tan(\theta + \theta)$, cannot distribute "tan."

9. $\tan\frac{1}{2}\theta = \dfrac{\sin\theta}{1 + \cos\theta}$　　　　*Identity* 29

$$= \dfrac{\sin\theta}{1 + \cos\theta} \cdot \dfrac{1 - \cos\theta}{1 - \cos\theta}$$

$$= \dfrac{\sin\theta(1 - \cos\theta)}{1 - \cos^2\theta}$$

$$= \dfrac{\sin\theta(1 - \cos\theta)}{\sin^2\theta}$$

$$= \dfrac{1 - \cos\theta}{\sin\theta}$$

11. $\dfrac{2\tan\frac{1}{2}\theta}{1 + \tan^2\frac{1}{2}\theta} = \tan\theta \neq \dfrac{2}{1 + \tan\theta}$

13. a. $180° < 2\theta < 360°$; III or IV　　**b.** $45° < \frac{1}{2}\theta < 90°$; I

15. a. I or IV　**b.** I or II　　**17.** II

19. a. $\frac{\sqrt{2}}{2}$　**b.** $\sqrt{2} - 1$　　**21. a.** 1　**b.** $\frac{1}{2}$

23. $\cos\theta = \frac{\sqrt{5}}{5}$; $\sin\theta = \frac{2\sqrt{5}}{5}$; $\tan\theta = 2$

25. $\cos\theta = \frac{\sqrt{10}}{10}$; $\sin\theta = \frac{3\sqrt{10}}{10}$; $\tan\theta = 3$.

27. $\cos\theta = \frac{1}{2}$; $\sin\theta = \frac{\sqrt{3}}{2}$; $\tan\theta = \sqrt{3}$

29. $\cos 2\theta = \frac{7}{25}$; $\sin 2\theta = \frac{24}{25}$; $\tan 2\theta = \frac{24}{7}$

31. $\cos 2\theta = \frac{119}{169}$; $\sin 2\theta = \frac{-120}{169}$; $\tan 2\theta = \frac{-120}{119}$

33. $\cos 2\theta = \frac{-31}{81}$; $\sin 2\theta = \frac{20\sqrt{14}}{81}$; $\tan 2\theta = \frac{-20\sqrt{14}}{31}$

35. $\cos\frac{1}{2}\theta = \frac{3}{\sqrt{10}}$ or $\frac{3}{10}\sqrt{10}$; $\sin\frac{1}{2}\theta = \frac{1}{\sqrt{10}}$ or $\frac{1}{10}\sqrt{10}$; $\tan\frac{1}{2}\theta = \frac{1}{3}$

37. $\cos\frac{1}{2}\theta = \frac{1}{26}\sqrt{26}$; $\sin\frac{1}{2}\theta = -\frac{5}{26}\sqrt{26}$; $\tan\frac{1}{2}\theta = 5$

39. $\cos\frac{1}{2}\theta = \frac{\sqrt{7}}{3}$; $\sin\frac{1}{2}\theta = \frac{\sqrt{2}}{3}$; $\tan\frac{1}{2}\theta = \frac{\sqrt{14}}{7}$

41.

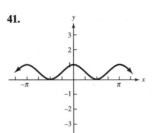

43.

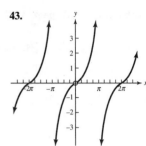

45.

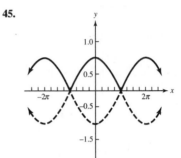

47. a. $M \approx 3.9$

　　b. $M = \sqrt{2} + \sqrt{6}$

　　$\left(\text{This is } \dfrac{2}{\sqrt{2 - \sqrt{3}}} \text{ after it has been simplified.}\right)$

　　c. $\theta = 2\sin^{-1} r^{-1}$

49. $x = \frac{1}{32}(256)^2\sin 90° \approx 2{,}048$ ft

51. $2\sin\frac{\alpha}{2}\cos\frac{\alpha}{2} = \sin\left(\frac{\alpha}{2} + \frac{\alpha}{2}\right)$

　　　　　　　　$= \sin\alpha$

53. $\dfrac{2\tan\theta}{1 + \tan^2\theta} = \dfrac{2\dfrac{\sin\theta}{\cos\theta}}{1 + \dfrac{\sin^2\theta}{\cos^2\theta}}$

$$= \dfrac{\dfrac{2\sin\theta}{\cos\theta}}{\dfrac{\cos^2\theta + \sin^2\theta}{\cos^2\theta}}$$

$$= \dfrac{2\sin\theta}{\cos\theta} \cdot \dfrac{\cos^2\theta}{1}$$

$$= 2\sin\theta\cos\theta$$

$$= \sin 2\theta$$

55. $\tan\frac{1}{2}\theta = \dfrac{\sin\frac{1}{2}\theta}{\cos\frac{1}{2}\theta}$

$$= \dfrac{\pm\sqrt{\dfrac{1 - \cos\theta}{2}}}{\pm\sqrt{\dfrac{1 + \cos\theta}{2}}}$$

$$= \pm \sqrt{\frac{1 - \cos\theta}{2} \cdot \frac{2}{1 + \cos\theta} \cdot \frac{1 - \cos\theta}{1 - \cos\theta}}$$

$$= \pm \sqrt{\frac{(1 - \cos\theta)^2}{1 - \cos^2\theta}}$$

$$= \pm \frac{\sqrt{(1 - \cos\theta)^2}}{\sqrt{\sin^2\theta}}$$

$$= \frac{1 - \cos\theta}{\sin\theta}$$

57. $\sin 2\theta \sec\theta = 2\sin\theta\cos\theta \cdot \dfrac{1}{\cos\theta}$

$$= 2\sin\theta$$

59. $\sin 2\theta = \dfrac{|\overline{CD}|}{|\overline{CO}|} = \dfrac{|\overline{CD}|}{1} = |\overline{CD}|$

In $\triangle ADC$, $\sin\theta = \dfrac{|\overline{CD}|}{2\cos\theta}$, so $2\sin\theta\cos\theta = |\overline{CD}| = \sin 2\theta$.

61. a. $|\overline{AO}| = |\overline{BO}| = r$, so $\alpha = \beta$ (base angles in an isosceles triangle). Also, $\alpha + \beta + \gamma = 180°$, and $\gamma = 180° - \theta$, so

$$\alpha + \alpha + (180° - \theta) = 180°$$
$$2\alpha - \theta = 0$$
$$\alpha = \tfrac{1}{2}\theta$$

b. $\sin\theta = \dfrac{|\overline{BD}|}{|\overline{BO}|}; \cos\theta = \dfrac{|\overline{OD}|}{|\overline{BO}|};$

$$\tan\alpha = \frac{|\overline{BD}|}{|\overline{AD}|} = \frac{|\overline{BD}|}{|\overline{AO}| + |\overline{OD}|} = \frac{|\overline{BD}|}{r + |\overline{BO}|\cos\theta}$$

$$= \frac{|\overline{BO}|\sin\theta}{r + |\overline{BO}|\cos\theta} = \frac{r\sin\theta}{r + r\cos\theta} = \frac{\sin\theta}{1 + \cos\theta}$$

Thus, $\tan\tfrac{1}{2}\theta = \dfrac{\sin\theta}{1 + \cos\theta}$

Problem Set 4.5, page 213

1. $\cos 11° - \cos 59°$ **3.** $\sin 94° - \sin 46°$
5. $\tfrac{1}{2}\cos 2\theta - \tfrac{1}{2}\cos 6\theta$ **7.** $\tfrac{1}{2}\cos 2\theta + \tfrac{1}{2}\cos 4\theta$
9. $2\sin 53.5°\cos 10.5°$ **11.** $2\sin 22.5°\cos 7.5°$
13. $2\cos 4x\cos 2x$ **15.** $2\cos 7y\cos 2y$
17. $\sqrt{2}\sin\left(\theta + \tfrac{\pi}{4}\right)$ **19.** $\sqrt{2}\sin\left(\tfrac{\theta}{2} + \tfrac{3\pi}{4}\right)$ **21.** $2\sin\left(\pi\theta + \tfrac{2\pi}{3}\right)$

23. $\dfrac{\sin\theta + \sin 3\theta}{2\sin 2\theta} = \dfrac{2\sin 2\theta\cos\theta}{2\sin 2\theta}$

$$= \cos\theta$$

25. $\dfrac{\sin\theta + \sin 3\theta}{4\cos^2\theta} = \dfrac{2\sin 2\theta\cos\theta}{4\cos^2\theta}$

$$= \frac{\sin 2\theta}{2\cos\theta}$$

$$= \frac{2\sin\theta\cos\theta}{2\cos\theta}$$

$$= \sin\theta$$

27. $\dfrac{\sin 5\theta + \sin 3\theta}{\cos 5\theta + \cos 3\theta} = \dfrac{2\sin 4\theta\cos\theta}{2\cos 4\theta\cos\theta}$

$$= \frac{\sin 4\theta}{\cos 4\theta}$$

$$= \tan 4\theta$$

29. $\dfrac{\cos 5\omega + \cos\omega}{\cos\omega - \cos 5\omega} = \dfrac{2\cos 3\omega\cos 2\omega}{-2\sin 3\omega\sin(-2\omega)}$

$$= \frac{2\cos 3\omega\cos 2\omega}{2\sin 3\omega\sin 2\omega}$$

$$= \cot 3\omega\cot 2\omega$$

$$= \frac{1}{\tan 3\omega}\cot 2\omega$$

$$= \frac{\cot 2\omega}{\tan 3\omega}$$

31. $\dfrac{\sin x + \sin y}{\cos x + \cos y} = \dfrac{2\sin\left(\dfrac{x+y}{2}\right)\cos\left(\dfrac{x-y}{2}\right)}{2\cos\left(\dfrac{x+y}{2}\right)\cos\left(\dfrac{x-y}{2}\right)}$

$$= \tan\left(\frac{x+y}{2}\right)$$

33. $\cos^2\tfrac{\theta}{2} - \sin^2\tfrac{\theta}{2} = \cos 2\left(\tfrac{\theta}{2}\right)$
$$= \cos\theta$$

35. $y = \cos\theta - \sin\theta$
$$= \sqrt{2}\sin\left(\theta + \tfrac{3\pi}{4}\right)$$
$(h, k) = \left(-\tfrac{3\pi}{4}, 0\right);$
$a = \sqrt{2}; b = 1; p = 2\pi$

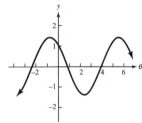

37. $y = 2\sin\theta - 3\cos\theta$
$$\approx \sqrt{13}\sin(\theta + 5.3)$$
$(h, k) = (-5.3, 0);$
$a = \sqrt{13}; b = 1, p = 2\pi$

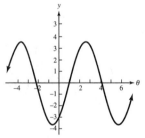

39. $y = 8\sin\theta + 15\cos\theta$
$$\approx 17\sin(\theta + 1.1)$$
$(h, k) = (-1.1, 0);$
$a = 17; b = 1; p = 2\pi$

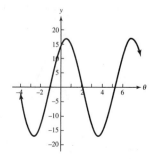

41. $0, \pi, \frac{\pi}{4}, \frac{3\pi}{4}, \frac{5\pi}{4}, \frac{7\pi}{4}$ **43.** $0, \frac{\pi}{4}, \frac{\pi}{2}, \frac{3\pi}{4}, \pi, \frac{5\pi}{4}, \frac{3\pi}{2}, \frac{7\pi}{4}$ **45.** $\frac{\pi}{2}$

47. Identity 32: $2 \sin \alpha \cos \beta = \sin(\alpha + \beta) + \sin(\alpha - \beta)$

$2 \sin \alpha \cos \beta = 2 \cos \beta \sin \alpha$
$\qquad = \sin(\beta + \alpha) - \sin(\beta - \alpha)$ which is identity 33
$\qquad = \sin(\alpha + \beta) + \sin(\alpha - \beta)$ which is identity 32

49. $2 \sin \alpha \cos \beta = \sin(\alpha + \beta) + \sin(\alpha - \beta)$
$\qquad\qquad\quad = \sin x + \sin y$

where $x = \alpha + \beta$ and $y = \alpha - \beta$, which is the system of equations

$$\begin{cases} x = \alpha - \beta \\ y = \alpha + \beta \end{cases}$$

Solve this system of equations to obtain $\alpha = \dfrac{x + y}{2}$ and

$\beta = \dfrac{x - y}{-2}$ and substitute these values into the original equation:

$$2 \sin\left(\frac{x + y}{2}\right) \cos\left(\frac{x - y}{-2}\right) = \sin x + \sin y$$

$$2 \sin\left(\frac{x + y}{2}\right) \cos\left(\frac{x - y}{2}\right) = \sin x + \sin y \quad \text{which is identity 36}$$

51. $y = 3 \sin x + 3 \cos x = 3\sqrt{2} \sin(x + \alpha)$
Amplitude is $3\sqrt{2}$ and period is 2π.

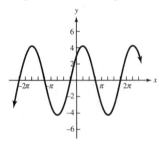

53. $y = \sqrt{2}\left(\sin\frac{x}{2} + \cos\frac{x}{2}\right) = 2 \sin\frac{1}{2}(x + 2\alpha)$
Amplitude is 2 and period is 4π.

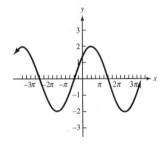

55. $y = 2 \sin(1{,}906\pi x)\cos(512\pi x)$

57. a.   **b.**

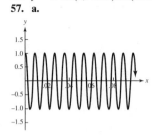 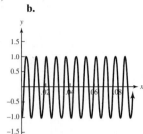

c.

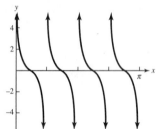

d. $y = 2 \sin(234\pi x)\sin(10\pi x)$

59. $y = \cot 4x$

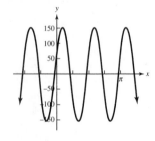

61. a. $v = \dfrac{480}{\pi} \sin 4\left(t + \frac{\pi}{24}\right)$

b. $a = \dfrac{480}{\pi}$, so the max/min values are
$\pm a = \pm \dfrac{480}{\pi}$
$\approx \pm 152.79$

Chapter 4 Sample Test, page 216

1. $\dfrac{\sec^2\theta + \tan^2\theta + 1}{\sec\theta} = \dfrac{\sec^2\theta + (\sec^2\theta - 1) + 1}{\sec\theta}$

$\qquad\qquad\qquad\quad = \dfrac{2\sec^2\theta}{\sec\theta}$

$\qquad\qquad\qquad\quad = 2\sec\theta$

2. $\dfrac{\sqrt{2} - \sqrt{6}}{4}$

3. $\dfrac{2\tan\frac{\pi}{6}}{1 - \tan^2\frac{\pi}{6}} = \tan\left(2 \cdot \frac{\pi}{6}\right)$

$= \tan\frac{\pi}{3}$

$= \sqrt{3}$

4. $-\sqrt{\dfrac{1 + \cos 240°}{2}} = \cos 120° = -\frac{1}{2}$

5. $\dfrac{\csc^2\alpha}{1 + \cos^2\alpha} = \dfrac{\csc^2\alpha}{\csc^2\alpha}$

$= 1$

6. $\dfrac{\sin^2\beta - \cos^2\beta}{\sin\beta + \cos\beta} = \dfrac{(\sin\beta + \cos\beta)(\sin\beta - \cos\beta)}{\sin\beta + \cos\beta}$

$= \sin\beta - \cos\beta$

7. $\dfrac{1}{\sin\gamma + \cos\gamma} + \dfrac{1}{\sin\gamma - \cos\gamma} = \dfrac{\sin\gamma - \cos\gamma + \sin\gamma + \cos\gamma}{\sin^2\gamma - \cos^2\gamma}$

$= \dfrac{2\sin\gamma}{\sin^2\gamma - \cos^2\gamma} \cdot \dfrac{\sin^2\gamma + \cos^2\gamma}{\sin^2\gamma + \cos^2\gamma}$

$= \dfrac{2\sin\gamma(\sin^2\gamma + \cos^2\gamma)}{\sin^2\gamma - \cos^4\gamma}$

$= \dfrac{2\sin\gamma}{\sin^4\gamma - \cos^4\gamma}$

8. $\dfrac{1 - \tan^2 3\theta}{1 - \tan 3\theta} = \dfrac{(1 - \tan 3\theta)(1 + \tan 3\theta)}{1 - \tan 3\theta}$

$= 1 + \tan 3\theta$

9. $\dfrac{\sin 5\theta + \sin 3\theta}{\cos 5\theta - \cos 3\theta} = \dfrac{2\sin 4\theta\cos\theta}{-2\sin 4\theta\sin\theta} = -\cot\theta$

Answers to Problems 10–12 vary; choose a counterexample.

10. False; let $\theta = 1$. Then $\tan\theta + \csc\theta \approx 2.7$ and $\cot\theta \approx 0.6$.
11. False; let $t = 1$. Then $(\sin t + \cos t)^2 \approx 1.9 \neq 1$.
12. False; let $t = 1$. Then $\cos t \tan t \csc t \sec t \approx 1.85 \neq 1$.
13. $\frac{\pi}{3}, \frac{2\pi}{3}, \frac{4\pi}{3}, \frac{5\pi}{3}$ **14.** $1.25, \frac{3\pi}{4} \approx 2.36, 4.39, \frac{7\pi}{4} \approx 5.50$
15. $0.31, 2.83$ **16.** $\frac{\pi}{6}, \frac{\pi}{3}, \frac{2\pi}{3}, \frac{5\pi}{6}, \frac{7\pi}{6}, \frac{4\pi}{3}, \frac{5\pi}{3}, \frac{11\pi}{6}$

17.

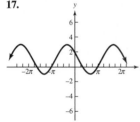

18.

19.

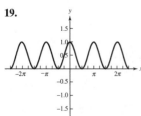

20.

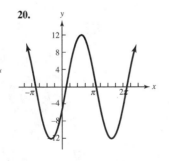

21. $\cos\theta = \frac{1}{10}\sqrt{10}; \sin\theta = \frac{3}{10}\sqrt{10}; \tan\theta = 3$ **22.** $2\sqrt{2}$
23. $|P_\alpha P_\beta| = \sqrt{(\cos\beta - \cos\alpha)^2 + (\sin\beta - \sin\alpha)^2}$
$= \sqrt{2 - 2(\cos\alpha\cos\beta + \sin\alpha\sin\beta)}$
24. $|AP_\theta| = \sqrt{2 - 2\cos\theta}$
25. Let P_α and P_β be defined as in Problem 23. Then,

$$|P_\alpha P_\beta| = \sqrt{2 - 2(\cos\alpha\cos\beta + \sin\alpha\sin\beta)}$$

The angle between rays through P_α and P_β is $\alpha - \beta$, so

$$|P_\alpha P_\beta| = \sqrt{2 - 2\cos(\alpha - \beta)} \quad \text{from Problem 22.}$$

Thus,

$$\sqrt{2 - 2\cos(\alpha - \beta)} = \sqrt{2 - 2(\cos\alpha\cos\beta + \sin\alpha\sin\beta)}$$
$$2 - 2\cos(\alpha - \beta) = 2 - 2(\cos\alpha\cos\beta + \sin\alpha\sin\beta)$$
$$\cos(\alpha - \beta) = \cos\alpha\cos\beta + \sin\alpha\sin\beta$$

26. $\cot\theta$ **27.** $-\frac{24}{25}$ **28.** $2\sin\left(\frac{h}{2}\right)\cos\left(\frac{2x + h}{2}\right)$
29. $\frac{1}{2}\sin 4\theta + \frac{1}{2}\sin 2\theta$ **30.** $13\sin(\theta + 5.9)$

Miscellaneous Problems, Chapters 1–4, page 217

1. $y = \sin x$ **3.** $y = \tan x$

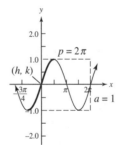

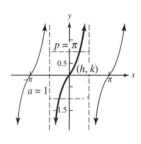

The dark portions are the principal values.

5. $y = \csc x$

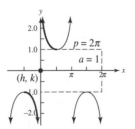

The dark portions are the principal values.

7. $y = \sin\frac{\pi}{20}x$

9.

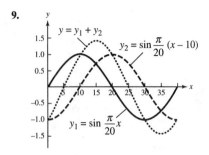

11. $-\frac{7}{25}$ **13. a.** $\frac{\sqrt{3}}{2}$ **b.** $\frac{\sqrt{2}}{2}$ **c.** $-\sqrt{3}$

15. a. $-\frac{\sqrt{3}}{3}$ **b.** $\frac{1}{2}$ **c.** $\frac{1}{4}$ **17. a.** $\frac{7}{25}$ **b.** $\frac{24}{25}$ **c.** $\frac{24}{7}$

19. a. $\frac{7}{24}$ **b.** $\frac{25}{7}$ **c.** $\frac{25}{24}$ **21.** π **23.** $0, 0.62$ **25.** $\frac{\pi}{6}$

27.
$$\frac{\sec u}{\tan^2 u} = \frac{\dfrac{1}{\cos u}}{\dfrac{\sin^2 u}{\cos^2 u}}$$
$$= \frac{1}{\cos u} \cdot \frac{\cos^2 u}{\sin^2 u}$$
$$= \frac{\cos u}{\sin^2 u}$$
$$= \frac{\cos u}{\sin u} \cdot \frac{1}{\sin u}$$
$$= \cos u \csc u$$

29.
$$\sec(-\theta) = \frac{1}{\cos(-\theta)}$$
$$= \frac{1}{\cos \theta}$$
$$= \sec \theta$$

31.
$$\frac{\sin \beta}{\cos \beta} + \cot \beta = \frac{\sin \beta}{\cos \beta} + \frac{\cos \beta}{\sin \beta}$$
$$= \frac{\sin^2 \beta + \cos^2 \beta}{\cos \beta \sin \beta}$$
$$= \frac{1}{\cos \beta \sin \beta}$$
$$= \sec \beta \csc \beta$$

33.
$$\sin \alpha = \sin\left(2 \cdot \tfrac{1}{2}\alpha\right)$$
$$= 2 \sin \tfrac{1}{2}\alpha \cos \tfrac{1}{2}\alpha$$

35.
$$\sin 2\theta \sec \theta = 2 \sin \theta \cos \theta \cdot \frac{1}{\cos \theta}$$
$$= 2 \sin \theta$$

37.
$$\frac{\csc \gamma + 1}{\csc \gamma \cos \gamma} = \frac{\csc \gamma}{\csc \gamma \cos \gamma} + \frac{1}{\csc \gamma \cos \gamma}$$
$$= \frac{1}{\cos \gamma} + \frac{1}{\csc \gamma \cos \gamma}$$
$$= \sec \gamma + \frac{\sin \gamma}{\cos \gamma}$$
$$= \sec \gamma + \tan \gamma$$

39.
$$\frac{\cot 3\theta - \sin 3\theta}{\sin 3\theta} = \frac{(\cot 3\theta - \sin 3\theta)\sin 3\theta}{\sin^2 3\theta}$$
$$= \frac{\dfrac{\cos 3\theta}{\sin 3\theta} \cdot \sin 3\theta - \sin^2 3\theta}{\sin^2 3\theta}$$
$$= \frac{\cos 3\theta - (1 - \cos^2 3\theta)}{\sin^2 3\theta}$$
$$= \frac{\cos^2 3\theta + \cos 3\theta - 1}{\sin^2 3\theta}$$

41. $(\sin \alpha \cos \alpha \cos \beta + \sin \beta \cos \beta \cos \alpha)\sec \alpha \sec \beta$
$$= \frac{\sin \alpha \cos \alpha \cos \beta + \sin \beta \cos \beta \cos \alpha}{\cos \alpha \cos \beta}$$
$$= \sin \alpha + \sin \beta$$

43. $\tan \alpha = \tan[(\alpha + \beta) - \beta]$
$$= \frac{\tan(\alpha + \beta) - \tan \beta}{1 + \tan(\alpha + \beta) \tan \beta}$$

45. $\dfrac{\sqrt{6} + \sqrt{2}}{4}$ **47.** $\sin \tfrac{1}{2}(270°) = \dfrac{\sqrt{2}}{2}$ **49.** $\tfrac{1}{6} \sec \theta$

51. $10 \sec \theta$ **53.** $8 \tan \theta$

55. $(\cos 30°, \sin 30°) = \left(\dfrac{\sqrt{3}}{2}, \dfrac{1}{2}\right)$

57.

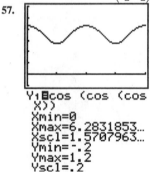

Y1■cos (cos (cos
X))
Xmin=0
Xmax=6.2831853...
Xscl=1.5707963...
Ymin=-.2
Ymax=1.2
Yscl=.2

Maximum values at
$x = 0, \pi, 2\pi$
and minimum values at
$x = \pi/2, 3\pi/2$

59. $\phi + \alpha + (180° - \beta) = 180°$ so $\phi = \beta - \alpha$
$$\tan \phi = \tan(\beta - \alpha)$$
$$= \frac{\tan \beta - \tan \alpha}{1 + \tan \beta \tan \alpha}$$
$$= \frac{m_2 - m_1}{1 + m_1 m_2}$$

Problem Set 5.1, page 228

1.
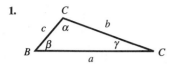

$A(c \cos \beta, c \sin \beta)$, $B(0, 0)$, and $C(a, 0)$

5. 54°   7. 49°   9. 22°   11. 13   13. 10   15. 17
17. $\alpha = 52.1°$; $\beta = 115.0°$; $\gamma = 12.9°$; $a = 14.2$; $b = 16.3$; $c = 4.00$
19. $\alpha = 125°$; $\beta = 41°$; $\gamma = 14°$; $a = 5.0$; $b = 4.0$; $c = 1.5$
21. ambiguous case; $\alpha = 90.8°$; $\alpha' = 25.0°$; $\beta = 57.1°$; $\beta' = 122.9°$; $\gamma = 32.1°$; $a = 98.2$; $a' = 41.6$; $b = 82.5$; $c = 52.2$
23. ambiguous case; $\alpha = 47.0°$; $\beta = 57.8°$; $\beta' = 122.2°$; $\gamma = 75.2°$; $\gamma' = 10.8°$; $a = 10.2$; $b = 11.8$; $c = 13.5$; $c' = 2.61$
25. No solution because the sum of the two smaller sides is not greater than the larger side.
27. $\alpha = 55°$; $\beta = 85°$; $\gamma = 41°$; $a = 4.3$; $b = 5.2$; $c = 3.4$
29. $\alpha = 25.8°$; $\beta = 139.4°$; $\gamma = 14.8°$; $a = 214$; $b = 320$; $c = 126$
31. $\alpha = 94.3°$; $\beta = 37.3°$; $\gamma = 48.4°$; $a = 140$; $b = 85.0$; $c = 105$
33. $\alpha = 61.0°$; $\beta = 29.0°$; $\gamma = 90.0°$; $a = 36.9$; $b = 20.5$; $c = 42.2$
35. The distance is 249 mi.   37. He is 5.39 mi from the target.
39. The cable is 42.8 ft.
41. The angle with the 8-ft side is 56°, and with the 10-ft side it is 41°.
43. $d = 2.25$; $q = 1.88$; $p = 2.13$; $\angle D = 68.1°$; $\angle P = 61.2°$; $\angle Q = 50.7°$
45. 28 in.   47. The perimeter is 4,605 ft.
49. It will take the boat 49 min to reach Avalon.
51. The original heading was 210° (S30°W), and the correct original heading was 225° (S45°W). The corrected heading after one hour is 244° (243.6374°).
53. The pilot must increase the airspeed to 182 mi/h.
55. Let B be in standard position. Then $C(a, 0)$ and $A(c \cos \beta, c \sin \beta)$. Find the distance between A and C:
$$b^2 = (c \cos \beta - a)^2 + (c \sin \beta - 0)^2$$
$$= c^2 \cos^2\beta - 2ac \cos \beta + a^2 + c^2 \sin^2\beta$$
$$= c^2(\cos^2\beta + \sin^2\beta) + a^2 - 2ac \cos \beta$$
$$= a^2 + c^2 - 2ac \cos \beta$$
57. From the law of cosines, $b \cos \gamma = \dfrac{a^2 + b^2 - c^2}{2a}$ and $c \cos \beta = \dfrac{a^2 + c^2 - b^2}{2a}$, so $b \cos \gamma + c \cos \beta =$
$$\frac{a^2 + b^2 - c^2}{2a} + \frac{a^2 + c^2 - b^2}{2a} = \frac{2a^2}{2a} = a. \text{ Thus,}$$
$a = b \cos \gamma + c \cos \beta$.

Problem Set 5.2, pages 238

5. one solution; 2 sig figs   7. two solutions; 2 sig figs
9. one solution; 3 sig figs   11. one solution; 2 sig figs
13. two solutions; 2 sig figs   15. no solution
17. cannot be solved since the sum of the given angles is not less than 180°
19. $\alpha = 25°$; $\beta = 110°$; $\gamma = 45°$; $a = 10$; $b = 23$; $c = 17$
21. $\alpha = 70°$; $\beta = 85°$; $\gamma = 25°$; $a = 85$; $b = 90$; $c = 38$
23. $\alpha = 120°$; $\beta = 53°$; $\gamma = 7°$; $a = 310$; $b = 280$; $c = 43$
25. $\alpha = 18.3°$; $\beta = 54.0°$; $\gamma = 107.7°$; $a = 107$; $b = 276$; $c = 325$
27. ambiguous case; $\alpha = 27.2°$; $\alpha' = 152.8°$; $\beta = 21.9°$; $\gamma = 130.9°$; $\gamma' = 5.3°$; $a = 10.8$; $b = 8.80$; $c = 17.8$; $c' = 2.20$
29. $\alpha = 98.0°$; $\beta = 28.7°$; $\gamma = 53.3°$; $a = 114$; $b = 55.0$; $c = 92.0$
31. $\alpha = 85.2°$; $\beta = 38.7°$; $\gamma = 56.1°$; $a = 148$; $b = 92.7$; $c = 123$
33. cannot be solved unless at least one side is given
35. cannot be solved since the sum of the two smaller sides is not larger than the third side
37. ambiguous case; $\alpha = 25°$; $\beta = 101°$; $\beta' = 29°$; $\gamma = 54°$; $\gamma' = 126°$; $a = 1.1$; $b = 2.6$; $b' = 1.3$; $c = 2.1$
39. The height of the building is about 260 ft.
41. The distance to the target is 291 km.
43. Luke's craft is 4,318 m from the first observation point and 6,815 m from the second.
45. The angle of elevation is 19.0°.
47. The ski lift is about 4,253 ft long.
49. *Sputnik* is 220 mi above point B.
51. Drop a perpendicular from C and let h be the length. From the definition of the trigonometric functions,
$$\sin \alpha = \frac{h}{b} \text{ and } \sin \beta = \frac{h}{a}$$
Solve for h and equate: $b \sin \alpha = a \sin \beta$
Divide by ab: $\dfrac{\sin \alpha}{a} = \dfrac{\sin \beta}{b}$

Problem Set 5.3, page 247

1. $K = \frac{1}{2}bh$ 3. $K = \frac{1}{2}bc \sin \alpha$; $K = \frac{1}{2}ac \sin \beta$; $K = \frac{1}{2}ab \sin \gamma$
5. Use the law of cosines to find an included angle and then use the formulas for two angles and an included side.
7. $K = \frac{1}{2}\theta r^2$ 9. 19.6 sq units   11. 13.6 sq units
13. 44 sq units   15. 37 sq units   17. 83 sq units
19. 680 sq units   21. 560 sq units   23. $\frac{25}{4}\sqrt{3}$ sq units
25. 6.4 sq units   27. 35 sq units   29. 54.0
31. ambiguous case; 58.2 sq units and 11.3 sq units
33. ambiguous case; 2,150 sq units and 911 sq units
35. 180 sq units   37. no triangle formed
39. 20,100 sq units   41. 72 in.²   43. 14.1 m²
45. 26 in.² $\left(\frac{25}{3}\pi\right)$ 47. Area is about 1.30 in.².
49. Area is 2.38 in.².

51. **a.** $K = \frac{1}{2} \sin \theta \cos \theta$ or $K = \frac{1}{4} \sin 2\theta$

b. $K = \frac{1}{2} \tan \theta$ **c.** $K = \frac{1}{2} \theta$

53. The apothem is 0.81 in.

55. The radius is about 7.82 cm.

57. Total $= 220,447 \text{ ft}^2 \approx 5.06$ acres. The cost is about $225. (Actually $227.73, if you round according to the rules we have set up for this book; however, in practice, this treatment would certainly be figured at 5 acres.)

59. The volume is about $21,000 \text{ ft}^3$ or 760 yd^3.

61. **a.** $V = h(\text{AREA OF BASE})$

$= h[(\text{AREA OF SECTOR}) - (\text{AREA OF TRIANGLE})]$

$= h\left[\frac{1}{2}\theta r^2 - r^2 \sin\frac{\theta}{2}\cos\frac{\theta}{2}\right]$

$= hr^2\left(\frac{\theta}{2} - \sin\frac{\theta}{2}\cos\frac{\theta}{2}\right)$

$b = r \sin\frac{\theta}{2}$

$h = r \cos\frac{\theta}{2}$

$\text{AREA OF TRIANGLE} = 2\left(\frac{1}{2}bh\right)$
$= r^2 \sin\frac{\theta}{2}\cos\frac{\theta}{2}$

b. 117 in.^3

Problem Set 5.4, page 259

5.

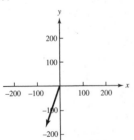

7.

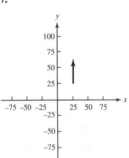

9.

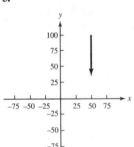

11.

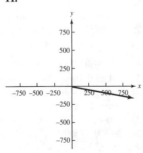

13.

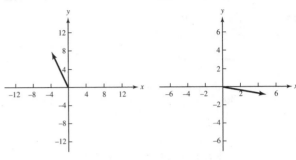

15.

17. a vector with a direction S37°E and a magnitude of 5.0

19. a vector with heading 110° and a magnitude of 220

21. a vector with heading 220° and a magnitude of 162

23. The horizontal component has a magnitude of 10 and a vertical component of 17.

25. The horizontal component has a magnitude of 110 and a vertical component of 57.

27. The horizontal component has a magnitude of 337 (in the negative direction) and the vertical component has a magnitude of 402.

29. The plane is traveling at 240 m/h with a course of 260°.

31. The horizontal component is 2,020 ft/s and the vertical component is 648 ft/s.

33. In two hours it has traveled 363 mi north.

35. The angles are 34° and 146°.

37. Ambiguous case; it will be there in 24 min or 1 hr 59 min.

39. The resultant force has magnitude 288 lb, and the log moves in a path that makes an angle of 23.5° with the 200-lb force and an angle of 11.5° with the 100-lb force.

41. The direction of the boat is N51.7°W, and it is traveling at 33.7 knots.

43. The planes are 374 mi apart.

45. The boats are 67.2 mi apart.

47. The tension on $\overline{AC}$ is 326 lb, and on $\overline{BC}$ the tension is 321 lb.

49. A 52-lb force would keep the barrel from sliding down the incline. The barrel is pushing against the plane with a force of 240 lb.

51. The angle of inclination is 37.4°.

53. The heaviest piece of cargo is 16,800 lb.

55. The tension on the cable is 69.8 lb. The horizontal force is 44.9 lb.

57. $\theta \approx 80.4°$

59. The track should be banked at an angle of 77°.

Problem Set 5.5, page 272

9. $\mathbf{v} = 4\sqrt{3}\mathbf{i} + 4\mathbf{j}$ **11.** $\mathbf{v} = 5\mathbf{i} - 5\mathbf{j}$

13. $\mathbf{v} = -8.6379\mathbf{i} - 5.9367\mathbf{j}$ **15.** $\mathbf{v} = 5\mathbf{i} + 7\mathbf{j}$

17. $\mathbf{v} = -13\mathbf{i}$ **19.** $\mathbf{v} = 4\mathbf{i} + 1,250\mathbf{j}$ **21.** 13

23. $\sqrt{34} \approx 5.8310$ **25.** $\sqrt{89} \approx 9.4340$
27. $\sqrt{122} \approx 11.0453$ **29.** not orthogonal   **31.** orthogonal
33. not orthogonal   **35.** -112; 10; 13; $-\frac{56}{65}$ **37.** 0; 8; 16; 0
39. -39; $3\sqrt{10}$; $\sqrt{29}$; $-\frac{13\sqrt{290}}{290}$ **41.** 1; 1; 1; 1 **43.** 0; 1; 1; 0
45. $75°$ **47.** $45°$ **49.** $90°$ **51.** $-\frac{8}{5}$ **53.** $\pm 3\sqrt{2}$
55. 31 ft-lb   **57.** $32 - 6\sqrt{2}$ ft-lb
59. $\mathbf{F}_1 = |\mathbf{F}_1|(\cos 31°\mathbf{i} + \sin 31°\mathbf{j})$, $\mathbf{F}_2 = |\mathbf{F}_2|(\cos 37°\mathbf{i} + \sin 37°\mathbf{j})$;
$\mathbf{W} = (|\mathbf{F}_1| + |\mathbf{F}_2|)[(\cos 31° + \cos 37°)\mathbf{i} + (\sin 31° + \sin 37°)\mathbf{j}]$
61. 325 ft-lb   **63.** 2,400 ft-lb   **65.** 2,600 ft-lb
67. 43,000 ft-lb
69. Let $\mathbf{v} = a\mathbf{i} + b\mathbf{j}$, $\mathbf{w} = c\mathbf{i} + d\mathbf{j}$; $(\mathbf{v} + \mathbf{w}) = (a + c)\mathbf{i} + (b + d)\mathbf{j}$;

$$
\begin{aligned}
(\mathbf{v} + \mathbf{w}) \cdot (\mathbf{v} + \mathbf{w}) &= (a + c)^2 + (b + d)^2 \\
&= a^2 + 2ac + c^2 + b^2 + 2bd + d^2 \\
&= (a^2 + b^2) + (c^2 + d^2) + 2(ac + bd) \\
&= |\mathbf{v}|^2 + |\mathbf{w}|^2 + 2(\mathbf{v} \cdot \mathbf{w})
\end{aligned}
$$

Chapter 5 Sample Test, page 275

1. a. $b^2 + c^2 - 2bc \cos \alpha$ **b.** $\dfrac{a^2 + c^2 - b^2}{2ac}$

c. $\dfrac{\sin \alpha}{a} = \dfrac{\sin \beta}{b} = \dfrac{\sin \gamma}{c}$

2. $\alpha = 29°$; $\beta = 109°$; $\gamma = 42°$; $a = 14$; $b = 27$; $c = 19$
3. ambiguous case; $\alpha = 57.0°$; $\beta = 74.5°$; $\beta' = 105.5°$;
$\gamma = 48.5°$; $\gamma' = 17.5°$; $a = 23.5$; $b = 27.0$; $c = 21.0$;
$c' = 8.42$
4. $\alpha = 18.3°$; $\beta = 112.4°$; $\gamma = 49.3°$; $a = 92.6$; $b = 273$; $c = 224$
5. 320   **6.** 51.7   **7.** 1,900
8. a. The magnitude is 7.1 and the direction is N47°W.
   b. The horizontal component is 2.8 and the vertical component is 3.5.
9. a. $\mathbf{v} = \mathbf{i} + \mathbf{j}$ **b.** $-4.4\mathbf{i} + 8.2\mathbf{j}$ **c.** $-6\mathbf{i} - 5\mathbf{j}$
10. a. $|\mathbf{v}| = \sqrt{13}$; $|\mathbf{w}| = \sqrt{2}$ **b.** 5   **c.** $11°$ **d.** $a = -\frac{3}{2}$
11. About 113 m of fencing is needed.
12. The area is 130,900 ft².
13. (ambiguous case) at 12:42 P.M. and at 4:40 P.M.
14. The force necessary to keep the boxcar from sliding down the hill is 0.2 ton (or about 420 lb).
15. It is about 430 ft.

Miscellaneous Problems, Chapters 1–5, page 276

1. a. $A(\cos \theta, 0)$; $P(\cos \theta, \sin \theta)$ **b.** $|\overline{OA}| = \cos \theta$
   c. $|\overline{PA}| = \sin \theta$ **d.** $|\overline{OB}| = \cos^2\theta$ **e.** $K = \frac{1}{2} \cos^3\theta \sin \theta$
3. a. $A(0, \sin \theta)$; $P(\cos \theta, \sin \theta)$ **b.** $|\overline{OA}| = \sin \theta$
   c. $|\overline{PA}| = \cos \theta$ **d.** $|\overline{OB}| = \sin^2\theta$ **e.** $K = \frac{1}{2} \sin^3\theta \cos \theta$
5. a. $\mathbb{R}$ **b.** $-1 \le y \le 1$ **c.** 1; 2π **d.** 1 **e.** $n\pi$
7. 3.81, 5.62

9. identity; $\tan \theta + \cot \theta = \dfrac{\sin \theta}{\cos \theta} + \dfrac{\cos \theta}{\sin \theta}$

$$
\begin{aligned}
&= \frac{\sin^2\theta + \cos^2\theta}{\cos \theta \sin \theta} \\
&= \frac{1}{\cos \theta \sin \theta} \\
&= \sec \theta \csc \theta
\end{aligned}
$$

11. 0.82, 2.32   **13.** 2.24, 4.05

15. $\dfrac{\tan(\alpha + \beta) - \tan \beta}{1 + \tan(\alpha + \beta)\tan \beta} = \tan(\alpha + \beta - \beta) = \tan \alpha$

17. 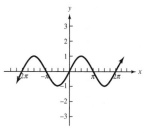   **19.**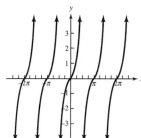

21.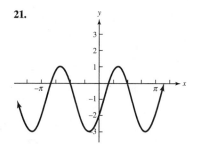

23. $\alpha = 23°$; $\beta = 58°$; $\gamma = 99°$; $a = 24$; $b = 52$; $c = 61$
25. $\alpha = 30°$; $\beta = 50°$; $\gamma = 100°$; $a = 30$; $b = 46$; $c = 59$
27. $\alpha = 50°$; $\beta = 70°$; $\gamma = 60°$; $a = 33$; $b = 40$; $c = 37$
29. $\alpha = 112°$; $\beta = 30°$; $\gamma = 38°$; $a = 15$; $b = 8.0$; $c = 10$
31. $\alpha = 82°$; $\beta = 54°$; $\gamma = 44°$; $a = 17$; $b = 14$; $c = 12$
33. $\alpha = 54°$; $\beta = 113°$; $\gamma = 13°$; $a = 7.0$; $b = 8.0$; $c = 2.0$
35. no solution
37. ambiguous case; $\alpha = 35.5°$; $\beta = 43.5°$; $\beta' = 136.5°$;
$\gamma = 101.0°$; $\gamma' = 8.0°$; $a = 14.5$; $b = 17.2$; $c = 24.5$; $c' = 3.49$
39. $\alpha = 21.3°$; $\beta = 108.7°$; $\gamma = 50.0°$; $a = 121$; $b = 315$; $c = 255$
41. $\alpha = 80°$; $\beta = 54°$; $\gamma = 46°$; $a = 11$; $b = 9.0$; $c = 8.0$
43. $\alpha = 20.7°$; $\beta = 67.8°$; $\gamma = 91.6°$; $a = 12.4$; $b = 32.5$; $c = 35.1$
45. no solution
47. $\alpha = 147.0°$; $\beta = 15.0°$; $\gamma = 18.0°$; $a = 49.5$; $b = 23.5$; $c = 28.1$
49. ambiguous case; $\alpha = 118.6°$; $\alpha' = 15.2°$; $\beta = 38.3°$;
$\beta' = 141.7°$; $\gamma = 23.1°$; $a = 117$; $a' = 34.9$; $b = 82.5$; $c = 52.2$
51. ambiguous case; $\alpha = 47.0°$; $\beta = 57.8°$; $\beta' = 122.2°$;
$\gamma = 75.2°$; $\gamma' = 10.8°$; $a = 10.2$; $b = 11.8$; $c = 13.5$; $c' = 2.61$

53. The tunnel is 979 ft.   **55. a.** $\beta \approx 12.1°$ **b.** $\beta \approx 20.6°$

57. The pilot will be 200 miles from Santa Rosa at 12:29 P.M. and again at 12:45 P.M.

59. It is 7,200 ft from the first tower and 6,220 ft from the second tower.

Problem Set 6.1, page 288

1. The number i is a number with the properties $i^2 = -1$ and $\sqrt{-a} = i\sqrt{a}$ if $a > 0$.

3. A complex number is simplified when it is written in the form $a + bi$ where a and b are simplified real numbers.

5. Every real number is a complex number, but not every complex number is a real number.

7. false; $i = \sqrt{-1}$

9. false; $(2 + 2i)(2 - 2i) = 4 - 4i^2 = 8$, not 0

11. a. $7i$ **b.** $2\sqrt{2}i$ **13. a.** $-\frac{36}{5}i$ **b.** $-\frac{3}{2}i$

15. a. $-2 + \sqrt{2} + \sqrt{2}i$ **b.** $-2 - \sqrt{3} + \sqrt{3}i$

17. a. $2 + 5i$ **b.** $-3 + 2i$ **19. a.** 68   **b.** 25

21. a. $5i$ **b.** $-2 - 8i$ **23. a.** $-i$ **b.** $-i$

25. a. $-i$ **b.** -1 **27. a.** $-8 - 6i$ **b.** $32 - 24i$

29. a. $-7 + 6\sqrt{2}i$ **b.** $-4 - 6\sqrt{5}i$ **31. a.** 12   **b.** 7

33. $1 - i$ **35.** $-1 + i$ **37.** $-\frac{3}{2} + \frac{3}{2}i$ **39.** $-2i$

41. $-\frac{1}{5} - \frac{3}{5}i$ **43.** $\frac{1}{5} - \frac{2}{5}i$ **45.** $-\frac{45}{53} + \frac{28}{53}i$ **47.** $-i$

49. $1 - 3i$ **51.** 0   **53.** $\pm\sqrt{2}i$ **55.** 0   **57.** $\frac{3}{2} \pm \frac{\sqrt{23}}{2}i$

59. 0   **61.** $1, -\frac{1}{2} \pm \frac{\sqrt{3}}{2}i$ **63.** $4.2429 + 4.2426i$

65.
$z_1 + z_2 = (a_1 + b_1i) + (a_2 + b_2i)$	*Substitution*
$= (a_1 + a_2) + (b_1 + b_2)i$	*Definition of addition*
$= (a_2 + a_1) + (b_2 + b_1)i$	*Commutative law for real numbers*
$= (a_2 + b_2i) + (a_1 + b_1i)$	*Definition of addition*
$= z_2 + z_1$	*Substitution*
$z_1z_2 = (a_1 + b_1i)(a_2 + b_2i)$	*Substitution*
$= (a_1a_2 - b_1b_2) + (a_1b_2 + b_1a_2)i$	*Definition of multiplication*
$= (a_2a_1 - b_2b_1) + (b_2a_1 + a_2b_1)i$	*Commutative law for real numbers*
$= (a_2 + b_2i)(a_1 + b_1i)$	*Definition of multiplication*
$= z_2z_1$	*Substitution*

Problem Set 6.2, page 298

9. false; $\theta = 0°$ or $180°$

11. false; it is plotted on the positive imaginary axis

13. false; 4 cis 250° is in Quadrant III

15.

a. $\sqrt{13}$
b. $2\sqrt{5}$
c. $\sqrt{61}$

17.

a. $\sqrt{29}$
b. $\sqrt{17}$
c. $\sqrt{2}$

19. a. 2 cis 30°   **b.** 2 cis 300°   **c.** 2 cis 240°

21. a. cis 90°   **b.** 2 cis 270°   **c.** 3 cis 90°

23. a. 4 cis 100°   **b.** 5 cis 246°   **25. a.** $\frac{3}{2} + \frac{3\sqrt{3}}{2}i$ **b.** $-5i$

27. a. $\frac{5}{4} - \frac{5\sqrt{3}}{4}i$ **b.** $-\frac{21\sqrt{2}}{10} + \frac{21\sqrt{2}}{10}i$ **29.** 15 cis 140°

31. $\frac{5}{2}$ cis 267°   **33.** 3 cis 130°   **35.** 81 cis 240°

37. 64 cis 120°   **39.** $-128 + 128i\sqrt{3}$; (256 cis 120°)

41. 3 cis 22°, 3 cis 112°, 3 cis 202°, 3 cis 292°

43. 3 cis 0°, 3 cis 120°, 3 cis 240°

45. $\sqrt[8]{2}$ cis 11.25°, $\sqrt[8]{2}$ cis 101.25°, $\sqrt[8]{2}$ cis 191.25°, $\sqrt[8]{2}$ cis 281.25°

47. 2 cis 40°, 2 cis 112°, 3 cis 184°, 2 cis 256°, 2 cis 328°

49. $1 + \sqrt{3}i, -2, 1 - \sqrt{3}i$

51. $-0.6840 + 1.8794i, -1.2856 - 1.5321i, 1.9696 - 0.3473i$

53. $1.9696 + 0.3473i, -0.3473 + 1.9696i, -1.9696 - 0.3473i, 0.3473 - 1.9696i$

55. $1, 0.3090 \pm 0.9511i, -0.8090 \pm 0.5878i$

57. $\pm 1, \frac{1}{2} \pm \frac{\sqrt{3}}{2}i, -\frac{1}{2} \pm \frac{\sqrt{3}}{2}i$

59. $\pm 2, \pm 2i$ **61.** $-1, \frac{1}{2} \pm \frac{\sqrt{3}}{2}i, -\frac{1}{2} \pm \frac{\sqrt{3}}{2}i$

63.

a. $e^{i\pi/4}$
c. $-e^{-i\pi}$
b. $2e^{-i\pi/6}$

65. $I \approx 3.31 + 0.39i$ **67.** 51 cis 45°

Problem Set 6.3, page 308

5. polar: $\left(4, \frac{\pi}{4}\right), \left(-4, \frac{5\pi}{4}\right)$; rectangular: $(2\sqrt{2}, 2\sqrt{2})$

7. polar: $\left(5, \frac{2\pi}{3}\right), \left(-5, \frac{5\pi}{3}\right)$; rectangular: $\left(-\frac{5}{2}, \frac{5\sqrt{3}}{2}\right)$

9. polar: $\left(\frac{3}{2}, \frac{7\pi}{6}\right), \left(-\frac{3}{2}, \frac{\pi}{6}\right)$; rectangular: $\left(\frac{-3\sqrt{3}}{4}, -\frac{3}{4}\right)$

11. polar: $(1, \pi), (-1, 0)$; rectangular: $(-1, 0)$

13. polar: $(0, 2\pi - 3)$, $(0, \pi - 3)$; rectangular: $(0, 0)$

15. $\left(2, \frac{2\pi}{3}\right)$ and $\left(-2, \frac{5\pi}{3}\right)$ **17.** $\left(2\sqrt{2}, \frac{5\pi}{4}\right)$ and $\left(-2\sqrt{2}, \frac{\pi}{4}\right)$

19. $(7.62, 1.17)$ and $(-7.62, 4.31)$ **21.** $\left(2, \frac{11\pi}{6}\right)$, $\left(-2, \frac{5\pi}{6}\right)$

23. $x^2 + y^2 - 4y = 0$ **25.** $x = 1$ **27.** $x^2 + 2y^2 = 2$

29. $x^2 + y^2 = \sqrt{x^2 + y^2} - y$

31.

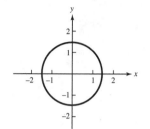

33.

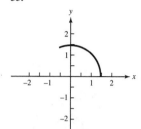

35.

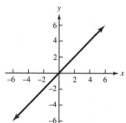

37.

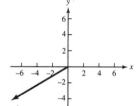

39. yes   **41.** no   **43.** yes   **45.** yes   **47.** no

49. no   **51.** no

61. **a.** United States   **b.** India   **c.** Greenland

d. Canada

63.

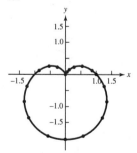

65. $a^2 = r^2 + R^2 - 2rR \cos(\theta - \alpha)$

Problem Set 6.4, page 321

5. **a.** four-leaved rose   **b.** lemniscate

c. circle (r is a constant)   **d.** sixteen-leaved rose

e. none (it is a spiral)   **f.** lemniscate

g. three-leaved rose   **h.** cardioid

7.

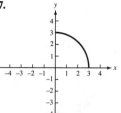

9.

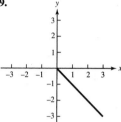

11.

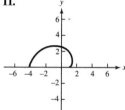

13.

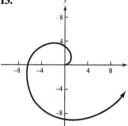

15.

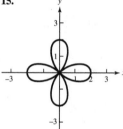

17.

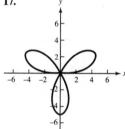

19.

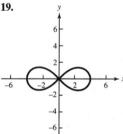

21.

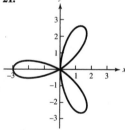

23.

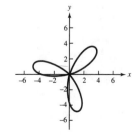

25. $r = \sin^2\left(\theta + \frac{\pi}{3}\right)$

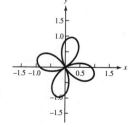

27. $r^2 = 16 \cos 2\left(\theta - \frac{\pi}{3}\right)$

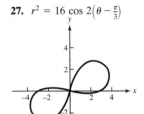

29.

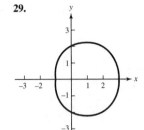

47.

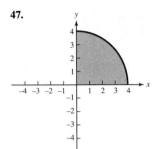

49.

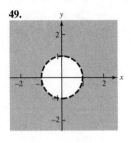

31.

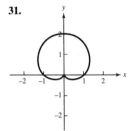

33.

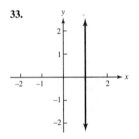

51.

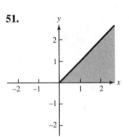

53.

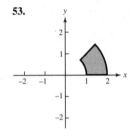

35.

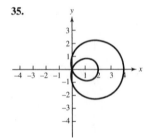

37.

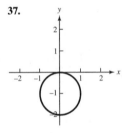

55.

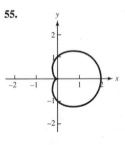

57.

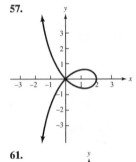

39.

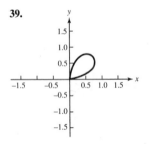

41.

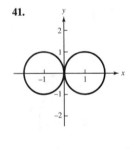

59.

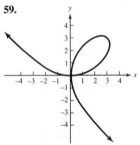

61.

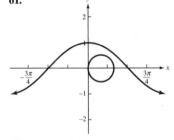

43.

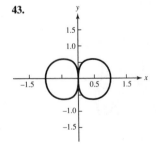

45.

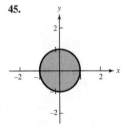

63.

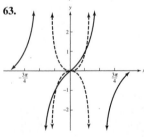

65.

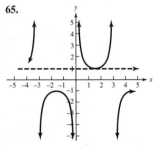

Chapter 6 Sample Test, page 324

1. a. $2 + 5i$ **b.** $-1 - 3i$ **c.** 8 **d.** $-21 - 20i$
 e. $-\frac{1}{2} + \frac{5}{2}i$
2. a. cis $154°$ **b.** $-128 - 128\sqrt{3}i$
3. a. $7\sqrt{2}$ cis $315°$ **b.** 3 cis $270°$ **c.** $-\sqrt{3} + i$
 d. $2\sqrt{2} - 2\sqrt{2}i$
4. a. $\sqrt{7}$ cis $165° \approx -2.5556 + 0.6848i$,
 $\sqrt{7}$ cis $345° \approx 2.5556 - 0.6848i$
 b. cis $0° = 1$; cis $90° = i$; cis $180° = -1$;
 cis $270° = -i$; also show these roots graphically.
5. $3, -\frac{3}{2} + \frac{3\sqrt{3}}{2}i, -\frac{3}{2} - \frac{3\sqrt{3}}{2}i$
6. a. $\left(3, \frac{4\pi}{3}\right), \left(-3, \frac{\pi}{3}\right)$ **b.** $\left(5, \frac{9\pi}{10}\right), \left(-5, \frac{19\pi}{10}\right)$
 c. $\left(6, \frac{2\pi}{3}\right), \left(-6, \frac{5\pi}{3}\right)$ **d.** $(5.3852, 1.1903), (-5.3852, 4.3319)$
7. a. no **b.** yes **c.** yes **d.** no
8. a. lemniscate **b.** cardioid **c.** five-leaved rose
 d. circle
9. a.
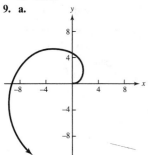
 b.
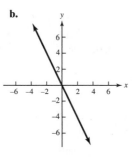
 c.
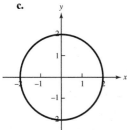
 d.
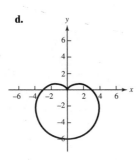
10. a.
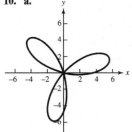
 b.
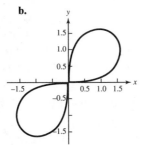

Miscellaneous Problems, Chapter 1–6, page 325

1. a.
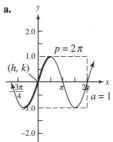
 b.
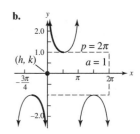
3. $y = \cos \theta$
5.
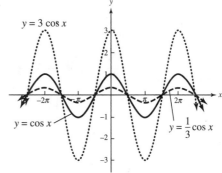
$y = \cos x$ is solid curve; answers vary; $y = a \cos x$ where a
is the amplitude.
7. $y = 2 \sin \pi x$ **9.** $y = \frac{1}{4} \cos \frac{\pi}{3}x$
11. $1 + \sin^2\theta = 1 + (1 - \cos^2\theta)$
 $= 2 - \cos^2\theta$
13. $\dfrac{(\sec^2\theta + \tan^2\theta)^2}{\sec^4\theta - \tan^4\theta} = \dfrac{(\sec^2\theta + \tan^2\theta)^2}{(\sec^2\theta - \tan^2\theta)(\sec^2\theta + \tan^2\theta)}$
 $= \dfrac{\sec^2\theta + \tan^2\theta}{1}$
 $= 1 + \tan^2\theta + \tan^2\theta$
 $= 1 + 2\tan^2\theta$
15. $\sec(-\theta) = \dfrac{1}{\cos(-\theta)}$
 $= \dfrac{1}{\cos\theta}$
 $= \sec\theta$
17. 1.57 **19.** 0.21 **21.** 7 cis $150°$ **23.** 16 cis $90°$
25. $\sqrt{13}$ cis $124°$ **27.** $-\frac{9}{2}\sqrt{2} + \frac{9}{2}\sqrt{2}i$ **29.** $-\frac{3}{2} - \frac{3\sqrt{3}}{2}i$
31. $4.5315 + 2.1131i$ **33.** -64 **35.** cis $228°$
37. cis $30°$, cis $150°$, cis $270°$
39. cis $0°$, cis $36°$, cis $72°$, cis $108°$, cis $144°$, cis $180°$, cis $216°$,
 cis $252°$, cis $288°$, cis $324°$
41. $2\sqrt[3]{4}$ cis $40°$, $2\sqrt[3]{4}$ cis $160°$, $2\sqrt[3]{4}$ cis $280°$

43.

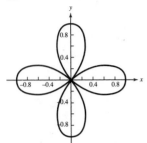

45.

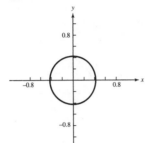

59.

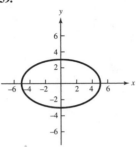

61.

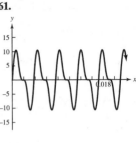

47.

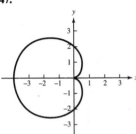

49.

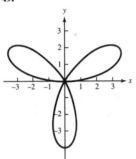

Cumulative Review for Chapters 4–6, page 328

23. $K = \frac{1}{2}bc \sin \alpha = \frac{1}{2}ac \sin \beta = \frac{1}{2}ab \sin \gamma$

25. $K = \sqrt{s(s-a)(s-b)(s-c)}$ where $s = \frac{1}{2}(a+b+c)$

27. The volume of a circular cone with base radius r and height h is $V = \frac{1}{3}r^2 h$. If the angle of elevation from the base to the vertex is α, then $V = \frac{1}{3}\pi r^3 \tan \alpha$.

29. $\mathbf{v} \cdot \mathbf{w} = 0$ **31.** $x = r \cos \theta, y = r \sin \theta$

33. $\theta = $ constant   **35.** $r = b \pm a \cos \theta$ or $r = b \pm a \sin \theta$

37. $r^2 = a^2 \cos 2\theta$ or $r^2 = a^2 \sin 2\theta$ **39.** A   **41.** E

43. B   **45.** B   **47.** A and C   **49.** A   **51.** B

53. C   **55.** B   **57.** C

59. a. rose curve   **b.** lemniscate   **c.** rose curve
     d. cardioid

61. a. circle   **b.** limaçon   **c.** cardioid   **d.** line

63. $\frac{\pi}{3}, \frac{2\pi}{3}, \frac{4\pi}{3}, \frac{5\pi}{3}, 0, \pi$

65. $\dfrac{\sin 3\theta + \cos 3\theta + 1}{\cos 3\theta} = \dfrac{\sin 3\theta}{\cos 3\theta} + \dfrac{\cos 3\theta}{\cos 3\theta} + \dfrac{1}{\cos 3\theta}$

$$= \tan 3\theta + 1 + \sec 3\theta$$

67. law of cosines; $a = 38$; $b = 42$, $c = 9.7$, $\alpha = 59°$, $\beta = 108°$, $\gamma = 13°$

69. law of sines; $a = 42$, $b = 52$, $c = 50$, $\alpha = 49°$, $\beta = 68°$, $\gamma = 63°$

71. no solution; the angle must be less than $180°$

73. right triangle (or law of cosines); $a = 3.0$, $b = 6.0$, $c = 5.2$, $\alpha = 30°$, $\beta = 90°$, $\gamma = 60°$

75. law of cosines; ambiguous case; $a = 55$, $b = 61$, $c = 53$, $c' = 13$, $\alpha = 57°$, $\beta = 68°$, $\beta' = 112°$, $\gamma = 55°$, $\gamma' = 11°$

77. polar form: $\left(-5, \frac{11\pi}{6}\right)$, $\left(5, \frac{5\pi}{6}\right)$; rectancular form $\left(-\frac{5\sqrt{3}}{2}, \frac{5}{2}\right)$

79. polar form: $(-3, 3.1416)$, $(3, 0)$; rectangular form: $(3, 0)$

81. $2\sqrt{2}$ cis $135°$, $2\sqrt{2}$ cis $315°$

51.

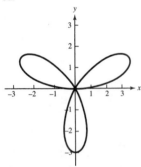

53.

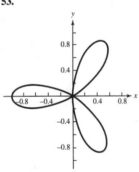

55.

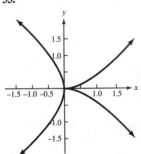

57.

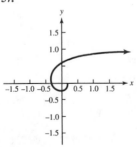

83.

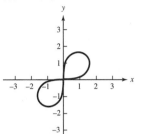

85.

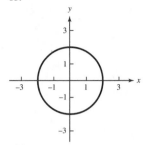

87. The height of the UFO is 0.3503 mi or 1,850 ft; the distance from the first city is 1.926 mi or 10,170 ft and from the second city it is 0.5363 mi or 2,832 ft.

89. The jetty is 5,028 ft.

Problem Set 7.1, page 342

1. An exponential function is a function of the form $y = b^x$ where $b \neq 1$, $b > 0$.

3. Suppose b is a real number greater than 1. Then, for any real number x, there is a unique real number b^x. Moreover, if h and k are any two rational numbers such that $h < x < k$, then $b^h < b^x < b^k$.

5. a. 27 **b.** 4 **c.** 16   **7. a.** $\frac{1}{8}$ **b.** $\frac{1}{27}$ **c.** $\frac{1}{16}$

9. a. $\frac{1}{25}$ **b.** $\frac{1}{81}$ **c.** $\frac{1}{16}$ **11. a.** 9 **b.** 729 **c.** 125

13. a. 20.08553692 **b.** 0.1353352832 **c.** 1.127496852
    d. 1.083287068

15. a. 2.716923932 **b.** 2.718268237

17. a.

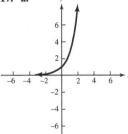

b.

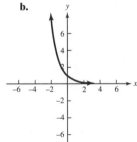

19. a.

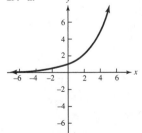

b.

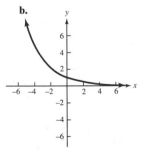

21. a.

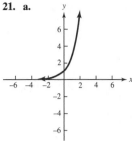

b.

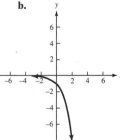

23.

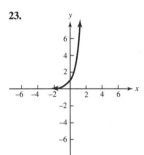

25.

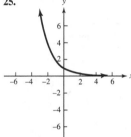

27.

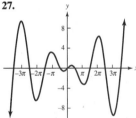

29.

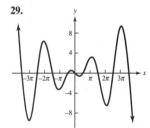

31.

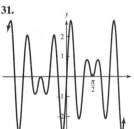

33.

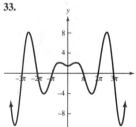

35.

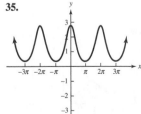

37.

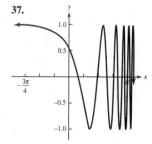

39.

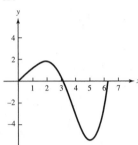

41. a. simple harmonic motion

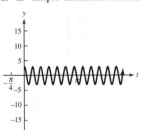

51.

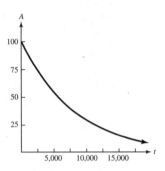

b. resonance

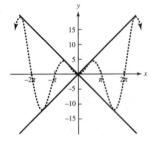

c. damped harmonic motion

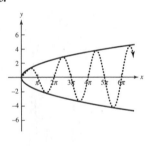

53. \$842.68   **55.** \$1.637.95

57.

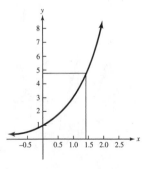

43.

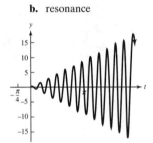

45.

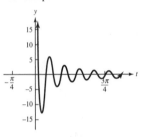

59. a. $\pi^{\sqrt{3}}$ is larger   **b.** $\pi^{\sqrt{5}}$ is larger   **c.** $(\sqrt{6})^{\pi}$ is larger

d. The following graph is a first approximation. Notice that the cursor is on the known value for N where the two expressions are the same size. We now zoom for the other value:

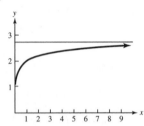

47.

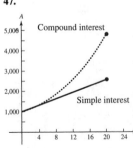

49. $3^{\sqrt{2}} \approx 4.7$

We can approximate the other value; namely, $N \approx 5.7$. This is why for $N = 3$ and 5, $\pi^{\sqrt{N}}$ is larger, but for $N = 6$, $(\sqrt{N})^{\pi}$ is larger. We would expect this to occur until $N > \pi^2$, in which case, the former expression is again larger.

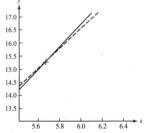

61. a.

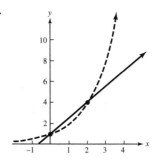

b. The secant line passing through $(0, 1)$ and $(2, 4)$ is shown; its slope is 1.5.

c. The secant line is shown here:

$$F(X) = 2^X$$

WANT MORE? **Y**	

$(X = 0)$

ΔX	$\dfrac{F(X+\Delta X) - F(X)}{\Delta X}$
.02815676	.69996
.01407838	.69654
.00703919	.69484
.00351959	.69399
.00175980	.69357
.00087990	.69336
.00043995	.69325
.00021997	.69320
.00010999	.69317
.00005499	.69316
.00002750	.69315
.00001375	.69315

d. The slope of the tangent line appears to be about 0.69315.

Problem Set 7.2, page 355

13. a. 1   **b.** 2   **c.** −3   **15. a.** 1   **b.** 3   **c.** −6
17. a. $x = \log_4 8$ **b.** $x = \log_5 10$ **c.** $x = \log_6 4.5$
19. a. $x = \ln 6$ **b.** $x = \ln 1.8$ **c.** $x = \ln 34.2$
21. a. $3 = \log 1{,}000$ **b.** $2 = \log_9 81$ **c.** $-1 = \ln \dfrac{1}{e}$
23. a. 0.63   **b.** 2.00   **c.** 3.00
25. a. 4.85   **b.** 2.00   **c.** 0.00
27. a. 0.82   **b.** 2.82   **c.** 1.99
29. a. 4.02   **b.** 111.79   **c.** 6.17

31. a. $\log_2 0.0056 = \dfrac{\ln 0.056}{\ln 2} \approx -7.480357457$

   b. $\log_{8.3} 105 = \dfrac{\ln 105}{\ln 8.3} \approx 2.199148599$

   c. $\log e^2 = \dfrac{\ln e^2}{\ln 10} \approx 0.8685889638$

   d. $\log e^8 = \dfrac{\ln e^8}{\ln 10} \approx 3.474355855$

33. a. 0.6666666667   **b.** 0.66666666667
35. a. 0.11111111111   **b.** 0.0833333333
37. a. 0.7796488803   **b.** 0.1679530909
39. a. 2.129072048   **b.** 1.855903615
41. a. −1.292029674   **b.** −2.481262426   **43.** 2
45. 2.305397424   **47.** 12.63153513
49. 0.53 (Note: 3.18 is not in the domain)   **51.** 3.00
53. 1.50   **55.** The plane is flying at about 9,190 ft.
57.

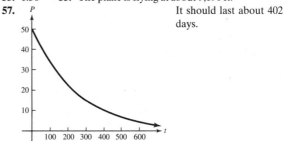

It should last about 402 days.

59. The cumulative number of deaths is 1,410,000.
61. The elapsed time is 7 yr.
63. $T = \dfrac{-E}{R \ln \dfrac{1}{\eta A}}$ **65.** $r = -\dfrac{1}{t} \ln \dfrac{I}{I_0}$ **67.** $h = \dfrac{t}{\ln \dfrac{A}{A_0}}$

Problem Set 7.3, page 369

1. a. x **b.** x **3. a.** 4.2   **b.** 3   **c.** x
5. Type I: The unknown is the logarithm, for example, $\log_2 \sqrt{3} = x$. Type II: The unknown is the base, for example, $\log_x 6 = 2$. Type III: The logarithm of an unknown is equal to a number, for example, $\ln x = 5$. Type IV: The logarithm of an unknown is equal to the logarithm of a number, for example, $\log_5 x = \log_5 72$.
7. true   **9.** false; natural logarithm is base e.
11. false; divide $\log N$ by $\log 5$.   **13.** true
15. false; $\ln \dfrac{x}{2} = \ln x - \ln 2$
17. false; $\log_b AB = \log_b A + \log_b B$
19. true   **21.** false; $\log_b A - \log_g B = \log_b \dfrac{A}{B}$
23. false; $\log N$ is negative when $N < 1$.
25.

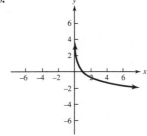

27.

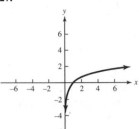

29.

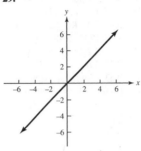

31.

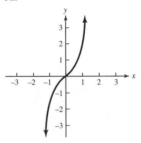

33.

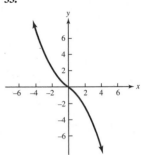

31. a. -1 **b.** 4 **c.** 3 **d.** -3

33. a. e^4 **b.** 14 **c.** 9.3 **d.** 109

35. a. $\ln 6$ **b.** $\ln \dfrac{4^3}{5^5}$ **c.** $\ln(2 \cdot 3^{-5})$

37. a. $\log \dfrac{x^3(x-3)}{x+3}$ **b.** $\log \dfrac{x-3}{x^6(x+3)}$ **c.** $\ln(x-2)$

39. a. 8 **b.** 10^8 **41. a.** $\frac{10}{3}$ **b.** 25

43. a. 2 **b.** $\dfrac{\sqrt{2}}{e^2}$ **45. a.** 2 **b.** e^2 **47.** 10^{10}

49. ± 1 **51.** 0   **53.** $\dfrac{3 \pm \sqrt{9 + 4e^2}}{2}$ **55.** 0

57. $\dfrac{\log 123 - 2\log 6}{\log 6 - 1}$ **59.** $\dfrac{2\ln 5 - 2\ln 6}{3\ln 6 - \ln 5}$ or $\dfrac{2\log 5 - 2\log 6}{3\log 6 - \log 5}$

61. a. 3.5 **b.** 6.2 **c.** $10^{-8.1} \approx 7.94 \times 10^{-9}$

63. a. The maximum typing rate is 80 wpm.

   b. $N = 80(1 - e^{-t/62.5})$

Problem Set 7.4, page 375

1. 3.6269   **3.** -0.7616 **5.** 1.8107   **7.** -4.9370

9. -0.9951 **11.** 11.5920   **13.** 0.4251   **15.** 1.0000

17. 1.6472   **19.** 1.6667   **21.** 0.8000   **23.** 1.9189

25. The graph is rising for all values of x.

27.   **29.**

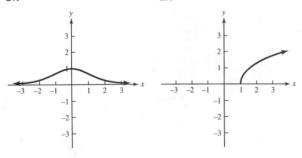

35. $\tanh x = \dfrac{\sinh x}{\cosh x}$

$= \dfrac{\dfrac{e^x - e^{-x}}{2}}{\dfrac{e^x + e^{-x}}{2}}$

$= \dfrac{e^x - e^x}{e^x + e^{-x}}$

37. $\dfrac{1}{\sinh x} = \dfrac{1}{\dfrac{e^x - e^{-x}}{2}}$

$= \dfrac{2}{e^x - e^{-x}}$

39. $\coth x = \dfrac{1}{\tanh x}$

$= \dfrac{1}{\dfrac{e^x - e^{-x}}{e^x + e^{-x}}}$

$= \dfrac{e^x + e^{-x}}{e^x - e^{-x}}$

$= \dfrac{e^x + \dfrac{1}{e^x}}{e^x - \dfrac{1}{e^x}}$

$= \dfrac{e^x + \dfrac{1}{e^x}}{e^x - \dfrac{1}{e^x}}$

$= \dfrac{\dfrac{e^{2x} + 1}{e^x}}{\dfrac{e^{2x} - 1}{e^x}}$

$= \dfrac{e^{2x} + 1}{e^{2x} - 1}$

41. $1 - \tanh^2 x = 1 - \left(\dfrac{\sinh x}{\cosh x}\right)^2$

$= 1 - \dfrac{\sinh^2 x}{\cosh^2 x}$

$= \dfrac{\cosh^2 x - \sinh^2 x}{\cosh^2 x}$

$= \dfrac{1}{\cosh^2 x}$

$= \operatorname{sech}^2 x$

43. $\sinh(-x) = \dfrac{e^{-x} - e^x}{2} = -\dfrac{e^x - e^{-x}}{2} = -\sinh x$

45. $\tanh(-x) = \dfrac{\sinh(-x)}{\cosh(-x)} = -\dfrac{\sinh x}{\cosh x} = -\tanh x$

47. $\sinh x \cosh y + \cosh x \sinh y$

$$= \frac{e^x - e^{-x}}{2} \cdot \frac{e^y + e^{-y}}{2} + \frac{e^x + e^{-x}}{2} \cdot \frac{e^y - e^{-y}}{2}$$

$$= \frac{e^x e^y - e^{-x}e^{-y}}{4} + \frac{e^x e^y - e^{-x}e^{-y}}{4}$$

$$= \frac{2(e^x e^y - e^{-x}e^{-y})}{4}$$

$$= \frac{e^{x+y} - e^{-(x+y)}}{2}$$

$$= \sinh(x + y)$$

49. $\tanh(x + y) = \dfrac{\sinh(x + y)}{\cosh(x + y)}$

$$= \frac{\sinh x \cosh y + \cosh x \sinh y}{\cosh x \cosh y + \sinh x \sinh y}$$

$$= \frac{\dfrac{\sinh x \cosh y}{\cosh x \cosh y} + \dfrac{\cosh x \sinh y}{\cosh x \cosh y}}{1 + \dfrac{\sinh x \sinh y}{\cosh x \cosh y}}$$

$$= \frac{\tanh x + \tanh y}{1 + \tanh x \tanh y}$$

51. $\cosh 2x = \cosh(x + x) = \cosh x \cosh x + \sinh x \sinh x = \cosh^2 x + \sinh^2 x$

53. The poles are located at $x = -20$ and $x = 20$ so the poles are 40 ft apart. When $x = 20$, $y = 45.10503861$, so each pole is about 45 ft tall.

55. From Problem 54, we estimate $b = 35$, and calculate (for $a = 46$) the sag to be about 13 ft (13.97011575 ft).

57.

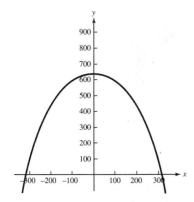

It looks as if the center of the coordinate system is on the ground directly below the center of the arch.

59. $\cosh^{-1} x = \ln(x + \sqrt{x^2 - 1})$

61. $\text{sech}^{-1} x = \ln(1 \pm \sqrt{1 - x^2}) - \ln x$; use $+$ for the principal value.

63. $\coth^{-1} x = \ln \sqrt{x + 1} - \ln \sqrt{x - 1}$

Chapter 7 Sample Test, page 378

1. a. 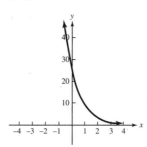 **b.**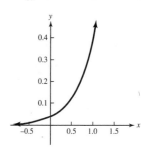

2. a. 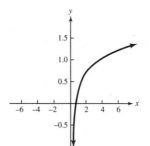 **b.**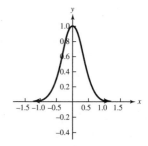

3. a. 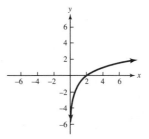 **b.**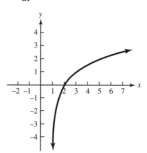

4. a. 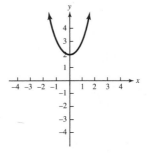 **b.**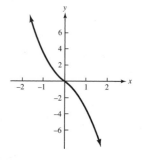

5. 3   **6.** 546   **7.** 4.3   **8.** 1.929418926

9. 0.9355541633   **10.** 1.117832865   **11.** 1,296

12. 223.6067978   **13.** 1.02020202   **14.** 5

15. $x = \log_{1+i}\left(\dfrac{A}{P}\right)$ **16.** $y = 14.8e^{x \ln 2.5}$

17. a. 3 years, 2 months   **b.** 9 years, 11 months   **c.** 14%

18. a. $k \approx -0.0261564974$ **b.** It would take about 101 h.

19. a. $f(\pi) \approx 0.9962720762$ **b.**

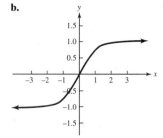

Miscellaneous Problems, Chapters 1–7, page 379

1. $\dfrac{e^{x+h} - e^x}{h}$ **3.** $\dfrac{5^{x+h} - 5^x}{h}$ **5.** $\dfrac{\ln(x+h) - \ln x}{h} = \dfrac{1}{h}\ln\dfrac{x+h}{x}$

7. a. -1.00 **b.** 3.00   **9. a.** 0.69   **b.** -2.08

11. a. 2.48   **b.** 1.78   **13. a.** 0.46   **b.** 0.55

15. $\sqrt{3} \approx 1.73$ **17.** $\sqrt{3} = 1.73$

19. a. $\mathbb{R}$ **b.** $y \geq 1$ **c.** minimum value is 1   **d.** $(0, 1)$

21. $y = 5\cos\frac{\pi}{2}x$ **23.** $y = 6\sin\frac{\pi}{2}(x+1)$ **25.** $y = 2\sec\frac{\pi}{4}x$

27. $y = 3\tan\frac{\pi}{6}x$ **29.** $y = 3\cos\frac{\pi}{2}x + 4$ **31.** rose curve

33. circle   **35.** cardioid   **37.** rose curve

39. lemniscate   **41.** rose curve   **43.** rose curve

45. cardioid   **47.** 1   **49.** $\frac{n\pi}{2}$ **51.** 243

53. ± 15.45 **55.** ± 19.91 **57.** 2.93

59. Note that $10^{\log(\cos x)} = \cos x$ and $10^{\log(1-\sin x)} = 1 - \sin x$, so

$$\dfrac{10^{\log(\cos x)}}{10^{\log(1-\sin x)}} = \dfrac{\cos x}{1 - \sin x}$$

$$= \dfrac{\cos x}{1 - \sin x} \cdot \dfrac{1 + \sin x}{1 + \sin x}$$

$$= \dfrac{\cos x(1 + \sin x)}{1 - \sin^2 x}$$

$$= \dfrac{\cos x(1 + \sin x)}{\cos^2 x}$$

$$= \dfrac{1 + \sin x}{\cos x}$$

$$= \dfrac{1}{\cos x} + \dfrac{\sin x}{\cos x}$$

$$= \sec x + \tan x$$

61. $1 + \cos^2 3\theta = 1 + (1 - \sin^2 3\theta)$

$$= 2 - \sin^2 3\theta$$

63. $(\tan 5\theta - 1)(\tan 5\theta + 1) = \tan^2 5\theta - 1$

$$= (\sec^2 5\theta - 1) - 1$$

$$= \sec^2 5\theta - 2$$

65. $(\sin\gamma - \cos\gamma)^2 = \sin^2\gamma - 2\sin\gamma\cos\gamma + \cos^2\gamma$

$$= 1 - 2\sin\gamma\cos\gamma$$

67. $(\sec 2\theta + \csc 2\theta)^2 = \left(\dfrac{1}{\cos 2\theta} + \dfrac{1}{\sin 2\theta}\right)^2$

$$= \left(\dfrac{\sin 2\theta + \cos 2\theta}{\cos 2\theta \sin 2\theta}\right)^2$$

$$= \dfrac{\sin^2 2\theta + 2\sin 2\theta\cos 2\theta + \cos^2 2\theta}{\cos^2 2\theta \sin^2 2\theta}$$

$$= \dfrac{1 + 2\sin 2\theta\cos 2\theta}{\cos^2 2\theta \sin^2 2\theta}$$

69. $\cot^4\theta$ **71.** $\sin^2\theta$ **73.** $\cot\theta\cos\theta$

75. a. $\sqrt{7}\,\text{cis}\,\frac{11\pi}{12}, \sqrt{7}\,\text{cis}\,\frac{23\pi}{12}$ **b.** $\frac{\sqrt{3}}{2} + \frac{1}{2}i, -\frac{\sqrt{3}}{2} + \frac{1}{2}i, -i$

77. $1 + \sqrt{3}\,i, -1 + \sqrt{3}\,i, -1 - \sqrt{3}\,i, 1 - \sqrt{3}\,i$

79.

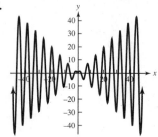

both; $x < 0$, damping
$x > 0$, resonance

81.

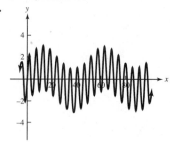

harmonic motion

83.

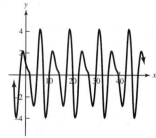

harmonic motion

85. 23 hr (23.17343162)
87. a. The surface area is 46 in.^{2}. **b.** $\theta \approx 3.9°$
c. The minimum surface area is 12.44 in.^{2}.

Appendix B, page 394

1. $4,914 \approx 4,900$ **3.** $5,517,840 \approx 5,520,000$
5. $6,624.48 \approx 6,620$ **7.** $378,193,764 \approx 380,000,000$
9. $40 \approx 40.0$ **11.** $7,714.00791 \approx 7,714.01$ **13.** 0.997
15. 20 **17.** 29.1 **19.** $-4,500$ **21.** $-4,533$
23. three **25.** three **27.** three **29.** 12.2 **31.** 31
33. 1,120 **35.** 0.264 **37.** 0.008353 **39.** 2.78
41. 0.310 **43.** 92.3 **45.** 0.24 **47.** 1.0 **49.** -0.379

Appendix C, page 409

1. a. $x^2 + 3x + 2$ **b.** $y^2 + y - 6$ **c.** $x^2 - x - 2$
d. $y^2 - y - 6$
3. a. $2x^2 - x - 1$ **b.** $2x^2 - 5x + 3$ **c.** $3x^2 + 4x + 1$
d. $3x^2 + 5x + 2$
5. a. $a^2 + 4a + 4$ **b.** $b^2 - 4b + 4$ **c.** $x^2 + 8x + 16$
d. $y^2 - 6y + 9$
7. a. $\frac{1}{3}(x + y)$ **b.** $\frac{1}{2}(x - y)$
9. a. $\frac{6 + 5x - x^2}{2x}$ **b.** $\frac{x^2 - 2}{x^2}$ **11.** 1
13. a. $m(e + i + y)$ **b.** $(a - b)(a + b)$ **c.** $a^2 + b^2$
d. $(a - b)(a^2 + ab + b^2)$

15. a. $(3x + 1)(x - 2)$ **b.** $(2y - 1)(3y - 2)$
c. $b(2a + 3)(4a - 1)$ **d.** $2(s + 3)(s - 8)$
17. a. $(x - y - 1)(x - y + 1)$ **b.** $4(x + 1)(x + 2)$
19. a. $(x + 2)(2x - 3)$ **b.** $(x - 4)(3x + 1)$
21. a. $(x + 8)(6x + 1)$ **b.** $(x - 8)(6x - 1)$
23. False; $(2x^2y^3)^2 = 4x^4y^6$
25. false; $(x + y)^2 = x^2 + 2xy + y^2 \neq (x^2 + y^2)$
27. false; need to get a common denominator
29. false; can only cancel factors
31. false; can only cancel factors
33. false; let $x = 3$ and $y = 4$.
35. $\frac{x^2 + x + 5}{(x + 1)(x - 4)}$ **37.** $\frac{x^2 + 7x - 10}{(x + 2)(x - 2)}$ **39.** $\frac{6x + 6y - 2}{x + y}$
41. $\frac{y + 6}{y + 3}$ **43.** $\frac{-2}{x + 7}$ **45.** $\frac{x^2 + 4x + 1}{(x + 3)(x + 4)}$
47. $\frac{7x}{(x - 2)(x + 3)}$ **49.** $\frac{x^2 + 14x + 1}{(x + 4)(x - 4)}$ **51.** $\frac{4}{3}, -2$
53. $\frac{1}{2}, \frac{5}{2}$ **55.** ± 3 **57.** $\pm \frac{1}{4}$ **59.** $-\frac{1}{6}$ **61.** $\frac{2}{5}, -\frac{5}{6}$
63. $2 \pm \sqrt{5}$ **65.** $3 \pm 2\sqrt{13}$ **67.** $7 \pm \sqrt{10}$
69. $1 \pm 3i$ (no solution over the reals) **71.** $3 \pm 2\sqrt{6}$
73. a. Answers vary; $x_0 = 3.001$
b. Answers vary; 27.009001
c. When x is close to 3, $\frac{x^3 - 27}{x - 3}$ is close to 27 (even though it is not defined at $x = 3$).
d. $\frac{x^3 - 27}{x - 3} = \frac{(x - 3)(x^2 + 3x + 9)}{x - 3} = x^2 + 3x + 9$, which is easier to calculate.

INDEX